Studies in Systems, Decision and Control

Volume 125

Series editor

Janusz Kacprzyk, Polish Academy of Sciences, Warsaw, Poland
e-mail: kacprzyk@ibspan.waw.pl

The series "Studies in Systems, Decision and Control" (SSDC) covers both new developments and advances, as well as the state of the art, in the various areas of broadly perceived systems, decision making and control- quickly, up to date and with a high quality. The intent is to cover the theory, applications, and perspectives on the state of the art and future developments relevant to systems, decision making, control, complex processes and related areas, as embedded in the fields of engineering, computer science, physics, economics, social and life sciences, as well as the paradigms and methodologies behind them. The series contains monographs, textbooks, lecture notes and edited volumes in systems, decision making and control spanning the areas of Cyber-Physical Systems, Autonomous Systems, Sensor Networks, Control Systems, Energy Systems, Automotive Systems, Biological Systems, Vehicular Networking and Connected Vehicles, Aerospace Systems, Automation, Manufacturing, Smart Grids, Nonlinear Systems, Power Systems, Robotics, Social Systems, Economic Systems and other. Of particular value to both the contributors and the readership are the short publication timeframe and the world-wide distribution and exposure which enable both a wide and rapid dissemination of research output.

More information about this series at http://www.springer.com/series/13304

Christian Berger-Vachon
Anna María Gil Lafuente
Janusz Kacprzyk · Yuriy Kondratenko
José M. Merigó · Carlo Francesco Morabito
Editors

Complex Systems: Solutions and Challenges in Economics, Management and Engineering

Dedicated to Professor Jaime Gil Aluja

 Springer

Editors
Christian Berger-Vachon
Lyon Neurosciences Research Center
  and Biomechanical
  and Impacts Laboratory
University Lyon 1
Villeurbanne
France

Anna María Gil Lafuente
Department of Economics
  and Business Organization
University of Barcelona
Barcelona
Spain

Janusz Kacprzyk
Systems Research Institute
Polish Academy of Sciences
Warsaw
Poland

Yuriy Kondratenko
Department of Intelligent Information
  Systems
Petro Mohyla Black Sea National University
Mykolaiv
Ukraine

José M. Merigó
Department of Management Control
  and Information Systems
University of Chile
Santiago
Chile

Carlo Francesco Morabito
Department of Civil, Energy, Environment
  and Materials Engineering
Mediterranean University of Reggio
  Calabria
Reggio Calabria
Italy

ISSN 2198-4182   ISSN 2198-4190   (electronic)
Studies in Systems, Decision and Control
ISBN 978-3-319-88875-0   ISBN 978-3-319-69989-9   (eBook)
https://doi.org/10.1007/978-3-319-69989-9

Printed on acid-free paper

This Springer imprint is published by Springer Nature
The registered company is Springer International Publishing AG
The registered company address is: Gewerbestrasse 11, 6330 Cham, Switzerland

Professor Jaime Gil Aluja

 This book represents a token of appreciation by our entire research and professional community to Prof. Jaime Gil Aluja for his great scientific and scholarly achievements, long-time service to many scientific and professional communities, notably those involved in economics and management and, more specifically, the foundations and applications of tools and techniques for dealing with uncertain, imprecise and incomplete information, notably by using fuzzy logic and possibility theory.

Through the contributions by prominent members of the community, this volume is intended to appreciate Prof. Gil Aluja's original and groundbreaking ideas, combined with his visionary initiatives, which have for decades inspired the scientific and scholarly communities from all over the world in numerous already well-established fields of economics and management sciences, and have initiated and triggered new research directions later pursued by many research groups all over the world. He has particularly motivated these communities by providing novel visions, paradigms, and finally tools and techniques, to improve the modeling of systems and decision processes in social sciences, particularly in economics and management. The essence of his research has always been a synergistic combination of theory and practice. On the one hand, he has contributed to the development of new perspectives and approaches, and tools and techniques for dealing with uncertain and imprecise information in decision sciences and economics, to handle complexity that is so characteristic for the real world facing social, economic, technological, etc., changes. Since Prof. Gil Aluja's works have always been based on realistic assumption, his results have found application for the solution of many practical problems.

More generally, if one looks at the contributions of Prof. Gil Aluja, one can clearly see him as a visionary scientist and scholar who has could, earlier than most of his peers, understand what is going to play an important role in science in the future, and then being able to put the pieces of ideas, tools, and techniques together in an innovative way that has initiated new things and added value. That is, in addition to his undisputable scientific and scholarly stature, he has also been a person who has always could "get things done", in the sense of practical implementation, which is a remarkable ability, quite rare among members of the academia.

Professor Gil Aluja's illustrious career that spanned over a couple of decades has been characterized by his uncompromised quest for scientific excellence and unquestionable integrity. He has been Full Professor at the University of Barcelona since 1967, holding a part-time research position between 1987 and 1998 at the University of Paris-Dauphine in France working on the development of tools for macroeconomic analyses.

He has published more than 200 scientific journal papers, 32 books about various aspects of the modeling and processing of uncertain information in economic, financial, and management sciences. These works have appeared in many languages exemplified by English, Spanish, Russian, French, Catalan, Italian and Romanian, and have helped reach a wider audience in many countries, both researchers, scholars, and professionals.

Professor Gil Aluja is the founder of the International Association for Fuzzy-Set Management and Economy (SIGEF), one of the main branches of the International Fuzzy Systems Association (IFSA). Through this association, he launched in 1995 the Fuzzy Economic Review journal which today is a well-established international journal. Moreover, since the 90s, he has been the main leader of the Association for Modeling and Simulation in Enterprises (AMSE) which publishes eight international journals. Additionally, he has been involved in many other national and international associations. Particularly, it is worth mentioning the important role he has had in the European Academy of Management and Business Economics (AEDEM) which publishes several international journals, including the European Research on Management and Business Economics journal which has been recently included in the Web of Science Core Collection through the Emerging Sources Citation Index.

The scientific and professional communities all over the world have greatly appreciated the efforts and achievements of Prof. Gil Aluja. He has been awarded a multitude of scientific and academic honors and awards. First, for many years he has served as President of the Royal Academy of Economic and Financial Sciences of Spain (RACEF—Real Academia de Ciencias Económicas y Financieras). His vision and persistence have helped establish the RACEF as one of the most prestigious and influential academies of sciences in the world, not only in economic and financial sciences. In addition to top scientists and scholars from all over the world, including a dozen of Nobel Prize winners in economics, as well as numerous political and business leaders have been elected providing an added value that has made it possible to contribute to the solution of great problems and even grand challenges facing many countries and regions around the world.

Twelve most prestigious national and international Academies from all parts of the world have appreciated the scientific stature, vision, leadership, and integrity of Prof. Gil Aluja and have elected him as honorary member. Moreover, he has received from 29 universities from all over the world the title of *doctor honoris causa* causa which has been the most prestigious title that can be granted by a university according to a long tradition of the European university system that has then been adopted practically by all universities around the world. These are just a few of awards, prizes, and other signs of a deep appreciation by the world community.

At the lower, though by no means less important level, Prof. Gil Aluja is a remarkable person who has been known for his consideration, generosity, respect, and modesty, that is characteristic of great human beings. This all has always been a projection of the good atmosphere around him, and has for sure contributed to his exceptional ability to inspire people around him. People have felt privileged because of his lessons and have fully appreciated how inspiring contacts with him have been, and how their professional development has been shaped by discussions and daily interactions with him. They also could feel his friendliness, integrity, and great personal qualities. This is what was also experienced by the editors of this volume.

It is clear that this volume, meant to be a token of appreciation for Prof. Gil Aluja by our entire community, is modest in comparison with his achievements in terms of providing inspiration and so many important research results, but especially in terms of loyal friendship and constant support. We are honored to have had the opportunity to prepare this volume.

Spring 2017
Christian Berger-Vachon
Ana María Gil Lafuente
Janusz Kacprzyk
Yuriy Kondratenko
José M. Merigó
Carlo Francesco Morabito

Preface

This volume is dedicated to Prof. Jaime Gil Aluja, President of the Royal Academy of Economic and Financial Sciences (RACEF—Real Academia de Ciencias Económicas y Financieras of Spain), a world famous scientific and scholarly corporation aimed at the promotion of broadly perceived economic and financial sciences as well as many related areas of science and technology exemplified by humanities and social sciences, organizational sciences, cognitive sciences, systems science, just to name a few. Among the RACEF members, there are many prominent scientists and scholars, including 10 winners of the Nobel Prize in Econimics, but also many financial and political leaders from all over the world.

By dedicating this book to Prof. Gil Aluja, we, the editors, on behalf of the entire research and professional community, wish to present a small token of appreciation for his great scientific and scholarly achievements, long-time service to many scientific and professional communities, notably those involved in economics, finance and management and, more specifically, the foundations and applications of tools and techniques for dealing with uncertain, imprecise and incomplete information, notably by using fuzzy logic and possibility theory. At a more personal level, we also wish to thank him for a constant support of what we and many of our students and collaborators have been undertaking for so many years.

The volume is divided into six parts that cover main issues related to the topic of the volume. Part I, "Basic Issues in Economic, Organization and Management Systems", deals with some basic and foundation issues in broadly perceived economic, organization, and management systems that are crucial for the understanding, analyses, and even shaping of the main aspects that concern our society and economy.

Finn E. Kydland, the winner of the Nobel Prize in Economics in 2004, in his paper "Does Policy Consistency Affect Economic Growth?" provides an in-depth analysis of some issues related to uncertainty in the economic and business environment the extent of which is extremely high in the present world, in particular over the last couple of decades. Many reasons can be quoted for this, notably a lack of clarity in economic policy making that exists in virtually all countries. It is obvious that all decisions that are meant to stimulate growth, innovation, investments, etc., require a clarity as to the policy environment, not only now but in the future, sometimes quite distant. This can include many aspects including tax policy, spending, and debt policy, extent of trade restrictions, just to name a few. It is contended that, taking into account what has been known in economic theory, the increase in policy uncertainty over the past decade has been normal and predictable to a large extent. This claim is illustrated by an analysis of the Euro zone, the United States, and China. The lessons learnt are summarized and some possible solutions for the future are pointed out.

Fernando Casado Juan ("Perception and Reality of the Spanish Economy") provides an insightful summary of the situation of the Spanish economy. First, the author starts by pointing out a positive state of the Spanish economy in recent years that has experienced the annual GDP growth of more than 3%. This positive situation is then analyzed, mainly by emphasizing as the main causes, paying a special attention to the strong performance of the external sector in the current global economic context. Moreover, the paper analyzes the key macroeconomic factors underpinning the Spanish economy, as well as the increase of the GDP, the public deficit, labor market, foreign investment, tourism, training, infrastructure, and company size. The author concludes his paper by highlighting the importance of the digitalization process on all aspects and areas of the economic and social life. The paper, though mainly focused on the Spanish economy, can also be very useful for people who are interested or involved in similar analyses related to other countries.

Mario Aguer Hortal ("The Virtual Company as a Value Generator in the New Economy") is concerned with the analysis of a new techno-economic paradigm, based on an interconnected combination of technological innovations which permit the company to cope in an environment that is increasingly dynamic and uncertain, and in which people more and more work together being physically in places which are distant from each other. This increasingly common work practice, and its underlying processes, result in the creation of virtual companies, factories, offices, and virtual communities. Basically, the virtual companies are those which conduct their activities in a virtual way, in part or even fully. The most important ones among them are those dedicated to logistic operations, commerce and trade, marketing, sales, design, etc. One can distinguish three creative components that are crucial for a virtual company: intellectual capital, information and technology. Needless to say that information plays in this context a crucial role, and various aspects related to information are analyzed, exemplified by the measuring of information quality. The same can be said about communication. The author then analyzes a very important aspect of the lack of dimensionality of the virtual company itself which partly results from the absence of physical dimensions. This makes classic procedures of, for instance, accounting or evaluations difficult to apply. Moreover, the proliferation of e-commerce and the Internet usage have had an enormous impact on all aspects of corporate activities. The author indicates potentials of results from the General Theory of Systems for the analysis and synthesis of such socioeconomic systems. In general, information and the intellectual capital within the company call for a correct evaluation method and require tools and techniques for this measuring of intangible assets. The virtual company's absence of dimensionality makes accounting and evaluation criteria different from the traditional ones. Virtual communication need to be objective, meaningful and valuable, referring to judgments and perceptions. This all poses a challenge for research and practice.

Camilo Prado-Román, Francisco Díez-Martín, and Alicia Blanco-González ("The Origin of the Legitimacy of Organizations and Their Determining Factors") are concerned with the analysis of to what extent personal characteristics have an influence on the assessment process of organizational legitimacy. For this purpose, a questionnaire is used which examines the effect of five personal characteristics on four types of legitimacy. Then, the regression analysis is applied on a sample of 258 individuals. The results obtained show that people with a higher social awareness are more prone to making decisions about organizations taking into account the moral, regulatory, and cognitive legitimacy. This type of assessment also occurs when the fear of receiving a social sanction increases. Likewise, in the perception of a higher economic risk inherent to the result of a decision, people are more likely to make decisions based on cognitive and pragmatic legitimacy. The results of this study extend the knowledge in the field of institutional theory about the origin of organizational legitimacy. It can also facilitate to improve the strategic planning of organizations by displaying the legitimacy preferences based on each person's profile.

José María de-Goñi-Oslé and Arturo Rodríguez-Castellanos ("A Model for the Management in Organizations Based on People and Knowledge: Aspects to be Considered in its Design") take as a point of departure that uncertainty and complexity in business and economic systems highlight a relevant role of people who that is crucial for the generation of intangible resources that are the source of business competitiveness. Therefore, the relevance of intangible assets and other aspects in the current society, especially those based on knowledge and people, makes them a strategic issue for the organizations. The broadly perceived intangibles are under research and study both in the academic and corporate communities. The authors consider some aspects related to the design and implementation of an efficient model for the management in organizations that is based on people and knowledge. First, an extensive review of the literature is provided, including analyses of models and their implementations. The results, especially from the observation of practical cases, suggest that there is an enormous difficulty in the timely implementation and then maintenance of these models, possibly due to difficulties that occur while evaluating their actual cost and benefits obtained. These issues are then analyzed and conclusions are provided.

Part II, "Decision Making and Systems Modeling", provides information on some important formal tools and techniques that can be used to represent uncertainty and imprecision, model systems under consideration, and determine proper decisions or strategies.

Joan Carles Ferrer Comalat, Xavier Bertran Roura, Salvador Linares Mustarós, and Dolors Cominas Coll ("Six Experimental Activities to Introduce the Theory of Fuzzy Set") are concerned with some important problems related to a proper presentation of the theory of fuzzy sets, the importance and potentials of which have been justified by many studies, in a way that can best convince a potential user. The authors have presented their highly innovative approach to the presentation of the main ideas underlying the concept of a fuzzy set by employing artworks, notably pictures exhibited in a gallery.

Galyna Kondratenko, Yuriy Kondratenko, and Ievgen Sidenko ("Fuzzy Decision Making System for Model-Oriented Academia/Industry Cooperation: University Preferences") discuss some important, and effective and efficient models of cooperation of universities and IT companies. Moreover, the authors propose a new hierarchic approach to the development of decision support systems (DSS) based on fuzzy logic that can be useful for solving the problems considered. Special attention is paid to the use of fuzzy logic in that DSS that is meant to be an advisor for choosing the most appropriate cooperation model for a certain department of a university that tries to become a partner for a certain IT company. The article features a hierarchic structure, results of rule bases, and DSS software based on an approximation of fuzzy systems with discrete output. It also contains the results of simulations of a process of developing the most rational model of cooperation of the "University—IT company" type.

Due to an oversight, the uncorrected version of the originally planned chapter "Towards the Convergence in Fuzzy Cognitive Maps Based Decision-Making Models" by Concepción et al. https://doi.org/10.1007/978-3-319-69989-9_8 was accepted; hence this chapter had to be removed.

Michał Jakubczyk, Bogumił Kamiński, and Michał Lewandowski ("Eliciting Fuzzy Preferences Towards Health States with Discrete Choice Experiments") are concerned with some issues related to health (quality and duration of life) that is clearly extremely important both from an individual and societal points of view. However, since people rarely choose between health states, their preferences are often not well-formed; moreover, the quality of life is frequently defined using imprecise terms. The authors propose to model preferences related to health states (precisely: disutilities of worsening health dimensions in the EQ-5D-5L descriptive system) as fuzzy, more specifically, each worsening is assigned an interval instead of a crisp number. They elicit such preferences with discrete choice experiment (DCE) data, using a maximum likelihood approach and bootstrapping to evaluate the estimation error. The authors' approach overcomes one of the nonintuitive features of the standard approach to DCE, in which even a clearly dominated alternative has a positive probability of being chosen, in that in the authors' model, if the disutility ranges do not overlap, the worse alternative will never be chosen. Also, the model is more consistent regarding the constant proportional trade-off condition in that the probability of a given health state being chosen in a pair will not change if durations are scaled proportionally which does not hold in the standard DCE model.

Silvia Bortot, Mario Fedrizzi, Michele Fedrizzi, Ricardo Alberto Marques Pereira, and Thuy Hong Nguyen ("The Soft Consensus Model in the Multidistance Framework") investigate the reformulation of the soft dissensus measure in relation with the notion of multidistance, recently introduced by Martín and Mayor, which is an extension of the classic concept of a binary distance obtained by means of a generalization of the triangular inequality. The new soft dissensus measure introduced by the authors is a particular form of the sum-based multidistance. This multidistance is constructed on the basis of a binary distance defined by means of a subadditive scaling function which in general puts more emphasis on small distances and attenuates large distances in preferences. The authors present a detailed study of the subadditive scaling function which is analogous but not equivalent to the one used in the traditional form of the soft consensus model.

Yamilis Fernández-Pérez, Ailyn Febles-Estrada, Carlos Cruz, and José Luis Verdegay ("Fuzzy Multi-Criteria Decision Making Methods Applied to Usability Software Assessment: An Annotated Bibliography") present a very interesting analysis of the main developments in the use of fuzzy multicriteria decision-making models for the analysis of usability of software assessment. Issue related to software development, maintenance, and use is very important and costly, and therefore it is essential to assess the impact on the sustainability. An important feature is here associated with the very meaning of the software usability, and one of crucial aspects is the ease of use which is difficult to define and hence measure. The authors first analyze main recent approaches employed for this purpose. Basically, these models have evolved from the use of conventional statistical techniques and soft computing methods. In this work, the authors present a critical account of a collection of annotated bibliography entries about the characteristics, attributes, and metrics used in the usability assessment, focused on the use of soft computing techniques.

F. González Santoyo, B. Flores Romero, A. M. Gil Lafuente, J. Flores Juan, and R. Chávez Rivera ("Production Systems Optimization Using Hierarchical Planning") consider a very interesting and important problem of production planning for lumber production which is relevant for many countries, including Mexico. The authors propose a new, effective, and efficient heuristic algorithm to solve the problem of hierarchical production planning in sawmills. The proposed solution is based on the mixed integer linear programming (MILP) formulation using Benders' decomposition and the Lagrangian relaxation techniques. The new method proposed is computationally efficient, better than methods applied, as shown on numerical examples.

Yuriy P. Kondratenko, Oleksiy V. Kozlov, Galyna V. Kondratenko, and Igor P. Atamanyuk ("Mathematical Model and Parametrical Identification of Ecopyrogenesis Plant Based on Soft Computing Techniques") present the development of a mathematical model and system for the parametric identification of the ecopyrogenesis (EPG) plant as a complex multi-coordinate control object using soft computing techniques. The synthesis procedure of the main parts of the EPG plant's mathematical model, including its fuzzy parametric identification system, adaptive-network-based fuzzy inference system for the calculation of the temperature of the multiloop circulatory system (MCS), and the Mamdani type fuzzy inference system for the determination of the reactor load level are shown. The analysis of computer simulation results in the form of graphs of static and dynamic characteristics of the EPG plant confirms that the complex neuro-fuzzy model proposed is a proper solution. The developed mathematical model with the parametric identification based on neuro-fuzzy technologies gives an opportunity to investigate the behavior of the given complex control object in steady and transient modes, in particular to synthesize and adjust the intelligent controllers of the multi-coordinate automatic control system of the EPG plant.

Part III, "Intelligent Data Analysis and Processing", provides an overview of some important and promising tools and techniques for broadly perceived data analysis that is crucial for all kinds of analyses, syntheses. and implementations in virtually all problems and setting encountered.

Andrzej Pownuk, Vladik Kreinovich, and Songsak Sriboonchitta ("Fuzzy Data Processing Beyond Min t-Norm") are concerned with some algorithmic issues related to the use of fuzzy data. The authors show that though the use of the minimum as an operation corresponding to "and", which is widely employed notably while using Zadeh's extension principle, can give good results, the use of other triangular norms (t-norms) can also be a viable and effective and efficient solution.

Leszek J. Chmielewski and Arkadiusz Orłowski ("Detecting Changes in Time Sequences with the Competitive Detector") discuss and then extend the concept of the competitive edge detector which, in the case of one-dimensional signals, can be denoted as the detector of changes. In such a detector, two approximators are used, one working on the "past" and one working on the "future" side of the considered data point. The difference of their outputs makes it possible to find the change of the value and the derivative of the signal. The new features introduced consist in performing a robust analysis and in adding an option to use a quadratic function as an approximator. A weighted voting of elementary subsets is used with weights related to the significance of a subset for the result. Results of change detection on test data as well as some real-life economic data are encouraging.

Vikas Singh and Nishchal K. Verma ("Deep Learning Architecture for High-Level Feature Generation Using Stacked Auto Encoder for Business Intelligence") consider a crucial problem in the present day science and technology which is related to huge data sets that are commonly encountered and have to be handled. The handling of such large amount of data by conventional machine learning algorithms is difficult because of their usually heterogeneous nature and large size. Deep learning, a new direction in machine learning, meant to deal with such heterogeneous nature and large size of data and to extract high-level representations of data through a hierarchical learning process, can be a viable and promising tool in this context. The authors propose a novel multilayer feature selection model with the conjunction of a stacked auto-encoder (SAE) to extract high-level features or representations, and eliminate lower level features or representations from data. The proposed approach is validated on the Farm Ads dataset and the result is compared with various conventional machine learning algorithms. The proposed approach has outperformed conventional machine learning algorithms for the given dataset.

Daniela Sánchez, Patricia Melin, and Oscar Castillo ("Comparison of Type-2 Fuzzy Integration for Optimized Modular Neural Networks Applied to Human Recognition") present optimization techniques for the Modular Neural Networks (MNNs) and their combination with a granular computing approach. A Firefly Algorithm (FA) and a Grey Wolf Optimizer (GWO) are developed to perform the MNN optimization. These algorithms perform the optimization of some parameters of the MNN such as the number of submodules, percentage of information for the training phase and number of hidden layers (with their respective number of neurons) for each submodule and learning algorithm. The MNNs are applied to human recognition based on the face, iris, ear, and voice. The minimization of the recognition error is the objective function. To combine the responses of the MNNs, different type-2 fuzzy inference systems are proposed and a comparison of results is performed.

Part IV, "Sustainability", is devoted to the discussion of various aspects of broadly perceived sustainability that is one of the most widely used terms in various scientific, political, economic, etc., discussions. Briefly speaking, it concerns various ways of how human societies, maybe more generally various systems, can survive, endure, and prosper in the conditions of global changes, overwhelming uncertainty, ecosystem degradation, social and political unrest, resource limitations, to just name a few.

Gorkhmaz Imanov ("Fuzzy Measure of National Sustainable Development Aggregate Index") proposes a fuzzy measure of a national sustainable development aggregate index (NSDAI) taking into account subindices of economic, social and environmental sustainability. The discussion is illustrated by examples corresponding to Azerbaijan. The author uses for measuring the subindices some elements of the intuitionistic fuzzy set, generalized entropy measure of intuitionistic fuzzy set, Zadeh's compositional rules of inferences, etc. Then, using the fuzzy method of forgotten effects proposed by Kaufman and Gil Aluja, relations between socioeconomic indicators are analyzed.

Massimiliano Ferrara and Bruno Antonio Pansera ("A Dynamic Game For A Sustainable Supply Chain Management") develop a dynamic game to allocate the corporate social responsibility (CSR) to the members of a supply chain. They propose a model of a three-tier supply chain in a decentralized state that includes the supplier, manufacturer, and retailer. For the analysis of the supply chain performance in the decentralized state and of the relationships between the members of the supply chain, the authors use the Stackelberg game and consider a hierarchical equilibrium solution for a two-level game. In particular, they formulate a model that involves multiple periods and propose a dynamic discrete Stackelberg game. An equilibrium point is obtained at which both the profits of members and the level of the CSR taken by the supply chains are maximized.

Ingrid Nineth Pinto López, Anna María Gil Lafuente, and Guillermo Sánchez Flores ("Pichat´s Algorithm for the Sustainable Regional Analysis Management: Case Study of Mexico") consider a multidimensional analysis for a group of regions with common similarities that can allow them an improvement of competence management, support its integration processes, as well as strengthen the social and economic development. The analysis of such groups of regions calls for the application of concepts, models, and algorithms that allow the analysis of ambiguous variables. The model proposed by the authors tries to contribute to an informed decision-making, getting closer to reality while identifying those regions that, because of their characteristics, share common features. The goal of this work is to design a process of grouping regions based on the analysis of sustainability indicators. An approximation method based on Pichat´s algorithm is proposed using sustainable development indicators of each region as a reference and suggesting the degree of similarity among the regions so that the determination of similar groups be possible.

Vicente Liern and Blanca Pérez-Gladish ("Companies' Selection Methods for Inclusion in Sustainable Indices: A Fuzzy Approach") are concerned with the selection of companies to be included in sustainability indices. Since sustainability indices involve concepts which are both numeric and nonnumeric, the use of fuzzy logic can be advantageous. Usually, these indices follow a three-step process to define sustainable investment universes. The first step consists of a sustainability assessment. In the second step, assets are rated based on the previously assessed sustainability scores and, final in the third step, the best assets are selected. This last step relies on the construction of a global score reflecting the performance of the assets in the main sustainability dimensions. The authors deal with the third step of the selection process. They review the aggregation process used by the sustainability indices to obtain overall sustainability scores and propose the use of flexible aggregation operators for the obtaining of a global score describing the sustainability degree of a firm that takes into account the characteristics of different dimensions to be aggregated. Assets are then ranked using this score from the most to less sustainable. The proposed approach is be compared with the three-step selection process applied by Euronext in their selection of companies in the Euronext Vigeo family of sustainability indices.

Part V, "Financial Analyses", deals with various aspects of financial analyses.

Alfonso M. Rodriguez ("About Formal Construction of Financial Analysis") discusses some basic and foundation aspects of broadly perceived financial analyses. Though the conventional financial mathematics boils down to financial calculi, an important financial phenomenon, the preference for liquidity, should always be present in the studies. The financial value considers, jointly with the monetary amount, its temporal deferral, i.e., its liquidity. Both of them must be formalized in a binary vector. The financing financial operations (FFOs) and investment financial operations (IFOs) are very different financial operations (FOs), with the FFO only pretending to get an interest, the financing price by its financial service. The IFO intends to get an investment yield, its economic result. The interest and investment yield are different economic magnitudes. The interest is reached by a financial market and it defines the financial equilibrium of the FFO (finance equivalence). The investment yield is an economic result attained at the financial disequilibrium of IFO. It is the reason why to investigate an investment yield as a financial equilibrium as an implicit interest, it is the grave financial mistake that IRR (Internal Rate of Return) commits, confusing investment yield with interest and an IFO with a FFO, with erroneous consequences to its definition and possible to investment decisions. The new methodological approach to financial mathematics makes it possible to revise conventional concepts as interest as the investment yield, financial productivity, financial profitability, etc., and to incorporate another unknown concepts as the financial degeneration, financial immunity, etc., and to know some other financial instruments.

Martín Iglesias Caride, Aurelio F. Bariviera, and Laura Lanzarini ("Stock Returns Forecast: An Examination by Means of Artificial Neural Networks") analyze—taking into account as a point of departure the validity of the efficient market hypothesis that has been under severe scrutiny since several decades but with the evidence against it being not conclusive. The authors discuss the possibility to use the artificial neural networks (ANN) as a model-free means to analyze the prediction power of past returns based on the knowledge of current returns. They analyze the predictability in the intraday Brazilian stock market using a backpropagation artificial neural network. The authors selected 20 stocks from the Bovespa index, according to different degrees of market capitalization, as a proxy for the stock size. They find that the predictability is related to the capitalization. In particular, larger stocks are less predictable than smaller ones.

Maciej Janowicz, Leszek J. Chmielewski, Joanna Kaleta, Luiza Ochnio, Arkadiusz Orłowski, and Andrzej Zembrzuski ("Persistent Correlations in Major Indices of the World Stock Markets") are concerned with the time-dependent cross-correlation functions that are calculated between returns of the major indices of the world stock markets. The authors consider one-, two-, and three-day shifts and find that, surprisingly enough, high, and persistent-in-time correlations are found among some of the indices. Part of those correlations can be attributed to the geographical factors, for instance, strong correlations between two major Japanese indices have been observed. The reason for other, somewhat exotic correlations, seems to be as much accidental as obvious. It seems that the observed correlations may be of a practical value in the stock market speculations.

Ezequiel Avilés Ochoa, Ernesto León Castro, Ana María Gil Lafuente, and José María Merigó Lindahl ("Forgotten Effects in Exchange Rate Forecasting Models") try to use the methodology of forgotten effects, hidden variables that influence the behavior of the forward exchange rate of US dollar—to—Mexican peso (USD/MXN) to then incorporate them into a structured model from the postulates of the theory of Purchasing Power Parity (PPP) and, thus, to reduce the forecast error. The authors discuss the problem if it is possible to decrease the prediction error of the PPP model by using the methodology of the forgotten effects to detect and include hidden or forgotten variables. It is found that the inclusion of hidden or forgotten variables in the PPP model decreased the forecast error for the exchange rate USD/MXN in 2015.

Antonio Terceño, M. Gloria Barberà-Mariné, Laura Fabregat-Aibar, and Maraia Teresa Sorrosal-Forradellas ("The Behaviour of Non-surviving Spanish Funds According to Their Investment Objectives") present an attempt to determine if the characteristics which define the non-surviving funds are different according to their investment objectives. The authors use the Self-Organizing Maps (SOMs) to cluster the mutual funds that disappeared in 2013, 2014, and 2015, based on the variables that define its survival capacity and, as a result, to analyze if these variables take similar values for all of them, or different values depending on the funds' investment objectives. The authors propose to analyze nine categories: bond funds, bond mixed funds, equity funds and equity mixed funds, distinguishing between those that invest their assets in national or international markets, and passive investing funds. Numerical results are shown for illustration.

Part VI, "Applications in Business and Technology", is an important part of the volume in which various applications of modern tools and techniques that stem from the area of broadly perceived intelligent systems are employed.

Igor Atamanyuk, Yuriy Kondratenko, and Natalia Sirenko ("Management System for Agricultural Enterprise on the Basis of Its Economic State Forecasting") present a new management system (MS) for an agricultural enterprise that is based on the use of economic state forecasting. The system makes it possible to estimate the results of the enterprise functioning in the future under the realization of certain reorganization acts (change of land resources, labor resources, fixed assets). The method for the calculation of forecasts of economic indices of agricultural enterprises on the basis of a vector polynomial exponential algorithm for extrapolation of the realizations of random sequences is proposed. The prognostic model makes it possible to estimate the results of enterprise functioning (to estimate future gross profit, gross production) after its reorganization, and it does not impose any restrictions on the forecast characteristics (linearity, stationarity, Markov behavior, monotonicity, etc.) and thus allows to fully take into consideration stochastic peculiarities of functioning of agricultural enterprises. The simulation results confirm a high efficiency of the introduced calculation method. The method can be implemented in a decision support systems for agricultural and nonagricultural enterprises with various sets of economic indices.

Marina Z. Solesvik ("Partner Selection in Green Innovation Projects") considers issues related to green innovation strategies, and in particular, focuses on the issues related to R&D strategic alliances aimed at developing green technologies in the maritime sector. Though still the managers often use their feelings to select partners from the prospective candidates, expensive R&D projects aimed at developing radical green innovations need a thorough preliminary analysis of collaborators. For this purpose, the author applies a fuzzy logic based formal concept analysis based approach to facilitate decision-making of management teams who are responsible for the selection of partners for collaborative green innovation projects.

Mario Versaci and Francesco Carlo Morabito ("Evaluation of Structural Integrity of Metal Plates by Fuzzy Similarities of Eddy Currents Representation") present an innovative practical application of the methodologies introduced by Gil Aluja and Zadeh in civil and electrical engineering. More specifically, they consider metallic plates which when biaxially loaded deform producing dangerous mechanical stresses that are not visible. Since the representation of such stress conditions by 2D images is extremely complex, the authors propose to generate suitable Eddy currents (ECs) images to translate the information content of mechanical stresses into representative electric signals that would be easier to show. By grouping the produced images in different classes related to different biaxial loads and in a single class all the images referring to plates in absence of loads, the evaluation of the integrity of a plate is transformed into a problem of classification/decision. This further step is carried out by means of some measure of fuzzy similarity between the 2D EC signal and the prototype classes. The attained performance is comparable to more established approaches that are commonly plagued by a higher computational load. The proposed methodology is also shown to be able to manage uncertainty in an application of relevant industrial interest.

C. Berger-Vachon, P. A. Cucis, E. Truy, H. Thai Van, and S. Gallego ("Cochlear Implants: Consequences of Microphone Aging On Speech Recognition") are concerned with a very important problem related to aging, specifically to the cochlear implants (CI) that are designed for the rehabilitation of a profound deaf-ness. One relevant aspect in this context concerns the microphone as an ongoing drift occurs over the time. The authors discuss the consequences of this for the speech recognition. A general population of CI users and subjects using a CI simulator is involved. Words are presented to the listeners in noise with a variable signal to noise ratio (SNR) and the percentage ranged from 0% to 100%. For the CI simulator, the drift is simulated from data coming from figures measured on regular hearing aids. The results are compared before and after cleaning the microphones. Moreover, in a subgroup of CI users, the replacement of the head filter protecting the microphone is done and the recognition percentages are compared with those coming from the standard "brush and blow" cleaning procedure. The results have been revisited and quantified after a curve fitting. The results indicate that the "CIS-like" coding schemes are less sensitive to aging than the n-of-m strategies, and the cleaning improves the recognition performance but not too much. Fur-thermore, the improvement mainly occurs in the middle of the SNR range where the noise was not too intense.

Brigitte Werners and Yuriy Kondratenko ("Alternative Fuzzy Approaches for Efficiently Solving the Capacitated Vehicle Routing Problem in Conditions of Uncertain Demands") deal with the analysis of fuzzy models and fuzzy approaches for an effective and efficient solution of the transportation and vehicle routing problems (VRP) with constraints on the capacity of the vehicles.The authors focus on the VRPs for marine bunkering tankers and on the planning and optimization of tanker's routes in the conditions of uncertain fuel demands at nodes. The triangular fuzzy numbers are proposed for the modeling of uncertain demands and the opti-mization problem is considered as a multicriteria problem with: (a) the minimiza-tion of the total length of planned routes, (b) the satisfaction of all orders at nodes (ships, ports), (c) the maximization of the total sale volume of unloaded fuel, (d) the minimization of the fleet size. Two alternative fuzzy approaches for efficiently solving such marine VRPs are discussed. The first alternative deals with the development of a multistage iterative heuristic procedure and the second alternative concerns the development of a fuzzy decision-making system for the evaluation of satisfaction values for uncertain order realizations. The solutions proposed are advantageous.

We would like to express our gratitude to all the authors for their interesting, novel, and inspiring contributions. Peer reviewers also deserve a deep appreciation, because their insightful and constructive remarks and suggestions have considerably improved many contributions.

And last but not least, we wish to thank Dr. Tom Ditzinger, Dr. Leontina di Cecco, and Mr. Holger Schaepe for their dedication and help to implement and finish this large publication project on time maintaining the highest publication standards.

Villeurbanne, France	Christian Berger-Vachon
Barcelona, Spain	Anna María Gil Lafuente
Warsaw, Poland	Janusz Kacprzyk
Mykolaiv, Ukraine	Yuriy Kondratenko
Santiago, Chile	José M. Merigó
Reggio Calabria, Italy	Carlo Francesco Morabito
Spring 2017	

Contents

Part I
Basic Issues in Economic, Organization and Management Systems

Does Policy Consistency Affect Economic Growth?

Finn E. Kydland

Abstract In the long run, nations prosper, in the sense of higher per-capita incomes, lower unemployment, lower poverty levels, and so on, the fewer the impediments are to steadily higher productivity, along with growing productive capacity to take advantage of the productivity growth. In large parts of the world today, however, the business environment is characterized by an extent of uncertainty that is unprecedented over the past several decades. The main reason is lack of clarity in economic policy making. In order to be properly informed and well founded, growth-promoting decisions, such as innovative activity, investment in new productive capacity, choice of new markets, and so on, require an assessment —an expectation—of the policy environment years into the future. Important policy dimensions are tax policy, spending and debt policy, extent of trade restrictions, and the regulatory environment in general.

On June 15, 2010, I had the honour of being inducted as a Member of RACEF. In connection with that spectacular event, my wife, Tonya, and I had the pleasure of meeting Jaume Gil, benefiting from his immense hospitality and cheerfulness. I'm delighted to have to opportunity to make a contribution to this volume in his honour.

In the long run, nations prosper, in the sense of higher per-capita incomes, lower unemployment, lower poverty levels, and so on, the fewer the impediments are to steadily higher productivity, along with growing productive capacity to take advantage of the productivity growth. In large parts of the world today, however, the business environment is characterized by an extent of uncertainty that is unprecedented over the past several decades. The main reason is lack of clarity in economic policy making. In order to be properly informed and well founded, growth-promoting decisions, such as innovative activity, investment in new productive capacity, choice of new markets, and so on, require an assessment—an

F. E. Kydland (✉)
Department of Economics, University of California, Santa Barbara
CA 93106-9210, USA
e-mail: finn.kydland@ucsb.edu

© Springer International Publishing AG 2018
C. Berger-Vachon et al. (eds.), *Complex Systems: Solutions and Challenges in Economics, Management and Engineering*, Studies in Systems,
Decision and Control 125, https://doi.org/10.1007/978-3-319-69989-9_1

expectation—of the policy environment years into the future. Important policy dimensions are tax policy, spending and debt policy, extent of trade restrictions, and the regulatory environment in general. In this chapter, I contend that, based on economic theory, the increase in policy uncertainty over the past decade was quite predictable. Contexts from various parts of the world, such as the euro zone, the United States, and China, are used as illustrations from which one can learn what may work and what is unlikely to.

In doing so, I'll draw primarily on two theoretical foundations: the aggregate production function, which is central to all of macroeconomics, including growth theory, and the finding that optimal government policy is time inconsistent. (Given that the intended target audience for this volume includes readers without a Ph.D. in economics, I'll spend a little extra time on the intuition of what the latter is all about.) For the growth of nations, I attribute a significant role to economic policy. I'm motivated in part by the tremendous economic differences, for decades, if not centuries, seen across the world, differences that are hard to explain without reference to the nations' political environments. But also, considering more specifically the most recent decade, and especially what led up to, and transpired after, the global financial crisis, my contention is that what has evolved economically in various parts of the world is sufficiently different that variations in the respective nations' economic policy surely must have played a significant role. A conclusion is that it's not clear the extent of uncertainty that has faced, and is still facing, those making growth-promoting decisions in the private sector will disappear any time soon.

1   Income Inequality Across the Globe

The disparity in nations' income levels is astounding. To be sure, measured income (or GDP) may provide a skewed picture of the associated welfare of the typical citizen of that nation. In some countries, an extensive informal sector means that many are better off than the statistics suggest. In other cases, the statistics on per-capita income may mask considerable inequality within that nation. Still, we have to take the statistics seriously, especially when the differences are large. Western European nations, along with countries such as the United States, Canada, Australia, Japan, Korea and Taiwan have annual per-capita incomes exceeding $30,000. Many nations hover in the $10,000–$20,000 range. And then we have all those poor nations, many in Africa, with per-capita incomes under $5,000, some even under $1,000.

Figures 1 and 2 display some of these contrasts. Figure 1 graphs per-capita real GDP in a diverse set of nations. (All the numbers are in constant 2005 dollars, adjusted for purchasing power parity.) Some are notable for rather healthy and steady growth rates, doing a good job getting closer to the United States and Canada, which one may think of as reasonable benchmark countries. These include Korea, Hong Kong, and Japan, although Japan has faltered in the past couple of decades. Some countries were at fairly similar levels to those three around the

1960s, but have since grown strikingly more slowly than those other countries. They include Argentina, Chile, and Mexico. Included in the figure are two nations that were created after the break-up of the Soviet Union, namely Azerbaijan and Kazakhstan. These are both resource-rich nations that seem to have put that wealth to reasonably good use. (Note that, as the graphs are not on log scale, a given steepness of a curve near the bottom of the chart corresponds to much higher annual rate of growth than does that same steepness near the top of the chart.) Finally, in light of the attention China gets for its economic size, it may surprise you to see its low *per-capita* GDP.

Figure 2 graphs a set of sub-Saharan nations. Observe the different scale on the vertical axis as compared with Fig. 1.

A way to organize our empirical knowledge about the growth of nations is in terms of the aggregate production function, which can be written as:

$$Y_t = A_t F(K_t, L_t),$$

in other words, a nation's aggregate output (say, as measured by real GDP) in any time period t is a function of its capital and labour inputs, but importantly enhanced over time if the technology, A_t, grows, as it should in a well-functioning economy.

Especially interesting from the perspective of this chapter is the fact that behind the production function lie decisions by millions of households and thousands of businesses, decisions that by their nature must be very much forward looking.

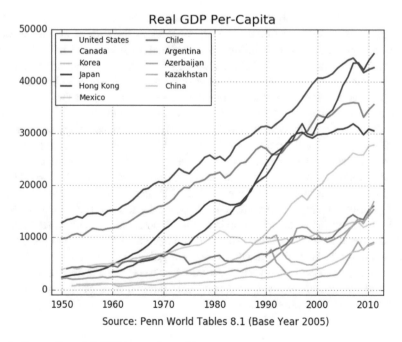

Fig. 1 Per-capita real GDP for several countries

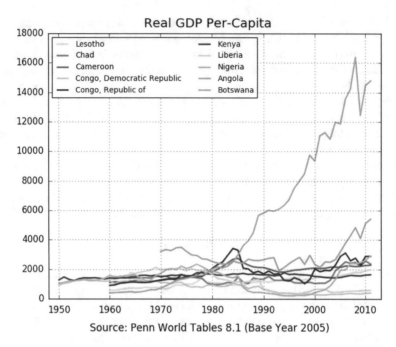

Fig. 2 Per-capita real GDP for African nations

These activities are costly as they take place, but the returns come over years, perhaps decades, into the future. The capital stock is the sum of non-depreciated factories, machines, office buildings, accumulated as investment over the years leading up to period t. Behind the technology level lie past innovative activities whose outcomes by their nature were uncertain and in principle informed by the decision makers' view about the future environment. In the case of the labour input, as a first approximation its growth is more or less in line with the nation's growth of its working-age population. But even here, a dynamic aspect plays an important role. It's desirable for the nation and for the workers themselves that, at least in the long run, L grows faster than the working-age population, as people become more skillful at market production because of improving education, on-the-job learning, and so on. In other words, the accumulation of human, as well as physical, capital are important for long-run growth. (As an anecdote, the last time I was in Kazakhstan to attend the Astana Economic Forum, I was told by government officials that because of their exceptional growth, it had become difficult to find people with the right skills to keep the growth momentum going. This, of course, is in some ways a benign problem for a nation. As I could observe myself over several years, at least at the university level, in recent years they've been busy improving and expanding their educational system.)

2 Time Inconsistency of Optimal Policy

There's a lot of knowledge available. Why can it not be transferred to any place in the world, if necessary with adaptation to local circumstances? What can account for the immense disparities we saw in Figs. 1 and 2? In large part, it comes down to the nature of nations' governments, political systems and institutions, the presence of corruption or lack thereof, the kinds of restrictions they place on economic activity and trade, and so on. (An interesting account of these issues can be found in [5].) But, perhaps surprisingly, there's a potential source of problem even for the more well-to-do nations. I'll refer to this problem as *the time inconsistency of optimal government policy* (see also [2]), sometimes referred to as *rules rather than discretion*. It is intimately related to the forwardlookingness, just pointed out, of growth-promoting decisions as they affect aggregate A, K, and L. In particular, the streams of future returns from such high-expense decisions are affected importantly by the tax and regulatory environments at each point in time.

For some intuition about the problem, for now let's imagine an "ideal" world as our abstraction. Suppose there were a way to quantify, using a mathematical formulation, the welfare of the nation's citizens, today and indefinitely far into the future. A benign policy maker could then select policy (including tax policy) for the indefinite future so as to maximize this expression, that is, citizens' welfare. We may call the result *the optimal policy*. Naturally, this policy would take into account the effects of its future portion on earlier private-economy decisions, such as the effect future capital-income taxes have on current investment. In general, such a policy would represent a prescription for what to do under various pre-specified circumstances (as governments aren't the *only* source of uncertainty in the world!)

As this policy is being implemented, suppose everything is moving along hunky-dory. But suppose also, after 5 years, say, the policy maker (or some hot-shot quantitative expert in his office) gets the bright idea of recalculating the optimal policy from that point on. That really ought to be unnecessary, as in our ideal world, the economic environment has not changed (in terms of the relations describing the motion of the economy) and the original calculated plan includes a blueprint for what to do also for the upcoming future. But suppose one goes ahead anyway. To everyone's surprise, perhaps (except those who understand time inconsistency!), a completely different policy path will be found than the continuation of the original plan. The implication seems to be, policy must change, in the interest of the nation's citizens. Not!!! Indeed, theory suggests that if the policy maker falls for the temptation to change, it would be bad, perhaps *very* bad, for society, as policy becomes much too focused on the short run at the expense of the long run.

What's the intuition for this inconsistency over time? Suppose we refer to year 0 as the year in which the original policy plan was determined. As mentioned, this policy takes into account, among other things, the effect policy (for example tax policy) from year 5 and beyond would have on private decisions (such as capital formation) in years 0, 1, 2, 3, and 4. But when year 5 arrives, those decisions for years 0–4 have already been made. So in the re-optimization in year 5, only the policy effects on investment and other private-economy decisions in year 5 and beyond would be taken into account, resulting in a completely different policy prescription than the continuation of the original plan. In particular, the new plan would surely suggest raising taxes on the income from capital that has already been built, perhaps with the promise of reducing taxes again in the future. It might suggest increasing the inflation rate significantly so that the real value of outstanding government debt issued at fixed nominal interest rates would be dramatically reduced, indeed almost to zero if they went so far as to create a hyperinflation—not uncommon, for example, in Latin America in the 1980s. The government might try to justify the change by arguing that it's facing a near-emergency situation. But potential investors would have to worry, Might the policy maker pull that one off again in the future? Even the mere uncertainty about whether or not it will happen again is enough to depress investment activity and therefore long-term growth.

Now you may ask (and it's a good question): If this is the situation under "ideal" circumstances, then what about the more realistic world where policy is selected under all kinds of political pressure from interest groups? Well, that should make it all the more likely that time inconsistency will rear its ugly head, especially in nations with weak institutions and poor property rights. In practice, the way this problem manifests itself is by making policy focused on the short run. Indeed, once one has understood this principle, an amazing number of developments in nations' economic histories can be seen in a new light.

Here's a paragraph from a comment (published in [1]) I made as an assigned discussant of a conference paper by James Alt on "The Evolution of Tax Structures":

> Since the government would like any change of policy to be considered a once-and-for-all change, it is natural that it would find an excuse (for example, an emergency such as a war) or another explanation in an attempt to make the new policy credible for the long run. In this sense, the above framework for thinking about changes in tax structures is complemented well by Alt's discussion on pp. 199–200. Policymakers might also argue that our understanding of the economic structure has changed and that a change of policy is necessary for that reason. It is hard to believe, however, that this could be done more than a couple of times and still achieve the intended "optimal" effects of the new policies. A likely eventual outcome is the time-consistent policy, which [2] showed could be quite suboptimal.

Doesn't this sound amazingly like what has happened in many parts of the world since the global financial crisis?!

3 Benefits of a Commitment Mechanism

An implication of this theory is that it is advantageous to shield policy making from political pressure—a kind of commitment mechanism to ensure that promised good policy will be carried out both currently and in the future. This principle has been understood and implemented by several nations in the arena of monetary policy, which is then carried out by central banks that are independent, to varying degrees, depending on the nation. For example, the Bundesbank in Germany for decades was regarded as the champion of consistency and transparency. The Federal Reserve Bank in the United States hasn't been far behind (although recently we've seen signs of erosion of that independence). The Bank of England was formally made independent in 1997. The central banks in Scandinavia generally have been made quite independent.

At the other end of the spectrum are central banks such as that of Argentina. One sign of a central bank's independence and consistency is that the president or chairman remains in that position for a considerable number of years. In the United States, for example, Janet Yellen is only the seventh chairperson since 1951. In contrast, over the 70-year period 1945–2015, Argentina's central bank had 56 presidents—an average of only 1.25 years per president. In the especially tumultuous year of 2002, the central-bank president changed three times!

Argentina is an interesting case in that it tried a different commitment mechanism —a currency board. In 1990, the nation was recovering from the decade of the 1980s —the so-called Lost Decade—which ended in hyperinflation, defaults on government debt, losses of pensions, and during which the nation's output per capita had dropped by over 20%. Figure 3 shows real GDP per working-age person over much of the post-World-War-II period. (Note that, unlike Figs. 1 and 2, this chart is on proportional (log) scale. In such a plot, constant growth rate will be represented by a straight line, for example that drawn, with the slope of the average growth rate over the entire period.) As newly elected president, Carlos Menem decided, in 1991, in order to raise the confidence among investors in his nation, to make the Argentine peso exchangeable one for one with the dollar, accumulating enough dollar reserves to make that policy seem credible. To the naked eye (as in Fig. 3), growth rates over the next half-dozen years look like this policy worked. But then, starting around 1998, it all fell apart. Output per capita fell again by over 20%, this time over a much shorter period of about 4 years, the peso had to be devalued, bank deposits were frozen, and all sorts of bad things transpired to the economy (Fig. 4).

The explanation usually given for the failure of this "commitment mechanism" is that, while Argentina seemed to have fixed its monetary problems, they forgot about fiscal policy, that is, tax and spending policy. Monetary policy cannot be completely separated from fiscal policy. For one thing, both are part of the same budget constraint. The provinces borrowed heavily, even in the seeming good times of the 1990s. When it became clear they wouldn't be able to pay back their debt, they came running to the federal government who had to bail them out. Federal debt ballooned and everything fell apart.

As an indication of how bad things can get when the "time-inconsistency disease" attacks, consider the behavior of Argentina's stock of business capital per working-age person (a per-capita picture would look essentially the same), which in a healthy nation ought to grow steadily over the long run. Its level peaked in 1982. As of 2008, the last year for which comparable data are available, this ratio was still about 10% below its 1982 level. It's a good bet that, even more than three decades after 1982, it's still lower.

An opposite example is Ireland over that same time period, from 1990 until the early 2000s. Starting in the 1960s and 1970s, Ireland had made secondary education

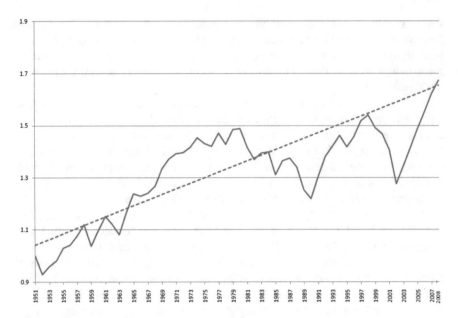

Fig. 3 Argentine log real GDP per working-age person

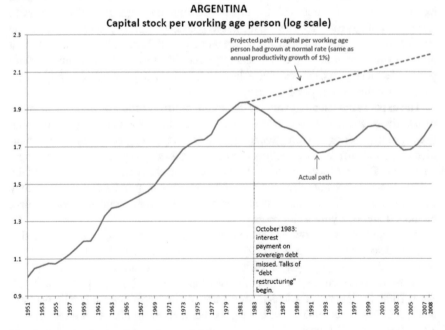

Fig. 4 Argentine log capital stock per working-age person

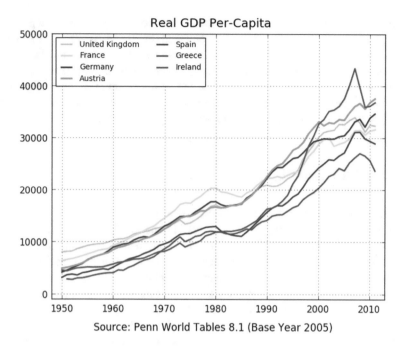

Fig. 5 Per-capita real GDP for European nations

free of charge. As a consequence, by 1990 the nation found itself with a potentially skillful work force, but not enough factories and equipment with which to put all of these skills to use. So the government decided to do their best to remove any uncertainty about future taxation. They announced that if you, Irish or foreigner, set up shop here, these will be your (not very high) tax rates in 1992, 1993, and so on, all the way to 2009. Of course there may have been other favorable factors as well. The bottom line is that Ireland grew spectacularly (Celtic Tiger), going, in the course of one decade, from being one of the lower per-capita-income countries in Western Europe to one of the very highest (see Fig. 5), surpassing Germany, the U. K., and France, for example.

But alas, this story does not have a completely happy ending, on account of policy makers panicking at the onset of the financial crisis in 2008. In the 2000s a debt-driven property boom had taken hold. When property values plunged in 2008, in part affected by what happened in the rest of the world, major Irish banks faced insolvency. The government made the highly questionable decision to bail them out, in the process saddling tax payers with a huge increase in government debt. This ending, however, does not take away from the lesson from the experience in the 1990s, when removal of uncertainty about the tax environment for the foreseeable future encouraged companies, foreign and domestic, to establish and expand productive capacity in Ireland.

Note the important role of fiscal policy in the adverse outcome in Argentina and in the good thing that happened in Ireland. It's hard, however, to see how the solution to the time-inconsistency problem for monetary policy—independence from political pressure—could be implemented in the fiscal arena. So how to commit to good fiscal policy? (I sometimes suggest that if some hot-shot young economist comes up with a solution to that problem, then 30 years later he may get to stand before the King of Sweden and accept an important prize!) At this point, at least, the case of Ireland seems too much like an aberration. How many nations would be likely to commit credibly to something analogous?

4   Rebuild Credibility?

An interesting question is: If a nation, such as Argentina, falls prey to the "time-inconsistency disease," how easy would it be to rebuild its credibility? The answer has to be, not easy at all. As already mentioned, Argentina, after the Lost Decade of the 1980s, seemed to grow at acceptable rates. It is possible, however, to "check the temperature" of a nation using a standard growth model as the "thermometer." As reported in [3], with the aid of the best available measurements, this "thermometer" showed that Argentina was "ill," that in light of its technology level it should have grown much faster even over the 1990–98 period. Especially the capital stock grew much more slowly than the model said ought to have been the case. This finding indicates that, in spite of President Menem's best intentions, the nation still suffered from severe lack of credibility among potential investors.

5   But the Mere Consistency of Policy Is Not Sufficient

It is important to emphasize that policy consistency is not *sufficient* for healthy growth. Notice the word *optimal* in the description of the basic time inconsistency problem. It won't be good if a country carries out a policy that's consistent, but *bad*. I will argue that China is such an example. Its economic policy certainly appears quite consistent. Admittedly, China has made important strides in its development, although, as seen in Fig. 1, its per-capita income is still low by international standards.

A stylized description of how nations grow in the long run is as follows. Entrepreneurial and innovative activity takes place, resulting in new and better ways of producing things, new production processes, new products, generally with the help of both private and public research and development. Factories, machines, and office buildings are needed to implement these innovative ideas. Workers are hired. Incomes grow. And so on. In order to facilitate all of this activity, however, a healthy banking system, or financial system, more generally, is important, as these costly decisions cannot be undertaken without the required funding.

As described in [6], in China banks are generally state-owned. These banks favor the state-owned companies. The state-owned enterprises (or, more generally, the large well-known companies) have easy access to credit and, at least until recently, to cheap labor. In the meantime, the less-established entrepreneurs with the really innovative ideas for products or ways of doing things have a hard time getting the necessary loans. They often have to save up in advance before they can implement their ideas. Naturally, activities that are relatively intensive in labor rather than capital are easier to finance. The overall result is a huge waste of resources. In other words, with the same use of resources, China could have grown substantially faster than their already high growth rates. It's a good bet that unless China opens up for more competition in the financial sector, this problem will eventually impede their long-run ability to grow at acceptable rates.

6 Some Comments on Recent Events

These ideas provide food for thought about what has been going on more recently in many areas of the world, including in the United States and Western Europe. Starting with the United States, Fig. 6 plots real GDP per capita post WWII. The straight line represents average growth 1947–2007 and is extended to the present. There are of course ups and downs about that straight line—what we call business cycles—but it does an amazing job in accounting for the long-run growth over these 60 years. The startling part, as further emphasized in Fig. 7, which magnifies the most recent time frame of Fig. 6, is how far below the trend line the economy fell in 2008 and after—by on the order of 12%. And worse, unlike prior recoveries, which were typically quite rapid once the bottom was hit, so far there's not been any sign of moving back towards the old trend. On the contrary, the two curves are still diverging more than a half-dozen years on.

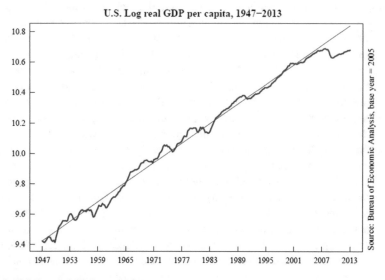

Fig. 6 U.S. log real GDP per capita

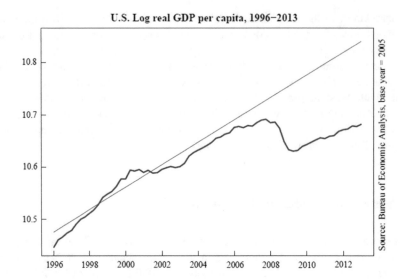

Fig. 7 U.S. log real GDP per capita, 1996–2013

Of course there are several factors contributing to the severity of this recession. One thing is remarkable: Unlike past recessions, the severe decline happened without an initial slowing of productivity. Another aspect has got some attention: The decline in consumption was relatively small by recession standards. The recession is largely investment driven.

As Zarazaga and I [4] show, a large portion of the recession can be accounted for as follows. Around 2008, the growth in the debt/GDP ratio, partly because of stimulus packages, partly for other reasons, started to gain attention in the press and elsewhere. Indeed, even before the financial crisis, the U.S. debt had been projected to rise substantially, largely as a consequence of the "baby boomers" retiring in ever larger numbers. The Bush tax-reduction law of 2001 already called for taxes to go back up starting Jan. 1, 2011. (As it turned out, this increase was postponed until 2013.) Suppose capital owners in 2009 were struck by the sentiment that taxes would rise in the future in order to keep the debt from growing further (say, as estimated by the bipartisan Congressional Budget Office). Suppose, to be specific in our model experiment, they thought the required tax increase would last for 10 years, starting in 2013. The time-inconsistency insight indeed suggests that capital income would be the main target. Our experiment, using a standard neoclassical growth model calibrated to the U.S. economy through 2007, with the magnitude of the tax increase calibrated according to the Congressional Budget Office's estimate, accounts for most of the decline in investment, about half of the decline in labor input, and it is the only explanation we're aware of that is consistent with consumption falling relatively little in such a severe recession. Moreover, the experiment indicates it could take a long time to move back to the vicinity of the old trend (whose slope, we show, ought even to be reduced somewhat because of demographic factors affecting the population and the composition of the work force). In summary, the culprit, in large part, seems to be a sentiment that developed soon after the financial crisis that capital-income taxes would have to rise in a few years' time, which, by the way, is in line with the insight from the time-inconsistency principle. Interestingly, when we modified our experiment to make all of the tax increase fall on labor income instead of capital income, then it didn't account at all for what has happened over these years.

In Europe, the euro zone, with its fixed exchange rate among a large number of countries, was conceived with seemingly little attention to enforceable fiscal rules to accompany the new monetary arrangement. As we know, some nations borrowed heavily and have had to be bailed out. Even *within* a nation such as Spain, much unchecked accumulation of debt by the provinces appear to have taken place. One is reminded of Argentina in the 1990s.

Since the failure of Greece, one often heard mentioned as potential additional problem nations Italy, Spain, Portugal, and Ireland. Let's get a sense of their backgrounds in terms of the main driving forces for sustainable growth: innovative activity and technological progress, as reflected in total factor productivity (TFP) and in labor productivity (output divided by hours worked). Figures 8 and 9 graph the logs of those two data series for each of the four nations. The average growth from 1960 to 1990 is indicated as a straight line and extended to the present. The shocking thing is that, for Spain, Portugal, and Italy, growth in both TFP and labor productivity basically came to a full stop in the early 1990s!

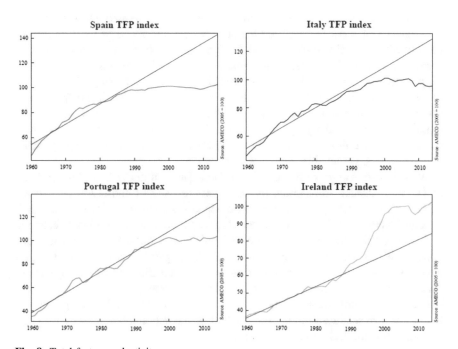

Fig. 8 Total factor productivity

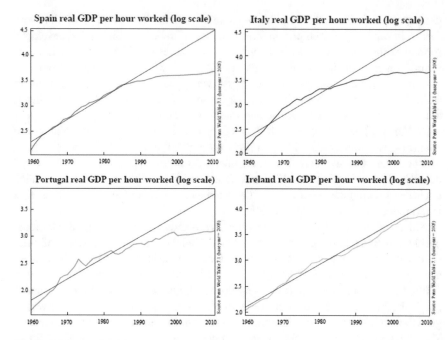

Fig. 9 Labour productivity

One might have suspected that the slowdown in these nations was partly a consequence of them having been tempted to take advantage of the low interest rates after joining the euro area and "live the good life." While there could be something to that, these charts show that these nations' problems are much more deep-seated and appear to date back to well before euro memberships. It's tempting to conclude that the attention, even blame, on the euro one has heard from many sources is only a "red herring" which, if anything, has distracted from dealing with more fundamental underlying structural problems. Until these three nations figure out how to make their respective curves in Figs. 8 and 9 turn back to significant positive slopes, sustainable growth will be lacking.

Now, in cooperation with Enrique Martínez-García of the Federal Reserve Bank of Dallas, I decided to entertain the idea that euro wasn't entirely a red herring. Although memberships in the euro zone actually started in the late 1990s, their announcements were made several years before, resulting in a significant decline in the country-risk components in these nations' interest rates. It turns out that this feature, and its subsequent repercussions, in large part can account for the productivity slowdown. In Spain, for example, one dramatic way in which it manifested itself was as follows: Suppose we divide economic activity into two sectors —tradable and nontradable. In the late 1980s, the relatively much more productive tradable-goods sector accounted for one-half of the nation's output. Twenty years later, this fraction had sunk to one-third.

For comparison in Figs. 8 and 9, I include the plots also for Ireland, a nation that after 2008 was sometimes mentioned in the same breath with these other three countries, in part because of the already mentioned build-up of its debt/GDP ratio. TFP displays an impressive pick-up in the 1990s, but then flattens out. Eventually, so does labor productivity. With its large debt, Ireland surely has its problems, but at least from a productivity standpoint, the situation looks much less dire than for the other three nations, as the flattening started only relatively recently and from a substantially higher level. These labor-productivity numbers for Ireland currently are on the order of 40–50% higher than those for the other three countries. This fact suggests that, if Ireland makes sure not to harm investors' confidence and credibility as a reliable nation in terms of its future policy environment, Ireland should be in good shape over the long run.

7 The Road Forward

It is probably fair to say that an unprecedented amount of uncertainty prevails about future economic policy in European Union countries as well as in the United States. In Europe, we have seen politicians reacting to short-run developments without any clear plan for next year and beyond. As has been argued in this chapter, this kind of uncertainty is bad for growth. One cannot fault potential innovators and investors in business capital if they choose to wait on the sidelines for a while. Worse, the insights from the time inconsistency literature give reasons to be pessimistic as to whether this uncertainty will be removed soon in any meaningful way.

As already suggested, fiscal policy is the key. But unlike with monetary policy, it is difficult to see how, with few exceptions, nations can commit to good long-run fiscal policy. Since the global financial crisis, it's too easy to claim that we're in a near-emergency situation and therefore to argue for a change of policy, which, if motivated primarily by short-run considerations, represents a turn for the worse. Or one may suggest that past modeling approaches have been at fault, that new approaches are needed, and that these models call for a change of policy.

A problem with time inconsistency, even for countries with strong institutions, is that although we understand the nature of the temptation, we don't know if, or when, the country might actually fall for that temptation. Hence, growth-promoting private-sector decisions that require looking far into the future are likely to be affected, to be postponed perhaps, or even completely discarded. The most shocking thing would be if even Ireland, with its relatively high productivity, not to mention well-deserved built-up credibility, were to fall for the temptation!

References

1. Kydland, F.E.: Implications of dynamic optimal taxation for the evolution of tax structures. Public Choice **41**, 229–235 (1983)
2. Kydland, F.E., Prescott, E.C.: Rules rather than discretion: the inconsistency of optimal plans. J. Polit. Econ. **85**, 473–491 (1977)
3. Kydland, F.E., Zarazaga, C.E.J.M.: Argentina's lost decade and the subsequent recovery puzzle. In: Kehoe, T.J., Prescott, E.C. (eds.) Great Depressions of the Twentieth Century, pp. 191–216. Federal Reserve Bank of Minneapolis (2007)
4. Kydland, F.E., Zarazaga, C.E.J.M.: Fiscal sentiment and the weak recovery from the great recession: a quantitative exploration. J. Monet. Econ. **79**, 109–125 (2016)
5. Parente, S.L., Prescott, E.C.: Barriers to Riches. The MIT Press (2000)
6. Zheng, S., Storesletten, K., Zilibotti, F.: Growing like China. Am. Econ. Rev. **101**, 196–233 (2011)

Perception and Reality of the Spanish Economy

Fernando Casado Juan

Abstract The current chapter on "Perception and reality of the Spanish economy" is a summary of the situation of the Spanish economy. The chapter starts with the positive evolution of the Spanish economy in recent years, which has seen annual GDP grow in excess of 3%. It analyses the main causes behind this, paying special attention to the strong performance of the External Sector in the current global economic context. At the same time, it analyses the key macroeconomic factors underpinning the Spanish economy, as well as developments in GDP, the public deficit, labour market, foreign investment, tourism, training, infrastructures and company size. The chapter closes by highlighting the importance that the digitalisation process will have in all areas of life, not only economic, but also across wider society. In summary, the chapter is not only useful for those interested in understanding the reality of the Spanish economy, but also for all those who want a primer on the issue.

It is a great pleasure for me to be able to participate in this book which has been prepared to honour His Excellency Dr. Jaime Gil Aluja on his 80th birthday in consideration of his magnificent academic career and his outstanding personal qualities. I have shared many years of collaboration with him which date back to my youth, both in University (his constant support from 1970 helped me to obtain the academic degrees that I have) as well on the governing bodies of the Institutions that we have both participated in, such as the Royal Academy of Economic and Financial Sciences or the EAE Business School, where he has always been a point of reference that I have sought to follow, in addition to offering me his friendship.

Data and information provided by Adecco, Bank of Spain, European Central Bank, International Monetary Fund, World Bank, World Trade Organisation, National Stock Exchange Commission, Ministry of Economy and Competitiveness, Ministry of Public Works, ICEX, Ministry of Housing, National Statistics Institute, National Telecommunications Observatory, Information Society and the research departments of companies belonging to the Competitiveness Business Council.

F. Casado Juan (✉)
Fundación E.A.E, Calvet 41 Entlo. 5º, 08021 Barcelona, Spain
e-mail: Fernando.casado@racef.es

© Springer International Publishing AG 2018

C. Berger-Vachon et al. (eds.), *Complex Systems: Solutions and Challenges in Economics, Management and Engineering*, Studies in Systems, Decision and Control 125, https://doi.org/10.1007/978-3-319-69989-9_2

1   Executive Summary

The current situation of the Spanish economy should be tackled by first starting with the state of competitiveness in Spain. In order to do so, it is necessary to distinguish between two concepts that nowadays are associated with the term "competitiveness" at the country level. On the one hand, there is so-called external competitiveness which refers to the position of an economy's goods and services in the global market. And on the other, the broader concept of country-competitiveness which relates not only to export performance but rather to the "combination of institutions, policies and factors that determine an economy's level of productivity".

With regard to the first concept, the performance of our economy can only be described as exceptional, especially during the crisis and subsequent recovery: since 2009, real goods exports rose by 42%, contributing to the correction of the record current account deficit accumulated in the pre-crisis years (from levels of close to 10% of GDP in 2007 to a surplus from 2013 onwards). Spanish export competitiveness is also reflected in Spain's relatively stable share of world trade over the last 15 years, compared to the declines registered in the majority of European countries.

This has been the result of the achievements made in terms of unit labour costs, with a cumulative reduction of close to 7% (compared to increases of over 5% in the main Euro Zone countries). It is also due to an effective policy of product and geographic diversification with an orientation towards higher value added projects and emerging regions, which has enabled our exporters to confront the worst years of the crisis from a stronger position. And, finally, an exporter base led by large, capable and innovative companies, which have been able to position themselves in key sectors such as infrastructures, telecommunications, the automotive sector or tourism, as well as in leading innovation activities such as the agroindustry or biotechnology.

But being more competitive requires addressing and reducing the still elevated productivity gap with the rest of Europe and especially in comparison with countries such as Germany. It is this point where the second competitiveness concept applies. If we pay attention to the diagnosis made by expert international organisations, Spain's position on country-competitiveness indicators is some distance away for what it should correspond to when looking at GDP per capita. While only 16% of countries have a higher GDP per capita than Spain, this percentage increases to 23% if we consider country-competitiveness in terms of the World Economic Forum ranking. The main areas for improvement are the excessive and complex regulatory environment, labour market inefficiencies, the suboptimal quality of the education system and the excessively fragmented business fabric.

2 Global Economy Context

Although the global economy has shown signs of recovery in recent years, it has done so in an environment of uncertainty and generalised wariness, which is not at all propitious for the main western economies. They are exposed to an international landscape that has suffered profound and intense changes in the last 3–4 years as a result not only of the current international economic crisis, but also due to other important factors.

In July the IMF revised down its forecast for growth in 2016 and 2017 by 0.1ppts to 3.1% and 3.4% respectively. This review is due to the uncertainty generated by Brexit. But during the start of 2016 economic activity was moderately stronger than expected, especially in emerging economies, with data continuing to move in line with expectations in July and August. And the risks surrounding the Chinese economy appear to have significantly diminished.

In terms of interest rates, and in spite of Yellen's optimistic speech on the American economy at Jackson Hole, the market only foresees one hike in Fed Funds rates during the next 18 months. It remains unclear whether the hike will come this December or in 2017. In this context, stock markets and emerging economy exchange rates remained stable in August. By contrast, Sterling has depreciated by 13% over the year, which could have global implications.

Oil prices in August remained at similar levels to last year, moving on the back of rumours of a potential freeze in OPEC crude oil production, which was reflected in strong movements in hedging of short positions. Although the data continue to show declines in crude oil production in US, the increase in inventories is explained by the rise in imports.

Against this backdrop it is worth highlighting the importance of the globalisation of markets and the economy in general. This globalisation has been substantiated by significant advances in technology and innovation, especially digitalisation, during the last decades, which has increased the need for companies' internal operations to function on the basis of economies of scale and on a global level. In other words, companies have to operate with a significant production base that enables reductions in unit labour costs and, at the same time, division of infrastructure costs across a wide range of markets. Both factors serve to highlight that those economies, and for that matter companies, that are committed to processes of internationalisation are also those that are most able to improve their efficiency and competitiveness—all of which are key aspects and a necessary condition to survive in a competitive international environment.

3   Relevant Aspects of the Spanish Economy

3.1   *Evolution of GDP*

Spain's economic recovery began in the third quarter of 2013 and has resulted in an increase in GDP of 9.4% to the end of 2016, meaning that Spain has practically recovered what was lost during the pronounced economic and financial crisis. Nonetheless, we should bear in mind that in order to maintain current growth rates (elevated in the context of numerous internal and external factors) over the medium-term, our country continues to need a series of structural reforms to improve efficiency and competitiveness and ultimately consolidate our labour market recovery.

The recovery has benefited from a relatively favourable international environment with world growth of around 3%, falling oil prices, geopolitical uncertainty in some of the main competitor countries for the tourism sector and ultra-expansive monetary policy. Other factors have also helped, including the release of pent-up demand, the end of the adjustment in sectors such as real estate, the recovery of net financial wealth, an expansive fiscal policy and reforms introduced in recent years.

The Spanish economy will expand for a fourth consecutive year in 2017, with GDP increasing by 2.3%. This implies a deceleration from rates of growth reached in 2015 (3.2%) and 2016 (3.1%). The recovery is holding up thanks to the continuation of various tailwinds which have supported growth in internal spending and exports. However, the impact of some of these tailwinds is beginning to wane. Furthermore, the need to continue with the fiscal adjustment and increased uncertainty regarding future political economy will restrict growth in domestic demand (Fig. 1).

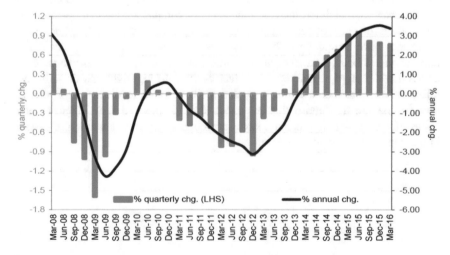

Fig. 1 Evolution of Spanish GDP 2008–16

3.2 Public Deficit

The public deficit closed 2015 at 5.1% of GDP. In the first quarter of 2016 the public deficit, excluding local corporations, reached 3.35% of observed GDP, this compares to a 2015 figure of 3.41%. The new deficit objectives set by the EC is 3.7% of GDP for 2016 as a whole and 2.5% in 2017.

According to the OECD report "Government at a Glance 2011", the weight of state revenues in Spain is relatively lower than elsewhere in the OECD, while regional and local spending have a much higher weight There appears to be a larger disparity between the capacity to decide revenues and spending than in other OECD countries, which could affect the responsible management and efficiency of public spending. This is an issue that is coming to fore in the current debate regarding the central and regional Government deficit ceilings (Fig. 2).

3.3 Public Debt

The Spanish economy has made firm progress in recent years in correcting the external imbalance: the current account has swung from registering a deficit of 9.6% of GDP in 2007 to an average surplus of 1.3% of GDP between 2013–2015. Despite the improvements that have been made, one should not forget that one of the major weaknesses of the Spanish economy continues to be the elevated stock of net international debt, which requires continued improvements in competitiveness and a sustained current account surplus over time in order to reach sustainable levels.

In 2016 Spain's net international investor position stood at −90.8% of GDP. This figure compares to 2000, when Spain registered a NIIP of around −35% of GDP, the maximum limit established in the European Commission's "Macroeconomic Imbalances Procedure". However, from 2001 onwards the NIIP deteriorated significantly, reaching a maximum of −97.7% of GDP in Q1 2015, one of the

Fig. 2 Evolution of GDP

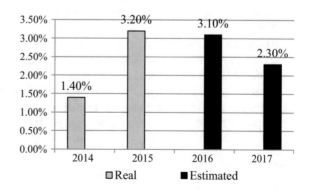

Fig. 3 Evolution of public
and private debt (% GDP)

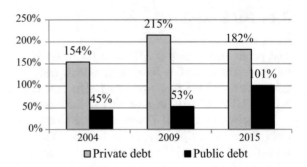

□ Private debt   ■ Public debt

highest net debt positions in the world. In fact, at the end of 2015, only the U S had
a higher net debt position in absolute terms (Fig. 3).

3.4   Labour Market

One of the factors that most affect the image of the Spanish economy is the current
unemployment rate. In order to remedy this situation, the most significant reform to
the Spanish labour market in recent times was introduced with the aim of aligning
the Spanish labour market with the rest of Europe in terms of flexibility and costs.
The 2012 Labour Market Reform addressed substantial issues that constituted a
clear handicap relative to our competitor countries and enabled unemployment to be
reduced to below 20%. The reform tried to simultaneously tackle cross-cutting
issues such as increasing companies' internal flexibility, reducing the elevated rate
of temporary employment by promoting permanent contracts, increasing job
security, eliminating administrative barriers, doing away with ultra-activity and
reforming public employment services, amongst others.

The labour market reform is expected to generate employment in the
medium-term through two main lines: firstly, through the reduction in dismissal
costs and lower rigidities in salary negotiations. Secondly, the reform could also
have an indirect impact, creating jobs through the increase in part-time employment, which in Spain represents 14% of employment, compared to an EU average
of 20%.

All of this should allow the current unemployment rate to be reduced to around
17.4% in the medium-term. The fight against fraud and increased control over
unemployment benefits should also help to lower current unemployment.

In terms of taxation, Spain's Social Security contribution rates are relatively
higher than the rest of the European Union. These are the equivalent of a tax on
employment and therefore limit hiring possibilities.

4 Key Aspects of the Spanish Economy

It is important to identify those arguments that help to defend Spain's position as a sustainable, credible and competitive economy, highlighting in particular Spain's external openness as one of the fundamental pillars of any modern economy, as well as the Spanish economy's attractiveness as a destination for Foreign Direct Investment and the competitiveness and leadership of Spanish companies abroad.

4.1 External Sector

Spanish exporters have advanced significantly, increasing from 100,000 to 150,000 between 2010 and 2015 with regular exporters (firms that export for four successive years) rising from 38,000 in 2012 to 48,000 in 2015.

It is important to highlight that this increase in exporting companies has spurred goods exports which have increased their weight in GDP from 15% in 2009 to 23% in 2015, eight points more. Likewise, the weight of services exports has increased from 7.5% in 2009 to 10% in 2015, meaning that the combined weight of goods and services exports in GDP has increased by ten percentage points since the crisis.

Since 2013, non-tourism services exports have practically doubled. It is particularly worth highlighting the strong performance of the information and IT sector, which has tripled, and of business services, such as consultancy or R&D. These sectors, together with transport-related services, accounted for more than 80% of non-tourism services exports in 2015.

Looking at how the current international backdrop is defined, we can see that company efficiency depends on operating with a global market base, where production levels are compensated by economics of scale and geographic distribution, resulting from the globalisation of markets and, therefore, an unstoppable internationalisation process. This necessary and sufficient condition to be able to compete globally, is being met by a quite a significant proportion of our companies, who are oriented and compelled towards international markets, with a high degree of international commitment.

A significant group of Spanish companies are well focused and positioned in international markets, with a high level of international commitment both to exports and direct investment, which represent a significant proportion of total sales. In fact, 60% of IBEX-35 companies' turnover is obtained from international markets. These companies are leaders in high-value added sectors with future growth potential, such as banking, fashion, retail, renewable energies, infrastructures, engineering, hotels, etc. Furthermore, destination markets are configured towards high growth potential zones such as Latin America, Eastern European countries, the Middle East and Asian countries.

In this regard, this regional diversification has not only reduced our dependence on the European Union but also ensured that the long-term international business of our companies is focused on high growth regions.

Large Spanish companies are also becoming beacon companies for SMEs, creating an important business cooperation network and serving as a source of competitive advantage. This phenomenon is one of the fundamental pillars that underpin our country's competitiveness.

The Spanish economy has made important progress in recent years in correcting the external imbalance: the current account has swung from registering a deficit of 9.6% of GDP in 2007 to an average surplus of 1.3% of GDP between 2013–2015 (Figs. 4, 5, 6, and 7).

Fig. 4 Exports per capita (euros)

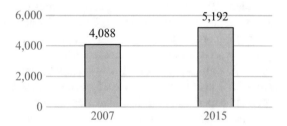

Fig. 5 Exports: diversification by destination (% total)

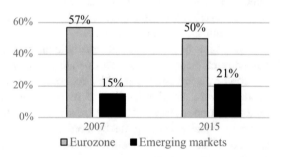

Fig. 6 Number of exporters ('000s)

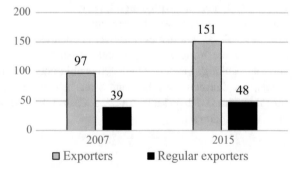

Fig. 7 Improvement in trade balance (2007–15) by type of good per capita (% GDP)

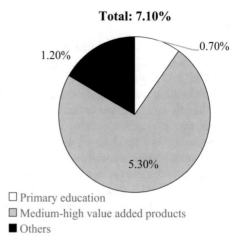

Total: 7.10%

0.70%

1.20%

5.30%

☐ Primary education
▨ Medium-high value added products
■ Others

4.2 Foreign Direct Investment (FDI)

Various factors are making Spain an attractive destination for FDI, but without doubt one of the most important reasons is the role that Spain plays as a business platform to one of the regions with the highest growth potential, Latin America, with 600 million inhabitants and a middle class which has expanded by 150 million people in the last decade. Furthermore, Spain serves as a landing point for Latin American companies in Europe, which brings together a combined market of 1.1 billion people.

Other factors explaining Spain's attractiveness for FDI include the increase in the number of foreign companies that decide to locate their R&D centres in our country, as well as the fact that Spain continues to be in the top 15 countries worldwide for brand image or reputation.

Spain, Switzerland and Germany are the only European countries to have two cities considered as preferred investor destinations, Spain also has one of the top 3 business schools and is the sixth EU country in terms of economic freedom, above France, Italy or Belgium. These are just a handful of data.

4.3 Tourism

The Spanish tourism sector currently leads the World Economic Forum's global rankings of tourism competitiveness. Among the factors contributing to the improvement in Spain's position from eighth in 2011 to first place is the improvement in the ranking of the cultural and natural resources endowment, which is reflected in the growing number of locations declared to be heritage of humanity, high interest intangible assets, sport facilities and the progressive increase in business tourism.

It goes without saying that tourism has been the key driver of Spanish economy growth, with tourism GDP growing by 3.7% in 2015, marking an acceleration from 0.6% in 2013. As a result, the tourism sector has contributed 0.5 additional points to Spanish GDP, increasing its share to 11.7% of total GDP or 124 billion euros.

This dynamic activity has translated to the labour market, with one out of every seven new job openings being in the tourism sector in 2015. In other words, this corresponds to 73,343 new jobs (5.5% more than in 2014), leading to the current figure of 1.4 million social security registrants.

Forecasts for the current year reinforce this favourable performance. As a result, tourism GDP will grow by 3.4%, a performance encompassing all Spanish destinations and subsectors in the tourism value chain.

4.4 Training

Training is one of the main levers for raising productivity given that it facilitates highly qualified and high performing workers. The absence of an education system focused on productive employability inhibits the educational quality of the labour force from an early stage.

Comparing the percentage of the employed adult population (25–64 years) that has participated in training activities recently (last 12 months), it can be seen that, while the EU-27 average is close to 42%, in Spain the proportion is 35.6%.

According to Eurostat, public unemployment support invested in training is extremely low in Spain, which ranks 19th within the EU-27.

Year after year the PISA reports place Spain below neighbouring countries and the World Economic Forum puts our education system in 107th place (a long way behind the OECD average). Only Malta beats Spain in terms of school dropout rates at the European level (which are above 30%).

In our system (and at all levels) there are a lack of values embodying work ethic and a culture of endeavour, something that is apparent in the high level of absenteeism. Achieving greater convergence in education towards leading countries in the field is vital to drive up our productivity, as well as to support labour integration, given that more years in education are associated with an increased chance of finding employment. Differences in education levels explain between 30 and 40% of the differences in productivity between OECD countries and the Spanish regions.

The outlook for job opportunities in the coming years points to a growing need for highly qualified professionals: more than half of the vacancies that will be created to 2025 (8.8 million in total) will require a high level of education, such as technical positions and directors.[1]

---

[1]Cedefop Spain [4].

4.5 Infrastructures

Spanish companies are leaders in international construction with foreign sales of 80 billion dollars, 13 companies among the 250 largest global contractors and a leading position in concessions having invested 50% of the capital acquired by the ten largest concessionaries.

This international presence is geographically well diversified, with a leading position in Latin America and an important position in North America and the rest of Europe, and progressively extending into other growth areas such as the Middle East or Asia with the awarding of flagship projects.

4.6 Company Size

One of the key levers for changing the productive model is to increase the size of our companies which are currently much smaller than comparable European countries. A convergence in the proportion of companies with more than 50 employees towards the average of Germany, France and Italy would imply needing 15,000 new companies in this range. This would in turn have a significant impact on our economy both in terms of employment and productivity, thanks—in part— to the significant carry-over effect that larger firms have on SMEs.

According to Eurostat estimations for 2014, the average number of workers per company in Spain is less than half that of the United Kingdom or Germany. The lower average company size is explained by the lower relative presence of companies with more than 50 employees[2] (which account for an estimated 0.8% of the total, compared to 3% in Germany and 1.9% in the United Kingdom). Thus in Spain there are 3 million companies,[3] of which, only 24,000 have more than 50 employees, and this figure falls to 3,800 with more than 250 employees.[4] This fact is not due to any greater dynamism of business in Spain, but rather structural weaknesses which impede growth and the consolidation of "young firms" in line with a normal company development cycle.

In addition, the concentration of employment in small and medium-size companies (more than 60% of Spanish employees work in companies with less than 50 people) has been embedded through a series of regulatory and fiscal incentives that can limit their subsequent medium-term development. The evidence shows that

---

[2]We consider the European Commission's definition of companies according to the number of employees. So, we consider micro-company (0–9 employees), small company (10–49), medium (50–249) and large (more than 250 employees).

[3]Including professionals and self-employed. If this figure is excluded, the number of firms is 1,246,381, according to Social Security statistics.

[4]Statistics of companies enrolled in the Social Security, July 2014.

these measures lead to cliff effects in company growth, encouraging informal labour and reducing tax revenues.

As a result of these stylised facts, one of the priorities for political economy in increasing long-term growth has to do not so much to do with increasing the number of firms themselves but rather creating an environment that is favourable to existing firms increasing their size and being able to become more productive, competitive and international (Figs. 8, 9, and 10).

Fig. 8 Innovative companies in Spain by size, 2010–2012 (%)

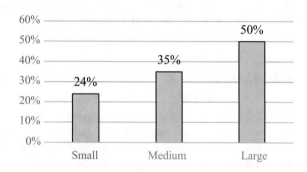

Fig. 9 Company productivity by size (US$/ employee)

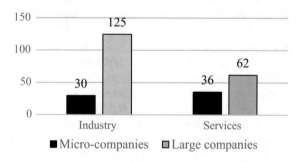

Fig. 10 Weight of exports in total company value-added by size

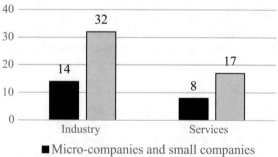

4.7   Digitalisation

The silent digital revolution that we are living is one of the most powerful and prominent mega-trends. The estimate that by 2018 there will be more than 4.5 billion smartphones, more than 2.5 billion social network users or more than 25,000 connected objects in the world bears witness to this. Indeed, the internet of things and all its implications are already a reality.

Digitalisation is a broad concept not just because it covers different technologies, but also because it depends on the willingness of countries to exploit the opportunities that these technologies offer to fit their description, this in turn depends on variables such as the level of qualification of the general population and of employees in particular, as well as the degree of development of different sectors, amongst other things.

Hence there exist numerous indexes and rankings to classify countries' degree of digitalisation which often respond to different concepts. These are set out in detail in chapter "A Model for the Management in Organizations Based on People and Knowledge: Aspects to be Considered in Its Design".[5] According to the different measures of the degree of digitalisation at the international level, Spain is among the top 25% and within the EU28 Spain it is practically in line with the European average, though still a long way from the level of leading countries in digitalisation.

As such, Spain has a long way to go if it aspires to become a truly digital economy and society. This means that it is important to focus on the axes of the digital economy, which will be key to the future:

1. Data Axis: analytics, filtering and visualisation, industry 4.0, the internet of things.
2. Cloud Axis: integration services, new computing techniques.
3. Digital Security and Privacy Axis.
4. Social Axis: open innovation, new media.

The development of the Digital Agenda, where more than 90% of measures are in implementation or completed and which affects different economic sectors, highlights the high extent of commitment by not only Spanish but also European authorities to increasing the degree of digitalisation at all levels.

This global push is already having a positive impact both in terms of global growth and employment creation. In the period from 2007 to 2010, digitalisation was responsible for 9.2 million jobs per year, accounting on average for a third of new jobs.[6] A year later, in 2011,[7] it is estimated that digitalisation explained 3% of

---

[5]Competitiveness Business Council [5].

[6]Strategy & (Formerly Booz&Company) [13].

[7]World Economic Forum [21].

the increase in total GDP and 15% of new jobs created, with developed countries enjoying the largest economic impact.[8]

In aggregated terms, reaching peak digitalisation levels in Spain in 2020 could result in GDP in that year being 3.2% higher,[9] which would imply 34.6 billion euros more GDP. Total employment would be 1.3% greater in 2020 compared to a scenario of no progress in digitalisation, which corresponds to nearly 250,000 new jobs.

However, while there is little room to doubt the benefits of digitalisation at the level of value creation for the country as whole, which is greater the more developed the level of digitalisation in the country, there are more discrepancies with regard to the impact on existing employment, although in global terms the impact on employment is expected to be positive.

Either way, job creation in the near future will require profiles that are completely different from what was required a few years ago.

New digital technologies as an education tool are an unprecedented innovation and have the potential to revolutionise how we understand education today. Living in a digital world implies new training needs in both the general population, as well as professionals working in this area and the role played by digital education will be key to individuals' employability. In this context, a 40% increase in the hiring of these types of professionals is expected for 2016[10] in our country.

5 Conclusions

It is undeniable that the Spanish economy faces a difficult but necessary change to its historic pattern of growth and that it is substituting the construction/debt cycle for a model based on diversification and internationalisation, and all of this must be done at a time of heightened market uncertainty. But it is also true that the economy takes on this enormous challenge from a solid base. Spain is in fact well placed at the international level, it is one of the largest European markets in term of income and size, its medium-term growth potential is above the EU's, it has leading and well diversified companies in key sectors, Spain possesses a strong and modern network of infrastructures and it remains a strategic spot to compete with certainty in a natural market of 1.1 billion inhabitants across Europe and Latin America.

But even so, the profound economic crisis has brought to light a series of accumulated imbalances, such as the overvaluation of real estate assets or a rigid labour market. The crisis has also led to a deterioration in public debt and has

---

[8]USA y Europe account for 30% of the gains in GDP but only 6% of employment (vs. their weight of 11% in global employment).

[9]Average of five selected econometrics analyses: Accenture Strategy (2015), Raul Katz. (Booz&Co), The Digital Single Market strategy [6], Digital Agenda for Spain (Ministry of Industry, 2013), Mckinsey "A labour market that works: connecting talent" (2015).

[10]Adecco (2016).

resulted in a weakening of the banking system, which has required significant restructuring.

Right now, having taken the decision to tackle these imbalances and having implemented appropriate measures to strengthen and consolidate the crisis-weakened financial system, Spain is now in a good position to improve its productivity levels, increase its share of exports and breathe confidence into consumers, businesses and investors.

Bibliography

1. Abad, J.: Causas y consecuencias de la reducida dimensión de la PYME en España. FAES (2014)
2. Besedes, T., Prusa, T.: The role of extensive and intensive margins and export growth. J. Dev. Econ. (2011)
3. CEBR: The economic impact of basic digital skills and inclusion in the UK (2015)
4. Cedefop Spain: Skill supply and demand up to 2025 (2015)
5. Competitiveness Business Council: Spain 2018 (2014)
6. European Commission: Europe's Digital Progress Report 2016. Digital Inclusion and Skills (2016)
7. Doménech, R., García, J.R., Ulloa, C.: Los efectos de la flexibilidad salarial sobre el crecimiento y el empleo. BBVA WP N° 16/05 (2016)
8. IEA: World Energy Outlook 2015. New Policies Scenario (2015)
9. IMF: Toil & Technology (2015)
10. Fundación España Digital: Competencias Digitales en España. Cómo mejorarlas (2015)
11. Garicano, L., Lelarge, C., Van Teenen, J.: Firm Size Distortions and the Productivity Distribution: Evidence from France. National Bureau of Economic Research, w18841 (2013)
12. González, M.J., Martín, C.: The internationalisation of Spanish SMEs: main developments and their determinants. Econ. Bull. (December, Bank of Spain) (2015)
13. Strategy & (Formerly Booz&Company): Maximizing the Impact of Digitization (2012)
14. OECD: Government at a Glance 2011 (2011)
15. OECD: Teaching and Learning International Survey (2013)
16. OECD: Education Policy Outlook 2015 (2015)
17. Schneider, F.: Size and Development of the Shadow Economy of 31 European Countries from 2003 to 2013. Johannes Kepler Universität Linz (2013)
18. VIER: Cambio educativo y productivo en España 1964–2013 (2013)
19. VIER: Early school leaving: early school leaving: an analysis of the Spanish case (2013)
20. VIER: Valoración económica y sostenibilidad de la red de infraestructuras (2014)
21. World Economic Forum: Digitization for Economic Growth and Job Creation (2013)
22. World Economic Forum: The Global Information Technology Report (2015)

The Virtual Company as a Value Generator in the New Economy

Mario Aguer Hortal

Abstract The primacy of information and intellectual capital within the company demand a correct evaluation and require creating techniques and appropriate methods for this measuring of intangible assets. Within virtual companies their intellectual qualities are more valuable than their material assets. The virtual company's absence of dimensionality makes accounting and evaluation criteria different from the traditional ones. Virtual communications need to be objective, meaningful and valuable, and can deal with: references to entities of reason, references and observations about perceptions, and references, observations and provisions about materials.

1 The Triangle Generator: Intellectual Capital, Information, Technology

"The new technology of information and communication form the new techno-economic paradigm, based on an interconnected combination of technological innovations which permit the company to cope in an environment which is increasingly dynamic and uncertain. That progress allows employees to work together even if distantly located, without any needs of being in the same place. The

I would like to express, with this chapter, my reward to the master, professor, person and friend Dr. Jaime Gil Aluja, and join the collective work realized as tribute to him for his wide dedication to University and to the financial world, in the act of teaching and researching with the utmost honesty and steadiness. A professor gifted with privileged intelligence, his work is going to be a milestone in the evolution of the Fuzzy Logic, in which he has been and continue being the great referent for teachers and graduated students.

Equally noteworthy is his great work as President of the Royal Academy of Economic and Financial Sciences of Spain (RACEF), whose prestige he has achieved raising to the utmost, during his mandate, thanks to its international dimension.

M. Aguer Hortal (✉)
Alta de Gironella, 58-60 5° 1ª, 08017 Barcelona, Spain
e-mail: rosablancoortiz@elcorteingles.es

© Springer International Publishing AG 2018
C. Berger-Vachon et al. (eds.), *Complex Systems: Solutions and Challenges in Economics, Management and Engineering*, Studies in Systems, Decision and Control 125, https://doi.org/10.1007/978-3-319-69989-9_3

creation of virtual companies, factories, offices and virtual communities is a result of this process." Virtual companies are those which realize their activity, either partially or completely in a virtual way. The most important ones among them are those dedicated to logistical operations, trade, marketing, sales, design, etc. using electronic tools.

It is clear that three creative components have taken part in the genesis of the virtual company: intellectual capital, information and technology.

Nobody doubts about the primacy of information in the world's economic work, when information is not only meant as an array of knowledge and expertise in the economic field, but also as an entity with which it is possible to give chance, color, sense, profitability to the objects we deal with. All this makes information a substantial part or even a whole with the objects which are managed.

The evidence that information is one of these objects, and clearly the main one, is remarked in all specialty publications disseminated and does not need to be especially underlined. Consider this evidence not only as the platform of some possible techniques of information evaluation itself, but mainly as a result of our own purpose of being able to evaluate information, meant as knowledge, that has been collected in a company, so that the evaluation makes us compile an inventory of the intellectual capital collected by the company and embodied by some specific people who service the company.

In the work *Calidad de la información y gestión del conocimiento* [17], by Kuan-Tsae Huang, Yang W. Lee and Richard Y. Wan it is implicitly suggested (pages 26 and following ones) a possible way to evaluate economic wealth generated by the correct information gathered inside a company. In fact, those authors show and insist on the dangers and costs entailed by the mistakes that a company might make in its daily management: the wrong reports about credits, mistakes concerning transports and journeys, sending of direct marketing media to the wrong groups, the imperfections in production processes, oversights and inaccuracies in the billing and in data management.

A whole chapter of this clear and recent work is devoted to explain how expensive is working within wrong information and data. Consequently, on the other way round it would be also useful to quantify how worth it is to have accurate information. And in case the information is embodied in a group of people, there is no obstacle to evaluate what kind of wealth is represented by these people inside the company or by the possible ad hoc works made on request.

Apart from this consideration of costs of errors, as a photo negative of the value of successes, we can benefit some lines approaching the measuring of information quality—the ever more popularized IQ—suggested by the three authors in the aforementioned work. They propose to introduce various parameters in order to evaluate the considered quality and, all at once, they suggest distinguishing 'what', 'how' and 'why' of the IQ. On one side, they also propose applying an IQ subjective metric, through the opinions of people who experience it, and an objective and independent metric based on the theory of quality control on the other. More specifically, the database which materializes the information is required to observe five kinds of integrity: the domain, the column, the entity, the referential ones and that defined by the user. The work offers interesting surveys and procedures focused on measuring and analyzing the quality of information.

According to Steward, it is necessary to distinguish between data, information, knowledge and wisdom/prudence while classifying intellectual capital. He follows stating that intellectual assets only exist and are worth caring within a strategy. This strategy has to take into deep consideration the problem of obsolescence of intellectual assets which are constantly getting outdated as far as they become completely useless. Obsolescence affects ideas themselves, steadily replaced by new ones on the way of progress, bodies and minds of those who bring them, and even the language and style they are expressed with. This last aspect is not usually remembered while studying obsolescence in general, but is an important part of it: a technical study, an internal circular, even a twenty years old civil contract can preserve something which is currently relevant, but they may also be written in a language that sounds old-fashioned and cannot cover present requirements.

Concerning the obsolescence of intellectual assets, it does not seem too rash to apply them a good deal of the rating methods used for machinery and tools, and for their regular replacement. Among the extensive bibliography concerning this subject it is worth mentioning Walter L. Smith's work, *Renewal theory and its ramifications* [31], which has been epoch-making.

The distinguished biologist Alfred J. Lotka bridged a gap, that can be useful for us, relating the ageing and substitution of living members with the replacement of machinery, as he sets out in his article *A contribution to the theory of self-renewing aggregates, with special reference to industrial replacement* [21]. Moreover, it also seems reasonable to quantify the validity of intellectual assets with techniques used in demography and in the study of animal populations in order to measure the structure of population by ages, physical integrity, work capacity, etc. The scientific journal *Biometrika* [4] has published a large number of works thereon.

The age-old problem of co-workers' ageing and initiatives is notoriously serious in those companies constituted exclusively on the basis of intangible assets, which are especially sensitive to obsolescence.

2 The Absence of Dimensionality of the Virtual Company

The conceptualization, not to mention the quantification, of the virtual company trips with a difficulty of an almost metaphysical dimension: the lack of dimensionality of the company itself. The absence of physical dimensions in the main part of its equipment and activities raise an objection: this is absolutely not the case of the desired isomorphism between the measuring criteria and the measured object. The classical procedures of accounting in running a business in traditional companies and of their evaluation are not applicable to virtual companies. In the first case, accounts are made dealing with quantifiable entities that, therefore, have an exact numerical dimension, while the virtual company itself lacks of measurable dimensions, apart from some aspects of its expenses and results which translate into money, but which do not reflect the total activity and company's volume either, and even less its past and future.

The concept and description of the so-called "infomediary" has recently consolidated as acting between producer and customer, conveying information about the customer to the former, so that helps the latter find the most suitable suppliers. With such a volatile and occasional activity the infomediary disrupts the commercial consolidated circuits: he steadily interferes with purchase and sales rhythms and therefore he continuously modifies the dimensions of production and consumption. Decades ago this was undoubtedly the sales agent's objective and modus operandi, or either the commission agent's or the travelling sales man's ones, and even the retailer's ones. But all these professions were practiced within quite strict routines and their limited physical scopes did not alter the production and consumption patterns.

The introductions of e-commerce and Internet possibilities, which are ever-growing, have enormously broadened the possibilities of these infomediaries, leading to dedicate them a wide bibliography, growing at every moment. Let us give the example of the work *Net worth: Shaping markets when customers make the rules* [16] by John Hagel III and Marc Singer translated to French with the title *Valeur sur le Net.* This volume details the informants-intermediaries' techniques with which they operate, and many of them may become companies dedicated to protect the consumer or to distribute products from the factories or even wholesale them, while they are commercializing the information they go on gathering about customers.

It is assumed that consumers are glad to provide this kind of information since they take advantage of a cheaper or more selective supply thanks to those companies' activity. At the same time they say they serve producer companies through three helpful effects: (1) The net effect used to raise the number of clients; (2) The effect of depreciation with which average costs are reduced by higher sales; (3) The learning effect, as companies discover new *know-hows* that reduce costs. Once again it should be noticed that it is difficult to program and quantify these effects insofar as they are due to infomediaries.

Even so, in the fore-mentioned study by Hagel and Singer, it is taken in consideration that the dynamic and successful infomediary may become a sized and large company in relatively few years, and the figure proposed is ten years, mentioning as reference case the success of the online bookstore Amazon, as well as successes brought by Internet to companies like American Express, Citicorp and America Online, but also to others whose sizes are well defined by the number of their subscribers and their business figures. The book concerned analyzes that the infomediary's operating expenses gradually reduce over the ten years mentioned, so that the main costs are devoted to engage customers and suppliers over the first year and decrease thereafter, even though it is clear that it is burdensome to acquire customers for the informediaries and they have to assign important amounts to the respective campaigns, unless they work in partnership with a traditional company that passes its own customers to them.

The General Theory of Systems with so many and wide applications in the socioeconomic analysis can, in some way, help us conceive the form and the basic properties of a virtual entity through the application of the research plan, which was devised by O. R. Young in 1964 [32] among other examples. Here is the list of concepts to analyze that he proposed to those who need to determine a social entity:

1. **Descriptive features**
1.1. Boundaries and borders
1.2. Integration and differentiation
1.3. Interdependency and independency
1.4. Subsystems
1.5. Is it an organistic or no-organistic system
1.6. Isolation or interaction
1.7. Centralization or decentralization

    1.7.1. Centralized system
    1.7.2. Lead sector
    1.7.3. Centralization
    1.7.4. Decentralization

1.8. Interaction sequence
1.9. Field
1.10. Open or close system
1.11. System of determined state
1.12. Equipurpose
2. **Regulation and preservation**
2.1. Balance

    2.1.1. Stable or unstable balance
    2.1.2. Stability
    2.1.3. Communication
    2.1.4 Feedback

        2.1.4.1. Negative feedback
        2.1.4.2. Positive feedback
        2.1.4.3. Finalist feedback

2.2. Homeostasis and self-regulation
2.3. Monitoring
2.4. Preservation of stability

The following reflection can be useful to achieve our aims: it is appropriate to remind that if we used the *levels scale* drawn by Kenneth Boulding to classify the systems, the company in general and therefore the virtual one, due to its attributes, would be on the eighth level, the social organizations' one, in between the *seventh,* human, and the *ninth*, abstract and transcendent.

In the collective work *Strategic human resource management* [10], E. Ferlie and A. Pettigrew address the issue of "Management through Networks", firstly stating that along the 20th century and until 1970, it seemed to prevail the form of the wide and vertical company which created standard products in large quantities, and to which was attributed the style of "Fordism" by some people, whereas thereafter the network style business organization seemed to forge its way ahead, with obvious consequences in the management.

The authors retake the phenomenon, which seems to belong to the eighties, concerning the growth of small companies, regional groups and high technology clusters, each of them connecting with its similar ones like a net. R. E. Miles y C. C. Snow talked about the "general transition from the inflexible pyramids to the more flexible net style structures". A very prolific point of the nets analysis is highlighted by N. Nohria and R. G. Eccles, who edited the collective work *Networks and organizations structures: Form and action* [24]. They point out that the networks approach focus on the importance of social relation patterns within the company, comprising informal connections and stepping aside traditional formalities. In such a way, market processes are revalued in emotional and social terms, and the moral and affective concepts pass on a higher and more powerful level than, those of the mere economic technique for instance.

These authors identify and mark that, on a different plan, various companies are reorganizing as smaller companies, coordinated in a lose way. An example of this is offered by some Italian textiles sectors which join through contracts more than by permanent structures. This is also true for medical and mutual entities in different countries. These phenomena are more or less lateral ones in relation to the one we are dealing with, and it is worth keeping them apart so not to confuse the concepts.

What is clear according to the mentioned Ferlie and Pettigrew is the usefulness of network forms of organization for some types of company, but not for others, especially those engaged with large quantities of productions, fixed objects and permanent systems. By contrast, the following assumptions seem to suggest the adoption of network forms of organization:

1. The increasing appetence of flexibility and learning, coping with the fact that many products or services are targeted to more and more specialized markets or reveal short products lifecycles.
2. The reduction of market uncertainty that many companies undertake fostering long-term stable relationships between firms; that is the case when they need to feel ever more reliant not only on external conditions of treatment, but also on intangible factors.
3. The joint management of production, by means of the tendency towards the fact that a group of independent companies work in close cooperation in order to operate flows of goods and services in agreed terms. This way they replace the old relations between large vertical companies with a method that allows saving IT and communications related expenditure among others.
4. A common high-technology basis, from which various companies, that uses sophisticated technology and similar staff, can take advantage from by mutual agreement in many aspects.

5. Management above cultural diversity, so that the network structure has adapted to the international nature of companies and the international business management overtaking boundaries and cultures. This subject has been especially studied by C. Barlett and S. Ghoshal, in their book *Managing across borders.*

The current trend of many companies organizing in networks raise some problems considered by N. Nohria and can be summarized in *four* outstanding points:

(a) *The issue of direction and power in the networks.* Starting by the undoubtable assumption that a network organization is less centralist and more diffused than a vertical one, it is clear that the network focal points entails a power. Therefore it is advisable to take into consideration that networks have position values with their own power.
(b) *The difficulty of organizing companies in network.* It is usually esteemed that the processes of networks organization, or that of reorganization of old business in networks, imply a high degree of skill and effort, which are even more complex in the area of the public sector.
(c) *Strategic alliances.* It is still to be clarified whether the theory of networks fosters strategic alliances a priori or if each of them needs to be considered within the whole network. Nothing prevent from forging partial partnerships on the way to a complex network of agreements.
(d) *The function of competence within the new networks.* It has been necessary to shape a new concept of competence that should reflect the present notion of the market and gather those factors which provide foundation and regulation to networks, fostering they be either institutions and business rules or vague precepts to guide actions. All this should be within the fluid and constantly reviewed nature of networks.

3 Communications of the Virtual Company as Socio-economic Relation and Object of Commerce

The above characterization sustains the classification of the virtual company as eminently placed in the area of social communication. In the traditional company the objectivity of the ways of working applied to goods, services and money is so clear that the relations sustained for its management lose relevance. On the other hand, in the virtual company relations constitute the main part of its activities and gain preeminence above the subject treated. This subject can be inexistent or, at least, impalpable or not able to be translated to physical or money related measures. This conceptual discordance between the relationship and its object force to consider each one's nature and features.

The virtual company's communications involve a praxis and a purposeful activity based on a group of references which become a system of data susceptible to analysis for what concerns whether they are complete or not. We know that in the present virtual world there are communicative practices which are acceptable and desirable while others are far from this. Even if we are not to enter such broad matter, it is clear that any communicative praxis, implicitly or explicitly, establishes value judgments about the reality communicated.

The object of commerce about which we communicate is neither part of the communicative systems in the virtual company nor among its components (players, expressions and instruments), but it is matter of the treatment between players. There cannot be business communication if there is not an object of commerce in a broad sense, even if there may be an object that does not exist in the physical meaning of the word, as we have said.

Benchmark languages are those which enable the object of commerce to enter the communication system. By benchmark languages we mean different kinds of vehicles depending on the point of view from which they are generated. They may be modulated energies (*physical languages*), perceptive stimuli (*expressions*), codified signals (*information*) and voluntary acts externalized through a combination of symbols (business communication for instance).

The communication engaged has to be objective, meaningful and valid to be considered as a socio-economic relationship created by the virtual company and worthy of its name. In accordance with *objectivity*, the benchmark languages have to correctly mark the designated benchmark. The objectivity of data results from the fact that they come from the benchmark item, they may be attributed to the object or it may be advocated that this object does not possess them. This objectivity depends on the fact that it owns the characteristic of being referenceable and, as such, is independent from the use that can be done with data in the communication. Do not forget the unavoidable existence of intermediation (instrumental and players) in the communication. The only not mediated objectivity would be the direct contact between player and object, which is not taken into account in the virtual company. The intermediation becomes an intentional action related to the realization of communication and it especially rests on the validity of data.

Meaningfulness requires a correct selection of baseline data according to the criteria used to consider the objects. There is a correct selection of baseline data when they can be legitimately attributed to the object and are also advisable to the criteria on which the communication is based. The meaningfulness of the baseline data is a quality related to their relevance and it is independent from the practical purposes aimed by the players.

Validity means that the array of data is enough to place the object in the context of other objects with which it relates. Objectivity and meaningfulness are needed to have validity, but they are not sufficient conditions. In the communication there may be a wrong use of baseline data, due to a mistake, as well as the illegitimate use of them, as consequence of an illegal or dishonest action; for instance when the company refers to the object providing insufficient or incomplete data, or deals with illegitimate matters, even when they are objective and meaningful matters.

Validity is then a quality of communication which depends on baseline data sufficiency to the communication recipient in order to correctly place its object in space, time and in relation with other objects.

We are going to use the term "object of commerce" in a conventional way to express the thing about which communication is produced in the virtual company. Its function is that of the material or ideal object of business communication within that virtual world.

The object of commerce is recalled in the communication as its own matter, but it remains external to it, that is to say, it is not part of the communicative system as players, expressions and instrumental are. This marks a substantial difference with the traditional company, where the object can be next to the players and can be touched by them.

As Manuel Martín Serrano says in *Teoría de la Comunicación* [22], the communication can deal with:

(a) *Entities of reason*: it is possible to refer to them.
(b) *Perceptions:* it is possible to refer to them and also observe them.
(c) *Materials*: it is possible to refer to them, observe and handle them as well.

Within all these communications it is worth highlighting that the traditional company basically focuses on—if not limits to—working with materials while the virtual one moves at ease, working with perceptions and entities of reason too.

References

1. Aguer, M.: La era de las organizaciones virtuales. Editorial Pirámide, Madrid (2005)
2. Aguer, M.: Evaluación de activos en las empresas virtuales. Editorial Pirámide, Madrid (2002)
3. Aguer, M.: La empresa virtual en el marco de la sociedad de la información. Publicaciones de la Real Academia de Ciencias Económicas y Financieras, Barcelona (2000)
4. Biometrika. Oxford University Press (2015)
5. Boulding, K.: General systems theory—the skeleton of science. Gen. Syst. (1956)
6. Castells, M.: La era de la información. Alianza Editorial, Madrid (2005)
7. Cuesta, F.: La empresa virtual. Revista Digital Universitaria **11**(10) (2010)
8. Eccles, R.G., Herz, R.H., Keegan, E.M., Phillips, D.: The Value Reporting Revolution. PriceWaterhouseCoopers, Wiley (2001)
9. Eicher, J.P., León Pérez, C.E.: Dirigir la organización virtual. Editorial Centro de Estudios Ramón Areces, Madrid (2002)
10. Ferlie, E., Pettigrew, A.: Strategic human resource management. In: Mabey, C., Salaman, G., Storey, J. (eds.). Sage, London (1998)
11. Gallego Vázquez, J.A., Galli Granada, R.A.: Todo lo que hay que saber de comunidades virtuales y redes sociales. Wolters Kluwer, España (2012)
12. Estallo, G., de los Ángeles, M: Empresa virtual: de la idea a la acción. Editor ESIC Editorial (2001)
13. Gil Lafuente, M.A.: El análisis de las inmovilizaciones en la incertidumbre (2004)
14. Gil Aluja, J., Kaufmann, A., Gil Lafuente, M.A.: La creatividad en la gestión de las empresas (1994)

15. Gupta, K.: Guía práctica para evaluar necesidades. Editorial Universitaria Ramón Areces, Madrid (2000)
16. Hagel, J., Singer, M.: Net Worth: Shaping Markets When Customers Make the Rules. Harvard Business School Press, McKinsey & Co. Boston (1999)
17. Huang, K.-T., Lee, Y.W., Wang, R.Y.: Calidad de la información y gestión del conocimiento. AENOR, Madrid (2000)
18. Jáuregui Machuca, K.L.: Formación virtual en las organizaciones. Editorial Académica Española (2011)
19. Jones, J.E., Eicher, J.P., Bearley W.: Dirigir la organización virtual. Editorial Universitaria Ramón Areces, Madrid (2002)
20. Kotler, P.: Los 10 pecados capitales del Marketing. Ediciones Gestión 2000, Madrid (2012)
21. Lotka, A.J.: A contribution to the theory of self-renewing aggregates, with special reference to industrial replacement. Ann. Math. Statist. X (1939)
22. Martín, M.: Teoría de la comunicación. MacGraw-Hill, Madrid (2007)
23. Miles, R.E., Snow, C.C.: Organizations new concepts for new forms. Calif. Manag. Rev. 28 (3) (1986)
24. Nohria, N., Eccles, R.G.: Networks and Organizations Structures: Form and Action. Harvard Business Scholl Press, Boston (1992)
25. Ortiz Fuertes, M., et al.: La contribución de las tecnologías de la información y comunicación a la creación de valor en la empresa. Universidad de Oviedo (1997)
26. Rodes, A., Moro, M.: Marketing Digital. Ediciones Paraninfo SA, Madrid (2014)
27. Timoteo, J.: Los intangibles en el valor de las empresas. Díaz de Santos, Madrid (2015)
28. Tirado, F., Domènech, M.: Lo social y lo virtual. Nuevas formas de control y transformación social. UOC (Universitat Oberta de Catalunya), Barcelona (2006)
29. Torres, Z., Cubillán, L.E.: El emprendedor virtual: una guía para tu éxito online (2015)
30. Viloria Vera, E.: De la empresa internacional a la empresa virtual (2005)
31. Smith, W.L.: Renewal theory and its ramifications. J. R. Stat. Soc. Ser. B XX (1958)
32. Young, O.R.: The impact of general systems theory on political science. Gen. Syst. IX (1964)

The Origin of the Legitimacy of Organizations and Their Determining Factors

Camilo Prado-Román, Francisco Díez-Martín,
Alicia Blanco-González and Alberto Prado-Román

Abstract The aim of this research consists in analysing to what extent personal characteristics have an influence on the assessment process of organizational legitimacy. For this purpose, a questionnaire was used which examines the effect of five personal characteristics on four types of legitimacy. Subsequently, regression analysis was applied on a sample of 258 individuals. The results have shown that the persons with higher social awareness are more prone to make decisions about organizations taking into account the moral, regulatory and cognitive legitimacy. This type of assessment also occurs when the fear of receiving a social sanction increases. Likewise in the perception of a higher economic risk inherent to the result of a decision, people are more likely to make decisions based on cognitive and pragmatic legitimacy. Future research projects may confirm if the results of this investigation are repeated among different activity sectors and different sociodemographic environments. Accordingly, it would be possible to design a conceptual framework where the legitimacy preferences of interest groups are explained by sectors and personal characteristics. The results of this study extend the knowledge in the field of Institutional Theory about the origin of organizational legitimacy and the causes which condition it. It also facilitates improving the strategic planning of organizations by displaying the legitimacy preferences based on each person's profile.

Keywords Legitimacy · Organizational legitimacy · Origin of legitimacy
Types of legitimacy · Sources of legitimacy · Institutional theory

C. Prado-Román (✉) · F. Díez-Martín · A. Blanco-González · A. Prado-Román
Universidad Rey Juan Carlos, Paseo de Los Artilleros, s/n, 28032 Madrid, Spain
e-mail: camilo.prado.roman@urjc.es

F. Díez-Martín
e-mail: francisco.diez@urjc.es

A. Blanco-González
e-mail: alicia.blanco@urjc.es

A. Prado-Román
e-mail: alberto.prado@urjc.es

© Springer International Publishing AG 2018
C. Berger-Vachon et al. (eds.), *Complex Systems: Solutions and Challenges in Economics, Management and Engineering*, Studies in Systems,
Decision and Control 125, https://doi.org/10.1007/978-3-319-69989-9_4

1   Introduction

At present, the legitimacy of an organization is considered as an essential resource for its survival [48]. This characteristic means that the most legitimate organizations are able to obtain better results [13]. Essentially because legitimacy favours better access to the required resources [9, 37]. These results suggest that the achievement of legitimacy is crucial for all organizations. In this line, previous research has indicated the strategies which lead to the enhancement of an organization's legitimacy [44] or on the other hand, its reduction [31].

A key topic to manage the legitimacy of organization consists in determining its origin. The legitimacy of an organization is determined by the persons who observe and assess them according to several social models [39]. However in spite of the role which individuals play in the creation of the legitimacy of organizations, they have been underestimated by the literature. This absence in the literature was made evident in the research by Suchman [44] and more recently by Bitektine [3]. Both suggest that the social context in which people operate is essential to understand the legitimacy of organizations. This can have major consequences for organizations. For example, many customers make their decisions based on purely pragmatic reasons (e.g. quality-price relation), while others place more importance on moral factors (e.g. reduction of environmental pollution). In the first case, the customers preferred companies with greater pragmatic legitimacy as opposed to the second case where they preferred companies with greater moral legitimacy. Why do these divergences occur? What causes some customers (or interest groups) to value organizations based on such diverse aspects? Understanding the existing relation between the sociodemographic characteristics of individuals and their preferences concerning the legitimacy of organizations can help managers to better address the cognitive needs of their evaluators and improve the efficiency of their management.

Based on the Social Judgement Theory [3] and Institutional Theory [33], this research aims to enhance the understanding of the process in which persons assess the legitimacy of organizations. To date, the majority of studies in this sector have focused on demonstrating legitimacy's importance to achieve the best results. They also point out the most suitable strategies for their achievement. However little research has shown an interest in developing in-depth analysis regarding the relation which exists between people's characteristics and the legitimacy of organizations. This would make it easier to determine what causes can condition the origin of the organizations' legitimacy. Therefore, we have defined the objective which consists in determining to what extent personal characteristics have an influence on the decisions based on the legitimacy of organizations.

For this purpose, the work has been organized into four parts as well as the introduction. The first section analyses the theoretical framework of the research. This section describes the concept of legitimacy and its importance for organizations. Likewise, it theoretically explains the functioning of the decision making process based on the legitimacy of organizations. Finally, it analyses the variables which influence this process and the research hypotheses are established. The paper's second section describes the methodology used to contrast the described hypotheses. The empirical study was carried out by means of a stepwise regression analysis on a sample of 258 students from the Business Administration degree at the Rey Juan Carlos University. The data was collected by means of a questionnaire which inquired about the personal characteristics and the type of assessment which each respondent used when shopping for clothes. The third section describes the achieved results and specifically describes which variables have an influence on the decision making based on the organizations' legitimacy. The paper's final section discusses the results and explains the scientific and managerial implications of the research. It also indicates the limitations and proposes future lines of research.

2 Theoretical Framework

2.1 Legitimacy and Its Types

The concept of legitimacy has created generalized interests and concerns among academic scholars of institutional theory [14, 33], the Resource Dependency Theory [36] and Organizational Ecology [43]. Its importance consists in its positive relation with the access to the required resources for the survival and growth of organizations [48]. The organizations which survive the longest are the ones which best adjust to the pressures from interest groups, act in an ethical way and display stability in the process of seeking competitive advantage [10]. The legitimacy status is a *sine qua non* condition to achieve easier access to resources and markets as well as ensure the long-term survival [5].

Legitimacy consists in a status which reflects the cultural alignment, the normative support or the accordance with the relevant rules and laws [41]. One of its most accepted definitions widely used by the literature considers it as "the generalized perception or assumption that an entity's activities are desirable, correct or appropriate, within any socially constructed system of rules, values, beliefs and definitions" [44]. It involves a judgement based on people's perception. To legitimize themselves, organizations attempt to obtain favourable judgements from their evaluators. In this way, the legitimate organizations have more probabilities to obtain a benefit from society in the form of a better access to resources. With this aim, they can implement actions which range from adjusting to the commonly established social models (isomorphism), to the manipulation of the environment in which they operate [35].

The study of the literature notes the existence of multiple types of legitimacy (See [3]). However, the majority of the research frequently uses three or four types. The origin of these types can be found in the works by Scott [41] and Suchman [44]. Both authors refer to three types of legitimacy, although they do not completely coincide. The former considers the existence of a regulatory, normative and cognitive legitimacy. The latter indicates a type comprised by pragmatic, moral and cognitive legitimacy. The analysis of these types permits combining them into four types of legitimacy.

Pragmatic legitimacy appears when the interest groups consider that their individual interests are fulfilled. The organizations obtain pragmatic legitimacy when several of their interest groups achieve their objectives through the organization. The support does not occur because they achieve major challenges such as major sales turnover, but because the interest groups observe that the organization is being receptive to their specific interests [12]. For example, when the shareholders' objective is to increase the value of the shares and this occurs, or when the customers desire products with a better quality and they are offered to them; also when governments aim to collect large taxes and they achieve this, or when society wants reductions in the pollution caused by the company and this takes place. This type of legitimacy represents a materialistic relation of power and dependency between the organization and their interest groups. It is worth mentioning that this type of legitimacy is also found in the literature under the name of normative legitimacy [26].

Moral legitimacy refers to the judgements about the morality which an organization reveals in their actions. Unlike pragmatic legitimacy where the result is what is important, moral legitimacy is more focused on the methods to achieve them. This involves a type of judgement in which the positive results obtained by an organization are not considered but rather the way to attain them. This type of legitimacy appears when individuals consider that the actions which they perform are "what must be done" [44]. Generally analysis is done by assessing if the techniques, procedures or organizational structure used to achieve the results are the most appropriate in a moral sense. The ethically motivated companies act with socially responsible behaviours because "this is the correct thing to do" [47]. The moral practices of an organization are motivated by the management values, the values of the organization itself, or even by the strategic interest (See [10]).

The regulatory legitimacy is derived from the regulations, rules, standards and expectations created by governments, accreditation associations, professional associations and other powerful organizations [48]. The organizations would obtain the regulatory legitimacy through the compliance with the rules. This type of legitimacy is useful, not only to prevent sanctions derived from the infringement of the rules but also to earn accreditations or licenses in order to develop a business activity. For example to obtain an administration contract, the entity must demonstrate that they do not possess any debts with the Tax Agency.

The cognitive legitimacy consists in the type of legitimacy founded on knowledge, rather than on the results or the methods to achieve it. Specifically, it arises from the knowledge which the evaluators possess about the organization. An organization achieves this type of legitimacy when the evaluators perceive that the organization is comprehensible. When they understand how the organization operates. Also when the evaluators assume that it is unthinkable "to do things" in another way (e.g. the option to pay with a card). The greatest cognitive legitimacy is accomplished when a new product, process or service is considered as established [1]. It is possible to obtain cognitive legitimacy by achieving that other legitimate organizations recommend you [48], adopting behaviours similar to the practices of the leaders in the sector [44], or even using communication tools such as *story-telling* [23].

2.2 The Legitimacy Assessment Process

The analysis of the above section highlights the presence of four different ways to assess the legitimacy of an organization. When a person assesses an organization's legitimacy, they aim to answer four different questions in relation to: Do the interest groups achieve their individual objectives through the organization? Does the organization act according to principles of social responsibility which are above the individual interests? Does the organization comply with the rules, regulations and laws? And does it understand and accept the organization's operation?

However what remains unclear is which of these questions does a person attempt to answer first. What order do people follow to provide an answer to these questions? This is one of the major problems with the assessment process of organizational legitimacy.

Generally, when people who make decisions are faced with assessments with many alternatives, they tend to simplify the complexity of the decision making process. They use relatively simple decision algorithms which allow them to first reduce the total number of alternatives. For example, to purchase a house, some people only focus on residences which have a swimming pool. Secondly after this easy simplification, they change the more complex algorithms which permit them to more precisely define their choice [32]. This reasoning is in harmony with the principle of cognitive economy [38]. This principle considers that people begin their assessment by using judgements based on categories and then develop more complex evaluations. Hence, the relations between organizations and the people who assess them would be the result of a process with two stages. With the aim to simplify complex decisions, the evaluators initially would classify the organizations on the basis of common features; they would subsequently select the most attractive organizations which share these characteristics [49].

In this way, the ideal assessment process of the organizations' legitimacy would begin with the assessment of the cognitive legitimacy, which would establish potential alternatives [3]. This (cognitive) assessment would have the aim to classify the organization as a member of a familiar and unproblematic category of organizations. Subsequently if this assessment fails or a more detailed examination is required for an organization, the evaluator would carry out an assessment using other types of legitimacy such as: regulatory, moral and pragmatic legitimacy. The latter analysis would be useful to finish selecting the organization about which they have the most knowledge and whose actions were the most desirable and acceptable for the evaluator.

However, we are aware that the assessment process of the organizations' legitimacy does not always take place in an ideal way. For example, numerous public organizations assess their suppliers, first considering the regulatory legitimacy. They only work with the suppliers which comply with the standards, regulations and laws. Postponing the assessments about the cognitive, moral or pragmatic aspects. In turn, many speculative investors do not manage to understand the operation of the companies in which they invest. Their assessment only considers the pragmatic legitimacy. They are only interested in the attainment of results which guarantee their investment goals. Likewise some consumers only purchase products from companies which are responsible with the environment, where it does not matter to them if the price of these products may be higher. In short, the relevance of a specific assessment method depends on the variables which have an influence on the evaluator.

2.3 Variables Which Have an Influence on the Legitimacy Assessment

There is no single factor which determines the prevalence of a legitimacy type when assessing an organization. Some evaluators are prone to perform assessments based on the pragmatic legitimacy and others on the cognitive type. The selection of an assessment method depends on the effect of multiple factors on the evaluators, as well as the importance of these factors for each evaluator. Indeed, people's social context determines the prominence of one assessment type over another [44]. This implies that the evaluator's social context can cause diverse effects in the assessment process of the organizations' legitimacy. Biteketine [3] suggests several variables which would influence the evaluator's judgements during this process: the visibility of the decision, the economic interests, the risk of receiving social sanctions, social awareness and trust.

An evaluator from a highly institutionalized sector (e.g. a public university) will be more interested in making decisions considering the regulatory and moral legitimacy as opposed to the cognitive or pragmatic legitimacy. For example, one of the most required terms in recent years to make a contract with a public entity consisted in demonstrating that the organization did not have any debt with the Public Administration (regulatory legitimacy). One explanation for the selection of this assessment type as opposed to other legitimacy types would be the visibility which this type of decision has in the most institutionalized sectors. In a highly institutionalized sector, the evaluator's decision involves a high visibility compared with social agents, due to the transparency required in the process. In this context, the specific assessment must show an expression of values, as well as the evaluator's conformity with the social rules. Hence, the evaluators with social awareness would be more prone to confide in this type of judgements.

An evaluator who is situated in an action scope whose decisions were not very visible to the remaining social agents would first select an assessment based on the pragmatic or cognitive legitimacy because their decision would not possess a social or public transcendence. Likewise, when there is perceived low risk of receiving social sanctions due to the decisions which they make, the evaluator will first select an assessment based on the pragmatic or cognitive legitimacy, because no one would be concerned about their decision.

H1: A higher visibility of the decision favours the assessments based on the regulatory and moral legitimacy.

H2: A higher visibility of receiving a social sanction favours the assessments based on the regulatory and moral legitimacy.

H3: An evaluator with greater social awareness favours the assessments based on the regulatory and moral legitimacy.

Economic interests can also influence the assessment process of organizational legitimacy. If the selection of the assessment type involves a low economic risk for the evaluator, said party will probably first select an evaluation based on the pragmatic, moral or regulatory legitimacy. In this context, the evaluator does not have anything to lose and can make a selection based on their individual values (pragmatic legitimacy), values (moral legitimacy) or the importance of adapting to the laws (regulatory legitimacy). Accordingly, when the assessment is associated with the probability of involving a high economic risk, the evaluator will use an assessment based on cognitive legitimacy. When an organization displays higher cognitive legitimacy this implies that the evaluator understands and accepts its operating method because he/she considers that this is the "best way to do things". This decision would have the objective to minimize the economic risk. In turn, the most distrustful evaluators probably prefer this type of assessment due to its lower risk [24, 29].

H4: A higher economic risk favours the assessments based on cognitive legitimacy.

H5: An evaluator with less confidence favours the assessments based on cognitive legitimacy.

Table 1 shows the variables which influence the decisions based on the legitimacy of organizations. Hence the influence of each variable will mean that different types of legitimacy are considered when making a decision. For example, the evaluator's higher social awareness, a greater likelihood to make an assessment considering the moral and/or regulatory legitimacy.

Table 1 Conditioning factors about the decisions based on the legitimacy

Variable	Type of legitimacy			
	Cognitive	Pragmatic	Moral	Regulative
Visibility of the decision	Low	Low	High	High
Social sanction risk	Low	Low	High	High
Social awareness	Low	Moderate	High	High
Economic risk	High	Moderate	Low	Low
Confidence	High	Moderate	Low	Low

Source Own elaboration

3   Methodology

3.1   Sample and Data Collection

The analysis of the relation between the personal characteristics and the decisions based on the legitimacy of organizations was carried out using a convenience sample. The sample was comprised by University students in the Business Administration and Management degree of the Rey Juan Carlos University in Madrid (Spain). This type of sample ensured that all the respondents shared a similar socio-cultural context, in reference to: geographical area (Fuenlabrada campus), age (between ages of 18 and 22), education (Business Administration). Previous investigations suggest that the University students have similar cultural aspects (See [30]). This is important because in the literature, it is considered that people's socio-cultural context is a determining factor when making decisions based on the types of legitimacy [3, 13, 44].

The empirical data for this research was collected from a total population of 614 University students during the 2015–2016 academic year. The data collection was performed by means of a questionnaire. Questionnaires have been described in business literature as a powerful and useful tool for data collection related to human characteristics such as: attitudes, thoughts and behaviour [20]. Since this research analyses these types of characteristics, the questionnaire was a natural choice with regard to the most practical way to control the opinion of a large group of people.

The questionnaire used was an electronic type. The Google Forms application was used as support for the electronic questionnaire. This application facilitates the preparation of electronic questionnaires, providing a link for their communication. The link of the electronic questionnaire was sent by e-mail during the first half of March 2016. In turn, in the development and sending process, procedures were used which have proven useful to improve the questionnaire response rate and to minimize the bias (See [21]): (1) The questionnaire was pre-tested by means of seven in-depth interviews with Business Administration students from the Rey Juan Carlos University and with three researchers, one from the University of Seville and two from the Rey Juan Carlos University; (2) subsequently during the first three days of March, a pilot test was carried out on the process to complete the electronic questionnaire, which made it possible to perfect its contents and structure based on the obtained reactions; (3) no identification was required to maintain the respondents' anonymity; (4) a phrase was included stating that the responses would be confidential and that the survey comprised part of a research project for the University; (5) the questionnaire's background was white; (6) a simple heading; (7) an image was shown in the e-mail of the questionnaire; (8) the word questionnaire was not mentioned in the e-mail's subject. The data was subsequently processed using the SPSS computer program.

The field work was carried out during the months of March and April 2016. The final sample was comprised by 258 valid questionnaires. The sample error was 4.65% with a confidence level of 95% ($Z = 1.96$; $p - q = 0.5$). The main

Table 2 Demographic features of the sample

	Frequency	%
Age		
18 years	68	26.4
19 years	42	16.3
20 years	57	22.1
21 years	30	11.6
22 years	61	23.6
Total	258	100
Gender		
Man	92	35.7
Woman	166	64.3
Total	258	100
Job		
No	205	79.5
Si	53	20.5
Total	258	100

Source Own elaboration

demographic characteristics of the students which composed the final sample are summarized in Table 2. The sample distribution suggests a homogeneous distribution among the respondents' ages. It also shows a higher number of females (64.3%). This fact is quite representative of Spanish Universities, where over half of the students are women [2]. In turn, 20.5% of the sample were working at the time of the survey.

3.2 Measurements and Variables

The majority of the empirical research about legitimacy has developed their own research instruments ad hoc. Previous investigations have used content analysis [13, 37], surveys [6, 8, 17, 22], interviews [40], case studies [19], and secondary data analysis [4, 7, 11, 12, 16].

The questionnaire used in this research was based on the theoretical and empirical literature about the management of legitimacy. The entire questionnaire has been meticulous and specially compared with the instruments proposed by other researchers, such as [3, 8, 18, 22, 45, 48]. Accordingly, the measurement of the legitimacy was carried out with four questions. Each question measures one of the four types of legitimacy described above (pragmatic, moral, cognitive and regulatory). In turn, the personal characteristics were measured with five questions: the decision's degree of visibility, the economic repercussions which the assessment could generate, the risk of receiving some type of sanction derived from the social environment, the level of social awareness and the degree of confidence concerning the assessed company. All the questions used in the questionnaire followed a 5-point Likert scale. In addition during the construction process, each variable was carefully compared with the works by Suchman [44], Scott [41] and Bitektine [3], with the aim to resolve any discrepancy which could arise between the different legitimacy measurements and the variables which have an influence on its selection.

In order to focus the respondents' attention on a single type of assessment, we decided to associate all the questions with a decision which the individuals of the sample commonly make. After a *focus group* with ten students, we decided that all the questions about legitimacy would be linked to shopping for clothes.

4   Results

In order to analyse the influence of people's characteristics on the assessment process of organizational legitimacy, we used stepwise regression analysis. This approach permits the comparison of the defined hypothesis by identifying several variables which make it possible to obtain a better adjustment.

Table 3 shows the average, standard deviation and correlations of all the variables used in the study. The results suggest that on average, the most relevant type of legitimacy for persons to assess an organization would be the following, in this order: (1) cognitive legitimacy; (2) pragmatic legitimacy, (3) moral legitimacy; and (4) regulatory legitimacy. Note that the pragmatic, moral and regulatory legitimacy achieve a similar average relevance. They were all below the average value of the employed scale. However, the cognitive legitimacy is granted a higher importance than the rest (3.47 points), which was above the average value of the scale.

Table 4 displays the results of the stepwise regression analysis. In order to analyse the influence which the personal characteristics have on the decisions based on legitimacy, four regression models were created, one for each legitimacy type. Among the five variables suggested by the literature which could have an influence on the selection of the legitimacy type to assess an organization, two proved to be insignificant: the visibility of the decision and the evaluator's confidence. Both variables have shown no significance in any of the four presented models. Hence, Hypotheses 1 and 5 were rejected.

The results suggest that a higher probability of receiving a social sanction has a significant and positive influence on the assessments made based on the regulatory and moral legitimacy. However they were also negative on the assessments based on cognitive legitimacy. At the same time, this characteristic does not produce any type of significant influence on the pragmatic legitimacy. This result confirms Hypothesis 2.

Hypothesis 3, which suggests that the evaluators with higher social awareness would increase the assessments based on the regulatory and moral legitimacy have been supported by the study's results. A significant and positive influence took place between the social awareness and moral, regulatory and cognitive legitimacy. In this case, the relative importance of the evaluator's social awareness on the moral

Table 3 Average, standard deviation and correlations

		Media	s.d	1	2	3	4	5	6	7	8
1	LegP	2.78	1.21								
2	LegM	2.61	1.11								
3	LegR	2.38	1.17								
4	LegC	3.47	1.29								
5	REco	1.94	1.12	0.20**	0.03	0.00	0.07				
6	Conf	3.41	0.92	0.16**	0.14*	0.10	0.12*	0.23**			
7	CSoc	3.31	1.12	0.04	0.49**	0.41**	0.32**	0.04	0.30**		
8	Vis	3.04	1.48	0.05	0.00	0.06	−0.02	0.04	0.06	0.00	
9	SanSoc	1.53	1.28	0.00	0.02	0.11	−0.17**	0.32**	0.04	−0.16**	0.03

Source Own elaboration

s.d standard deviation; *LegP* pragmatic legitimacy; *LegM* moral legitimacy; *LegR* regulative legitimacy; *LegC* cognitive legitimacy; *Vis* visibility of the decision; *SanSoc* social sanction risk; *REco* economic risk; *Conf* confidence; *CSoc* social awareness

*p < 0.05; **p < 0.01

Table 4 Regression analysis

Variable	Model I Pragmatic legitimacy		Model II Moral legitimacy		Model III Regulative legitimacy		Model IV Cognitive legitimacy	
	Beta	t	Beta	t	Beta	t	Beta	t
Economic risk	0.196	3.261**					0,123	1,981*
Confidence								
Social awareness			0.511	9.323***	0.448	7.940***	0.291	4.889***
Visibility of the decision								
Social sanction risk			0.108	1.975*	0.190	3.369***	−0,173	−2.755**
R square corrected	0.074		0.249		0.203		0.125	
Sample	258		258		258		258	

Source Own elaboration
*p < 0.05; **p < 0.01; ***p < 0.001

and regulatory legitimacy (beta = 0.51; 0.44) is significantly higher than that produced on the cognitive legitimacy (beta = 0.29). Likewise, this characteristic does not produce any type of significant influence on the pragmatic legitimacy.

The results also confirm Hypothesis 4. There is a significant and positive influence between the economic risk perceived by the persons and the assessments which they made on the organizations. In this case, a higher perceived risk would favour the assessments based on the cognitive and pragmatic legitimacy. In turn, there is no type of significant influence between the economic risk and the moral and regulatory legitimacy.

5 Discussion, Limitations and Future Lines of Research

5.1 Scientific Implications

Over the years, the study of organizational legitimacy has become one of the research fields with the highest growth in the business management sector [27, 46]. However, there are still major gaps to resolve concerning the legitimacy process of organizations. One issue consists in a better understanding of the process with which people assess the legitimacy of organizations.

This research has analysed to what extent people's characteristics influence the decisions based on the legitimacy of organizations. Five variables were examined for this purpose: the visibility of the decision, the economic interests, the risk of receiving social sanctions, the social awareness and the confidence. These variables form a part of the training model of the social judgements described by Bitektine [3]. The results suggest that not all the personal variables described in this model have a significant influence on the assessment of the legitimacy of organizations.

Social awareness has proven to be a very influential personal variable when assessing an organization's legitimacy. An evaluator with higher social awareness is more likely to assess organizations based on the moral and regulatory legitimacy and to a lesser degree, the cognitive type. This result is in accordance with the training model of social judgements described by Bitektine [3]. Taking into account the organization's morality and legality as well as their operation is the type of behaviour expected among persons with a higher social awareness. Recently, an advertising campaign conducted by the organization, Fashion Revolution in 2015 showed that the persons with social awareness would not purchase inexpensive clothing yet manufactured in unfair work conditions. Nevertheless, this is not always the case. Recent studies have shown that people who socially reject the underground economy have actively participated in it, making payments aware that the collecting party was not going to file taxes on it [28]. First performing a moral and regulatory assessment on an organization does not always prevail over the pragmatic legitimacy.

The fear of receiving a social sanction has also proven to be another of the most influential personal characteristics on the assessment of organizational legitimacy. When people have a greater fear of receiving a social sanction as the result of their assessment, they are more likely to make assessments considering the moral and regulatory legitimacy. Although they are less likely to make an assessment based on the cognitive legitimacy. This result is also expected. On possible explanation could be when a person considers that the result of his/her assessment (in this case, the purchased clothing), can create problems with their environment (e.g. social marginalization, source of irony), this person would automatically cease to make assessments of the organization based on the cognitive legitimacy. This type of assessment is actually not required because the environment itself informs this person not to support this organization or their products, since this organization's activity is not desirable within their social environment. Previous investigations have shown that the external validity which is obtained by means of support from the social environment is a determining factor to assess the legitimacy [45].

At the same time, no significant relation has been detected between the fear of receiving a social sanction and the development of assessments with a more pragmatic nature. A person who potentially has a major fear of receiving a social sanction would not make an assessment based on the possibility of achieving their individual objectives but rather based on the possibility of obtaining a morally and legally acceptable result from the social environment, which reduces the risk of receiving any social sanction. In short, the risk of receiving some type of social sanction means that the people who buy clothing seek acceptance more from the social environment than from the personal level.

Likewise, the results suggest that a person who perceives an economic risk inherent to the result of the decision which he/she will make, this individual is more likely to use an assessment based on the cognitive and pragmatic legitimacy. This implies that the perception of a greater economic risk leads to an increase of decisions based on cognitive and pragmatic aspects. This means, when the evaluator believes that he/she has high probabilities to obtain an economic loss, this person would choose the organization which does things better and functions better. Hence, this person could minimize the potential losses caused by this decision. This result would extend the research by Shepherd and Zacharakis [42], who showed that the customers of newly created companies are more likely to purchase products from these companies about which they have greater knowledge (cognitive legitimacy). On the other hand, the economic risk does not seem to have a significant influence on the performance of assessments based on the morality or regulatory compliance shown by the vendor. Faced with the likelihood of an economic risk, people do not take moral or regulatory aspects into account [25].

Finally, we observed that the variables: degree of confidence and visibility of the decision did not obtain the expected results. Both variables did not prove to be significant with any of the legitimacy types used in the organizations' assessment. With regard to the degree of confidence, the lack of significance could be explained due to the type of relation existing between the evaluator and the assessed party. In this research, we analyse a customer-business relation (C2B). In this case, we understand that the customers of clothing businesses do not develop a sufficiently significant personal relation with the vendors. This means, a relation which can generate a degree of confidence. On the other hand, the appearance of some degree of confidence does seem more probable among the relations of a business-business type (B2B). This leads us to believe that the degree of confidence represents a variable which could have an influence on the assessment of the legitimacy of organizations when a there is a B2B relation. In relation to the lack of significance between the visibility of a decision and the legitimacy, this could be explained based on the context in which the research has been applied. Shopping for clothes is a type of action with an uncontrollable degree of visibility. In general, it is difficult to conceal the clothing that one is wearing and who can see it. This would explain the lack of this variable's significance when selecting a specific type of legitimacy assessment in this sector.

5.2 Limitations and Future Research Projects

This research is an exploratory approach for the study of personal judgements and decision making based on the legitimacy of organizations. This fact involves several limitations, which in turn, reveal the path for future lines of research. The first limitation is related to the results of the research and its implications. Since this involves a convenience sample, the results cannot be transferred to the entire population. In any case, they can only be applied to the organizations whose activity was retail sales and more specifically, clothing sales. As well as the population comprised by persons with a similar context as the University students. Future research projects could develop the study defined here, among organizations which belong to multiple activity sectors. A *cross sectional* research project could confirm to what extent the personal characteristics influence the assessment process of the organizational legitimacy. This would be useful to contrast and complement the Social Judgement Theory [3]. In this line of study, new variables could be included in the model in order to discriminate the behaviours based on the sociodemographic aspects.

A second limitation can be found in reference to the research's purpose. Since the objective consisted in analysing the relation between the characteristics of individuals and the legitimacy assessment type which they perform, we have not taken into account another type of variables which could also influence the decision making process. Future research could consider another type of variables, not personal types, which could influence the legitimacy preferences when assessing an organization, and in this way, achieve a greater degree of explanation for the presented models.

A third limitation of the research consists in the measurement of legitimacy. In this study, the legitimacy has been measured by means of the individual perceptions of people regarding the desirability of the organization. This type of assessment is completely accepted by the literature. It has been used in numerous studies (e.g. [8,

22]). However, a more accurate measurement of the legitimacy would involve incorporating individual assessment for the external persons who have an influence on the individual. Hence, the assessment of the external persons (social environment, colleagues, bosses, family members) provide validity on the individual assessment [45]. Accordingly, it would be able to completely analyse the *reference locus* of the legitimacy (see [15]). Thus, the measurement of legitimacy could be improved, extending it to more assessment sources.

5.3 Managerial Implications

There are several managerial implications which can be drawn from this research. In light of the results described above, managers must understand that their customer profiles have an influence on the degree of importance which they grant to the company's legitimacy. Knowledge of the individual profiles associated with the types of legitimacy will allow managers to concentrate their strategic efforts on obtaining the legitimacy type most suitable for their customer profiles. Hence, the businesses whose customers show a major social awareness must concentrate their efforts on earning greater moral, regulatory and cognitive legitimacy. Not only demonstrating their "good performance", but also their "know-how". Without placing much importance on their degree of pragmatic legitimacy.

Likewise, when facing customer profiles with the fear of being socially penalized by the results of their decisions, they must focus their efforts on obtaining greater moral and regulatory legitimacy. Finally, when facing customers with an evident fear of economic risks, they must focus their efforts on earning cognitive and pragmatic legitimacy. This would accordingly reduce the sensation of risk.

Organizations can only manage to influence the desirability and the acceptance of their activities by working in an active way. In this sense, the literature contains numerous strategies to earn, maintain and recover legitimacy (See [44]). For example by affiliating with economic groups which provide experience and reputation [34].

References

1. Aldrich, H.E., Fiol, C.M.: Fools rush in? The institutional context of industry creation. Acad. Manag. Rev. Acad. Manag. **19**(4), 645–670 (1994)
2. Alonso-Almeida, M., Fernández de Navarrete, F.C., Rodriguez-Pomeda, J.: Corporate social responsibility perception in business students as future managers: a multifactorial analysis. Bus. Ethics Eur. Rev. **24**(1), 1–17 (2015)
3. Bitektine, A.: Toward a theory of social judgments of organizations: the case of legitimacy, reputation, and status. Acad. Manag. Rev. **36**(1), 151–179 (2011)
4. Blanco-González, A., Cruz-Suárez, A., Díez-Martín, F.: The EFQM model as an instrument to legitimise organisations. In: Peris-Ortiz, M., Álvarez-García, J., Rueda-Armengot, C. (eds.)

Achieving Competitive Advantage through Quality Management, pp. 155–169. Springer International Publishing (2015)

5. Brown, A.D.: Narrative, politics and legitimacy in an IT implementation. J. Manag. Stud. **35**(1), 35–58 (1998)

6. Chaney, D., Lunardo, R., Bressolles, G.: Making the store a place of learning: the effects of in-store educational activities on retailer legitimacy and shopping intentions. J. Bus. Res. (2016). http://doi.org/10.1016/j.jbusres.2016.04.104

7. Chen, J., Wu, H., Yao, X.: Status, legitimacy, and the presence of outside directors in China. Manag. Decis. **54**(5), 1205–1221 (2016)

8. Chung, J.Y., Berger, B.K., DeCoster, J.: Developing measurement scales of organizational and issue legitimacy: a case of direct-to-consumer advertising in the pharmaceutical industry. J. Bus. Ethics (2015). http://doi.org/10.1007/s10551-014-2498-8

9. Cruz-Suárez, A., Prado-Román, A., Prado-Román, M.: Cognitive legitimacy, resource access, and organizational outcomes. RAE-Revista de Administração de Empresas **54**(5), 575–584 (2014)

10. Cruz-Suárez, A., Prado-Román, C., Díez-Martín, F.: Por qué se institucionalizan las organizaciones. Revista Europea de Dirección y Economía de la Empresa **23**(1), 22–30 (2014)

11. Cruz-Suárez, A., Prado-Román, C., Escamilla-Solano, S.: Nations of entrepreneurs: a legitimacy perspective. In: Peris-Ortiz, M., Sahut, J.-M. (eds.) New Challenges in Entrepreneurship and Finance, pp. 157–168. Springer International Publishing (2015)

12. Deephouse, D., Carter, S.: An examination of differences between organizational legitimacy and organizational reputation. J. Manag. Stud. **42**(2), 329–360 (2005)

13. Deephouse, D.L.: Does isomorphism legitimate? Acad. Manag. J. **39**(4), 1024–1039 (1996)

14. Deephouse, D.L., Suchman, M.: Legitimacy in organizational institutionalism. In: Greenwood, R., Oliver, C., Suddaby, R., Sahlin-Andersson, K. (eds.) The Sage Handbook of Organizational Institutionalism, pp. 49–77. Sage Publications, London (2008)

15. Díez-Martín, F., Blanco-González, A., Prado-Román, C.: Measuring organizational legitimacy: the case of mutual guarantee societies. Cuadernos de Economía y Dirección de la Empresa **43**(junio), 115–144 (2010)

16. Díez-Martín, F., Blanco-González, A., Prado-Román, C.: Explaining nation-wide differences in entrepreneurial activity: a legitimacy perspective. Int. Entrepreneurship Manag. J. **12**(4), 1079–1102 (2016)

17. Díez-Martín, F., Prado-Román, C., Blanco-González, A.: Efecto del plazo de ejecución estratégica sobre la obtención de legitimidad organizativa. Investigaciones Europeas de Dirección y Economía de la Empresa **19**(2), 120–125 (2013)

18. Díez-Martín, F., Prado-Roman, C., Blanco-González, A.: Beyond legitimacy: legitimacy types and organizational success. Manag. Decis. **51**(10), 1954–1969 (2013)

19. Drori, I., Honig, B.: A process model of internal and external legitimacy. Organ. Stud. **34**(3), 345–376 (2013)

20. Easterby-Smith, M., Thorpe, R., Jackson, P.R.: Management Research, 4th edn. SAGE Publications, London (2012)

21. Edwards, P.J., Roberts, I., Clarke, M.J., Diguiseppi, C., Wentz, R., Kwan, I., Cooper, R., et al.: Methods to increase response to postal and electronic questionnaires. Cochrane Database Syst. Rev. (3) (2009)

22. Finch, D., Deephouse, D., Varella, P.: Examining an individual's legitimacy judgment using the value-attitude system: the role of environmental and economic values and source credibility. J. Bus. Ethics **127**(2), 265–281 (2015)

23. Garud, R., Schildt, H.A., Lant, T.K.: Entrepreneurial storytelling, future expectations, and the paradox of legitimacy. Organ. Sci. **25**(5), 1479–1492 (2014)

24. Gil Aluja, J.: La selección de inversiones en futuro incierto. Revista de economía y empresa **7**(17–18), 61–80 (1987)

25. Gil Aluja, J.: El riesgo y la incertidumbre en el momento actual. Revista de economía y empresa **8**(20–21), 7–28 (1988)

26. Greenwood, R., Suddaby, R., Hinings, C.R.: Theorizing change: the role of professional associations in the transformation of institutionalized fields. Acad. Manag. J. **45**(1), 58–80 (2002)
27. Haveman, H.A., David, R.J.: Ecologists and institutionalists: friends or foes? In: Greenwood, R., Oliver, C., Sahlin, K., Suddaby, R. (eds.) The Sage Handbook of Organizational Institutionalism, pp. 573–595. Sage, London (2008)
28. Iglesias Pérez, F.: Estudio de la economía sumergida: relación entre legitimidad y medidas para combatirla. Tesis Universidad Rey Juan Carlos (2014)
29. Kaufmann, A., Gil Aluja, J.: La programación de actividades por el método del semáforo. Revista europea de dirección y economía de la empresa **1**(1), 7–14 (1992)
30. Liñán, F., Urbano, D., Guerrero, M.: Regional variations in entrepreneurial cognitions: start-up intentions of university students in Spain. Entrepreneurship Reg. Dev. (Routledge) **23** (3–4), 187–215 (2011)
31. Maguire, S., Hardy, C.: Discourse and deinstitutionalization: the decline of DDT. Acad. Manag. J. **52**(1), 148–178 (2009)
32. Manrai, A.K., Andrews, R.L.: Two-stage discrete choice models for scanner panel data: an assessment of process and assumptions. Eur. J. Oper. Res. **111**(2), 193–215 (1998)
33. Meyer, J., Rowan, B.: Institutionalized organizations: formal structure as myth and ceremony. Am. J. Sociol. **83**(2), 340–363 (1977)
34. Mingo, S.: Entrepreneurial ventures, institutional voids, and business group affiliation: the case of two Brazilian start-ups, 2002–2009. Academia Revista Latinoamericana de Administración **26**(1), 61–76 (2013)
35. Oliver, C.: Strategic responses to institutional processes. Acad. Manag. Rev. **16**(1), 145–179 (1991)
36. Pfeffer, J., Salancik, G.R.: The External Control of Organization: A Resource Dependence Perspective. Harper & Row, New York (1978)
37. Pollack, J.M., Rutherford, M.W., Nagy, B.G.: Preparedness and cognitive legitimacy as antecedents of new venture funding in televised business pitches. Entrepreneurship Theory Pract. **36**(5), 915–939 (2012)
38. Rosch, E.: Principles of categorization. In: Rosch, E., Lloyd, B.B. (eds.) Cognition and Categorization, pp. 27–48. Lawrence Erlbaum Associates (1978)
39. Ruef, M., Scott, W.: A multidimensional model of organizational legitimacy: hospital survival in changing institutional environments. Adm. Sci. Q. **43**(4), 877–904 (1998)
40. Rutherford, M.W., Buller, P.F.: Searching for the legitimacy threshold. J. Manag. Inquiry **16** (1), 78–92 (2007)
41. Scott, W.R.: Institutions and Organizations. SAGE Publications (1995)
42. Shepherd, D., Zacharakis, A.: A new venture's cognitive legitimacy: an assessment by customers. J. Small Bus. Manag. **41**(2), 148–167 (2003)
43. Singh, J., Tucker, D., House, R.: Organizational legitimacy and the liability of newness. Adm. Sci. Q. **31**(2), 171–193 (1986)
44. Suchman, M.C.: Managing legitimacy: strategic and institutional approaches. Acad. Manag. Rev. **20**(3), 571 (1995)
45. Thomas, T.E.: Are business students buying it? A theoretical framework for measuring attitudes toward the legitimacy of environmental sustainability. Bus. Strategy Environ. **14**(3), 186–197 (2005)
46. Überbacher, F.: Legitimation of new ventures: a review and research programme. J. Manag. Stud. **51**(4), 667–698 (2014)
47. Wood, D.J.: Corporate social performance revisited. Acad. Manag. Rev. **16**(4), 691–718 (1991)
48. Zimmerman, M., and Zeitz, G.J.: Beyond survival: achieving new venture growth by building legitimacy. Acad. Manag. Rev. **27**(3), 414 (2002)
49. Zuckerman, E.W.: The categorical imperative: securities analysts and the illegitimacy discount. Am. J. Sociol. **104**(5), 1398–1438 (1999)

A Model for the Management in Organizations Based on People and Knowledge: Aspects to Be Considered in Its Design

José María de-Goñi-Oslé and Arturo Rodríguez-Castellanos

Abstract As Professor Gil-Aluja states, the progressive uncertainty and complexity in business and economic systems highlights the relevant role of people, basis for the generation of intangible resources that are the source of business competitiveness. Thus, the relevance of intangibles in the current society, especially those based on knowledge and people, makes them a strategic issue for the organizations, due to the competitive advantage that they can generate, and also by their impact on results. Intangibles are under research and study both in the scientific-academic and in the corporate spheres. In this work we intend to consider some aspects in the design and implementation of an efficient model for the management in organizations, based on people and knowledge. To do this, we review the literature on the matter, as well as apply to the observation of some relevant cases of implementation of related models in companies belonging to different territorial spaces. Results, especially from the observation of practical cases, suggest that there is an enormous difficulty in the phases of implementation and maintenance in time of these models, possibly due to the difficulties that emerge to evaluate and monetize adequately their actual cost and the benefits obtained.

1 Introduction

As Professor Gil Aluja [37] states, a thick fog covers the world; its effects respond to two names: "complexity" and "uncertainty". We found in a world each time more interrelated—"complex", but at the same time and surely by it, more uncertain. And this affects the overall scope, but it manifests itself most clearly in the economy [2], and also in companies [57].

Precisely Professor Gil Aluja had long ago the vision of this current reality, deciding to anticipate it and dedicate their intellectual efforts to find instruments,

J. M. de-Goñi-Oslé · A. Rodríguez-Castellanos (✉)
Facultad de Ciencias Económicas y Empresariales, Departamento de Economía Financiera II, Universidad del País Vasco, Avda. del Lehendakari Agirre, 83, 48015 Bilbao, Spain
e-mail: arturo.rodriguez@ehu.eus

© Springer International Publishing AG 2018
C. Berger-Vachon et al. (eds.), *Complex Systems: Solutions and Challenges in Economics, Management and Engineering*, Studies in Systems, Decision and Control 125, https://doi.org/10.1007/978-3-319-69989-9_5

especially from logic and mathematics then considered "non-conventional", enabling to deal with these new circumstances. Thus, together with Professor Arnold Kauffmann, developed mathematical instruments for taking economic and business decisions in conditions of uncertainty, through the logical of fuzzy subsets [50, 51], passing then to the application of other types of logics [35, 36], work that has been developed additionally by other researchers [40, 41]. Also, initiated the analysis of complex social and economic phenomena, through the investigation of forgotten effects [52], subsequently expanding the analysis to other logics [38].

The work of Professor Gil-Aluja on these subjects is very wide, and at the same time, of a great depth. In this chapter we adopt a position matching to highlight the relevance of progressive complexity and uncertainty when analyzing economic and business phenomena, and also to make management decisions, but hope, modestly, make an additional contribution in a way already pointed by Professor Gil-Aluja to indicate that "the evolution of the economic system will always be conditioned by the subjectivity inherent to the human being falling, again and again, in the decisions taken in its bosom" [39]. Finding us in the presence of social, economic and business systems simultaneously complex and uncertain, the subjective dimension oh the human being achieves special relevance. And, focusing on the business system, we understand that, increasingly, are people, with their knowledge and their values, basic sources for the development of intangible resources, company's innovation, and hence for competitiveness and value creation [61]. People-based management will allow companies to make facing the challenges raised by complexity and uncertainty.

Currently, terms such as *intellectual capital, intangibles or knowledge assets* are used interchangeably in the academic fields and in the business world. A boom is produced in this field, especially in the last two decades (at the end of the 20th century up to today's date), although these concepts are not new.

As shown by Bueno et al. [18], the concept of "intellectual capital" was introduced by the economist Franz List at the end of the 19th century (1885), defining it with a focus on a nation or the humanity as 'the accumulation of discoveries, inventions, efforts, etc., of previous generations'.

Therefore recognise the importance of intangibles, of knowledge, and by association, the people in the company—it should not be forgotten that knowledge is in the people—is not something new, and there is abundant literature on the matter. Thus, authors who follow the approach of Intellectual Capital (IC) insist on the importance of Human Capital—as the sum of training, skills and competencies of the people in the company—for innovation and business competitiveness [61].

The interest has notability raised in recent years by the field of intangible resources, knowledge or IC. Its growth is exponential. As example, more than 50% of the GDP of the major economies of the OECD in 1990 are based or are related to knowledge [55]; nevertheless, in the report of the World Bank in 2014, and relative to the first decade of the 21st century, that percentage increases to levels reaching 57%.

One of the reasons, possibly the most relevant, for the progressive importance of knowledge assets is that both physical and financial assets have become commodities. For this reason, at the present time a third stage in the research on intangible resources is taking place, which has led to more practice oriented studies, looking for the implementation and improvement of their management [24].

Despite the research carried out so far, companies do not have tools that allow them to measure the true extent that for value generation has the management that they carry out on their intangible resources [23]. Furthermore, despite the large amount of proposed models, reality is that in general there has been a scarce implementation of them. To this end, it is noteworthy that despite the interest in intangibles both in business and in the scientific or academic spheres, research may be somewhat away from the business world.

The fundamental basis of intangible resources or intellectual capital is knowledge (whether human, structural, technological or relational), which is linked to the human being, that is, to the person. In many cases, the biggest difficulty lies in the identification of intangible assets in an organization—or in a sector, or region or country, as well as to identify which the generate value, and which t are critical or strategic. Therefore, valuate or quantify them (monetize them is not always possible), and also to manage them, is often extremely difficult.

In this work, within the research project "A management model based on people and knowledge", we will come up some aspects to consider in the design of the mentioned model that we consider particularly relevant.

To do this, in the first place we present a review of the literature on the matter in recent decades (1980–2016). Later we consider some of the most striking cases in the application of models of this type, two of the Basque environment (Arteche Group and Irizar Group), two of the Spanish environment (Repsol and Bankinter) and two international (Xerox and Volvo IT). Lastly, the conclusions synthesize the positive and negative aspects of models, both in the theoretical approaches and in the practical realizations, in such a way that they allow identifying the most important aspects to be considered in the design of a model that can surpass the limitations encountered.

2 People, Intangibles, Knowledge and Innovation: A Literature Review

2.1 Intangible Resources, Intellectual Capital, Knowledge and Innovation

Our competitive environment demand continuous changes: the speed with which ideas born, compete and die, leads to the need to manage organizations very differently to as it was only 4 or 5 years ago. The need to generate new ideas more quickly has provided that the value of information and knowledge quote upward. Proof of this are the sectors that are directly related: telecommunications, internet, computing in general, training, etc. Therefore it is recognized the need to ensure that people accept to invest all its talent in the organization, with a level of participation and involvement much greater. In this line have emerged a series of management approaches that recognize the value of the existing knowledge in the people and seeking to promote it, structure it and make it operative and valuable for the company. Some of these approaches are: the Intangible Resources, Intellectual Capital (IC), Knowledge Management (KM), or Lifelong Learning.

According to [33], we can affirm that a generally accepted definition of "Intangibles" do not exists. In fact, it is an adjective that usually accompany different nouns such as assets, activities, resources, etc. However, the fact that, often, the adjective is substantivized, this is, is used as a noun, is a good proof of the difficulties that exist for try to establish a correct definition [21].

In any case, the wide variety of definitions of Intangibles that can be found in the literature has common elements [20]. In general, from an accounting perspective, "Intangibles" are defined as non-monetary sources of future economic benefits, without physical substance, controlled, or at least influenced by the company, as a result of past transactions and events—purchased, produced by the company, or obtained in any other manner—and that may or may not be sold separately from other assets of the company.

Considering that the difference between the market value and the accounting or book value is due only to the value of the intangible resources and capabilities is an error. As is indicated by García-Ayuso [32] and Andriessen [5], there are many factors that can influence the market, in addition to intangibles; for example, tangible and financial assets can are undervalued, or the stock can be altered by market anomalies. Moreover, the IC value is not only the sum of the isolated values of its elements, but that the relationships that take place among them and with the rest of resources also form part of IC and increase its value.

IASC [43], through IAS num. 38, defines "intangible assets" as identifiable and non-monetary assets, without physical substance, used in the production or supply of goods and services, or for lease to others, or for administrative reasons. In addition, the standard indicates that its cost should be able to be reliably measured so it can be recorded.

The above accounting standard establishes the criteria for a "resource" can be considered "active":

- It must be subject to control by the company, as a result of past events.
- It is likely that future economic benefits attributable to the asset are obtained.

The term "control" refers to both the ability to perceive economic benefits generated by the asset, as the ability to restrict competitive access to the same by the existence of legal rights. Therefore, the concept of "intangible asset" is more restrictive than the of "intangible resource", requiring compliance with a series of requirements collected in accounting standards. These requirements are not met in many cases by intangible resources, and much less by capabilities [21, 29].

Turning to the concept of "Intellectual Capital" or IC, this is defined as the combination of human, technological, organizational and relational resources—that is, intangible resources—of an organization. There are numerous studies that deal with IC and its elements [9, 11, 15, 27, 33, 68]. In fact, Intellectual Capital is nothing new, but it has been present since the time when the first seller established a good relationship with a client. Later, it was named *goodwill*. What has happened over the course of the last two decades is an explosion in certain key technical areas, including the information and communication technologies (ICT), which have provided new tools with which a global economy has been built. Many of these tools provide intangible benefits that now we take for granted, but that before did not exist, to the point that the organization cannot function without them. The property, or at least the control, over such tools provides competitive advantages, and therefore constitute an asset.

Organizational knowledge is measured through the Intellectual Capital of the organization. IC can be considered as the set of intangible resources of an organization which, though not necessarily be reflected in the traditional financial statements, generate value at present or have potential to generate it in the future [28]. Another definition of IC indicates that it is the sum and synergy of all the knowledge that brings a company, all the experience accumulated in its members, all what it has achieved in terms of relationships, processes, discoveries, innovations, market presence and influence in the community. For [26] IC can be divided into Human Capital and Structural Capital, which may be Organizational Capital or Client Capital. Subsequently, several authors have established the division of IC into Human Capital (HC), Structural Capital (SC)—that sometimes divides into Organizational Capital (SC) and Technological Capital (TC) or Innovation Capital —and Relational Capital (RC), which also divides sometimes into Business Capital —sometimes, Client Capital—and Social Capital [9, 15, 17, 28, 67].

"Knowledge" is not the same thing as "intelligence". Knowledge is "a combination of organized and structured ideas and information, taken by one or more human beings in a specific context and for a purpose" [60], while intelligence is what makes lack to create knowledge. Intelligence implies that there are skills to learn, transfer knowledge, reasoning, see what is possible, find new interpretations, generate alternatives and make wise decisions.

It is also clear that we differentiate between "knowledge" and "information"; knowledge is superior to the information, in the sense that it is more complex, is structured, has more dimensions than the information, and, perhaps most importantly, is a specific quality of human beings. Moreover, information is inert and static; however, knowledge, when being linked to the individual, has subjective elements, and is essentially dynamic.

To expand intelligence, IC is generated by creating new knowledge, that is, the "raw material" that allows people to innovate through creating new products, services, processes and management methods.

IC thrives on relationships with high level of integration. In fact, integration is the foundation of strategic advantages, because knowledge creation, innovation and customer collaboration depend on it. In business, those who want to develop intellectual capital and manage knowledge should be collaborating people, essential to create and transfer new knowledge and creating and implementing innovation.

Current times generate a new competitive dynamic [46], in which companies increasingly grant greater importance to intangible resources and capabilities when faced with competitors, so is widely recognized that the new knowledge, as well as its application to innovation, are key factors to achieve business success and to keep it [30, 31].

In this sense, in a competitive and dynamic environment like the present one, innovation (technological and non-technological) is becoming, increasingly, a key aspect of business competitiveness. However, given the close relationship between the organizational knowledge (expressed through IC, and especially HC) and the capacity of innovation [22, 65, 68, 72], the studies dedicated to the improvement of knowledge processes in companies are of great interest [3].

Knowledge is the differentiating factor between the organizations that generate wealth through innovation and which do not; by this reason, the decision makers in companies must develop all the strategies necessary for identify the features that orient the knowledge to constitute himself in a valuable asset. This is the aim of Knowledge Management. These aspects cause that HC gets to constitute the dimension of greater preeminence in IC [61].

The pressures that environment changes, globalization and the acceleration of products cycles are producing in business make it necessary to accumulate a high volume of knowledge so that, subsequently, have as a result all kinds of innovations. In the professional field, managers of companies needed instruments to assess the processes and outcomes of innovation activities. Furthermore, it is interesting to note that [71], in reviewing the literature, observes that innovation, among other concepts such as intangible assets—rooted tacit routines, core competencies and knowledge creation, has an important role in the consideration of the company's assets that create value, beyond the physical and financial resources.

2.2   Models for the Measurement of Intellectual Capital Applicable to Knowledge Management

Below is a brief review of models for measuring Intellectual Capital that we consider most relevant to the management of people and knowledge. A thorough review is impossible given the space available.

Balanced Scorecard [47–49]. It has been one of the pioneer models in the IC field; it includes in its analysis the management indicators that had been excluded traditionally in accounting because only the related ones to the financial aspect were included.. Thus, the model gets to be an effective tool for decision making [34, 62].

Skandia Navigator. Presented by Edvinsson [25] and Edvinsson and Malone [26]. This model emerged initially in 1994, as an annex to the annual report of the company Skandia, where pointed out the contribution of Intellectual Capital developed within the organization. It clearly identifies the difference between financial capital and Intellectual Capital of the company; its main objective was to create a tool to address the managerial decision-making process.

Technology Broker. Annie [11] is the proponent; part of the same concept that Skandia model, but includes both tangible as intangible resources. In this model qualitative indicators are elaborated through the development of a methodology to audit the information related to IC.

Model of the Canadian Imperial Bank [63]. It has been created to measure IC in the Imperial Bank of Canada; It make explicit the relationship between the processes of learning and IC; the indicators proposed are basically of learning [23, 66]; raises that intellectual Capital is made up of three elements: (a) Human Capital, (b) Structural Capital, (c) Client Capital. In addition, the Financial Capital is also considered.

University of Western Ontario Model [8, 9]. It carries out an analysis referring to the cause-effect relations between the basic dimensions of—IC HC, SC and RC— and its effect on business results. The success of the model is primarily due to the importance that has H C as a basic and fundamental resource for the achievement of the goals and indicators of performance desired and required by the company [34, 61]; also its importance within an organizational culture focused on intelligence can be reaffirmed, because it allows to structure knowledge-based organizations thanks to the development of that culture.

Intangible Assets Monitor [69, 70]. It is presented by Karl-Erick Sveiby; intangible assets are considered as central aspect in its approach. In this order of ideas, according to [10], the design of the model can be synthesized in the following way: (a) Human Capital: has like relevant element the competences of the people, based on the capacity to act in different situations; it includes experience and education; (b) Structural Capital: is subdivided in two subcategories: "Internal Structure", alluding to the elements created by the company, but which are the property or the possession of the organization, such as patents, processes, models, information systems and organizational culture; and "External Structure", focused on the relationships with customers and suppliers; aspects such as reputation of the organization, image of the company and brand names are taken into account in this subcategory. The basic concepts of this model are similar in some respects to those of the Skandia Navigator, because it shows what is called the "Invisible Balance". Sveiby proposes three types of indicators for the measurement and evaluation of these resources, within each of the categories and subcategories: (a) indicators of growth and innovation, (b) efficiency indicators and (c) stability indicators. All it jointly allows to arrive at the so-called Intangible Assets Monitor.

Dow Chemical Model. According to [64], this model arose due to the failure of a scheme allowing to assess the management of intangible assets in Dow Chemical Company. The established methodology is based on the process of measuring and management of invisible assets and their impact on financial activity. This becomes relevant because it is a company that managed to develop a method for coding and managing portfolios of patents [56]. The model is structured on three pillars: (a) Human Capital, related to the skills to generate solutions for the clients, (b) Organizational Capital, referring to the capabilities of the organization to encode and use knowledge and, (c) Client Capital, based on the attention and service to the client (client understanding, generation of loyalty and franchise management). All these resources generate value for the company. The indicators vary according to the type of organic structure; thus, the model of management of intellectual assets consists of six basic steps such as: Business Strategy, Valuation of Competitors, Classification of the Own Advantages, Pricing, Investment and Management of the Portfolio.

Model of Competence-Based Strategic Management: Intangible Capital [12]. Another approach to assess IC is the so called "Strategic Management (SM) of the company". Bueno [12] stated that Intellectual Capital is the basis of the process of Strategic Management. Thus, identifies IC with core competencies. The model is based on three essential elements: (a) those of technological origin, (b) those of organisational origin and (c) those of a social nature. The main generic competences are present in this model, namely: (a) attitudinal, (b) aptitudinal, and (c) appraisal; all referred or related with the members of the organization, as foundations of the work of the company and estimates of what the company is capable of doing [64].

Nova Model (1999) [19]. Created by the company Nova Care and developed by the Knowledge Management Club of the Valencian Community in Spain, it is a model that is applicable to any type of company; it allows not only measure, but also make IC management processes; it is aimed to foster the growth of companies; it is divided into four elements: (a) Human Capital, (b) Organization Capital, (c) Social Capital and (d) Innovation and Learning Capital. When deepening a little more in it, authors as [56] argue that the essence of the model lies basically in the management by competencies; likewise, in regards to the indicators, [58] placed them within the range of process dynamics.

Model for the Structure of Intellectual Capital—Intellect (Euroforum). [13] coincides with the approach of [9] in the sense of focus the model on three types of capital related: (a) Human Capital, (b) Structural Capital, (c) Relational Capital. Although already they have been considered previously, next we display a brief elucidation of each one of them:

Human Capital. It refers to the capabilities and commitments that make part of the knowledge of the people; competencies, ability to innovate and improve, motivation and commitment, are articulated therefore. In this regard, [13] states that both tacit and explicit knowledge, possessed by individuals, are useful for the organization. It is clear that this accumulation of knowledge is acquired through socialization, education, training and communication processes.

Structural Capital. It is the own knowledge of the organization. It arises whereas the knowledge possessed by individuals and teams is made explicit, codified, systematized and internalized through a formal process that operates through the creation of routines. To register this type of capital bibliographies, documentary compilations, databases, management systems, development of new technologies, patents, and the culture and values of the organization can get together.

Relational Capital. As its name suggests, it is the set of successful relationships that the organization maintains with clients and suppliers, as well as the value that has managed to develop through compliance with political, social, economic and environmental issues for the State and society itself, with which a value is generated that potentiates the procurement of new clients.

Intellectus Model [15, 17]. Developed in the Centre for Research on the Knowledge Society of the Autonomous University of Madrid, from the traditional dimensions of IC—HC, SC and RC—[9], difference in SC two sub-dimensions, Technological Capital and Organizational Capital, and within RC also developed other two sub-dimensions, in particular Business and Equity Capital. On that basis, a comprehensive system of indicators is constructed.

The *Updated Intellectus Model* [17], on the basis of the initial model, part of the reflection on the dynamic role played by both processes of entrepreneurship and innovation, like they which lead to the I + D function. These considerations place this proposal in the determination of a new logic in the analysis of the composition of intangible resources. Of this form a "Map" of elements and variables for each one of the five basic initial capitals settles down, for which also identifies a set of "Accelerators", which is completed with the incorporated with the new Capital of Entrepreneurship and Innovation.

Model for the Valuation of Intellectual Capital [54]. These authors define separate items for each type of capital, and from them, they indicate its specific measurement. In some cases it is obtained more than one factor for each dimension. At the same time, within the three-dimensions traditional, each of these dimensions has been divided into various sub-components. Thus, within HC they propose subcomponents such as the ability of employees and their satisfaction; SC includes culture and organisational processes, information systems and intellectual property; and in RC the subcomponents of clients and partners.

Despite the diversity of proposed models, they display more coincidences of which one might think at first, since, in general, they consider that it exists a series of not financial aspects and intangible assets that affect the economic performance of the organization, which can be grouped in the dimensions that are commonly considered that they conform to IC and that, in general, it is considered to be the three usual ones of Human Capital, Structural Capital and Relational Capital, which contribute to the success of the activities, projects and programmes of an organization and, ultimately, to increase its value.

2.3 Indicators for the Measurement of Intellectual Capital

The identification of IC has aroused in the organizations a great interest to develop a process that allows them to know in what level are its components, so that these contribute that companies can be developed in an open and global economy, act in a socially responsible way and achieve a competitiveness held in time, based on intangible resources that they possess or they get to acquire at a specific time. Given the youth of the thematic one, the way to identify and measure IC in companies is not exempt of difficulties, not only because of the new management culture that implies, but because of the diversity of models and approaches, which can lead to confusion. For this reason, it is necessary to ask for a panel of indicators that serve as a guide and reference for management.

"Indicators" can be defined as numerical values that provide measures to assess the quantitative and/or qualitative performance of a system, an individual or an organization. In general, refers to "management indicator" as an empirical instrument that allows representing, to a certain extent, the theoretical dimension of a practical variable [1]. Specifically, IC measurement indicators are instruments for the evaluation of the organizations' intangible resources, expressed in different units of measurement [14, 16]. It should be made clear that the indicators are not a goal in themselves, but that are descriptions that require to be interpreted and evaluated according to standards. Thus, for the thoroughness their concept, principles of use, characteristics and classification criteria are due to need.

Some authors such as [6] emphasize the process of measuring IC, presenting it as the most important wealth of the organization, reason why its measurement is a vital topic. Thus, this author displays a frame confronting different factors such as finance, processes, market and Human Capital, with which he stipulates a basic process which begins by the creation of the vision and the identification of core competencies and key success factors; with this it is arrived at the identification of typical IC indicators in organizations.

Although there is still no homogeneous criteria perfectly delimited for the valuation of intangible resources, the use of indicators for their measurement and management has become a common practice. The absence of general criteria has led to the emergence of panels of indicators of different types. So, it is necessary to propose a set of principles for the use of IC indicators in order to homogenise its practical application by users and experts, and refine intangibles' measurement. They are also necessary homogeneous and universal criteria that allow economic agents to interpret and make comparisons between companies. In this regard, important efforts are being made to design general indicators according to a set of common guidelines, materialized in basic principles and features.

With base in the above considerations, the use of IC indicators should be governed by three basic principles: *permanence in time*, *levels of aggregation* and *transparency*.

Permanence in time implies that the set of indicators must be interrelated and balanced over time, so that it could constitute a systematic management tool. The information should be assessable and shared in time and space. The stability in time of the indicators allows gathering measurement experiences, testing its use and integrating them as organizational routines, which can get to reach the force of habits and be internalized as spontaneous mechanisms within the organization.

Aggregation levels refer to the form of representation of indicators at different levels of information (breakdown) and in different units of measurement for each one of the different levels, in order to provide information as thoroughly as possible on the basis of the most general to the most particular, depending on the organizational structure of each company—the whole organization and functional departments, either matrix, subsidiaries, business units and departments, according to the existing organizational structure. Aggregation also allows reducing the number of calculation operations, since the addition of the minimum levels of aggregation will automatically provide overall levels. It is also possible to classify the measurements, at all times, since a minimum level of detail, up to a maximum level.

Transparency is essential to convey confidence and credibility, both to the markets as the individuals—read employees.

The fact that companies have begun to carry out measurements and reporting of intellectual capital is very positive. In this respect, and given that the first steps have been taken from the business world and not from accounting regulators, each company has produced its own report without following "generally accepted" principles and rules—as it is the case of principles and accounting standards, so that each report is unique. As a result of this situation, it is very difficult to compare IC reports between companies, even between those of the same country, and in that sense the necessity of standards becomes decisive.

Once established the concept, the principles of use, the characteristics and the types of IC indicators, with the aim of elaborate concrete indicators, we have came to consider each of the IC dimensions, since, in the practice, such indicators are referred to these. Therefore, we have addressed both the measurement of HC, SC and RC.[1]

IC management, besides to identify the organization's intangible resources and detect the activities and initiatives that are affecting their performance [53], is due mainly to transform Human and Relational Capital in Structural Capital useful to the organization. But to achieve this, IC must first be measured in the three dimensions that comprise it (human, structural and relational), which is not an easy task but necessary, since their importance is reflected in the organizations' capital structure. Intangible assets, such as patents and intellectual property rights—which, generally, make up a small part of SC—tend to be easily measurable and therefore countable as organization's assets, but for the rest of the intangible resources that is not usually possible, reason why the above mentioned methods arose.

3 Some Relevant Cases

Since the mid 90s of the last century, as a result of the intense globalization of markets, constant technological change, and also changes in the production and organization structures, the international context acquires greater relevance in the economic and competitive aspects. Technological innovations and, largely, the related to the information and communication technologies (ICT) are those that have facilitated of more intense way the changes in all social dimensions—

---

[1]The lists of indicators are not included here by reasons of space, but they can be provided on request.

economic, political, cultural and organizational—by its connector, structuring, diffuser and innovative character. Result of all this is a reality that is increasingly characterized by the reduction of the importance of manufacture, the increased presence of services, the emergency and consolidation of ICT, new ways of managing organizations, the relevance of the professional and technical occupations, the importance of the control of technology and the need to innovate continuously. Innovation appears like a condition *sine qua non* for competitiveness and sustainable development, but its process of creation and development is cumulative, uncertain and of socio-economic, as well as technological, nature.

Therefore, after examining different models of IC measurement and management, we believe necessary present, although briefly, diverse cases, both of our close environment, that is, the Basque Country (Arteche and Irizar), as also of the Spanish area (Repsol and Bankinter) and the international environment (Xerox and Volvo IT), in which they have been applied, in varying degrees and in different ways, models of this type, in order to detect both the advantages and disadvantages of their application to the business practice.

3.1 Arteche Group

Today this business group is an international reference in the electricity sector, with teams working in more than 150 countries, with companies in Europe, America, Asia and Oceania, and a service consisting of more than 80 technical and commercial offices, whose business philosophy is based on the development of a multicultural team integrated by professionals committed to the objectives of the group, capable of providing efficient response to any challenge, and who share their knowledge in a free and active form in a climate of trust and participation, with the aim of creating value for the organization, for the collectives involved in the business, and for the social environment.

The transformation process started in 1995 at the parent company, Electrotécnica Arteche Hermanos Sociedad Anónima[2] (EAHSA), based in Mungia (Bizkaia), has crystallized into a corporate identity reflected in an operation style that oriented the company towards new cultures or ways to make based on trust in people and in their potential [42, 45]. The system related to knowledge management in EAHSA is organized internally. Access to management information through panels of indicators are developed to overcome the limitations that access though the traditional accounting systems provided, because this information was disseminated in hindsight, it was generic and too abstract, did not allow to follow the upcoming action events nor to see them in detail by workshops, products, etc., nor allowed the involvement of every person in the running of the company's work. The operation's

---

[2]In Spanish legislation, "Sociedad Anónima" is a joint-stock company.

style of Arteche Group orients the company towards new cultures or ways to make, based on trust in people and in their potential [4].

3.2   Irizar Group

Headquartered in Ormaiztegi (Gipuzkoa), and with legal form of a cooperative, is the leading company in Spain and second in Europe on manufacture and equipment of luxury coaches; its commitment is "knowledge to do" as a basis for innovation. Irizar set up innovation as a goal and KM as a method to achieve it, based, on the one hand, in collective learning and, on the other hand, in teams and multidisciplinary projects, involving people with a high degree of commitment.

The Irizar model was established in 1992 under the leadership of Koldo Saratxaga. With him, Irizar focused its efforts on the production of luxury coaches, while approached the international market, but above all he revolutionize organizational model to management based on people united by the challenge of satisfying the client. Irizar is not using the word "management" referring to issues having to do with people, because it is understood that people are not managed, but that are served and animated to achieve both individual and common objectives. Irizar pursues that multidisciplinary teams share experiences that generate innovation and creativity and which in turn contribute to new knowledge [73]. By that, they maintain that when in most of the other companies speak of KM like a competitiveness tool, they refer to that there is to think how to cause that the high part of the company's pyramid acquires more knowledge so that generates more, reason why they do not get to have the scope that it has in Irizar.

The final result is a horizontal organization, with self-managed work teams in which the active participation of all the people turns out key to achieve profitability and to progress jointly. These teams have a few very specific objectives to be achieved in a given period of time. One of the team's members is responsible for leading it, a role which has already assumed a significant percentage of the workforce. At Irizar they do not like to talk about "knowledge management", because they understand that knowledge belongs to the people and "them it is not possible to be managed". For the same reason either speech of "human resource management", but to "share experiences". All workers, including those in the production plant, attend a very high number of meetings. In them, the main objectives are to define which the best practices for a specific task are or to solve problems that exist within the teams. The communication between employees takes place not only through the meetings, but also through a computer system that collects all the relevant information and making unnecessary workers to report to a superior one. The lack of "seniors" allows workers to "turn the client in Chief". The objective is that, with base in the creativity, the individual autonomy and the mutual commitment, each employee understands the client and can add value to him, identifying the causes of the creation of competitive advantages. The model is completed with an equitable distribution of a percentage of the profits.

3.3   Repsol

This company occupied the fifth place in 2008 in the European Make Awards (behind BP, British Broadcast, Heineken and Nokia). They define KM as a means to get the objective that to convert individual knowledge into a shared good accessible for the whole of the organization, to the object to share ideas, manage the learned lessons and facilitate the transmission of best practices, focusing on the establishment of virtual communities and social networks. Also mechanisms have settled down that allow to determine the causes of the success or failure of each one of the undertaken actions ("*post mortem* analysis"). Conceptual maps, knowledge maps and technological surveillance systems are other tools implemented in this organization. The analysis that company carries out about the *cost of the not-knowledge* is especially remarkable [59].

3.4   Bankinter

It is a banking entity in whose annual reports, from 1999 to the present time, are collected a systematized and plentiful set of data on IC and its measurement in the organization. Given its small size with respect to its more direct competitors (BBVA, Bank Santander, Caixabank, etc.), they have arrived at the conviction that its competitiveness is tied fundamentally to the intangible resources. Therefore, they interpret that the value of the company is determined largely by IC; for that reason it is necessary to have the most complete information about intangible resources relevant to the company and try to homogenize it. Thus, the entity has specific indicators over IC dimensions, providing information in their annual reports of the indicators relating to HC, SC and RC [7].

3.5   Xerox

If there is a mythical company in terms of knowledge management, this is, without a doubt, Xerox. In the early 90s of the last century, Jonh Seely Brown, CEO of the PARC (Xerox Palo Alto Research Center) established a group of specialists (with anthropologists among them) to try to improve the printer repair service offered to its clients. The group discovered that the technicians obtained a great effectiveness thanks to the information they shared while they were talking next to the coffee machine. This anecdotal fact, true or not, has gone down in history as the driving force behind the development of KM at Xerox. The true fact is that in 2001 this company had implanted a system (Eureka) able to accredit and compensate employees (recognized already by then as *knowledge workers*) who share their knowledge. The effort focused on developing a platform where technicians could record the detected problems and the applied solutions—they pride themselves on having more than fifty thousand of these documented solutions, as well as developing expert systems for the resolution of problems able to leverage the vast amount of available information.

The consequence was a 10% of improvement in productivity in the first two years, which has been kept at the time (in 2010 was estimated at 5–10%, about 10 million USD). The ability to evaluate the return on investment in KM, consequence of a proper integration with financial services, is another of the differential aspects of this company. Communities of practice have always been another of Xerox strengths, along with mechanisms to ensure the motivation of employees. This last aspect is critical in an organization that recognizes the interaction among people as the main source of knowledge, and gives to coffee machine almost divine powers. To get it they have implemented economic reward and recognition systems and it has been paid special attention to the commitment of the people in charge, always reluctant to their teams spend too much time away from their daily work tasks. Some of the objectives pursued by Xerox with their model are:

- To make emerge employees' knowledge and to be able to encode it.
- Locate the most appropriate means to decode the knowledge that is shared by the entire organization.
- Allow quick and easy access to knowledge.
- Encourage employees to consider knowledge as an exclusive advantage for their personal development, share it, and promote the creation of new solutions and strategies to improve the business.

3.6 Volvo IT

Volvo Information Technology (Volvo IT), a service company of information technologies (IT) belonging to the Volvo Group, devoted four months during 1998 to implement a recommendation system based on intelligent agents and in studying its use. At this time the Volvo Intranet consisted of 450 Internet servers and counted on approximately 400,000 documents, which were mostly presentations of the departments, reports, projects, FAQs, and online help material. The study was done on 48 users who agreed to participate; the way in which they are motivated was based on the assumption that the tested prototype would provide them with specific information at a minimal effort. People were invited to attend an introductory meeting whose main purpose was to explain the application design and how to use it. Additionally, users were requested to maintain an informal record of the events that deserved to be considered as important, then collect this information. Users ranged from technicians and systems developers to content providers and administrators. All of them were computer users experienced with access to Intranet-connected personal computers. Additionally, the system had a function called Community, basically oriented to users to have a directory to locate colleagues with similar roles in the organization. In order to obtain this, the system would match the descriptions of the jobs provided by the users.[3]

4 Conclusions

As a culmination of this work, following a review of the literature, as well as the business cases, we consider that some conclusions can be deduced.

With regard to intangibles, IC and KM, the advances conducted by the academic proposals and research are very important; however, that effort seems poorly compensated with effective applications in the business world. These are exceptional cases, and many of them lack scientific and academic rigour. Therefore, despite the relevance perceived by companies with respect to their intangible resources, its management has not received the necessary attention. Evidence suggests that the business efforts devoted to the identification and management of intangible assets are weak in general.

In this respect, a double problem is detected: As academic research runs the risk of being away from the business world, also in business there are attitudes and behaviours in taking decisions and designs strategies that are far away, or not consider, rigour provided by academic and scientific activity. It must be taken into account that many companies define its strategy properly, but a large part of them fail in its implementation. It is likely that mutual interaction and support between both business and academic worlds will reduce this high proportion of failures.

Also, it is clear that people's capabilities and skills of are the primary source of innovation and value creation in the company, although those potentials should be integrated in a favourable organizational culture, and under a suitable leadership.

It is interesting to note that the literature is extensive in terms of intangibles as "assets"—"positive" value drivers, but very small with respect to "intangible" liabilities—"negative" value drivers, although it is true that they can be considered as the first, but with negative value. What happens is that in practice the simultaneous existence in the company of both intangible positive and negative value drivers tends to produce in many cases a "net effect" of value close to zero with what both assets and liabilities disappear, because the positive effect of the firsts is offset by the negative impact of the seconds.

Also, when considering people in developments that are carried out with respect to IC and KM areas, the complexity of human personality in emotional or psychological aspects seems to be little valuated.

Currently a situation, at least, paradoxical is taking place: Many managers know what Knowledge Management is, and also are aware that they need that their organization is impregnated of that new approach, but often they do not know by

---

[3]For both Volvo's and Xerox's cases, see Innovación Tecnológica, Calidad, Gestión del Conocimiento [44].

where to start. This circumstance occurs as a result of the existence of much literature approaches about "what", but very little on "how". All it has led to the sale of methodologies which, under the name of "Knowledge Management" include on many occasions interesting topics such as building of competitive advantages, innovation, alignment of the company with its clients and markets, excellence in processes, competitive intelligence, etc., but who do not consider the fundamental element of KM, that is, that knowledge lies in people.

In the model that we intend to develop in our research project, as well as the fundamental aspects collected in the literature, and in the largest number of possible experiences, they will have to be considered,, in a special way:

- The close linkage between the knowledge of people and business innovation.
- The involvement and commitment of the human team with the project in the process of model's implementation.
- Mutual collaboration with the near scientific and research community, and vice versa.
- The relevance, in the process facilitator team, of aspects relative to the integrity of the person in the field of ethics and morality. In other words, we talk about "people" not only as bearers of knowledge, but also of ethical values.

As Professor Gil-Aluja underlines, the progressive complexity and uncertainty of the economic and business systems make highlight the role of the people, each time more relevant. Only a people-based business management will allow companies engage successfully the challenges of the new realities.

References

1. AECA: Indicadores para la gestión empresarial Documento Num 17. AECA, Comisión de Contabilidad de Gestión, Madrid (1998)
2. Aguer-Hortal, M.: New horizons in the evolution of economic science In ¿Hacia dónde va la ciencia económica?, pp. 45–59. Publicaciones de la Real Academia de Ciencias Económicas y Financieras, Barcelona (2016)
3. Alegre-Vidal, J.: La Gestión del Conocimiento como motor de la innovación: lecciones de la Industria de Alta Tecnología para la empresa. Jaume I University Press, Castellón de la Plana (2004)
4. Alvarado, C.: Arteche - Historia de los hechos empresariales 1946–2006. University of Deusto, Bilbao (2008)
5. Andriessen, D.: Making Sense of Intellectual Capital. Butterworth-Heinemann, Burlington, MA (2004)
6. Arbonies, A.L.: Conocimiento para Innovar: La Sociedad del Conocimiento, 2nd edn. Díaz de Santos, Madrid (2006)
7. Bankinter: Memoria Anual. bankinter.com/www2/.../es/inf.../memoria (1999–2015)
8. Bontis, N.: There's a price on your head: managing intellectual capital strategically. Bus. Q. **60**(4), 41–47 (1996)
9. Bontis, N.: Intellectual capital: an exploratory study that develops measures and models. Manage. Decis. **36**(2), 63–76 (1998)

10. Bontis, N.: Assessing knowledge assets: a review of the models used to measure intellectual capital. Int. J. Manage. Rev. **3**(1), 41–60 (2001)
11. Brooking, A.: Intellectual Capital. Core Asset for the Third Millennium Enterprise. International Thomson Business Press, London, (1996)
12. Bueno, E.: El capital intangible como clave estratégica en la competencia actual. Boletín de Estudios Económicos **53**, 207–229 (1998)
13. Bueno, E.: Perspectivas sobre dirección del conocimiento y capital intelectual. University Institute Euroforum Escorial, Madrid (2000)
14. Bueno, E. (dir.): Metodología para la elaboración de indicadores de capital intelectual. Documentos Intellectus Num 4. Knowledge Society Research Centre, Madrid (2003a)
15. Bueno, E. (dir.): Model for the measurement and management of Intellectual Capital: Intellectus Model. Intellectus Documents Num 5, Knowledge Society Research Centre, Madrid (2003b)
16. Bueno, E. (dir.): Methodology for the design of Intellectual Capital indicators (Updated Version). Intellectus Documents Num 4, Knowledge Society Research Centre, Madrid (2004)
17. Bueno, E. (dir.): Modelo Intellectus de medición, gestión e información del capital intelectual (Nueva versión actualizada). Documentos Intellectus Num 9/10, CIC-IADE (UAM), Madrid (2012)
18. Bueno, E., Salmador, M.P., Merino, C.: Génesis, concepto y desarrollo del capital intelectual en la economía del conocimiento: Una reflexión sobre el modelo Intellectus y sus aplicaciones. Estudios de Economía Aplicada **26**(2), 43–63 (2008)
19. Camisón, C., Palacios, D., Devece, C.: Un nuevo modelo para la medición del capital intelectual: el modelo Nova. ACEDE X Congress, Oviedo (2000)
20. Cañibano, L., García-Ayuso, M., Sánchez, M.P.: Accounting for intangibles: a literature review. J. Account. Lit. **19**(1), 102–130 (2000)
21. Cañibano, L., García-Ayuso, M., Chaminade, C. (eds.): MERITUM Project: Guidelines for Managing and Reporting on Intangibles. Airtel Móvil Foundation, Madrid (2002)
22. Černe, A., Jaklič, M., Škerlavaj, M., Ülgen, A., Polat, D.: Organizational learning culture and innovativeness in Turkish firms. J. Manage. Organ. **18**(1), 193–219 (2012)
23. Coduras, O.: Estudio sobre el estado del arte de los intangibles de la empresa. Institute for Intangible Assets, Madrid (2006)
24. Dumay, J., Garanina, T.: Intellectual capital research: a critical examination of the third stage. J. Intellect. Cap. **14**(1), 10–25 (2013)
25. Edvinsson, L.: Developing intellectual capital at Skandia. Long Range Plan. **30**(1), 366–373 (1997)
26. Edvinsson, L., Malone, M.S.: Intellectual Capital: Realising Your Company's True Value by Finding its Hidden Brainpower. Harper Business, New York (1997)
27. Edvinsson, L., Sullivan, P.: Developing a model for managing intellectual capital. Eur. Manage. J. **14**(4), 356–364 (1996)
28. Euroforum: Medición del capital intelectual. Modelo Intelect Euroforum, Madrid. http://gestiondelconocimiento.com/modelo_modelo_intelec.htm (1998)
29. Fincham, R., Roslender, R.: The Management of Intellectual Capital and its Implications for Business Reporting. Research Committee of The Institute of Chartered Accountants of Scotland, Edinburgh (2003)
30. Galende, J.: Analysis of technological innovation from business economics and management. Technovation **26**(3), 300–311 (2006)
31. Galende, J.: The appropriation of the results of innovative activity. Int. J. Technol. Manage. **35**(1/2/3/4):107–135 (2006b)
32. García-Ayuso, M.: Intangibles. Lessons from the past and a look into the future. J. Intellect. Cap. **4**(4), 597–605 (2003)
33. García-Merino, J.D.: Una propuesta metodológica para la valoración de los intangibles empresariales. Doctoral Dissertation, University of the Basque Country, Bilbao (2015)

34. García-Zambrano, L.: Intangible resources, competencies and business results: evidences for Basque and Spanish companies. Doctoral Dissertation, University of the Basque Country, Bilbao (2016)
35. Gil-Aluja, J.: Elements for a theory of decision in uncertainty. Springer Science & Business Media, Dordrecht (1999)
36. Gil-Aluja, J.: La pretopología en la gestión de la incertidumbre. Publicaciones de la Universidad de León, León (2002)
37. Gil-Aluja, J.: El mundo en el que es imposible volver atrás. Revolución, Evolución e Involución en el futuro de los Sistemas Sociales, pp. 15–20. Publicaciones de la Real Academia de Ciencias Económicas y Financieras, Barcelona (2014)
38. Gil-Aluja, J., Gil-Lafuente, A.: Algoritmos para el tratamiento de fenómenos económicos complejos. Ramón Areces, Madrid (2007)
39. Gil-Aluja, J.: Algunas reflexiones sobre el futuro de la investigación económica. Anales del Curso Académico 2014–2015, pp. 67–82. Publicaciones de la Real Academia de Ciencias Económicas y Financieras, Barcelona (2015)
40. Gil-Lafuente, A.: La matemática no numérica en el futuro de la ciencias económicas. Nuevos horizontes científicos ante la incertidumbre de los escenarios futuros, pp. 97–129. Publicaciones de la Real Academia de Ciencias Económicas y Financieras, Barcelona (2015)
41. Gil-Lafuente, A.: The economic sciences faced with new realities, from the classic mechanic to the multi-varied logic. ¿Hacia dónde va la ciencia económica?, pp. 67–96. Publicaciones de la Real Academia de Ciencias Económicas y Financieras, Barcelona (2016)
42. Goyarzu, A., Igarza, R.M., Martínez, U.: Grupo ARTECHE. Gestión del conocimiento: más allá de la tecnología informática, el desarrollo personal. In: Zarrabeitia, J. (Coord.) Empresas Avanzadas en Gestión. Cluster del Conocimiento and Ediciones PMP, Bilbao (1999)
43. IASC (International Accounting Standards Committee): International Accounting Standard, no 38. Intangible Assets, IASC, London (1998)
44. Innovación Tecnológica, Calidad, Gestión del Conocimiento. www.calidadytecnologia.com/ .../Gestion-Conocimiento-Mejores-Empresas-en-la-Gestión-del-Conocimiento (2014). Accessed 16 May 2014
45. Jiménez, J.L.: Generación del conocimiento: más allá de la tecnología informática, el desarrollo personal. In VV. AA. (2001) Entorno empresarial del Siglo XXI y gestión del conocimiento. IV Jornadas Internacionales del Cluster del Conocimiento. Cluster del Conocimiento and Ediciones PMP, Bilbao, pp. 54–58 (2001)
46. Johnson, W.H.A.: Leveraging intellectual capital through product and process management of human capital. J. Intellect. Cap. **3**(4), 415–429 (2002)
47. Kaplan, R.S., Norton, D.P.: The balanced scorecard—measures that drive performance. Harv. Bus. Rev. **70**(1), 71–79 (1992)
48. Kaplan, R.S., Norton, D.P.: The Balanced Scorecard: Translating strategy into action. Harvard Business School Press, Boston (1996)
49. Kaplan, R.S., Norton, D.P.: Strategy Maps. Converting Intangible Assets into Tangible Outcomes. Harvard Business School Press, Boston (2004)
50. Kaufmann, A., Gil-Aluja, J.: Introducción de la teoría de los subconjuntos borrosos a la gestión de las empresas. Milladoiro, Santiago de Compostela (1986)
51. Kaufmann, A., Gil-Aluja, J.: Técnicas operativas de gestión para el tratamiento de la incertidumbre. Hispano Europea, Barcelona (1987)
52. Kaufmann, A., Gil-Aluja, J.: Modelos para la investigación de efectos olvidados. Milladoiro, Santiago de Compostela (1988)
53. Marr, B.: What are Key Performance Questions? Management White Paper, Advanced Performance Institute, London. www.ap-institute.com (2008)
54. Moon, Y.J., Kym, H.G.: A model for the value of intellectual capital. Can. J. Adm. Sci./Revue Canadienne des Sciences de l'Administration **23**(3), 253–269 (2006)
55. OECD: The Knowledge-Based Economy. OECD, Paris (1996)
56. Osorio, M.: El Capital Intelectual en la gestión del conocimiento. ACIMED **11**(6), 113–128 (2003)

57. Poch-Torres, S.: Reflexiones sobre riesgos e incertidumbres en el nuevo paradigmaempresarial. Anales del Curso Académico 2014–2015, pp. 83–87. Publicaciones de la RealAcademia de Ciencias Económicas y Financieras, Barcelona (2015)
58. Pomeda, J., Merino, C., Murcia, C.: Towards an intellectual capital report of Madrid: new insights and developments. Paper presented at the Conference on the Transparent Enterprise. The Value of Intangibles. Madrid, 25–26 Nov (2002)
59. Repsol: Memoria Anual. © Repsol. www.repsol.com (2000–2013)
60. Rodríguez-Castellanos, A.: Gestión del Conocimiento y Finanzas: una Vinculación Necesaria. Publicaciones de la Real Academia de Ciencias Económicas y Financieras, Barcelona (2002)
61. Rodríguez-Castellanos, A.: Las personas, fuente de innovación y de creación de valor en la empresa. Publicaciones de la Real Academia de Ciencias Económicas y Financieras, Barcelona (2015)
62. Sáez Vegas, L.: Cuadro de Mando Integral. In: Planificación y control empresarial. University of the Basque Country, Bilbao (2013)
63. Saint-Onge, H.: Tacit knowledge: the key to the strategic alignment of intellectual capital. Strategy Leadersh. 24(2), 10–14 (1996)
64. Sánchez, P., Chamichade, C., Olea, M.: Management of intangibles: an attempt to build a theory. J. Intellect. Cap. 1(4), 188–209 (2000)
65. Santos, H., Figueroa, P., Fernández, C.: The influence of human capital on the innovativeness of firms. Int. J. Bus. Econ. 29(2), 463–474 (2010)
66. Soret, I., de Pablos, C., Montes, J.L.: Medición de capital intelectual y ventajas competitivas: Una aplicación para la iniciativa ECR (Efficient consumer response). Esic Mark. 137, 65–106 (2010)
67. Stewart, T.A.: Intellectual Capital: The New Wealth of Organizations. Doubleday, New York (1997)
68. Subramaniam, M., Youndt, M.A.: The influence of intellectual capital on the types of innovative capabilities. Acad. Manage. J. 48(3), 450–463 (2005)
69. Sveiby, K.E.: The New Organisational Wealth—Managing and Measuring Knowledge-Based Assets. Berrett-Koehler, San Francisco (1997)
70. Sveiby, K.E.: The intangible assets monitor. J. Hum. Resource Costing Account. 2(1), 73–97 (1997)
71. Swart, J.: Intellectual capital: disentangling an enigmatic concept. J. Intellect. Cap. 7(2), 136–159 (2006)
72. Wu, S.H., Lin, H.Y., Hsu, M.Y.: Intellectual Capital, dynamic capabilities and innovative performance of the organization. Int. J. Technol. Manage. 39(3–4), 279–296 (2007)
73. Zarrabeitia, J.: Irízar. La reingeniería como modelo de gestión, un proyecto basado en las personas. In Zarrabeitia J (Coord.) Empresas Avanzadas en Gestión. Cluster del Conocimiento and Ediciones PMP, Bilbao (1997)

Part II
Decision Making and Systems Modeling

Six Experimental Activities to Introduce the Theory of Fuzzy Sets

**Joan Carles Ferrer Comalat, Xavier Bertran Roura,
Salvador Linares Mustarós and Dolors Corominas Coll**

Abstract During 2016, on the occasion of the 50th anniversary of the publication of Lotfi Zadeh's article Fuzzy Sets, which gave rise to the theory of fuzzy sets, and within the framework of Catalonia's Science Week, the Barcelona Museum of Science "CosmoCaixa" organized an exhibition at the museum in collaboration with several Catalan Universities and institutions to present the results of 50 years of research in the field of fuzzy set theory. The inaugural speech at the exhibition was delivered by Professor Jaime Gil Aluja and entitled "The Spanish school of fuzzy economy", in which he brilliantly presented the large audience with an overview of the work done in recent years in the field of applying fuzzy logic methods to the economy. As a tribute to the person who taught numerous research groups and introduced them to the application of fuzzy logic methods to economics and business management, in this paper we present a sample of the six activities recreated in the first exhibition area. These activities were aimed at introducing visitors to the foundations and ideas that gave rise to fuzzy set theory and this new approach to reasoning.

1 Introduction

In classic set theory, every element of a referential set will belong or not belong to a subset of that set.

J. C. F. Comalat (✉) · X. B. Roura · S. L. Mustarós · D. C. Coll
Faculty of Economics and Business Sciences, Department of Business Administration,
University of Girona, C/ Universitat de Girona, 10, 17071 Girona, Catalonia, Spain
e-mail: joancarles.ferrer@udg.edu

X. B. Roura
e-mail: xavier.bertran@udg.edu

S. L. Mustarós
e-mail: salvador.linares@udg.edu

D. C. Coll
e-mail: dolors.corominas@udg.edu

© Springer International Publishing AG 2018
C. Berger-Vachon et al. (eds.), *Complex Systems: Solutions and Challenges
in Economics, Management and Engineering*, Studies in Systems,
Decision and Control 125, https://doi.org/10.1007/978-3-319-69989-9_6

Aristotle (*De interpretatione*, Chapter IX) believed that the statement "A sea battle will be fought tomorrow" could not be classified either within the subset of true statements or in that of false statements. His reasoning can be summarized as follows: if the statement is true at the time it is spoken nothing can prevent a sea battle tomorrow. If it is false, the same is true, or in other words, there will be no way to start a sea battle tomorrow. Accepting, then, that the sentence is true or false at the time it is pronounced would mean that our future has already been determined and will be determined forever more, and this does not make sense, as with future events there is the possibility of them happening or not.

Aristotle's reflections won him acknowledgement as one of the first visionaries of the problems associated with classic theory.

Zadeh [12] posited a linguistic context for a new set paradigm that opened the door to new techniques used in many industrial applications today, such as control systems for focusing video cameras or vehicle braking [4, 6, 7].

Fuzzy set theory has established itself in fields as diverse as artificial intelligence, medicine, geology and business management [1–3, 8–10], and studies using the theory can currently be counted in tens of thousands, while there would still seem to be huge research possibilities in this new field.

It was in 1993, when the University of Girona's Department of Economics had implemented a doctoral program on mathematical techniques applied to accounting and business management. that our teacher, Professor Carles Cassú, first contacted Professor Jaume Gil Aluja, then full professor at the University of Barcelona, to teach a monographic course on fuzzy logic applied to economics and business management. The year following this first meeting, the first congress of the International Society of Fuzzy Economics and Management (SIGEF) was held in Reus and attended by several professors from the University of Girona. This was when we made our initial contact with the fuzzy logic research groups that had founded SIGEF from the University of Barcelona and Rovira i Virgili University, and many other universities from around the world that were present at that first congress.

From this moment onwards, Professor Gil Aluja collaborated in numerous seminars and meetings held at the University of Girona to promote research in the field he is so passionate about. As a result of his perseverance and generosity, his teachings have permeated our University and allowed many researchers in our group to continue working in the field he introduced us to, based on the assumption that black and white coexist alongside a varied array of greys, and in which fuzzy set theory is essential for development. In economics, due to the human factor that permeates many of the attitudes and decisions inherent in economic activity and the social sciences, this vision is fundamental in establishing new perspectives of analysis for economic problem-solving.

As a result of Professor Gil Aluja's work and collaboration with different research centres, the authors of this article continue to work in collaboration with our colleagues from different universities on applications of fuzzy logic for improving economic models and management techniques.

These collaborations bore singular and very successful fruit in the aforementioned activity last year, and the authors of this article presented our work as a tribute to Professor Gil Aluja as the person who was able to transmit his enthusiasm in fuzzy methodology research to us.

Given that 2015 marked 50 years since the publication of the first article that popularized what has come to be called fuzzy set theory, the Barcelona Science Museum, in collaboration with the University of Girona, Rovira i Virgili University, the Open University of Catalonia, the Royal Academy of Economic and Financial Sciences and the International Society of Fuzzy Economics and Management organized a joint exhibition to commemorate the event. Based on objects, images and experiences, the exhibition "Explore: 50 years of fuzzy ideas" was designed to illustrate to all publics some results of research in fuzzy methodologies at the technical and business levels. To this end, the exhibition was structured into three complementary sections, as shown in Figs. 1 and 2.

First, upon entering the room there was the discovery zone (Tables 1–6, Zone A). This presented a series of six experimental activities related to fuzzy set theory concepts that were specially designed to attract the attention of children and arouse curiosity in adults and prepare them to understand the paradigmatic leap in logical principles on which fuzzy set theory is based.

The second section, or scientific linear zone (Showcases 7–11, Zone B), was for exhibiting academic and practical works in fuzzy logic. The first showcase showed the academic birth of Zadeh's theory and illustrated its extremely broad application with various research books. The second showcase showed several objects from the field of engineering that incorporated fuzzy control systems. To help visitors understand the basic idea of how these systems are operated, a piece of software was presented together with the objects, created expressly for the exhibition, which simulated the operation of a fan governed by fuzzy rules. The third showcase showed a collection of objects that demonstrated how the word fuzzy has been incorporated into modern-day society to identify works by plastic artists and musicians. Thus, visitors found jewels, design objects, books of poems and music CDs that contain the word fuzzy in their name. The section also contained an adult workshop consisting in sorting various sentences into groups of true and false. Visitors could see that many of the sentences were extremely difficult or impossible to classify, thus fuzzy logic was endowed with meaning from philosophy. Finally, as shown in Fig. 3, the last showcase presented academic works developed in the field of economics by Professors Kaufmann and Gil Aluja. Our aim here was to pay homage to the two men who spread the theory to the universities that jointly organized the event.

Finally, the third section comprises the main idea reflection zone (Zone C), illustrated in Fig. 4, presented the collection of pictorial works "Fuzzy Art" by Barcelona artist Queralt Viladevall. The pictorial representations of everyday fuzzy scenes captured by the artist were intended to portray, in a subtle and captivating way, that the coexistence of being and non-being and the principle of gradual simultaneity are present in our daily lives. The explanations of the principles of fuzzy logic depicted through the paintings completed the ideas opened in the first zone whose theoretical and practical possibilities were reflected in the second.

Fig. 1 Overview of the
venue with zones and activity
tables

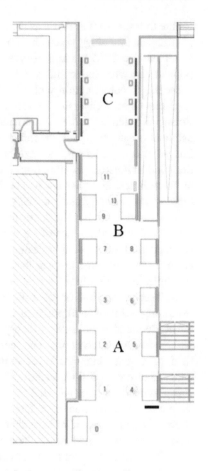

This text provides a detailed account of the first six experimentation zones the
visitor encountered at the beginning of the exhibition, which were designed to
introduce them to the foundations of set theory.

The text has been structured as follows. First of all, although there are interesting
bibliography about the theoretical foundations about fuzzy sets [5, 11], in order to
contextualize the experiments we provide a section on the foundations of intuitive
set theory and another on the foundations of fuzzy set theory. In the next section we
detail each of the six experiments included in the first section of the exhibition. This
section includes a total of six subsections explaining how each experiment is related
to fuzzy set theory. Finally, there is a conclusions section and the bibliography.

Fig. 2 Shot of the venue

2 Classic Set Theory

The current intuitive idea of the classic set, commonly found in dictionaries and encyclopedias, is that a set is a "well-defined collection of objects, called elements, with a common characteristic that allows it to be precisely established whether a reference element belongs to the given set or not". This wording is a simplification of Cantor's idea of sets established in 1872 in the first known attempt to formalize modern set theory: "a set is the gathering together in one of certain and distinct objects of our intuition, perception or thought, called the elements of the set, each clearly differentiated and differentiable from one another".

A set can be determined, or by extension expressed, by listing each and every one of its elements; or by comprehension, through stating a property that serves to determine all the elements of the set.

To express that an element is part of the set, we use the symbol of belonging $\in$ from the Greek letter epsilon, introduced by Peano. To express that an element is not part of the set, the symbol $\notin$ is used.

A set with no element is called "empty set" and is represented by the symbol borrowed from the Norwegian alphabet $\emptyset$, following the representation posited by André Weil.

Similarly, the current intuitive idea of a subset is that "it is a well-defined collection of objects from a referential set".

It is also possible to determine a subset of a set either by extension or by comprehension.

A very convenient way of depicting subsets while illustrating definitions is to represent them by Venn diagram drawings.

Fig. 3 Fuzzy research in the field of business

Thus, for example, Fig. 5 shows the representation of a subset A.

The quadrilateral in which the circle is located represents the "universe" of discourse and constitutes all possible elements. The circle identified as A represents the set and inside it are all the elements that comprise it.

The part of the quadrilateral outside the circle, which is a dark colour, represents the existing elements that do not belong to the set A. This part of the drawing is called "complementary to A" and is usually denoted as A′ in classic set theory. The predicate that identifies it is "not A".

Fig. 4 Fuzzy thinking through art

Fig. 5 Subset within a
universe

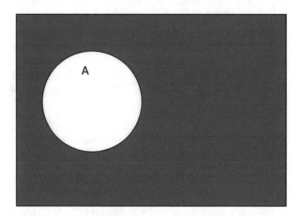

Figure 5 helps to clarify two characteristic principles of classic set theory known as the "excluded third principle" (an element of discourse is either in A or A', i.e. there are no other options beyond these two) and the "non-contradiction principle" (if the discourse element is in A, then it cannot be in A', or if the discourse element is in A', then it cannot be in A, i.e., it cannot be in both at the same time). It is possible to unify both principles by means of the expression: "every element of E belongs either to A, or, if not, then it must belong to non-A".

The term "classic predicates" is used for those predicates that divide a reference set into two complementary subsets fulfilling the two principles above. Subsets characterized by a classic predicate are called "precise, crisp or ordinary subsets".

Consequently, given a reference frame E, a subset A characterized by a classic predicate can be determined by comprehension through the following expression:

A = {elements of the reference frame E that fulfill the predicate P}[1]

There is a practical way of representing a subset based on listing all the elements of the referential set together with a one-to-one assignment of the values 1 or 0 depending on whether the element is in the subset or not.

Formally, let $E = \{x_1, x_2, x_3, x_4, ...\}$ a reference set and $A = \{x_i, x_j, ...\}$ a subset of it. The subset A is identified by the following expression:

$$A = \{(x_1, \mu_A(x_1)), (x_2, \mu_A(x_2)), (x_3, \mu_A(x_3)), (x_4, \mu_A(x_4)), ...\}$$

where $\mu_A(x_i)$ is number 1 if the element x_i is in A, or it is 0 if the element is not in A.

The expression of the subset can be abbreviated using the following notation:

$$A = \{(x, \mu_A(x))/x \in E\}$$

The above identification allows us to define what is known as the characteristic function or indicating function of subset A, represented by μ_A, as the function of the reference set on the set $\{0,1\}$ defined by:

$$\mu_A : E \to \{0, 1\}$$

$$x \to \mu_A(x) = \begin{cases} 1 & \text{if x belongs to A} \\ 0 & \text{if x does not belong to A} \end{cases}$$

One interesting observation is that, in the representation using characteristic functions, for every element x of a reference frame E, the following equality is fulfilled:

$$\mu_A(x) = 1 - \mu_{A'}(x) \tag{1}$$

An important aspect of set theory is being able to work with more than one subset of the same reference set. With Venn diagrams it is possible to represent different subsets and visually check which different types of situations can arise.

For example, the illustration in Fig. 6 shows two different subsets of the same reference frame E.

The common area between the two circles is the subset comprising the referential elements in A and B. This new subset is called "A intersection B" and its notation is A ∩ B.

---

[1]In the present work we have used the usual key symbol notation posited by Cantor and Zermelo to enclose the elements of the set.

Fig. 6 Sets A and B of a
reference frame

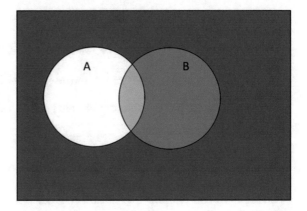

The subsets A, B and A ∩ B are represented using characteristic functions as
follows:

$$A = \{(x, \mu_A(x))/x \in E\}$$
$$B = \{(x, \mu_B(x))/x \in E\}$$

and

$$A \cap B = \{(x, \mu_{A \cap B}(x))/x \in E\}$$

where we observe that the characteristic function of the intersection subset can be
represented by the expression:

$$\mu_{A \cap B}(x) = \text{minimum}(\mu_A(x), \mu_B(x)) \tag{2}$$

later used by Zadeh to define the intersection of fuzzy subsets.

If we look again at Fig. 6, the area formed by the two circles, determined by the
illustration in Fig. 7, represents the subset comprising the elements of the reference

Fig. 7 The union set of
subsets

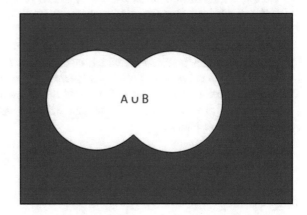

frame that are either in A or B. This new subset of the reference frame is called "A union B" and its notation is A ∪ B.

The subset A ∪ B is represented using characteristic functions as follows:

$$A \cup B = \{(x, \mu_{A \cup B}(x))/x \in E\}$$

where the characteristic function of the union subset can be represented by:

$$\mu_{A \cup B}(x) = \text{maximum}(\mu_A(x), \mu_B(x)) \tag{3}$$

the Operator that was later used to model the union of fuzzy subsets.

Based on the symbology presented here, in classic set theory the principle of the excluded third and the principle of non-contradiction are formulated by means of the following two expressions:

$$A \cup A' = E \tag{4}$$

$$A \cap A' = \varnothing \tag{5}$$

3 Foundations of Fuzzy Set Theory

The fact that an element of a set does or does not fulfill a property is the basis of binary logic, which was inherited from the Greek world and has guided our scientific and technological development. However, many examples can be found in nature that would be difficult to classify according to whether they fulfill a property or not.

In [13], Professor Lotfi Zadeh published the article "Fuzzy Sets". Using a new nomenclature for the concept of "vagueness" and allowing characteristic functions to take values in the interval [0,1], Zadeh initiated a new and fruitful theoretical field in mathematics known as "Fuzzy Set Theory".

Below, we set out the ideas posited by Zadeh that extended the possibilities of classifying subsets without modifying their previously developed inherent structure.

As we have seen, in classic set theory it is assumed that in an ordinary subset it can be clearly determined whether any element x of the reference set does or does not fulfill a certain predicate P, which characterizes the subset. A predicate that cannot classify a collection of objects into two disjointed sets according to whether the objects meet a condition or not is called the fuzzy predicate.

The subset that identifies each element with the value of fulfilling the fuzzy predicate is represented as:

$$\tilde{A} = \{(x_1, NA_1), (x_2, NA_2), (x_3, NA_3), \ldots\},$$

where the numbers NA_i are real values between 0 and 1 which represent the degree of belonging to the set according to the fuzzy predicate defining the subset.

The number that for each element of a subset identifies the level at which it fulfills a fuzzy predicate is called the element's degree of belonging to the fuzzy subset.

There are infinite ways of defining the complementary, union and intersection between fuzzy subsets so that they continue to fulfill the properties of defining the union, intersection and complementary in extreme cases where the elements have only a degree of belonging of 0 or 1 and are therefore ordinary subsets. Generally, the decision maker determines them according to the convenience of the situation to be modelled. The most common tend to be those defined by Zadeh in his seminal article:

$$\tilde{A}' = \left\{ \left(x, \mu_{\tilde{A}'}(x) \right) \right\} \text{ con } \mu_{\tilde{A}'}(x) = 1 - \mu_{\tilde{A}}(x) \tag{6}$$

$$\tilde{A} \cap \tilde{B} = \left\{ \left(x, \mu_{\tilde{A} \cap \tilde{B}}(x) \right) \right\} \text{ with } \mu_{\tilde{A} \cap \tilde{B}}(x) = \text{minimum} \left(\mu_{\tilde{A}}(x), \mu_{\tilde{B}}(x) \right) \tag{7}$$

$$\tilde{A} \cup \tilde{B} = \left\{ \left(x, \mu_{\tilde{A} \cup \tilde{B}}(x) \right) \right\} \text{ with } \mu_{\tilde{A} \cup \tilde{B}}(x) = \text{maximum} \left(\mu_{\tilde{A}}(x), \mu_{\tilde{B}}(x) \right) \tag{8}$$

We conclude this section by emphasizing that this new conception of sets does not fulfill the principle of the excluded third and the principle of non-contradiction, principles which are fulfilled in classic set theory.

4 Six Experimental Activities to Show Fuzzy Set Theory

In the knowledge that most visitors to the CosmoCaixa Barcelona Science Museum are families and that we were competing with a number of highly attractive alternatives (Amazon rainforest, planetarium, …), the first part of the exhibition was designed to capture the attention of children so that they would enter with their parents. It was hoped that once their attention had been caught and their curiosity aroused, and while they waited for their children to finish the activities, the parents would want to visit the rest of the exhibition. The methodology and dynamics used in the first zone were based on guidance from CosmoCaixa experts in order to make the exhibition as interactive as possible and fit with centre's philosophy. A monitor from the Museum was always on hand to help children with the activities and motivate them to discover strange objects.

4.1 First Activity. The Dollhouse Game

The first activity in Zone A, shown in Fig. 8, consisted in locating "strange" objects in a house. A fully-furnished Victorian dollhouse was used, as seen in Fig. 9. Some of the objects in the house were duplicated, adding some modifications to make them appear like the furniture normally found in the house.

Thus, visitors found beds you cannot sleep in, tables you cannot work or eat at, lamps that due to their construction do not project light even when lit, sinks with holes in them that allow the water to drain directly to the floor and never fill up, and seats you cannot sit in, like the one in Fig. 10.

The new objects were constructed using the 3D design software "Rhinoceros" and printed using the "Entres Up Box" printer. They were painted by hand using plastic paints to imitate the style of the initial base objects.

Once printed and painted, the objects formed a natural part of the scene, as can be seen in Fig. 11.

The panel accompanying the table invited the visitor to reflect on why an object receives a common name. In the panel, under the title of non-functional objects, it read: "Is there a piece of furniture in the house that does not fulfill its function properly? Maybe a chair you cannot sit in, a table you cannot eat at or a bed it would be too uncomfortable to sleep in? Find the strange objects!"

Fig. 8 Dollhouse with fuzzy objects

Fig. 9 Victorian dollhouse

Fig. 10 Sofa with extremely large arms in dining room

Fig. 11 Lamp with useless light function in dining room

The aim of the activity was for visitors to question the fact that we sometimes include an object within a generic set of objects such as "chairs", "tables", "lamps", etc., even though the object does not entirely fulfill the principal function that it is generally agreed an object must fulfill to be labelled with the word that identifies said category of objects.

4.2   Second Activity. The Transport Game

The second activity was to classify the means of transport (land, air or water) associated with different vehicles. Figure 12 presents a detailed photograph of the table.

On the table were objects that could be classified in only one of the categories, such as cars or trains, but also objects that could be classified in several categories, such as a submarine plane or a train boat. This allowed visitors to work on subset intersection. However, the main vehicle employed to generate debate was a train engine in the form of a plane without wheels pulling a group of wagons with wheels behind it. Each element of the train could be classified as a single means of transport. Interestingly, when we formed the train, it became unclassifiable. As one part was land and one part air transport, classifying this "train" entailed considering that for this set it may be necessary to accept, given that it comprised some elements that fulfilled a characteristic and others that did not, that its classification as a single land, air or water type of transport was not completely possible, even if it was not altogether impossible. Consequently, the activity used a concrete example to question that a predicate would divide the referential set of means of transports into

Fig. 12 Different objects for transportation

two complementary subsets: land transport and non-land transport. This questioned the principle of the excluded third.

The panel accompanying this activity and inviting visitors to reflect on the difficulty of having to choose a group for each object was entitled "Surprising Transportation". On it, beneath the image shown in Fig. 13 it read: "Guess which means of transport (land, air or water) each vehicle belongs to! Maybe the vehicles are not what they seem!"

Fig. 13 Surprising transportation

4.3   Third Activity. The Ordering Game

In this section, as shown in Fig. 14, apart from a 3D object printer, there were three collections of objects that visitors had to order from not-being to being. The main idea of this activity was to make visitors think about the possibility that the intermediate objects had a part of each extreme object and that the extremes were clearly classifiable into one of the two categories the objects moved between. The activity introduced the idea of change between two states and justified why Zadeh accepted that the function of an element belonging to a subset could take real values between 0 and 1, inclusive, while also serving to demonstrate the sorites paradox, so familiar to fuzzy logic researchers, to the general public.

All of the objects in the three series were created by the company Grogthobjects. Grogthobjects is a professional company dedicated to creating 3D material, one of whose members is 3D designer Jordi Bayer, who is also an associate professor at the University of Girona and advised the Mathematics unit belonging to the University of Girona's Enterprise Department on all processes with regard to manufacturing the 3D objects in the exhibition.

The image in Fig. 15 shows one of these collections in detail.

The main aim of the printer creating fuzzy 3D objects in this zone was to arouse curiosity regarding how the objects in the exhibition were created.

Fig. 14 Activity zone 3

Fig. 15 From cube to face collection

The panel accompanying the activity was entitled "From not-being to being". It read: "Can you order the following collections of objects progressively from not-being to being? Do you notice the difficulty in deciding when to say that we definitively have the being?"

4.4 Fourth Activity. The Spoon and Fork Classification Game

In this zone, as Fig. 16 shows, visitors had to group a number of objects into the categories of spoons or forks. The activity was designed to reinforce all the concepts seen in the three previous zones.

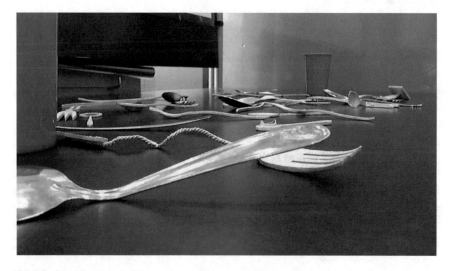

Fig. 16 Classification of objects

Fig. 17 A robot resembling the robot "Wall-e" classifies objects into the sets of spoons and forks. In the Disney film of the same name, the robot momentarily goes into a classification loop when it finds a knife and cannot decide whether to classify it into only one of the sets of spoon or fork

Creating 3D pieces was intended to complicate the classification options.

The panel accompanying this activity was entitled "Spoons, forks or neither one nor the other". Under the image shown in Fig. 17, it read: "Place the following objects inside the cups for spoons or forks. Have you found any of them difficult or impossible to classify?"

4.5 Fifth Activity. *Now You See It, Now You Don't Game*

In this activity, represented in Fig. 18, a collection of objects served to reflect on the fact that being and not-being can coexist at the same time, even if it does not seem so. Thus, the "principle of non-contradiction" was placed in doubt by showing objects that could and could not be considered to belong to a set at the same time.

The section featured a large cylindrical mirror for anamorphic drawings provided by the Girona Film Museum (Tomàs Mallol Collection) and several small mirrors for transforming drawings that initially seem to have no known form or forms, as well as a collection of thaumatropes, some lent by the University of Girona's Business Administration Department and some acquired in the gift shop of the aforementioned Girona Film Museum.

Fig. 18 Detail of table from activity 5

The section also offered the possibility of doing a workshop to create a thaumatrope based on the figure of a hummingbird feeding on a flower's nectar. The illustrations of the thaumatrope were done by the illustrator Queralt Viladevall Valldeperas and can be seen in the image shown in Fig. 19. When spinning, the thaumatrope showed a hummingbird feeding on a flower. Although this cannot be perceived when the thaumatrope is not moving, our brain makes it true when it is.

The panel presenting the activity was entitled "Being and not-being". It read: "Play with these devices and discover how they work! See how being and not-being can coexist at the same time! Create your own!"

Fig. 19 Hummingbird

4.6   Sixth Activity. The Creating Continuity Game

In Activity 6, represented by the image in Fig. 20, visitors could see how our brain endows a series of static moments with continuity. Thus, it made perfect sense to accept that an object can switch continuously from not-being to being, justifying Zadeh's acceptance that the function of an element belonging to a subset could take all real values between 0 and 1.

On the table visitors could find praxinoscopes, zoetropes, folioscopes of drawings and photographs and a *toupie fantoche* with a plinth ceded by the Girona Film Museum (Tomás Mallol Collection), all of which wowed visitors with their optical effects.

In this section, visitors could do a second artistic workshop that consisted in colouring the film of a zoetrope represented in Fig. 21. Visitors painted a green monster opening and closing its mouth on the film, simulating a biting motion. When the film was put on a device and spun, visitors found that our brain produces the feeling of continuity from a group of static images.

The panel accompanying this last activity was entitled "Creating continuity". It read: "Have you realized that our brain generates the illusion of continued motion from a succession of independent and static images? Create your own films!"

Fig. 20 Detail of table from activity 6

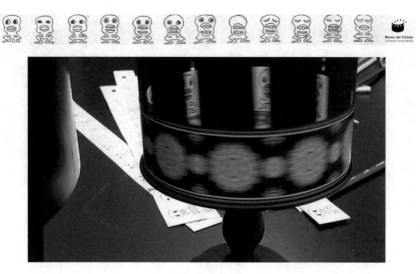

Fig. 21 Film from activity 6

5 Conclusions

CosmoCaixa Barcelona is a science museum located in a Modernist building in Barcelona. The building was originally a home for the blind designed by Josep Domènech Estapà. It was remodeled and expanded in 1979–1980 by Jordi Garcés and Enric Soria to accommodate the "La Caixa" Foundation Science Museum, and a later reform by the architects Robert and Esteve Terrades turned the space into the current CosmoCaixa. The new museum was opened on 25 September, 2004. In 2006, the European Museums Forum gave it the European Museum of the Year Award and it is currently considered one of the best in Europe.

The present article has described six activities from the exhibition "50 years of fuzzy ideas", which was held in the above museum and invited visitors to reflect on the principles of fuzzy research.

The activities demonstrated the ideas that an object can partially present a property and that our thinking accepts the idea of continuity. Accepting these ideas was key if visitors were to then accept the transition from the characteristic function of classical set theory, based on the values identified by "0" or "1", to the belonging function in fuzzy set theory, with values defined in the continuous interval [0,1].

The activities also worked on the absence of the principle of the excluded third and the principle of non-contradiction, helping to construct a first means by which society might understand the philosophy about fuzzy set theory. This was done by fostering citizen participation in issues related to fuzzy logic research and innovation, thereby also promoting scientific education.

We would like to emphasize that the aim of the experimental activities was to surprise visitors with strange objects that might confuse them and break their

Fig. 22 Twilight. Queralt Viladevall Valldeperas

bivalent classification regarding the level of truth of a proposition so that in the other zones of the exhibition they would have an open mind to accept the validity of principles on which fuzzy set theory is based. In Zone C, which contained paintings, different principles were formulated and their possible validity shown via everyday scenes. By way of example, the "twilight" painting in Fig. 22 demonstrated the idea that at twilight it is part day and part night. Thus, Gil Aluja's principle of gradual simultaneity could be perfectly comprehended through an image from everyday life.

Finally, we would like to highlight the fact that the activities formed part of one of the first exhibitions of fuzzy set theory to be held in a science museum anywhere in the world. Given that such museums are characterized as spaces dedicated to stimulating citizens in favour of scientific knowledge, scientific method and scientific opinion, this presentation of fuzzy set theory to society and its explanation in this article has given us the opportunity to pay the best possible tribute to our beloved teacher and friend Professor Jaime Gil Aluja. In addition to his extraordinary scientific work, Professor Gil Aluja has based much of his work and career on raising awareness about fuzzy set theory on the most diverse of forums and in various fields of the social sciences in the most enjoyable and didactic way possible. His excellent scientific work and his relationship with our University over many years led the University of Girona to appoint him doctor *honoris causa* in a ceremony held at the University's main hall on 21 October, 2004. We believe that the aforementioned exhibition and description of the same in the present article is the best homage that our research group can pay to him in recognition of his work teaching and disseminating fuzzy set theory in the different fields of the social sciences.

References

1. Barro, S., Marin, R.: Fuzzy Logic in Medicine. Physica-Verlag (2001)
2. Demicco, R., Klir, G.: Fuzzy Logic in Geology. Academic Press (2003)
3. Kaufmann, A., Gil Aluja, J.: Introducción de la Teoría de los Subconjuntos Borrosos a la Gestión de las Empresas. Editorial Milladoiro, Santiago de Compostela (1986)
4. Kaufmann, A., Gupta, M.M.: Fuzzy Mathematical Models in Engineering and Management Science. Elsevier Science Publisher, The Netherlands (1988)
5. Lazzari, L., Machado, E., Perez, R.: Los conjuntos borrosos: Una introducción. Cuadernos de Cimbage (2), 1–25 (1999)
6. Tanaka, K.: An Introduction to Fuzzy Logic for Practical Applications. Springer, Berlin (1997)
7. Terano, T., Asai, K., Sugeno, M.: Fuzzy Systems Theory and Its Applications. Academic Press, Boston (1987)
8. Torra, V.: Fonaments d'Intel·ligència artificial. Fundació UOC (2007)
9. Trillas, E. (ed.): Fundamentos e introducción a la Ingeniería Fuzzy. Omron Electronics S. A. (1994)
10. Trillas, E.: La inteligencia artificial. Temas de Debate, Madrid (1998)
11. Trillas, E., Alsina, C., Terricabras, J.M.: Introducción a la lógica borrosa. Ariel, Barcelona (1995)
12. Zadeh, L.A.: Fuzzy sets. Inf. Control **8**, 338–353 (1965)
13. Zadeh, L.A.: Nacimiento y evolución de la lógica borrosa, el soft computing y la computación con palabras: Un punto de vista personal. Psicothema **8**(2), 421–429 (1996)

Fuzzy Decision Making System for Model-Oriented Academia/Industry Cooperation: University Preferences

Galyna Kondratenko, Yuriy Kondratenko and Ievgen Sidenko

Abstract This paper discusses the effective models of cooperation of universities and IT companies, as well as the hierarchic approach towards projecting certain decision making support systems (DSS) based on fuzzy logic. Special attention is paid to fuzzy DSS as an advisor in choosing the most appropriate cooperation model for a certain department of universities eager to become partners within the frames of future cooperation with a certain IT company. The article features hierarchic structure, results of rule bases and DSS software based on the approximation of fuzzy systems with discrete output. It also contains the results of imitational DSS modeling based on the elaborated DSS developing the most rational model of cooperation for a university party of the cooperation of the "University—IT company" type.

Keywords Decision support system · Fuzzy logic · Membership function
Linguistic term · Rule base · University-industry cooperation

1 Introduction

Scientific potential of higher educational institutions is a fundamental basis for the introduction of personal scientific achievements and developments (modern theories, inventions, up-to-date scientific projects etc.) into the spheres of industry, IT, and national economy altogether. The ability to create world-level competitive technologies concerning IT-engineering can be substantially geared up by

G. Kondratenko · Y. Kondratenko (✉) · I. Sidenko
Department of Intelligent Information Systems, Petro Mohyla Black Sea National University,
68-th Desantnykiv Street 10, Mykolaiv 54003, Ukraine
e-mail: yuriy.kondratenko@chmnu.edu.ua

G. Kondratenko
e-mail: halyna.kondratenko@chmnu.edu.ua

I. Sidenko
e-mail: ievgen.Sidenko@chmnu.edu.ua

© Springer International Publishing AG 2018
C. Berger-Vachon et al. (eds.), *Complex Systems: Solutions and Challenges in Economics, Management and Engineering*, Studies in Systems, Decision and Control 125, https://doi.org/10.1007/978-3-319-69989-9_7

increasing the level of interaction between the departments of universities and industrial producing companies. For instance, the formation of universities and IT companies dealing with IT development and introduction into academic industrial consortiums (AICs) opens up vast opportunities of Science-to-Business (S2B) and Business-to-Science (B2S) directions involving new technologies of programming and hardware, as well as increasing intellectual potential of both modern IT-companies and universities [1–3].

2   Related Works and Problem Statement

The problem of choosing the most efficient university-industry cooperation (UIC) model with an IT-company is something a certain university faces as a potential partner at the initial phase of cooperation. The analysis of asserted literature allows us to define a wide circle of basic factors influencing the UIC model within the frames of an AIC, for instance, the following could be listed: IT-students level of knowledge, their extent of participation in international exchange programs, their cooperation with existing IT-companies, the students' grades, levels of IT company staff's experience, educational potential of the said IT company, IT-certification of its lecturers, business workshops held at universities, startup experience, the amount of grants distributed to finance the scientific research processes, level of articles published etc. [3–5]. Based on the listed factors both the university and the company must search for the best partnership considering the most rational cooperative model from the corporate interest point of view. This task is hard due to the fact that certain factors can only be represented with vague and fuzzy evaluations, and the search process is connected with fuzzy information based on multidimensional fuzzy dependencies [6–10].

Intellectual DSS developed based on up-to-date methods, technologies and approaches towards system analysis, prognostication, fuzzy logic, neural networks, AI can contribute to the efficiency increase in cooperation. Involvement of these methods into the development of modern DSS allows to process vast amounts of various data on a brand new level of interaction between a human operator (HO) and a computer system [4, 5, 7, 10–12].

In order to choose a rational cooperation model within the frames of an AIC hierarchically-organized DSS based on fuzzy logic discrete output is considered to be appropriate. Previous surveys, research and analysis of the existing successful experience of AIC cooperation prove that modern solution to the choice of cooperative model problem features 4 alternative models E_i, $(i = 1 \ldots m)$ to choose from where $(m = 4)$ is the optimal one. In this case the E_1 corresponds to the A1 model (interaction between a university and an IT-company aimed at education and knowledge enrichment, experiential exchange and staff preparation for an IT company); E_2 corresponds to A2 model (structuration and maintenance of certification of cooperation results); E_3 corresponds to B model (creation of joint scientific research center and the development of joint scientific projects); E_4

corresponds to C model (creation of student scientific research clusters and creation of independent enterprises aimed at business and startup implementation). The efficient selection of the cooperation model depends on the chosen $x_j, (j = 1, 2, \ldots, n)$ factors characterizing each of the partners of the future AIC between a university and an IT company [3, 6, 7].

The implementation of the bespoken models of cooperation requires considering and processing vast amounts of input data based on the analysis of previous experience of cooperation between the parties of interest, advantages and directions of development, scientific and educational levels of the future AIC participants, level of students' and lecturers' involvement etc. A poorly chosen cooperation model, as well as the inconsideration of certain cooperation requirements within the frames of the AIC can lead to unexpected and unaccounted consequence, for instance, a considerable loss of intellectual and\or material resources, decrease of educational and qualification levels of the lecturers and specialists, lesser potential and ability of creative approach [3–9].

The aim of current research is development and approbation of an intellectual DSS based on fuzzy logic allowing the university to choose the most rational cooperation model to establish an AIC, considering all the preferences of the university as potential partner, basic values of its activity and scientific potential of the IT department of the university.

3   Analysis of Existing Methods and Approaches for Choosing the Model of Cooperation Between the University and the IT-Company

The choice of a proper cooperation model between a university and an IT company is quite a complicated process for many reasons, for instance, multiple factors of evaluation, complexity of preliminary consideration of all possible stages of decision making, absence of any possibility to define the values of certain input factors, insufficient knowledge of up-to-date IT technologies' peculiarities or insufficient material base etc. [3, 7].

In order to choose the most efficient (appropriate) cooperation model one might use the following [13–19]: (a) Delphi method; (b) pairwise comparisons method; (c) Saati method of analytic hierarchy process etc.

The Delphi method is based on a thoroughly developed procedure of consequential individual questioning of the experts by making them fill in the survey questionnaire. It is accompanied by constant updating of the experts on the results of the processing before the answers are obtained. The expertise lasts a few rounds until most of the experts acquire the appropriate result according to their collective judgment. The median of final answers of all the experts is considered to be the ultimate evaluation [16, 21]. This method is undergoing constant improvement due to the ability of combining it with other methods. New modifications of the Delphi method provide increased comprehensive methodology, rapid and exact acquiring of joint expert evaluations.

The pairwise comparisons method means direct involvement of the expert evaluating the aims. According to this method the aims are being compared in all possible combinations. Each pair has a best goal chosen from. The processing of evaluation matrix allows determining the values of aims, thus distinguishing their relative importance levels [19, 20].

Saati method of analytic hierarchy process (AHP) is a calculations tool of system approach towards the complicated problems of decision making. It allows the decision maker to find the option of solving the problem, to find such an alternative which is the most appropriate according to his understanding of the essence of the problem and given requirements towards its solution [14–17, 19]. Building a hierarchical structure of the task allows to analyze all the possible necessary and sufficient elements of hierarchy. The models available for selection within the frames of the AIC are A1, A2, B, C [3, 6, 7] as well as their combinations A1&A2, A2&B, B&C etc.

The methods mentioned above have certain disadvantages: necessity of experts' evaluations accordance calculation; limited levels of hierarchy and pairwise comparison matrix size; necessity to maintain constant contact with the experts for questioning, necessity to update the DSS structure along with the change of input vector coordinates etc. [18].

The fuzzy logic inference method allows to perform a multifactorial evaluation of possible cooperation level and a choice of the appropriate model (A1, A2, B, C) for the university cooperation with the IT company without limits to the number of input coordinates and production rules. It is very convenient to use because of easily adjustable parameters and availability of decision making under the circumstances of uncertainty. The development of fuzzy logic selection of DSS algorithms for cooperation under uncertainty circumstances is one of the most prospective direction in modern IT sphere, considering the formation of input expert data with a high level of indetermination and variable structure of input coordinates vector [6, 7, 20–24].

4 The Structure of Fuzzy DSS for Choosing the Expedient UIC Model for University Department

According to the experience gained by the experts in projecting specific fuzzy systems for different purposes, the one-level structure of DSS under conditions of high-dimensional $X = \{x_j\}, j = 1 \ldots n$ input coordinates (factors, values) decreases the sensitivity of fuzzy rules towards the alternation of input coordinates $x_j, (j = 1, 2, \ldots, n)$. First of all this is linked to the complexity of establishing appropriate fuzzy rules for the implementation of all the possible relations between the input and output parameters of the system $y_k = f(x_1, x_2, \ldots, x_n), k = 1 \ldots K$ [7].

Let us observe in detail the procedure of appropriate cooperation model definition the authors propose for the university and its IT department considering its preferences towards the future partner within the frames of the AIC. The authors have developed a separate module of DSS to choose the necessary UIC model between the department and the IT company including 17 input $X = \{x_j\}, j = 1 \ldots 17$ coordinates and one output y, that are interconnected by fuzzy dependencies

$$y_k = f(x_1, x_2, \ldots, x_{17}), k = \overline{1, 7}$$

of the corresponding rule bases of 7 fuzzy subsystems FSS1, FSS2,..., FSS7 (Fig. 1). The input variables $X = (x_1, x_2, \ldots, x_{17})$ are x_1—level of scientific value of masters' and bachelors' thesis papers (MTP and BTP respectfully); x_2—practical implementation of BTP and MTP; x_3—accordance of BTP and MTP to the direction of research; x_4—level of IT-experience among the students of the department; x_5—participation of the students in the international exchange programs; x_6—level of interaction between students and IT companies; x_7—students' grades; x_8—level of the departments research work; x_9—number of patents; x_{10}—number of grants; x_{11}—level of published scientific research at university; x_{12}—amount of scientific publications; x_{13}—category and rating of the university; x_{14}—level of IT-certification among the lecturers; x_{15}—level of held business-courses and workshops; x_{16}—experience in organization of student companies; x_{17}—experience in managing a joint collective of people to implement IT-related projects [6, 7].

The principle of DSS approximation for choosing the appropriate UIC model lies within using the fuzzy logical equations based on the knowledge matrix (Table 1) or a system of fuzzy logic statements (1) [12, 20].

Fuzzy logical statements (1) obtained from the knowledge matrix (Table 1) [20].

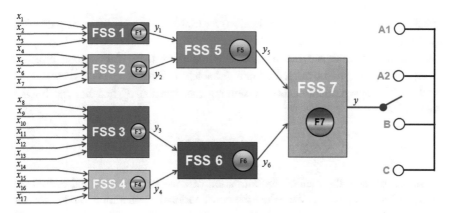

Fig. 1 Fuzzy DSS structure for choosing the appropriate UIC model for the university department within the framework of cooperation with the IT company

Table 1 Knowledge matrix $y=f(x_1, x_2, \ldots, x_n)$

Number of combination	x_1	x_2	...	x_i	...	x_n	y
11	a_1^{11}	a_2^{11}	...	a_i^{11}	...	a_n^{11}	d_1
12	a_1^{12}	a_2^{12}	...	a_i^{12}	...	a_n^{12}	
...	...	...	...	...	...	...	
$1k_1$	$a_1^{1k_1}$	$a_2^{1k_1}$	...	$a_i^{1k_1}$	...	$a_n^{1k_1}$	
...				...		...	
$j1$	a_1^{j1}	a_2^{j1}		a_i^{j1}		a_n^{j1}	d_j
$j2$	a_1^{j2}	a_2^{j2}		a_i^{j2}		a_n^{j2}	
...				...		...	
jk_j	$d_1^{jk_j}$	$d_2^{jk_j}$		$d_i^{jk_j}$		$d_n^{jk_j}$	
...				...		...	
$m1$	a_1^{m1}	a_2^{m1}		a_i^{m1}		a_n^{m1}	d_m
$m2$	a_1^{m2}	a_2^{m2}		a_i^{m2}		a_n^{m2}	
...				...		...	
mk_m	$a_1^{mk_m}$	$a_2^{mk_m}$		$a_i^{mk_m}$		$a_n^{mk_m}$	

$$
\begin{aligned}
&\text{IF } \left(x_1 = a_1^{11} \text{ AND } x_2 = a_2^{11} \text{ AND } \ldots \text{ AND } x_n = a_n^{11}\right) \text{ OR} \\
&\left(x_1 = a_1^{12} \text{ AND } x_2 = a_2^{12} \text{ AND } \ldots \text{ AND } x_n = a_n^{12}\right) \text{ OR } \ldots \\
&\text{OR } \left(x_1 = a_1^{1k_1} \text{ AND } x_2 = a_2^{1k_1} \text{ AND } \ldots \text{ AND } x_n = a_n^{1k_1}\right) \text{ THEN } y = d_1 \text{ ELSE} \\
&\text{IF } \left(x_1 = a_1^{21} \text{ AND } x_2 = a_2^{21} \text{ AND } \ldots \text{ AND } x_n = a_n^{21}\right) \text{ OR} \\
&\left(x_1 = a_1^{22} \text{ AND } x_2 = a_2^{22} \text{ AND } \ldots \text{ AND } x_n = a_n^{22}\right) \text{ OR } \ldots \\
&\text{OR } \left(x_1 = a_1^{2k_2} \text{ AND } x_2 = a_2^{2k_2} \text{ AND } \ldots \text{ AND } x_n = a_n^{2k_2}\right) \text{ THEN } y = d_2 \text{ ELSE} \\
&\ldots \qquad \ldots \qquad \ldots \qquad \ldots \qquad \ldots \\
&\text{IF } \left(x_1 = a_1^{m1} \text{ AND } x_2 = a_2^{m1} \text{ AND } \ldots \text{ AND } x_n = a_n^{m1}\right) \text{ OR} \\
&\left(x_1 = a_1^{m2} \text{ AND } x_2 = a_2^{m2} \text{ AND } \ldots \text{ AND } x_n = a_n^{m2}\right) \text{ OR } \ldots \\
&\text{OR } \left(x_1 = a_1^{mk_m} \text{ AND } x_2 = a_2^{mk_m} \text{ AND } \ldots \text{ AND } x_n = a_n^{mk_m}\right) \text{ THEN } y = d_m
\end{aligned}
\tag{1}
$$

where a_i^{jk} is a linguistic term (evaluation) i variable (x_i) to evaluate the j decision (d_j) according to the k rule [20, 25].

It allows to calculate the membership functions (MF) for various types of decisions d_j, $j = \overline{1, m}$ provided the input variables are fixed at x_i, $i = \overline{1, n}$ for a fuzzy system. The approximation task is defining the decision d^* that has the biggest value of MF:

$$
\mu^{d^*}\left(x_1^*, x_2^*, \ldots, x_n^*\right) = \max_{j = \overline{1, m}}\left(\mu^{d_j}\left(x_1^*, x_2^*, \ldots, x_n^*\right)\right).
$$

Let us observe the method for formation of knowledge matrix and processing fuzzy information while choosing the most rational cooperation model for the department of the university with IT-company based on the first subsystem FSS1 $y_1 = f_1(x_1, x_2, x_3)$ of the fuzzy DSS (Fig. 1).

Let us define the input and output coordinates of the system along with their characteristic parameters [7].

Input linguistic variables:

- X_1—level of scientific value of masters' and bachelors' thesis papers (MTP and BTP respectfully): range of vary—[0 100], number of linguistic terms (LT)—3 ("low", "medium", "high"), shape of MF—triangular [25–27];
- X_2—BTP and MTP's practical implementation value:): range of vary—[0 100], number of LTs—3 ("low", "medium", "high"), shape of MF—triangular;
- X_3—correspondence of BTPs and MTPs with the research direction of the department: range of vary—[0 100], number of LTs—3 ("low", "medium", "high"), shape of MF—triangular.

Output linguistic variable:

- y—BTP and MTP level evaluation: range of vary—[0 100], number of LTs—5 ("low"—L, "lower than medium"—LM, "medium"—M, "higher than medium"—HM, "high"—H), shape of MF—triangular.

Let us form the models for all the triangular LTs for the evaluation of input and output variables of the subsystem $y_1 = f_1(x_1, x_2, x_3)$, as well as the rule base and correspondent knowledge matrix. We choose the following parameters for the triangular LT models $.3em\_A = (a_1, a_0, a_2)$ for all the variables x_1, x_2, x_3, y_1:

- X_1—$low = (0, 0, 50)$; $medium = (0, 50, 100)$; $high = (50, 100, 100)$;
- X_2—$low = (0, 0, 50)$; $medium = (0, 50, 100)$; $high = (50, 100, 100)$;
- X_3—$low = (0, 0, 50)$; $medium = (0, 50, 100)$; $high = (50, 100, 100)$;
- X_4—$\begin{cases} L = (0, 0, 25); \ LM = (0, 25, 50); \ M = (25, 50, 75); \\ HM = (50, 75, 100); \ H = (75, 100, 100) \end{cases}$.

LT graphic representation for the variables x_1, x_2, x_3 and y_1 is at Fig. 2.

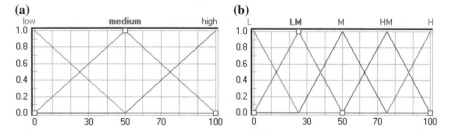

Fig. 2 LT with triangular shape of MF for variables x_1, x_2, x_3 (**a**) and y_1 (**b**)

Table 2 Rule base for subsystem $y_1 = f_1(x_1, x_2, x_3)$

Number of rule	x_1	x_2	x_3	y_1
1	Low	Low	Low	L
2	Low	Low	Medium	L
3	Low	Low	High	LM
4	Low	Medium	Low	L
5	Low	Medium	Medium	LM
6	Low	Medium	High	M
7	Low	High	Low	LM
8	Low	High	Medium	LM
9	Low	High	High	M
10	Medium	Low	Low	L
11	Medium	Low	Medium	LM
12	Medium	Low	High	M
13	Medium	Medium	Low	LM
14	Medium	Medium	Medium	M
15	Medium	Medium	High	HM
16	Medium	High	Low	M
17	Medium	High	Medium	HM
18	Medium	High	High	HM
19	High	Low	Low	LM
20	High	Low	Medium	M
21	High	Low	High	M
22	High	Medium	Low	M
23	High	Medium	Medium	HM
24	High	Medium	High	HM
25	High	High	Low	HM
26	High	High	Medium	H
27	High	High	High	H

Let us form a rule base (Table 2) and a knowledge matrix (Table 3) based on formerly gained expert data and knowledge, using the previously developed LT models with triangular shape of MF for x_1, x_2, x_3, y_1 variables [7, 20].

Knowledge matrix (Table 3) [20, 28, 29] is formed according to the rule base (Table 2) by combining the rules according to the output value $y_1 \in \{L, LM, M, HM, H\}$.

When the user enters data $X^* = \left(x_1^*, x_2^*, x_3^*\right)$, the best possible decision for the subsystem $y_1 = f_1(x_1, x_2, x_3)$ is defined. On the next hierarchy level (Fig. 1) it is represented by the input coordinate $y_1 \in \{L, LM, M, HM, H\}$ of the corresponding subsystem FSS5 $y_5 = f_5(y_1, y_2)$.

Therefore, fuzzy logical statements (1) can be represented as the fuzzy logical equations (2) by triangular MF (Fig. 2) and a knowledge matrix (Table 3):

Table 3 Knowledge matrix for subsystem $y_1 = f_1(x_1, x_2, x_3)$

Number of rule and combination		x_1	x_2	x_3	y_1
1	11	Low	Low	Low	**L**
2	12	Low	Low	Medium	
4	13	Low	Medium	Low	
10	14	Medium	Low	Low	
3	21	Low	Low	High	**LM**
5	22	Low	Medium	Medium	
7	23	Low	High	Low	
8	24	Low	High	Medium	
11	25	Medium	Low	Medium	
13	26	Medium	Medium	Low	
19	27	High	Low	Low	
6	31	Low	Medium	High	**M**
9	32	Low	High	High	
12	33	Medium	Low	High	
14	34	Medium	Medium	Medium	
16	35	Medium	High	Low	
20	36	High	Low	Medium	
21	37	High	Low	High	
22	38	High	Medium	Low	
15	41	Medium	Medium	High	**HM**
17	42	Medium	High	Medium	
18	43	Medium	High	High	
23	44	High	Medium	Medium	
24	45	High	Medium	High	
25	46	High	High	Low	
26	51	High	High	Medium	**H**
27	52	High	High	High	

$$\mu^L(x_1, x_2, x_3) = \left(\mu^{low}(x_1) \wedge \mu^{low}(x_2) \wedge \mu^{low}(x_3)\right) \vee$$
$$\left(\mu^{low}(x_1) \wedge \mu^{low}(x_2) \wedge \mu^{medium}(x_3)\right) \vee \left(\mu^{low}(x_1) \wedge \mu^{medium}(x_2) \wedge \mu^{low}(x_3)\right) \vee$$
$$\left(\mu^{medium}(x_1) \wedge \mu^{low}(x_2) \wedge \mu^{low}(x_3)\right),$$
$$\mu^{LM}(x_1, x_2, x_3) = \left(\mu^{low}(x_1) \wedge \mu^{low}(x_2) \wedge \mu^{high}(x_3)\right) \vee$$
$$\left(\mu^{low}(x_1) \wedge \mu^{medium}(x_2) \wedge \mu^{medium}(x_3)\right) \vee \left(\mu^{low}(x_1) \wedge \mu^{high}(x_2) \wedge \mu^{low}(x_3)\right) \vee$$
$$\left(\mu^{low}(x_1) \wedge \mu^{high}(x_2) \wedge \mu^{medium}(x_3)\right) \vee \ldots \vee \left(\mu^{high}(x_1) \wedge \mu^{low}(x_2) \wedge \mu^{low}(x_3)\right),$$
$$\ldots \qquad \ldots \qquad \ldots \qquad \ldots \qquad \ldots$$
$$\mu^H(x_1, x_2, x_3) = \left(\mu^{high}(x_1) \wedge \mu^{high}(x_2) \wedge \mu^{medium}(x_3)\right) \vee$$
$$\left(\mu^{high}(x_1) \wedge \mu^{high}(x_2) \wedge \mu^{high}(x_3)\right).$$

$$(2)$$

Let us assume the user entry data $x_1^* = 60$; $x_2^* = 30$; $x_3^* = 90$, concerning the first subsystem $y_1 = f_1(x_1, x_2, x_3)$ for BTP and MTP level evaluations. Based on the direct (vertical) LT model with triangular shape of MF (3) we fuzzificate the input coordinates FSS1, defining the membership degrees of vector components for X^* towards the corresponding LTs (Fig. 2) [25–29]:

$$\mu_{.3em \sim A}(x) = \begin{cases} 0, & \text{for } (x < a_1) \cup (x > a_2) \\ \dfrac{x - a_1}{a_0 - a_1}, & \text{for } a_1 \leq x \leq a_0, \text{ if } a_1 \neq a_0 \\ \dfrac{a_2 - x}{a_2 - a_0}, & \text{for } a_0 < x \leq a_2, \text{ if } a_0 \neq a_2 \\ 1, & \text{other cases} \end{cases} \tag{3}$$

The result of the fuzzification is the following:

$$x_1^*: \mu_{low}(60) = 0; \ \mu_{medium}(60) = 0,8; \ \mu_{high}(60) = 0,2;$$
$$x_2^*: \mu_{low}(30) = 0,4; \ \mu_{medium}(30) = 0,6; \ \mu_{high}(30) = 0;$$
$$x_3^*: \mu_{low}(90) = 0; \ \mu_{medium}(90) = 0,2; \ \mu_{high}(90) = 0,8.$$

Then the calculated membership degrees substitute to the system of Eq. (2) and we obtain the LT vector of membership degrees for the LT output signal $\mu^{d_j}(60, 30, 90) \in \{0; 0, 2; 0, 4; 0, 6; 0\}$, $d_j \in \{L, LM, M, HM, H\}, j = \overline{1,5}$ (4), where

$$\mu^L(60, 30, 90) = (0 \wedge 0, 4 \wedge 0) \vee \ldots \vee (0, 8 \wedge 0, 4 \wedge 0) = 0;$$
$$\mu^{LM}(60, 30, 90) = (0 \wedge 0, 4 \wedge 0, 8) \vee \ldots \vee (0, 2 \wedge 0, 4 \wedge 0) = 0, 2; \tag{4}$$
$$\mu^M(60, 30, 90) = (0 \wedge 0, 6 \wedge 0, 8) \vee \ldots \vee (0, 2 \wedge 0, 6 \wedge 0) = 0, 4;$$

$$\mu^{HM}(60, 30, 90) = (0, 8 \wedge 0, 6 \wedge 0, 8) \vee (0, 8 \wedge 0 \wedge 0, 2) \vee \ldots$$
$$(0, 8 \wedge 0 \wedge 0, 8) \vee (0, 2 \wedge 0, 6 \wedge 0, 2) \vee (0, 2 \wedge 0, 6 \wedge 0, 8) \vee \ldots$$
$$(0, 2 \wedge 0 \wedge 0) = (0, 6 \vee 0 \vee 0 \vee 0, 2 \vee 0, 2 \vee 0) = 0, 6;$$
$$\mu^H(60, 30, 90) = (0, 2 \wedge 0 \wedge 0, 2) \vee (0, 2 \wedge 0 \wedge 0, 8) = 0.$$

Below is the matrix implementation of the approximation procedure for $y_1 = f_1(x_1, x_2, x_3)$ with discrete output (Table 3), as a result of transformations performed upon a system of fuzzy logic equations (2), (4) using the t-norm ($\wedge$) MIN and s-norm ($\vee$) MAX operators. A combined set of rules with appropriate membership degrees in the universal and numerical forms are in the matrix implementation [20, 25].

№	x_1	x_2	x_3			
1	$\mu^{low}(x_1)$	$\mu^{low}(x_2)$	$\mu^{low}(x_3)$	}min		
... ...					}max	
10	$\mu^{medium}(x_1)$	$\mu^{low}(x_2)$	$\mu^{low}(x_3)$	}min	(0)	
3	$\mu^{low}(x_1)$	$\mu^{low}(x_2)$	$\mu^{high}(x_3)$	}min		
... ...					}max	
19	$\mu^{high}(x_1)$	$\mu^{low}(x_2)$	$\mu^{low}(x_3)$	}min	(0,2)	}max
6	$\mu^{low}(x_1)$	$\mu^{medium}(x_2)$	$\mu^{high}(x_3)$	}min		
... ...					}max	
22	$\mu^{high}(x_1)$	$\mu^{medium}(x_2)$	$\mu^{low}(x_3)$	}min	(0,4)	
15	**0,8**	**0,6**	**0,8**	}min (0,6)		
17	**0,8**	**0**	**0,2**	}min (0)	}max	
18	**0,8**	**0**	**0,8**	}min (0)		(0,6)
23	**0,2**	**0,6**	**0,2**	}min (0,2)	(0,6)	
24	**0,2**	**0,6**	**0,8**	}min (0,2)		
25	**0,2**	**0**	**0**	}min (0)		
26	$\mu^{high}(x_1)$	$\mu^{high}(x_2)$	$\mu^{medium}(x_3)$	}min		
27	$\mu^{high}(x_1)$	$\mu^{high}(x_2)$	$\mu^{high}(x_3)$	}min	}max (0)	

We define the best decision d^* if $X^* = \left(x_1^*, x_2^*, x_3^*\right)$, where, for instance, $x_1^* = 60$; $x_2^* = 30$; $x_3^* = 90$, for the first subsystem FSS1 $y_1 = f_1(x_1, x_2, x_3)$. The best is the decision, whereupon

$$\mu^{d^*}(X^*) = \max_{j=\overline{1,5}} \left(\mu^{d_j}(60, 30, 90) \in \{0; 0, 2; 0, 4; 0, 6; 0\}\right) = 0.6$$

Since $\mu^{d^*}(X^*) = 0.6$ is in accordance with decision d_4^*, the BTP and MTP evaluation level y_1 corresponds to the LT $\{HM\}$—"higher than medium".

According to the method mentioned above we define the best decision (UIC model for the university department) for the resulting subsystem $y = f_7(y_5, y_6)$, where y_5—general educational level of the students, y_6—level of research\business orientation of university lecturers, y—UIC model for the university department. Knowledge matrix (partial set of rules) for the corresponding subsystem is at Table 4.

Table 4 Partial set of rules of the knowledge matrix for subsystem $y = f_7(y_5, y_6)$

№ of rule and combination	1	2	6	16	3	4	5	... ...	17	21	22
	11	12	13	14	21	22	23		28	29	210
y_5	L	L	LM	HM	L	L	L		HM	H	H
y_6	L	LM	L	L	M	HM	H		LM	L	LM
y	**A1**	**A1**	**A1**	**A1**	**A2**	**A2**	**A2**		**A2**	**A2**	**A2**
№ of rule and combination	9	10	13	14	15	18	19	20	23	24	25
	31	32	33	34	35	36	41	42	43	44	45
y_5	LM	LM	M	M	M	HM	HM	HM	H	H	H
y_6	HM	H	M	HM	H	M	HM	H	M	HM	H
y	**B**	**B**	**B**	**B**	**B**	**B**	**C**	**C**	**C**	**C**	**C**

Thus, based on the authors proposed fuzzy DSS (Fig. 1) exists a choosing of appropriate UIC model within the cooperation with IT company using discrete logical inference. The advantage of this approach is that the DSS result is a specific decision $y \in \{A1, A2, B, C\}$ (Table 4). In addition, when the values of membership degrees $\mu^{d^*}(X^*)$ are similar, a decision making human can choose several variants of decisions (combined models), for example $A1\&A2$, $A2\&B$, $B\&C$ etc. It allows more accurately choose the appropriate UIC model and reduce the time of result calculation by eliminating the need for defazzification as in the case in systems with continuous logical inference [30–33].

Therefore, the UIC model for the IT department of Petro Mohyla Black Sea National University within the framework of cooperation with IT company is the combined model $A1\&A2$ by models $A1$ and $A2$ (Table 4), because $\mu^{d^*}(X^*) = \max_{j=1,4} \left(\mu^{d_j}(60, 30, 90, \ldots, 10) \in \{0, 5; 0, 5; 0, 2; 0\} \right) = 0.5$ that corresponds to the LTs (decisions) $\{A1, A2\}$. Intermediate results of discrete logical inference are shown below:

$$y_1 = f_1(x_1, x_2, x_3) \Rightarrow y_1 \in \{HM\}, \mu^{HM} = 0.6, X^* = (60, 30, 90)$$
$$y_2 = f_2(x_4, x_5, x_6, x_7) \Rightarrow y_2 \in \{HM\}, \mu^{HM} = 0.7, X^* = (70, 60, 80, 75)$$
$$y_3 = f_3(x_8, x_9, \ldots, x_{13}) \Rightarrow y_3 \in \{LM\}, \mu^{LM} = 0.6, X^* = (45, 30, 25, 40, 15, 4)$$
$$y_4 = f_4(x_{14}, \ldots, x_{17}) \Rightarrow y_4 \in \{L, LM\}, \mu^L = 0.5, \mu^{LM} = 0.5, X^* = (10, 20, 15, 10)$$
$$y_5 = f_5(y_1, y_2) \Rightarrow y_5 \in \{HM\}, \mu^{HM} = 0.6, Y^* = (HM, HM)$$
$$y_6 = f_6(y_3, y_4) \Rightarrow y_6 \in \{L, LM\}, \mu^L = 0.5, \mu^{LM} = 0.5, Y^* = (LM, \{L, LM\})$$
$$y = f_7(y_5, y_6) \Rightarrow y \in \{A1, A2\}, \mu^{A1} = 0.5, \mu^{A2} = 0.5, Y^* = (HM, \{L, LM\})$$

So, having on the inputs of the resulting subsystem $y = f_7(y_5, y_6)$ corresponding decisions $Y^* = (HM, \{L, LM\})$, a nonzero value of membership degrees will have two rules $Rule16: \{HM, L\}$ and $Rule17: \{HM, LM\}$. The result is two models A1 and A2.

In contrast to the continuous logical inference developed DSS with discrete logical inference (output signal) allows choosing several models (as at Petro Mohyla Black Sea National University).

5 Conclusions

The necessity of development of intellectual DSS in the AIC sphere can also be explained by the increase in complexity of processing scattered, incomprehensive or controversial data. At the project and implementation stage for the DSS there is a number of certain methodological and technical problems the developers stumble upon directly. For instance, in Ukraine we can name the following problems: absence of conceptual integrity and correspondence between certain traits and methods of engineering knowledge; lack of certified experts in the given area; low adaptive ability of the existing software; absence of the technical and economical values for the efficiency of such systems; the empiric nature of the tool selection procedure and testing, as well as the absence of unified criteria [2, 4–7, 11].

It is of paramount importance for now to develop the scheme of selection of partnership models based on the developments in the field of multifactorial evaluation of future cooperation levels between universities and IT-companies. Implementation of the fuzzy logic-based DSS gives an opportunity to choose the best model for the "University—IT company" consortium development from the point of view of the existing successfully operating consortiums and successful results of cooperation between universities and IT companies [7].

The performed analysis of the existing methods and approaches towards the choice of the appropriate UIC model for the IT department of the university with an IT company shows that upon increasing the input DSS parameters and the necessity of input coordinates vector there emerges a necessity to apply the intellectual methods and approaches based on fuzzy logic principles [20, 21].

The authors have represented the theoretically-methodological approach towards the hierarchic organization of DSS for choosing UIC model between universities and IT companies within the frames of the AIC with the implementation of the processing procedures of unclear expert data by using the triangular shape LTs [7, 20, 25].

Structure and rule base for a multilevel fuzzy logic-based DSS have been developed. They give an opportunity to present the decision making processes in the hierarchically organized DSS and graphically visualize them to a larger extent [16].

Besides, the analysis performed by the authors of this article on the materials of an existing successful innovational cooperation of academic educational institutions and IT companies [7–10] confirms that the creation of various "University-IT company" type entities and consortiums on purpose of finding solutions to the current and potential problems based on mutual work in the IT and Internet-communications' sphere is an extremely prospective direction for those who want to increase the efficiency of higher education field as well. In particular, the National Aerocosmic University "Kharkiv Aviation Institute" named after M. E. Zhukovskiy, Odessa National Polytechnic University, Yuriy Fedkovych Chernivtsi National University, Chernihiv State University, Petro Mohyla Black Sea National University, Institute of Cybernetics of National Academy of Sciences of Ukraine and others are members of such international academic-industrial consortia, which includes universities and IT-companies from Great Britain, Spain, Italy, Portugal, Ukraine and Sweden [3–6]. This consortium is created to develop and implement models of cooperation between universities and industry (IT companies) such as A1, A2, B and C within the project TEMPUS-CABRIOLET 544497-TEMPUS-1-2013-1-UK-TEMPUS-JPHES "Model-oriented approach and Intelligent Knowledge–Based System for Evolvable Academia-Industry Cooperation in Electronics and Computer Engineering" (2013–2017).

References

1. Drozd, J., Drozd, A.: Models, methods and means as resources for solving challenges in co-design and testing of computer systems and their components. In: The Ninth International Conference on Digital Technologies, Zhilina, Slovak Republic, 29–31 May, pp. 225–230 (2013). https://doi.org/10.1109/DT.2013.6566307
2. Kazymyr, V.V., Sklyar, V.V., Lytvyn, S.V., Lytvynov, V.V.: Communications management for academia-industry cooperation in IT-engineering: training. In: Kharchenko, V.S. (ed.) Chernigiv-Kharkiv: MESU, ChNTU, NASU "KhAI" (2015) (in Ukrainian)
3. Kharchenko, V.S., Sklyar, V.V.: Cooperation between universities and IT-industry: some problems and solutions. J. Kartblansh **3–4**, 43–50 (2014) (in Russian)
4. Kondratenko, Y., Simon, D., Atamanyuk, I.: University curricula modification based on advancements in information and communication technologies. In: Ermolayev, V. et al. (eds.) Proceedings of the 12th International Conference on Information and Communication Technologies in Education, Research, and Industrial Application. Integration, Harmonization and Knowledge Transfer, vol. 1614, ICTERI'2016, CEUR-WS, Kyiv, Ukraine, 21–24 June, pp. 184–199 (2016)
5. Kondratenko, Y.P.: The role of inter-university consortia for improving higher education system. In: Smithee, M. (ed.) Proceedings of Phi Beta Delta, vol. 2, issue 1, pp. 26–27. Honor Society for International Scholars, USA (2011)
6. Kondratenko, Y., Kharchenko, V.: Analysis of features of innovative collaboration of academic institutions and IT-companies in areas S2B and B2S. J. Tech. News **1**(39), 15–19 (2014) (in Ukrainian)
7. Kondratenko, Y.P., Kondratenko, G.V., Sidenko, Ie.V., Kharchenko, V.S.: Cooperation models between universities and IT companies, decision-making systems based on fuzzy logic. monograph. In: Kondratenko, Y.P., (ed.) Kharkiv: MESU, PMBSNU, NAU "KAI" (2015) (in Ukrainian)
8. Lytvynov, V.V., Kharchenko, V.S., Lytvyn, S.V., Saveliev, M.V., Trunova, E.V., Skiter, I.S.: Tool-Based Support of University-Industry Cooperation in IT-Engineering. Chernigiv, ChNTU (2015). (in Ukrainian)
9. Starov, O., Kharchenko, V., Sklyar, V., Khokhlienkov, N.: Startup company and spin-off advanced partnership via web-based networking. In: Proceedings of the University-Industry Interaction Conference, Amsterdam, May, pp. 115–124 (2013)

10. Starov, O., Sklyar, V., Kharchenko, V., Boyarchuk, A., Phillips, C.: A student-in-the-middle approach for successful university and business cooperation in IT. In: Proceedings of the University-Industry Interaction Conference, Barcelona, Spain, April, pp. 193–207 (2014)
11. Trunov, A.N.: An adequacy criterion in evaluating the effectiveness of a model design process. Eastern-Eur. J. Enterp. Technol. 1 **4**(73), 36–41 (2015)
12. Trunov, A.: Recurrent approximation as the tool for expansion of functions and models of operation of neural networks. Eastern-Eur. J. Enterp. Technol. 5 **4**(83), 41–48 (2016)
13. Blokhin, L.N., Osadchiy, S.I., Bezkorovainyi, Y.N.: Technology of structural identification and subsequent synthesis of optimal stabilization systems for unstable dynamic objects. J. Autom. Inf. Sci. **39**(11), 57–66 (2007)
14. Chang, D.Y.: Applications of the extent analysis method on fuzzy AHP. J. Eur. J. Oper. Res. **95**, 649–655 (1996)
15. Cheng, R.W., Chang, C.-W., Lin, H.-L.: A fuzzy ANP-based approach to evaluate medical organizational performance. J. Int. Manag. Sci. **19**, 53–74 (2008)
16. Kondratenko, Y.P., Sidenko, Ie.V.: Decision-making based on fuzzy estimation of quality level for cargo delivery. In: Zadeh, L.A., et al. (eds.) Recent Developments and New Directions in Soft Computing. Studies in Fuzziness and Soft Computing, vol. 317, pp. 331–344. Springer International Publishing, Switzerland (2014). https://doi.org/10.1007/978-3-319-06323-2_21
17. Laarhoven, V., Pedrych, W.: Fuzzy extension for Saaty's priority theory. J. Fuzzy Sets Syst. **11**, 229–241 (1983)
18. Messarovich, M.D., Macko, D., Takahara, Y.: Theory of Hierarchical Multilevel Systems. Academic Press, New York (1970)
19. Narasimha, B., Chen, N.: Effect of imprecision in specification of pair-wise comparisons on ranking of alternatives using fuzzy AHP. J. AMCIS 238–243 (2001)
20. Rotshtein, A.P.: Intellectual Technologies of Identification: Fuzzy Logic, Genetic Algorithms, Neuron Networks. UNIVERSUM, Vinnitsa (in Russian)
21. Gil-Aluja, J.: Investment in Uncertainty. Kluwer Academic Publishers, Dordrecht, Boston, London (1999)
22. Gil-Lafuente, A.M., Merigo J.M.: Decision making techniques in political management. In: Lodwick, W.A., Kacprzhyk, J. (eds.) Fuzzy Optimization. Studies in Fuzziness and Soft Computing, vol. 254, pp. 389–405. Springer, Berlin, Heidelberg (2010)
23. Osadchiy, S.I., Kalich, V.M., Didyk, O.K.: Structural identification of unmanned supercavitation vehicle based on incomplete experimental data. In: IEEE 2nd International Conference on Actual Problems of Unmanned Air Vehicles Developments, Kiev, Ukraine, 15–17 October, pp. 93–95 (2013). https://doi.org/10.1109/APUAVD.2013.6705294
24. Palagin, A.V., Opanasenko, V.N.: Reconfigurable computing technology. J. Cybern. Syst. Anal. (Springer, New York) **43**, 675–686 (2007)
25. Piegat, A.: Fuzzy Modeling and Control. Springer, Heidelberg (2001)
26. Zadeh, L.A.: Fuzzy sets. J. Inf. Control **8**(3), 338–353 (1965)
27. Zimmerman, H.J.: Fuzzy Set Theory. Kluwer, Boston (1991)
28. Shebanin V., Atamanyuk I., Kondratenko Y., Volosyuk Y.: Application of fuzzy predicates and quantifiers by matrix presentation in informational resources modeling. perspective technologies and methods in MEMS design. In: Proceedings of the International Conference MEMSTECH-2016. Lviv-Poljana, Ukraine, 20–24 April, pp. 146–149 (2016). https://doi.org/10.1109/MEMSTECH.2016.7507536
29. Kondratenko, Y.P.: Robotics, automation and information systems: future perspectives and correlation with culture, sport and life science. In: Gil-Lafuente, A.M., Zopounidis, C. (eds.) Decision Making and Knowledge Decision Support Systems. Lecture Notes in Economics and Mathematical Systems, vol. 675, pp. 43–56. Springer International Publishing, Switzerland (2015). https://doi.org/10.1007/978-3-319-03907-7_6
30. Drozd, J., Drozd, A., Maevsky, D., Shapa, L.: The levels of target resources development in computer systems. In: Proceedings of the IEEE East-West Design & Test Symposium, Kiev, Ukraine, pp. 185–189 (2014)

31. Kondratenko, Y.P., Klymenko, L.P., Al Zu'bi, E.Y.M.: Structural optimization of fuzzy systems' rules base and aggregation models. J. Kybernetes **42**(5), 831–843 (2013). doi:10.1108/K-03-2013-0053
32. Lodwick, W.A., Kacprzhyk, J. (eds.): Fuzzy optimization. In: Journal of Studies in Fuzziness and Soft Computing, vol. 254. Springer, Berlin, Heidelberg (2010)
33. Setnes, M.: Simplification of fuzzy rule bases. In: Proceedings of the International Conference EUFIT, Aachen, Germany, pp. 1115–1119 (1996)

RETRACTED CHAPTER: Towards the Convergence in Fuzzy Cognitive Maps Based Decision-Making Models

Leonardo Concepción, Gonzalo Nápoles, Isel Grau, Koen Vanhoof and Rafael Bello

Abstract Roughly speaking, decision-making can be defined as the process to select a decision (or group of decisions) among a set of possible alternatives in a given decision activity. Most real-life problems are unstructured in nature, often involving vagueness and uncertainty. This makes difficult to apply exact models, being necessary to use approximate methods based on Soft Computing techniques. In recent years, Fuzzy Cognitive Maps have been used in designing Decision Support Systems due to their capability for explaining the underlying reasoning process. This includes the development of learning methodologies for adjusting the inherent parametric requirements. Less attention has been given to the map convergence and its implications in the decision process. In this paper, we study the convergence issues of Fuzzy Cognitive Map based models used in decision-making. More explicitly, we present a learning procedure that allows improving the network convergence by preserving the ordinal relation between the alternatives. In this learning algorithm, the direction and intensity of causal relations cannot be altered since they comprise the system semantic. Numerical simulations show the practical usability of theoretical contributions proposed in this paper, when solving decision-making problems.

Keywords Decision-making · Fuzzy cognitive maps · Convergence

The original version of the book was revised: For detailed information please see Erratum. The erratum to the book is available at https://doi.org/10.1007/978-3-319-69989-9_32

L. Concepción (✉) · G. Nápoles · I. Grau · R. Bello
Universidad Central "Marta Abreu" de Las Villas, Santa Clara, Cuba
e-mail: lcperez@uclv.cu

G. Nápoles · K. Vanhoof
Hasselt University, Diepenbeek, Belgium

© Springer International Publishing AG 2018
C. Berger-Vachon et al. (eds.), *Complex Systems: Solutions and Challenges in Economics, Management and Engineering*, Studies in Systems, Decision and Control 125, https://doi.org/10.1007/978-3-319-69989-9_8

1 Introduction

Decision-making problems have become an active research area due to their impacts in solving real-world problems. Roughly speaking, decision-making process could be defined as the task of determining and selecting the most adequate action that allows solving a given problem. This scheme is supported by the knowledge concerning the problem domain allowing justifying the selected decision.

- More details about decision-making
- Brief presentation of cognitive mapping
- Brief categorization of learning algorithms
- Convergence issues and problem formulation
- Presentation of the proposal and paper's goals

The rest of the paper is organized as follows: in following Sect. 2 the theoretical background of FCM is described. Here we point out some aspects concerning the map inference process using continuous threshold functions. In Sect. 3 we introduce the proposed hybrid model consisting in three main steps: (i) the computation of positive, negative and boundary regions, (ii) the construction of the map topology, and (iii) the map exploitation using the similarity class of the target instance. Section 4 provides numerical simulations illustrating the behavior of our algorithm. Finally, conclusions and further research aspects are discussed in Sect. 5.

2 Fuzzy Cognitive Maps

Fuzzy Cognitive Maps (FCM) are a suitable knowledge-based tool for modeling and simulation. From a connectionist perspective, FCM are recurrent networks with learning capabilities, consisting of nodes and weighted arcs. Nodes are equivalent to neurons in connectionist models and represent variables, entities or objects; whereas weights associated to connections denote the *causality* among such nodes. Each link takes values in the range $[-1, 1]$, denoting the causality degree between two concepts as a result of the quantification of a fuzzy linguistic variable, which is often assigned by experts during the modeling phase. The activation value of neurons is also fuzzy in nature and regularly takes values in the range $[0, 1]$. Therefore, the higher the activation value of a neuron, the stronger its influence over the investigated system, offering to decision-makers an overall picture of the systems behavior.

Without loss of generality, a FCM can be defined using a 4-tuple (C, W, A, f) where $C = \{C_1, C_2, C_3, \ldots, C_M\}$ is a set of M neurons, $W: (C_i, C_l) \rightarrow w_{il}$ is a function which associates a causal value $w_{il} \in [-1, 1]$ to each pair of nodes (C_i, C_l), denoting the weight of the directed edge from C_i to C_l. The weigh matrix $W_{M \times M}$ gathers the system causality which is often determined by experts, although may be computed using a learning algorithm. Similarly, $A\colon (C_i) \to A_i$ is a function that associates the activation degree $A_i \in \mathbb{R}$ to each concept C_i at the moment $t(t = 1, 2, \ldots, T)$. Finally, a transformation function $f\colon \mathbb{R} \to [0, 1]$ is used to keep the neuron's activation value in the interval $[0, 1]$. Following Eq. (1) portrays the inference mechanism using the vector $A^{(0)}$ as the initial configuration. The inference stage is iteratively repeated until a hidden pattern or a maximum number of iterations T is reached.

$$A_i^{(t+1)} = f\left(\sum_{j=1}^{M} w_{ji} A_j^{(t)} + A_i^{(t)} \right), i \neq j \tag{1}$$

The most used threshold functions are: the bivalent function, the trivalent function, and the sigmoid variants. It should be stated that authors will be focused on Sigmoid FCM, instead of discrete ones. It is motivated by the benchmarking analysis discussed in Ref. [11] where results revealed that the sigmoid function outperformed the other functions by the same decision model. Therefore, the proper selection of this threshold function may be crucial for the system behavior. From [12] some important observations were concluded and summarized as follows:

- Binary and trivalent FCM cannot represent the degree of an increase or a decrease of a concept. Such discrete maps always converge to a fixed-point attractor or limit cycle since FCM are deterministic models (expand here)
- Sigmoid FCM, by allowing neuron's activation level, can also represent the neuron's activation degree. They are suitable for qualitative and quantitative tasks, however, may additionally show chaotic behaviors.
- Mathematical definition of fixed-point, limited cycle, chaos
- Relation between convergence and symmetric weights.

3 Related Work on FCM Convergence

- Previous research on FCM convergence (including our results)
- Towards the end, we must introduce the motivation and challenges, namely, *to improve the convergence of FCM-based models used in decision-making without modifying the system modeling.*

4 Converge of Decisions in FCM-Based Systems

- New definitions about convergence.

5 The Proposed Learning Algorithm

• In this section we detail the proposed learning algorithm.

6 Numerical Simulations

• Describe the procedure to generate the artificial maps
• Key goals of simulations and parameter settings
• Discussion of results and remarks

In case of the RCN model the similarity threshold ε is fixed to 0.9. This process is performed by "trial and error" although we could use a learning method as was stated in the previus section. However, in the present paper the authors prefer to be focused on the methodology to deal with decision-making problems (Fig. 1).

In case of the RCN model the similarity threshold is fixed to 0.9. This process is performed by "trial and error" although we could use a learning method as was stated in the previus section. However, in the present paper the authors prefer to be focused on the methodology to deal with decision-making problems.

From Table 1 we can conclude that...

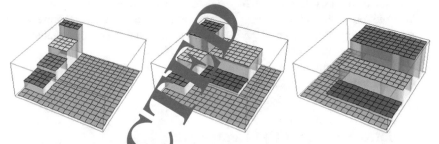

Fig. 1 Activation value of decision neurons for different scenarios. (i) a single positive region is activated, (ii) two positive regions are activated, (iii) only boundaries regions are activated

Table 1 Example

| | Predicting the most important decision | | | | | Predicting the correct order in decisions | | | | |
Study	MLP	BN	DT	FCM	RCN	MLP	BN	DT	FCM	RCN
E1	97.11	95.63	94.26	99.47	99.50	94.38	93.12	87.29	96.27	95.11
E2	92.06	91.37	89.39	93.74	93.29	82.40	80.25	77.59	88.72	90.32
Av	94.72	93.50	91.82	96.60	95.89	88.39	86.68	82.44	92.45	92.71

7 Conclusions

- Concluding remarks, summary of key findings, future work.

Eliciting Fuzzy Preferences Towards Health States with Discrete Choice Experiments

Michał Jakubczyk, Bogumił Kamiński and Michał Lewandowski

Abstract Most people consider health (quality and duration of life) as important but since we rarely choose between health states, our preferences are often not well-formed; moreover, the quality of life is frequently defined using imprecise terms (e.g. *moderate difficulties doing usual activities*). Therefore, we propose to model preferences towards health states (precisely: disutilities of worsening health dimensions in the EQ-5D-5L descriptive system) as fuzzy: each worsening is assigned an interval instead of a crisp number. We elicit such preferences with discrete choice experiment (DCE) data, using a maximum likelihood approach and bootstrapping to assess the estimation error. For example, the disutility of moderate difficulties doing usual activities was estimated as lying in the interval (0.018; 0.206). Pain/discomfort and anxiety/depression are associated with greatest upper bounds of disutilities and largest fuzziness (longest ranges). Our approach dispenses with one of the non-intuitive features of the standard approach to DCE, where even a clearly dominated alternative has a positive probability of being chosen; in our model, if the disutility ranges do not overlap, the worse alternative will never be chosen. Also, our model is more consistent regarding the constant proportional trade-off condition: the probability of a given health state being chosen in a pair will not change if durations are scaled proportionally; something that is not true in the standard DCE model.

Keywords Fuzzy modelling · Discrete choice experiment · Health-related quality of life · Utility · Preference elicitation · Multiple-criteria decision making

M. Jakubczyk (✉) · B. Kamiński · M. Lewandowski
Decision Analysis and Support Unit, SGH Warsaw School of Economics, Warsaw, Poland
e-mail: michal.jakubczyk@sgh.waw.pl

B. Kamiński
e-mail: bogumail.kaminski@sgh.waw.pl

M. Lewandowski
e-mail: michal.lewandowski@sgh.waw.pl

© Springer International Publishing AG 2018
C. Berger-Vachon et al. (eds.), *Complex Systems: Solutions and Challenges in Economics, Management and Engineering*, Studies in Systems, Decision and Control 125, https://doi.org/10.1007/978-3-319-69989-9_9

131

1 Introduction

Fuzzy sets have been introduced by [32] as a tool to formally define the imprecision inherent in some concepts, e.g. *a set of all tall men* or *a number much greater than 10*. Over the years, fuzzy sets (and related notions, as fuzzy logics) proved to be an effective way of describing imprecision and working with it: e.g. defining new concepts based on pre-existing ones (a set of tall *and* well-dressed men), conducting algebraic operations (a number around 5 *times* a number around 3), or verifying the truth value of complex statements (does being tall *imply* being good at basketball?). In many real-life situations, there is no clear-cut border between two opposing notions, such as being tall and not being tall or feeling good or not felling good. Therefore, it is natural to adopt the principle of gradual simultaneity: "*a proposal can at one and the same time be true and false, on the condition that a degree is assigned to its truth and a degree to its falseness*" ([17], pp. xxx). Hence, fuzzy logic and fuzzy arithmetic are more appropriate to model uncertainty inherent in many aspects of everyday life than the traditional Boolean logic and deterministic arithmetic are.

As a consequence of the above, fuzzy modelling has proved to be very useful in many economic and engineering applications. One important area of applications is that of controlling complex systems, such as vehicle routing problems [23], providing support for the business and management processes [18], or for the preparation and selection of investment, as well as handling the necessary stock of investment equipment [19].

People's preferences (over a set of alternatives) might also be perceived as a system: a collection of items (opinions about which of two alternatives is better) that are interdependent (e.g. via transitivity). This system may be very complex due to several reasons. First, it may be very large, if the set of alternatives contains many elements; also, the individual elements—decision alternatives—may be complicated, e.g. described by many attributes in the multiple-criteria decision making context. Second, it may not be given a priori, even to the person concerned; only inspecting parts of the system—e.g. when the person is asked to make a choice—may make the subject discover/form these preferences. Third, this system may be changing under inspection, and discovering preferences in one part may motivate the person to revise and change some previously established preferences.

In the present paper, we seek to verify whether fuzzy sets can be successfully used to model preferences towards health states in the multiple-criteria context. In this setting, the set of alternatives is enormous, comprising various combinations of health state quality and life duration. People, typically, are not used to making choices between health states: we might consult a physician to improve quality or longevity of our lives, but not to face trade-offs. (Some of us may have experience with choosing between staying healthy or falling victim to a pleasant but unhealthy activity, such as smoking.) Comparing health states requires counterfactual thinking involving hypothetical situations that may be very hard to imagine: what it would be like to be confined to bed or to feel extreme pain, even though you have no relevant experience.

However difficult, evaluating preferences towards health states is crucial in making informed decisions which health technologies to finance with public money. This preference elicitation in applied research is more and more often done by conducting discrete choice experiments (DCEs): presenting to a group of respondents a number of pairwise choice problems between two health descriptions. Then, in the standard approach, we assume that individual choices result from some true, underlying (crisp) preferences plus a stochastic error term. The underlying preferences may be estimated by econometric modelling (see, e.g. [3]).

The starting point in the present paper is the assumption that the underlying preferences are not crisp: the decision maker is in principle unable to say precisely how much more important the mobility is (not being confined to bed) versus the life duration, and such trade-off coefficient should rather be treated as fuzzy. The idea that preferences towards health should be modelled with fuzzy sets has been proposed before. Jakubczyk and Kamiński [21] present a model of fuzzy willingness-to-pay/willingness-to-accept (WTP/WTA) for health. The idea is that people rarely set a clear-cut number such that crossing it changes their preference from *definitely buy* (*sell*) to *definitely don't buy* (*don't sell*). Jakubczyk and Golicki [22] show how to elicit preferences towards various aspects of health with data collected using the time trade-off method (TTO, presented in the next section). Here, we present the first (to the best of our knowledge) application of fuzzy preferences in health based on DCE.

In Sect. 2, we discuss the concept of health-related quality of life in more details and discuss standard preference elicitation techniques. In particular, we present the standard approach to modelling DCE data and discuss some disadvantages further motivating our study. This also allows us to introduce the standard notation used in crisp approach to modelling utility of health states. Then, in Sect. 3 we formally present our model of fuzzy preferences, focusing on key differences with the standard model. We want to focus on the general idea and keep the model simple (e.g. neglecting heterogeneity of respondents). In Sect. 4 we present the dataset, technical details of the estimation process (e.g. our approach to calculating the estimation errors), and final results. In Sect. 5, we discuss our results in a wider context, also of some relevant (non-fuzzy) models present in the literature on utility modelling. The last section concludes.

2 Health Related Quality of Life and Standard Techniques of Preference Elicitation

Health care is largely financed with public resources; hence, the public regulator must decide which drugs should be reimbursed. In order to make such decisions in an informed way, we need to measure and compare the effectiveness of available technologies (how much health is bought), accounting for the effectiveness being reflected in improving the quality or longevity of life. Typically, in this health technology assessment process, the amount of health is measured with quality-adjusted life years (QALYs), i.e. number of years of life in the full health [30, 31].

In the QALY model, it is assumed that spending T years in a health state Q yields a utility of $u(Q) \times T$ QALYs, where $u(Q)$ is an index value of state Q, $u(Q) \leq 1$ (and T can be treated as discounted years, if needed). We normalize the scale by setting u(full health) $= 1$ and u(dead) $= 0$ (then we don't need to artificially consider for how long a person is dead). When health state changes, the streams of QALYs are added. The QALY model has been founded axiomatically [6, 7, 27] and is widely referred to and used when making actual public decisions [11].

Determining index values requires defining health states first. Usually, the EQ-5D-3L (or an EQ-5D-5L) descriptive system is used [8, 20]—a health state is defined as a combination of descriptions in five dimensions: mobility (MO), self-care (SC), usual activities (UA), pain/discomfort (PD), and anxiety/depression (AD). In each dimension, a person can be at exactly one of three (3L) or five (5L) levels. In EQ-5D-5L, slightly simplifying, these levels range from *no problems* to *extreme problems*, and are denoted by integer numbers 1–5 (see also Table 1). Hence, the full health is coded as 11111, and the worst possible health state as 55555. For example, state 21134 would denote a person who (MO) has slight problems in walking about, (SC) has no problems washing or dressing himself/herself, (UA) has no problems doing his/her usual activities, (PD) has moderate pain or discomfort, and (AD) is severely anxious or depressed. How the levels are defined leaves room for subjectivity (what are the usual activities?) and imprecision (where is the boundary between slight and moderate problems?). This observation motivates using fuzzy sets to model preferences over EQ-5D-5L system of health states.

Various methods have been proposed to assign index values to health states, the two most popular being TTO and DCE. In TTO the respondent is implicitly asked to find, for a given health state Q, such level of T^* (by testing various levels in a predefined sequence) that she would be indifferent between 10 years in Q or T^* years in 11111, both followed by immediate death [10]. Within the QALY model, this immediately translates to $u(Q) = \frac{T^*}{10}$ (a modified protocol—*lead-time TTO*—is used, if the respondent reveals $u(Q) \leq 0$, cf. [2, 9]). With TTO, usually only a subset of all (3125 for 5L) health states is used, and a linear model is estimated to explain $u(Q)$, where $Q \neq 11111$, as a linear function of dummies denoting the worsening of individual dimensions (plus an error term, accounting for other factors and, e.g., panel structure of the dataset):

$$u(Q) = 1 - \alpha_0 - \sum_{i=1}^{5} \sum_{j=2}^{5} \alpha_{i,j} d_{i,j}(Q) + \epsilon, \tag{1}$$

where αs are parameters to be estimated, and $d_{i,j}(Q) \in \{0, 1\}$ are dummy variables, equal to 1, if in Q the dimension i is at level j. Observe that for all $i \in \{1, \ldots, 5\}$ we have $\sum_{j=2}^{5} d_{i,j}(Q) \leq 1$. We assume that the levels "1" for different dimensions represent no worsening of utility (constituting a part of full health, 11111), and so $\alpha_{i,1}$ are not introduced. The intercept, α_0, is usually used and, in economic parlance, represents a possible aggregate complementarity of dimensions, as the first worsening (i.e. departing from 11111) triggers a larger disutility than subsequent worsenings.

In a DCE, the respondent is presented a sequence of choices, each between two health states lasting for some duration (e.g. 33322 for 8 years vs 21222 for 4 years, immediate death is often used as an option; also a version without duration is used sometimes). In each pair, the respondent selects the preferred alternative; DCE is considered to put less cognitive burden on a respondent than TTO. As index utility is not directly observed in DCE, additional modelling assumptions are needed. For example, [3] assume that the utility of T years in state Q is given as[1]:

$$u^*(Q, T) = T \left(\beta_0 + \sum_{i=1}^{5} \sum_{j=2}^{5} \beta_{i,j} d_{i,j}(Q) \right), \tag{2}$$

where βs are parameters. When two profiles are compared, (Q_1, T_1) vs (Q_2, T_2), then the probability of the former being selected is assumed to be equal to:

$$P((Q_1, T_1) \succ (Q_2, T_2)) = \frac{\exp(u^* (Q_1, T_1))}{\exp(u^*(Q_1, T_1)) + \exp(u^*(Q_2, T_2))}. \tag{3}$$

Estimated parameter values α and β in Eqs. 1 and 2, respectively, cannot be directly compared (the latter are defined on a latent scale), and so βs need to be anchored to be interpreted in the QALY scale, as αs [28].

As observed by [3], the specification of $u^*(Q, T)$ is consistent with the QALY model in that T enters the equation in a linear way. Then, the constant proportional trade-off, one of the consequences of the QALY model [5], is met: if $u^*(Q_1, T_1) = u^*(Q_2, T_2)$ (and so each would be chosen with 50% probability), then $u^*(Q_1, \lambda \times T_1) = u^*(Q_2, \lambda \times T_2)$, for any $\lambda \geq 0$. Nevertheless, if $u^*(Q_1, \lambda \times T_1) > u^*(Q_2, \lambda \times T_2)$, and so the former would be chosen more often than 50%, then increasing both durations by the same factor $\lambda > 1$ would result in the first profile being selected more often. Even though the constant proportional trade-off condition is often criticized and probably violated [1], it is important that the standard model specification used in DCE in principle cannot fully comply with this condition, even if trying to.

There is an additional difficulty with the standard way the DCE data are modelled. Consider two health profiles, (Q_1, T) and (Q_2, T) (identical duration, T, for simplicity), such that the latter is Pareto-dominated, i.e. the dimensions in Q_2 are equal or worse than in Q_1 (and at least one dimension is strictly worse). Under the standard specification, it is still possible (probability > 0%) that the dominated profile will be selected; a non-intuitive result, unless we want our model to takes into account the respondent making mistakes, i.e. selecting an option that is clearly inferior. The model presented in the Sect. 3 fully resolves the constant proportionality violation and partially solves the Pareto-dominance violation. This result strengthens and extends the motivation of the proposed model provided above.

---

[1]The general idea is presented. The original formulas and notation are slightly changed.

Jakubczyk and Golicki [22] suggested a way to estimate health state utilities as fuzzy numbers based on TTO data (without the need to introduce any changes to the TTO protocol to account for fuzziness). Here, we plan to show how DCE data can be used for this purpose. In the next section, we introduce our model; the general idea is to replace parts of Eq. 1 with fuzzy parameters and to propose a novel way of calculating the probability of choice based on fuzzy utilities.

3   The Model

In the standard approach, as described in the previous section, the utility of a health state is treated as a crisp number, and so are the utility losses attributed to individual dimensions/levels. In our model, we acknowledge that quality of life is an inherently imprecise notion and the respondent is not capable of perceiving the utility loss in a precise way. Instead, we assume the utility of a health state is a fuzzy set, $\widetilde{u(Q)}$, with membership function denoted by $\mu_{u(Q)}$ (or μ_u, if no particular state is being discussed; we also drop the tilde for clarity).

We assume $\mu_u(\cdot)$ to be non increasing, $\mu_u(x) = 1$ for low enough x, and $\mu_u(x) = 0$ for $x > 1$. We interpret $\mu_{u(Q)}(x)$ to denote respondent's conviction that the utility of state Q is at least x. We use this 'at least' interpretation, following how [21] model the fuzzy WTP/WTA. We also find this interpretation more natural: it might be difficult to assign any strictly positive conviction to the statement that the utility of health state Q is *exactly* x.

If $\mu_{u(Q)}(0) = 1$, then the respondent is fully confident that Q is better than being dead, i.e. the respondent is fully confident that Q yields at least zero utility (which is equivalent to being dead); if $\mu_{u(Q)}(0) = 0$, then the respondent is fully confident that Q is worse than dead; and in the remaining case the respondent cannot clearly determine how Q compares to dead. If $\mu_{u(Q)}(1) > 0$, then the respondent at least partially agrees that a given state does not really entail any utility loss (compared to full health), etc.

In order to further simplify the model, we assume that $\mu_{u(Q)}(x)$ is fully determined by two parameters: $L(Q)$ and $H(Q)$, $L(Q) \leq H(Q)$. Specifically, $\mu_{u(Q)}(x) = 1$ for all $x \leq L(Q)$, $\mu_{u(Q)}(x) = 0$ for all $x > H(Q)$, and $0 < \mu_{u(Q)}(x) < 1$ for all $L(Q) < x < H(Q)$. We also assume that $\mu_{u(Q)}(x)$ is linear for $L(Q) < x < H(Q)$ and continuous everywhere if $L(Q) < H(Q)$. $[L(Q), H(Q)]$ is the range, within which the respondent is not entirely sure where to locate the true utility; the respondent has crisp opinion, only if $L(Q) = H(Q)$.

We assume that the values $L(Q)$ and $H(Q)$ are given as linear equations, similar to Eq. 1:

$$L(Q) = 1 - \sum_{i=1}^{5} \sum_{j=2}^{5} h_{i,j} d_{i,j}(Q), \qquad (4)$$

$$H(Q) = 1 - \sum_{i=1}^{5} \sum_{j=2}^{5} l_{i,j} d_{i,j}(Q). \tag{5}$$

Notice that we do not introduce any error term at this stage and we drop the constant term for simplicity. Parameters $l_{i,j}$ and $h_{i,j}$ denote the bounds for the utility loss related to the dimension i being worsened to the level j (relative to level 1 for this dimension). We use parameters h to define L (and vice versa: parameters l to define H), because of the subtraction in the equations above. Parameters $l_{i,j}$ and $h_{i,j}$ can be interpreted in the following way: the respondent is fully confident that worsening the dimension i to the level j does not decrease the utility of life by more than $h_{i,j}$ and by less than $l_{i,j}$, and cannot rule out the decrements in between.

We assume that the utility of living for $T > 0$ years in health state Q has a membership function of the following form (with a slight abuse of notation for $u(\cdot)$):

$$\mu_{u(Q,T)}(x) = \begin{cases} 1, & \text{for } x \leq L(Q)T \\ \text{linearly decreasing}, & \text{for } L(Q)T \leq x \leq H(Q)T , \\ 0, & \text{for } H(Q)T \leq x \end{cases} \tag{6}$$

i.e. the bounds are simply multiplied by T, in analogy to the standard, crisp QALY model.[2] As already implicitly suggested above, we assume the membership function of full health (11111 health state) for T years satisfies $\mu_{u(11111)}(\cdot) = \mathbb{1}_{(-\infty,T]}(\cdot)$, and the membership function of being dead (does not depend on T): $\mu_{u(\text{dead})}(\cdot) = \mathbb{1}_{(-\infty,0]}(\cdot)$.

In Fig. 1, we present the examples of membership functions of four health descriptions (with duration). The gray membership function represents a health state that is clearly considered to be worse than dead (membership decreasing to 0 for negative

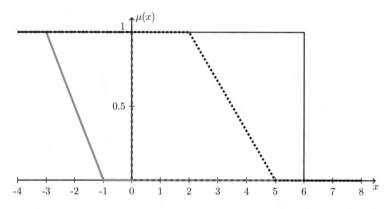

Fig. 1 Four examples of decision alternatives as fuzzy sets; a gray, dashed, dotted, and thin membership function (jumps in membership functions drawn as continuous lines for visibility)

---

[2]In a degenerate case $L(Q) = H(Q)$, the membership function jumps discontinuously from 1 to 0.

utilities). The dashed line represents the membership function of immediate death
(to increase readability also the vertical drop has been drawn). The dotted line represents a health state that is clearly better than dead (membership equal to 1 at 0). The
thin solid line represents six years in full health, where there is no fuzziness about
the total utility.

We now need to assume how the description of the alternatives is translated into
a choice in a single DCE task. We assume that when comparing two health descriptions, the respondent compares (or behaves as she would be comparing) two mem-
bership functions $\mu_{u(Q_1,T_1)}$ and $\mu_{u(Q_2,T_2)}$. Several qualitatively different situations can
be obtained, which we describe without using too many formalities, presenting the
intuition.

In the first step, the decision maker quantifies the advantage of (Q_1, T_1) over
(Q_2, T_2), denoted by $\delta_{(Q_1,T_1),(Q_2,T_2)}$, and vice versa:

$$\delta_{(Q_1,T_1),(Q_2,T_2)} = \sup_{x\in\mathbb{R}} \left(\mu_{u(Q_1,T_1)}(x) - \mu_{u(Q_2,T_2)}(x) \right). \tag{7}$$

The above expression can be interpreted as a conviction that there is some utility
level provided by (Q_1, T_1) and not by (Q_2, T_2) (when using the bounded sum AND
operator). For example, when $H(Q_1)T_1 \leq L(Q_2)T_2$, then $\delta_{(Q_2,T_2),(Q_1,T_1)} = 1$, demonstrating the respondent being fully convicted that (Q_2, T_2) is better (the left panel of
Fig. 2).

Furthermore, notice that when $L(Q_1)T_1 \leq L(Q_2)T_2$ and $H(Q_1)T_1 \leq H(Q_2)T_2$, then
$\delta_{(Q_1,T_1),(Q_2,T_2)} = 0$, demonstrating that the respondent would not agree that (Q_1, T_1)
is better. If additionally, $H(Q_1)T_1 > L(Q_2)T_2$, then $\delta_{(Q_2,T_2),(Q_1,T_1)} < 1$, showing lack
of full conviction (the middle panel of Fig. 2). If the two membership functions of
(Q_1, T_1) and (Q_2, T_2) cross (the right panel of Fig. 2), then $0 < \delta_{(Q_1,T_1),(Q_2,T_2)} < 1$ and
$0 < \delta_{(Q_2,T_2),(Q_1,T_1)} < 1$.

We then calculate the net advantage of (Q_1, T_1) over (Q_2, T_2) as $\Delta_{(Q_1,T_1),(Q_2,T_2)} =$
$\delta_{(Q_1,T_1),(Q_2,T_2)} - \delta_{(Q_2,T_2),(Q_1,T_1)}$. We shall use a simplified notation, Δ only, when the

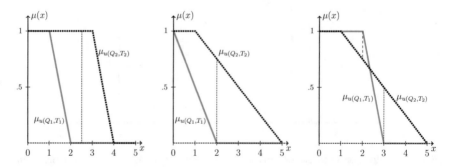

Fig. 2 Calculating the advantage of one fuzzy alternative over another based on their membership
functions. In each panel a solid and a dotted membership function is drawn. Thin, solid, vertical
lines in all panels represent $\delta_{(Q_2,T_2),(Q_1,T_1)}$; a dashed line (right panel) represents $\delta_{(Q_1,T_1),(Q_2,T_2)}$

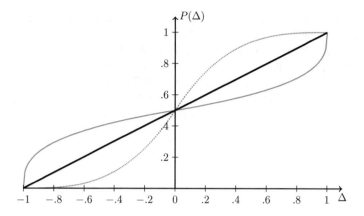

Fig. 3 The relation between the assessed advantage of one profile over another (Δ) and the probability of it being chosen ($P(\Delta)$). Black solid lines represents the relation as estimated from the data ($\rho = 0.989$), i.e. almost linear. Gray lines represent relations for other values: $\rho = 0.333$ (solid) or $\rho = 3$ (dotted)

states are clear from the context. Notice that $-1 \leq \Delta \leq 1$ and $\Delta_{(Q_1,T_1),(Q_2,T_2)} = -\Delta_{(Q_2,T_2),(Q_1,T_1)}$.

Finally, we have to assume how the advantage of one profile over another (Δ) is transformed into the probability of this profile being chosen, $P = P(\Delta)$. We assume intuitively $P(\cdot)$ to be non-decreasing, $P(0) = 0.5$ and $P(\Delta) = 1 - P(1 - \Delta)$ to guarantee symmetry. In general, we would not assume $P(-1) = 0$ and $P(1) = 1$ to allow some probability of error, even if only for technical reasons. In the present paper, we assume parametrically $P(\Delta) = \frac{(\Delta+1)^\rho}{2}$ for $\Delta \leq 0$, for some $\rho > 0$, to be estimated. Then $P(\Delta)$ for $\Delta > 0$ is calculated as $P(\Delta) = 1 - P(-\Delta)$. In Fig. 3, we present examples how Δ will impact P for various ρs. Additionally, as was suggested above, we clamp $P(\Delta)$ to the interval $[0.00001, 0.99999]$ to allow for respondent's error even in case of dominance.

To sum up, in our model we need to estimate 41 parameters: 5 dimensions times 4 non-trivial levels, lower and upper bounds for each, plus one parameter specifying the $P(\cdot)$ function. In the next section, we shortly discuss the technical approach we took to estimation and the results on the dataset we used.

4   Estimation and Results

4.1   Dataset

We used the data from a DCE predictive competition organized by the International Academy for Health Preference Research (IAHPR) and funded by The EuroQol

Group, available for general public and described in the IAHPR website (http://iahpr.
org/eq-dce-competition/, as of 16 Nov, 2016).

The dataset consists of 4074 respondents from the USA, each making a choice
between two health states in 20 pairs, selected out of 1560, described in EQ-5D-5L
descriptive system. Each state was associated with its duration. Four time units were
used (days, weeks, months, and years; same unit for both states in each pair), but
notice that in our model units in which the time is measured cancel out.

In the estimation process, we aggregated individual responses, for each pair of
health states and durations noting the number of comparisons and the number of
times the first profile was selected.

4.2 Estimation Process

We used maximum likelihood estimation, using the following constraints: (1) $l_{i,j}$
and $h_{i,j}$ must be non-decreasing with j for every i, (2) $h_{i,j} \geq l_{i,j}$ for every i and j.
Because the optimization problem is constrained and not-smooth, we used a custom
optimization routine in Julia [4] developed by us. In order to obtain the distribution
of estimators, we used bootstrapping. Then, the 95% confidence interval (CI) was
obtained with percentile method.

Observe that it would be straightforward to use the same setting in the Bayesian
fashion by assuming a priori distributions on model parameters (l, h, and ρ) and
calculating the posterior distributions with the described-above probabilistic model.
It would then be natural to use a hierarchical model and account for respondents
heterogeneity (assuming individual l, h, and ρ are drawn from some population-wise
distribution). This is left for further research.

4.3 Results

We present the estimates in Table 1, point estimates along 95% CI. The value of ρ
was found to be 0.989, with 95% CI $= (0.905; 1.063)$. The equality of some parameters (e.g. h for MO2 and MO3) results from the monotonicity constraints put on
the parameters, as described above. Still, the ranges, ($l_{i,j}, h_{i,j}$), for consecutive levels
(i.e. for j and $j + 1$) overlap, as there was no constraint requiring otherwise.

We used the estimated values of l and h to calculate the estimated fuzzy utilities
of all states possible in EQ-5D-5L. We present the results in Fig. 4, ordered by the
average of $L(Q)$ and $H(Q)$. As many as 20% states have $L(Q) < -1$, while only 8
states have $H(Q) \leq 0$. That shows that the fuzziness regarding the utility of a health
state is very large for bad states (not surprising looking at the $h_{i,j} - l_{i,j}$ for large js).
Importantly, the utility values lower than -1 are not inconsistent with the QALY
model—the utilities can be arbitrarily negative. Some methods are unable to elicit

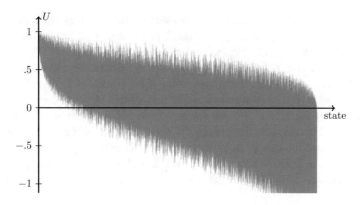

Fig. 4 Fuzzy value set (fuzzy utilities assigned to health states), ordered by the mid-point of a range. Figure trimmed at the utility equal to −1 for clarity

values less than −1: for example, standard lead-time TTO, possibly requiring special treatment in modelling [14].

The observation that virtually all $H(Q) > 0$ agrees with the intuition that it is discomforting for most people to declare a health state as being worse than dead. Still, very many states have $L(Q) < 0$, showing that often (i.e. for many states) the respondents have difficulty in ascertaining that a given state is better than dead.

5 Discussion

We have shown how to model the utilities of health states as fuzzy, rather than crisp, numbers, and how these utilities may result from the components defining health states, i.e. parameters assigned to individual health dimensions and their worsening. In order to estimate these parameters using standard DCE data, i.e. information on the frequency of a specific health state being chosen from a pair of health states (with some durations), we had to propose a model how the fuzzy description of alternatives is transformed into the relative advantage of each alternative in a pair and—subsequently—into the probability of each alternative being chosen.

The function transforming the relative advantage into the probability was defined with a parameter ρ (see Fig. 3); our estimation yielded $\rho \approx 1$, which proves that the fuzzy measure of conviction that one alternative is better than the other requires almost no transformation when converted into the observables, i.e. the frequency of a given choice. We find that very promising, as it may suggest that our fuzzy model may be close to the real decision making process, and in this sense the model is ontologically grounded.

A natural next step would be to extend standard DCE questions to ask also about the conviction of the respondent towards a declared choice [12]. Discovering how this (observed) conviction relates to the (observed) frequency of choice and (unobserved, modelled) fuzzy-model parameter Δ might shed new light on the decision making process in the decision problems with multiple criteria.

Our model partially solved the problem of a possible violation of a Paretodominance, present with standard modelling of DCE data. If $H(Q_1) < L(Q_2)$, and so Q_1 was found decidedly worse than Q_2, then (when both are considered for an equal duration) the former will be selected with 100% probability in our model, while standard specification underlying DCE modelling would accept some strictly positive chance of the Pareto-dominated outcome. An example of such a pair according to our results in Table 1 would be 11311 and 11141: severe pain or discomfort were found to have a decidedly bigger disutility than moderate problems doing usual activities.

Table 1 Estimation results. Description of dimensions/levels shortened for brevity

Dimension/level	Description	l (95%CI)	h (95%CI)
MO2	Slight problems in walking about	0.034 (0.002; 0.045)	0.215 (0.215; 0.220)
MO3	Moderate problems ...	0.034 (0.009; 0.045)	0.215 (0.215; 0.220)
MO4	Severe problems ...	0.200 (0.153; 0.204)	0.500 (0.458; 0.543)
MO5	Unable to walk about	0.320 (0.281; 0.341)	0.601 (0.569; 0.642)
SC2	Slight problems washing or dressing	0.000 (0.000; 0.000)	0.186 (0.173; 0.188)
SC3	Moderate problems ...	0.026 (0.005; 0.070)	0.278 (0.192; 0.301)
SC4	Severe problems ...	0.116 (0.066; 0.168)	0.388 (0.385; 0.412)
SC5	Unable to wash or dress myself	0.208 (0.180; 0.247)	0.530 (0.530; 0.600)
UA2	Slight problems doing usual activities	0.018 (0.005; 0.018)	0.138 (0.127; 0.149)
UA3	Moderate problems ...	0.018 (0.007; 0.018)	0.206 (0.190; 0.224)
UA4	Severe problems ...	0.138 (0.117; 0.193)	0.355 (0.320; 0.381)
UA5	Unable to do usual activities	0.138 (0.123; 0.209)	0.389 (0.362; 0.421)
PD2	Slight pain or discomfort	0.000 (0.000; 0.000)	0.296 (0.289; 0.296)
PD3	Moderate ...	0.071 (0.030; 0.090)	0.296 (0.289; 0.296)
PD4	Severe ...	0.210 (0.202; 0.251)	0.546 (0.512; 0.659)
PD5	Extreme ...	0.266 (0.248; 0.304)	0.771 (0.758; 0.937)
AD2	Slightly anxious or depressed	0.031 (0.000; 0.032)	0.120 (0.108; 0.120)
AD3	Moderately ...	0.091 (0.046; 0.100)	0.242 (0.217; 0.290)
AD4	Severely ...	0.091 (0.051; 0.132)	0.701 (0.701; 0.811)
AD5	Extremely ...	0.138 (0.126; 0.157)	0.727 (0.713; 0.822)

Still, the ranges of disutilities for consecutive levels in a single dimension overlap. This means that in our model the respondent is not 100% probably to, for example, choose 11311 over 11411. This proves that efforts should now be directed at separately considering, in the model, worsenings in various dimensions (more generally: attributes, criteria) and worsenings in a single dimension (attribute).

Some of the parameters were estimated equal, e.g. h for MO2 and MO3. That resulted from constraints (requiring MO3 value to be greater or equal). This also proves there is a room for improvement, either regarding the model specification or the data to be used for estimation. Ideally, the parameters should respect the intuitive monotonicity constraints by themselves.

In light of the above discussion, it is clear that our model should be treated as an initial attempt to apply fuzzy thinking to multiple criteria decision making in DCE context. We are aware of no other such approach ([12], mentioned above accounted for the conviction of choice, but not in a formal setting). We are also aware of no other fuzzy model in the context of health state comparison, except for the work of [22] on TTO. We find it important that our model, even if loaded with ad hoc assumptions at this stage, was constructed formally from the notion of fuzzy sets, via addition (to aggregate worsening in various dimensions), multiplication (to account for duration), and using the bounded sum AND NOT operator to define δ. In the next step we might reformulate the model, working on its predictive validity (ignored for now). For example, above we stressed the ability of our model to consistently represent the assumption of constant proportional trade-off (contrary to what standard DCE modelling can do). We feel this capability adds credibility to the whole general approach; still, the assumption itself is most likely false and so improving the predictive validity will require dropping it and incorporating duration in a non-linear way.

Another change would be to allow more flexibility regarding the shape of the membership function; for example, replace the linearity assumption with an S-shaped curve. More specifically, for the case where $L(Q) < H(Q)$, we propose to a use a membership function that is relatively flat at the edges of the interval $[L(Q), H(Q)]$ and steep in the middle. This particular shape is motivated by possible range effects that might be present in the evaluation of utility values. This idea derives from [24], who propose the range-dependent utility model and find that there is a range-dependent function defined over the range of stimulus values, i.e. the interval between the lowest and the highest stimulus arising in a given context (in their case, it is the range of monetary prizes arising in a given risky gamble; in the present paper, it is the interval $[L(Q), H(Q)]$ of utility values). We hypothesize that this change might improve descriptive accuracy of the model.

An additional appealing feature of our model is that its predictions (probabilities of choice) are continuous in the description of the decision alternatives, as long as at least one alternative is actually fuzzy (i.e. its membership function is continuous and does not drop discontinuously from 1 to 0). Then, changing the duration infinitesimally will only result in a small change of the probability. That does not hold when two alternatives are perceived in a crisp way (which we find intuitive): the outcome of choosing between full health for T or full health for $T + \eta$ years will depend discontinuously on η around 0.

We also see the possibility to improve our model by assuming a hierarchical approach to parameters, i.e. directly acknowledging the respondents differ, and we should rather be estimating the population-wise average of some distribution.

The general difficulty in modelling preferences towards health states seems to lie in the fact that these preferences are not observable from choice data, but rather they are expressed by the decision maker who contemplates them by performing a counterfactual reasoning inside his or her head. This makes the alleged imprecision of preferences possibly more pronounced and justifies using a non-standard fuzzy approach, as the one postulated in the paper. There are, however, two classic models in the literature that could possibly be used to address similar challenges, and we discuss them here in order to make a comparison to the fuzzy model presented here, as well as to possibly drive further research.

First, [25] propose to use semi-orders to describe preferences. This idea is motivated by Weber's ([29]) law from psychophysics, which states that the minimal change in a measurable stimulus that would be noticed (*just noticeable difference*, *jnd*), ΔS, is proportional to the level of that stimulus, S. Let $>$ be a binary relation on a set of alternatives X. This relation is interpreted as strict preference, where $I = (> \cup >^{-1})^c$ ($>^{-1} = \{(x, y) : (y, x) \in >\}$) denotes no preference in either direction, or indifference. Luce proposes a utility representation of the following form: $x > y$ if and only if $u(x) - u(y) > 1$, for a utility function $u : X \to \mathbb{R}$. Putting it in words, in order to have a strict preference for one alternative over another, one needs the utility difference to be greater than 1, which corresponds to the *jnd* from Weber's law. This model has been generalized by [15] and [16] by defining an *interval relation*. The resulting representation allows a variable, instead of a constant, *jnd*, denoted by $\delta(y)$ and may be stated as: $x > y \iff u(x) - u(y) > \delta(y)$.

The semi-order or the interval-relation models might fit our context in the following way: the respondent may only feel one health description as better than another if the improvement in some aspects (e.g. duration) is large enough. If the difference is non-zero but small, the decision maker might have difficulties in making the choice in DCE (i.e. may be reluctant to choose either of the two options). The alternatives for which the level of utility differences falls short of the *jnd* level would correspond to $\delta < 1$ in our model (Eq. 7).

Note that semi-orders allow intransitive choices. It may well be that the differences in utility values between x and y and between y and z are both smaller than the *jnd* level, but the difference between x and z exceeds it. In this case, $x \sim y$ and $y \sim z$, but $x > z$, violating transitivity of indifference. Instead of questioning transitivity, the probabilistic utility model[3] of [26] relaxes completeness. The idea is that the decision maker may have trouble in saying which of the two alternatives she prefers. It should be stressed that the decision maker has no trouble in telling apart the two alternatives as psychical stimuli; the assumed uncertainty pertains rather to the decision maker's ability to separate alternatives as to preference. Suppose that X is the set of all alternatives and x, y are two elements of X. Luce assumes that there

---

[3]This model is originally meant for decisions under uncertainty, but we can confine attention to sure alternatives, because they constitute a subset of all alternatives.

exists an objective probability $P(x, y)$ that a given decision maker will prefer x to y; if this probability is always either 1 or 0, we are back in the deterministic utility case. Otherwise, an individual may not be completely sure as to what she prefers. The idea is that when forced to do so, the decision maker may express preference of say x over y, but if asked many times, she might switch her preference some of the time. The frequency of choosing x over y is on average equal to the probability $P(x, y)$. Having defined $P(x, y)$, Luce introduces an auxiliary preference relation $\gtrsim$ in X, defined as follows: $x \gtrsim y$, whenever $P(x, z) \geq P(y, z)$ and $P(z, y) \geq P(z, x)$ hold for every $z \in X$. Obviously, $\gtrsim$ is transitive by construction but in general it might not be complete. Luce thus introduces an axiom that forces such comparability, i.e. for all $x, y \in X$, either $x \gtrsim y$, or $y \gtrsim x$. By imposing transitivity right from the start and then assuming completeness of the above form, Luce weakens comparability demands as compared to the usual and reverse strategy of imposing completeness first and then questioning transitivity. In this approach, a utility function $u : X \to \mathbb{R}$ is assumed instead of obtained in a representation. Given such utility function, the representation states that there is a function $P^* : \mathbb{R} \to [0, 1]$, such that $P(x, y) = P^*[u(x) - u(y)]$, for $x, y \in X$. This finding connects nicely with the definition of Fechner ([13]) subjective scale sensation, which is an extension of the Weber's law mentioned before. Using Luce's axioms one can even pin down the exact functional form of $P(x, y)$.

The model of fuzzy preferences presented in this paper as well as the probabilistic utility model described above lead to a probabilistic specification of choice between a given pair of alternatives. This general idea is common to both approaches. However, the way these probabilities are formed in both models differs substantially. Based on revealed preference, the latter model assumes preferences as probabilistic and imposes axioms that lead to a certain functional form of the implied probabilities. The model of fuzzy preferences, on the other hand, does not assume probabilistic preferences as primitives, but, instead, introduces a fuzzy utility value for a given alternative. Based on the assumptions imposed on the way these fuzzy utility values are modelled (viz. membership functions and their functional form) and compared (via δ and Δ), the probability of choosing one alternative over another is derived as a result. The main focus of our model has been on its applicability. In fact, as demonstrated in the paper, our model offers a ready-to-use procedure of estimating the choice probabilities. The disadvantage of the model proposed by Luce lies in its non-operational character. As stated before, probabilities of the form $P(x, y)$, for $x, y \in X$ are primitives of the model and may be difficult to estimate.

6 Conclusion

Health state description seem to inherently involve imprecision. Hence, modelling preferences towards health states should also account for this element; preferably in a formal way, in order to allow precise description and elicitation. We showed this can be done with a fuzzy model. Interestingly, the parameters of our model can be estimated with standard data collected in health preference research, not requiring any changes in the protocol. The model has promising formal properties but further work should focus on studying and improving its predictive validity.

Acknowledgements The research was financed by the funds obtained from National Science Centre, Poland, granted following the decision number DEC-2015/19/B/HS4/01729.

References

1. Attema, A.E., Brouwer, W.B.: On the (not so) constant proportional trade-off in TTO. Qual. Life Res. **19**, 489–497 (2010)
2. Attema, A.E., Versteegh, M.M., Oppe, M., Brouwer, W.B., Stolk, E.A.: Lead time TTO: leading to better health state valuations? Health Econ. **22**, 376–392 (2013)
3. Bansback, N., Brazier, J., Tsuchiya, A., Anis, A.: Using a discrete choice experiment to estimate health state utility values. J. Health Econ. **31**, 306–318 (2012)
4. Bezanson, J., Edelman, A., Karpinski, S., Shah, V.B.: Julia: a fresh approach to numerical computing (2014). arXiv.org/1411.1607
5. Bleichrodt, H., Johannesson, M.: The validity of QALYs: an experimental test of constant proportional tradeoff and utility independence. Med. Decis. Mak. **17**, 21–32 (1997)
6. Bleichrodt, H., Quiggin, J.: Characterizing QALYs under a general rank dependent utility model. J. Risk Uncertain. **15**, 151–165 (1997)
7. Bleichrodt, H., Wakker, P., Johannesson, M.: Characterizing QALYs by risk neutrality. J. Risk Uncertain. **15**, 107–114 (1997)
8. Brooks, R., De Charro, F.: EuroQol: the current state of play. Health Policy **37**, 53–72 (1996)
9. Devlin, N., Tsuchiya, A., Buckingham, K., Tilling, C.: A uniform time trade off method for states better and worse than dead: feasibility study of the 'lead time' approach. Health Econ. **20**, 348–361 (2011)
10. Dolan, P., Gudex, C., Kind, P., Williams, A.: The time trade-off method: results from a general population study. Health Econom. **5**, 141–154 (1996)
11. Drummond, M.F., Scupher, M.J., Torrance, G.W., O'Brien, B.J., Stoddart, G.L.: Methods for the Economic Evaluation of Health Care Programmes. Oxford University Press (2005)
12. Elrod, T., Chrzan, K.: The value of extent-of-preference information in choice-based conjoint analysis. In: Gustafsson, A., Herrmann, A., Huber, F. (eds.) Conjoint Measurement, pp. 209–223. Springer, Methods and Applications (2000)
13. Fechner, G.: Elemente der Psychophysik (2 Vols) (1860). Breitkopf and Hartel. Vol. 1 trans, by Adler, H.E. (1966)
14. Feng, Y., Devlin, N., Shah, K., Mulhern, B., van Hout, B.: New methods for modelling EQ-5D-5L value sets: an application to English data. Health Economics & Decision Science (HEDS) Discussion Paper Series, University of Sheffield (2016)
15. Fishburn, P.C.: Intransitive indifference in preference theory: a survey. Oper. Res. **18**(2), 207–228 (1970)
16. Fishburn, P.C.: Interval graphs and interval orders. Discret. math. **55**(2), 135–149 (1985)
17. Gil-Aluja, J.: Elements for a Theory of Decision in Uncertainty. Springer Science+Business Media Dordrecht (1999)
18. Gil-Aluja, J.: Fuzzy Sets in the Management of Uncertainty. Springer, Berlin Heidelberg (2004)
19. Gil-Aluja, J.: Investment in Uncertainty, vol. 21. Springer Science & Business Media
20. Herdman, M., Gudex, C., Lloyd, A., Janssen, M., Kind, P., Parkin, D., Bonsel, G., Badia, X.: Development and preliminary testing of the new five-level version of EQ-5D (EQ-5D-5L). Quality of life research: an international journal of quality of life aspects of treatment, care and rehabilitation **20**, 1727–1736 (2011)

21. Jakubczyk, M., Kamiński, B.: Fuzzy approach to decision analysis with multiple criteria and uncertainty in health technology assessment. Ann. Oper. Res. (2015). https://doi.org/10.1007/s10479-015-1910-9
22. Jakubczyk, M., Golicki, D.: Estimating the impact of EQ-5D dimensions as fuzzy numbers with hierarchical Bayesian modelling of regular TTO data. In: EuroQol Plenary Meeting (2016)
23. Kondratenko, G.V., Kondratenko, Y.P., Romanov, D.O.: Fuzzy Models for Capacitive Vehicle Routing Problem in Uncertainty. In: Proceedings of 17th International DAAAM Symposium "Intelligent Manufacturing and Automation: Focus on Mechatronics & Robotics", pp. 205–206 (2006)
24. Kontek, K., Lewandowski, M.: Range-dependent utility. Manag. Sci. (forthcoming) (2016)
25. Luce, R.D.: Semiorders and a theory of utility discrimination. Econometrica, 178–191 (1956)
26. Luce, R.D.: A probabilistic theory of utility. Econometrica, 193–224 (1958)
27. Miyamoto, J.M., Wakker , P.P., Bleichrodt, H., Peters, H.J.M.: The Zero-Condition: a simplifying assumption in QALY measurement and multiattribute Utility. Manag. Sci. 44 (1998)
28. Rowen, D., Brazier, J., van Hout, B.: A comparison of methods for converting DCE values onto the full health-dead QALY scale. Med. Decis. Mak. 35, 328–340 (2015)
29. Weber, E.: De tactu. Koehler, Leipzig (1834)
30. Weinstein, M.C., Torrance, G., McGuire, A.: QALYs: the basics. Value Health 12, S5–S9 (2009)
31. Whitehead, S.J., Ali, S.: Health outcomes in economic evaluation: the QALY and utilities. British Medical Bulletin 96, 5–21 (2010)
32. Zadeh, L.: Fuzzy Sets. Inf. Control 8, 338–353 (1965)

The Soft Consensus Model in the Multidistance Framework

Silvia Bortot, Mario Fedrizzi, Michele Fedrizzi,
Ricardo Alberto Marques Pereira and Thuy Hong Nguyen

Abstract In the context of the soft consensus model due to (Fedrizzi et al. in Journal international journal of intelligent systems 14:63–77, 1999) [27], (Fedrizzi et al. in New mathematics and natural computation 3:219–237, 2007) [28], (Fedrizzi et al. in Preferences and Decisions: models and applications, studies in fuzziness and soft computing Springer, Heidelberg, pp. 159–182, 2010) [30], we investigate the reformulation of the soft dissensus measure in relation with the notion of multidistance, recently introduced by Martín and Mayor (Information processing and management of uncertainty in knowledge-based systems. Theory and methods, communications in computer and information science, springer, heidelberg, pp. 703–711 2010) [43], Martín and Mayor (Fuzzy sets and systems 167:92–100 2011) [44]. The concept of multidistance is as an extension of the classical concept of binary distance, obtained by means of a generalization of the triangular inequality. The new soft dissensus measure introduced in this paper is a particular form of sum-based multidistance. This multidistance is constructed on the basis of a binary distance defined by means of a subadditive scaling function, whose role is that of emphasizing small distances and attenuating large distances in preferences. We present a detailed study of the subadditive scaling function, which is analogous but not equivalent to the one used in the traditional form of the soft consensus model.

Keywords Multidistances · Dissensus measures · Soft consensus model

1 Introduction

The notion of consensus is central to decision making models involving the aggregation of individual preferences. We can distinguish essentially two complementary readings of the consensus concept. In general terms, it refers to the consensual

S. Bortot (✉) · M. Fedrizzi · M. Fedrizzi · R.A. Marques Pereira · T.H. Nguyen
Department of Economics and Management, University of Trento, Via Inama, 5,
38122 Trento, Italy
e-mail: silvia.bortot@unitn.it

© Springer International Publishing AG 2018
C. Berger-Vachon et al. (eds.), *Complex Systems: Solutions and Challenges in Economics, Management and Engineering*, Studies in Systems, Decision and Control 125, https://doi.org/10.1007/978-3-319-69989-9_10

preference resulting from the aggregation scheme, whether or not preference aggregation is formulated as an iterative consensus reaching process. More specifically, the notion of consensus refers to the construction of consensus (dissensus) measures, which express the level of agreement (disagreement) present in the collective profile of individual preferences.

In general, consensual aggregation models involve some form of explicit or implicit averaging of the individual preferences. In the context of aggregation theory, comprehensive reviews of averaging functions can be found in [2, 7, 21, 33, 34].

In our approach we are primarily interested in the class of aggregation schemes which are based on consensus (dissensus) measures, often constructed on the basis of some binary distance acting pairwise on the individual preferences.

In this respect, the recent literature on the use of penalty functions in aggregation [3–6, 17, 18, 20] provides a suggestive framework in which to describe the interrelation between aggregation functions and consensus (dissensus) measures.

Further interesting investigation on the construction and applications of consensus (dissensus) measures can be found in [1, 8–11, 13–16, 19, 23–26, 46, 47, 49–53].

In the tradition of the fuzzy approach to consensus in the aggregation of individual preferences [31, 32, 35, 36, 40, 41], the soft consensus model was originally proposed in [37–39] and later reformulated in [27–30]. The soft consensus model is based on a dissensus measure constructed from pairwise square differences, composed with a subadditive scaling function (substituting the linguistic quantifiers in the original version of the model), whose role is that of emphasizing small (attenuating large) preference differences by means of a smooth thresholding effect.

In this paper we wish to revisit the soft consensus model and investigate the formulation of the soft dissensus measure in relation with the notion of multidistance, recently introduced in [22, 42–45, 48]. The concept of multidistance is as an extension of the classical concept of binary distance, obtained by means of a generalization of the triangular inequality.

With respect to the traditional soft consensus model, here the idea is to construct a new multidistance dissensus measure directly from the pairwise absolute value differences and the subadditive scaling function, keeping the traditional character of the soft dissensus measure but avoiding the square differences in the functional form.

The paper is organized as follows. In Sect. 2 we briefly review the soft consensus model and the construction of the traditional soft dissensus measure. In Sect. 3 we review the basic notions regarding multidistances and in Sect. 4 we introduce the new multidistance dissensus measure, with a detailed study of the subadditive scaling function. Finally, in Sect. 5 we present some concluding remarks and notes on future research.

2 The Soft Consensus Model

In this section we present a brief review of the traditional soft consensus model in the formulation introduced in [27]. Our point of departure is a set of individual fuzzy preference relations. If $A = \{a_1, \ldots, a_m\}$ is a set of decisional alternatives and $I = \{1, \ldots, n\}$ is a set of individuals, the fuzzy preference relation R_i of individual i is given by its membership function $R_i : A \times A \to [0, 1]$ with

$$
\begin{aligned}
R_i(a_k, a_l) &= 1 && \text{if } a_k \text{ is definitely preferred over } a_l \\
R_i(a_k, a_l) &\in (0.5, 1) && \text{if } a_k \text{ is preferred over } a_l \\
R_i(a_k, a_l) &= 0.5 && \text{if } a_k \text{ is considered indifferent to } a_l \\
R_i(a_k, a_l) &\in (0, 0.5) && \text{if } a_l \text{ is preferred over } a_k \\
R_i(a_k, a_l) &= 0 && \text{if } a_l \text{ is definitely preferred over } a_k,
\end{aligned}
$$

where $i = 1, \ldots, n$ and $k, l = 1, \ldots, m$. Each individual fuzzy preference relation R_i can be represented by a matrix $[r_{kl}^i]$, $r_{kl}^i = R_i(a_k, a_l)$ which is commonly assumed to be reciprocal, that is $r_{kl}^i + r_{lk}^i = 1$. Clearly, this implies $r_{kk}^i = 0.5$ for all $i = 1, \ldots, n$ and $k = 1, \ldots, m$.

The general case $A = \{a_1, \ldots, a_m\}$ for the set of decisional alternatives is discussed in [27, 28]. Here, for the sake of simplicity, we assume that the alternatives available are only two ($m = 2$), which means that each individual preference relation R_i has only one degree of freedom, denoted by $x_i = r_{12}^i$.

In the framework of the soft consensus model, assuming $m = 2$, the degree of dissensus between individuals i and j as to their preferences between the two alternatives is measured by

$$
V_{ij} = g((x_i - x_j)^2) \qquad i, j = 1, \ldots, n \tag{1}
$$

where $g : [0, 1] \to \mathbb{R}$ is a scaling function defined as

$$
g(u) = \frac{1}{\alpha} \ln \left(\frac{1}{1 + e^{-\alpha(u-\beta)}} \right) \qquad u \in [0, 1]. \tag{2}
$$

In the scaling function formula above, $\beta \in (0, 1)$ is a threshold parameter and $\alpha \in (0, \infty)$ is a free parameter which controls the polarization of the sigmoid function $g' : [0, 1] \to (0, 1)$ given by

$$
g'(u) = \frac{1}{1 + e^{\alpha(u-\beta)}} \qquad u \in [0, 1]. \tag{3}
$$

In the network representation of the soft consensus model [27], each decision maker $i = 1, \ldots, n$ is represented by a pair of connected nodes, a primary node (dynamic) and a secondary node (static). The n primary nodes form a fully connected subnetwork and each of them encodes the individual opinion of a single decision maker. The n secondary nodes, on the other hand, encode the individual opinions originally declared by the decision makers, denoted $s_i \in [0, 1]$, and each of them is connected only with the associated primary node.

The iterative process of preference change corresponds to the gradient descent optimization of a cost function W, depending on both the present and the original network configurations. The value of W combines a measure V of the overall dissensus in the present network configuration with a measure U of the overall change from the original network configuration.

The various interactions involving node i are modulated by interaction coefficients whose role is to quantify the strength of the interaction. The consensual interaction between primary nodes i and j is modulated by the interaction coefficient $v_{ij} \in (0, 1)$, whereas the inertial interaction between primary node i and the associated secondary node is modulated by the interaction coefficient $u_i \in (0, 1)$. In the soft consensus model the values of these interaction coefficients are given by the derivative g' of the scaling function according to

$$v_{ij} = g'((x_i - x_j)^2) \qquad i, j = 1, \ldots, n \tag{4}$$

$$v_i = \sum_{j(\neq i)=1}^{n} v_{ij}/(n-1), \quad u_i = g'((x_i - s_i)^2) \qquad i = 1, \ldots, n. \tag{5}$$

The average preference $\bar{x}_i$ of the context of individual i is given by

$$\bar{x}_i = \frac{\sum_{j(\neq i)=1}^{n} v_{ij} x_j}{\sum_{j(\neq i)=1}^{n} v_{ij}} \qquad i = 1, \ldots, n \tag{6}$$

and represents the average preference of the remaining decision makers as seen by decision maker $i = 1, \ldots, n$.

The construction of the cost function W that drives the dynamics of the soft consensus model is as follows. The individual dissensus cost is given by

$$V_i(\boldsymbol{x}) = \sum_{j(\neq i)=1}^{n} V_{ij}/(n-1) \qquad i = 1, \ldots, n \tag{7}$$

and the individual opinion changing cost is

$$U_i(\boldsymbol{x}) = g((x_i - s_i)^2) \qquad i = 1, \ldots, n. \tag{8}$$

Summing over the various decision makers we obtain the collective dissensus cost V and inertial cost U,

$$V(\boldsymbol{x}) = \frac{1}{4} \sum_{i=1}^{n} V_i(\boldsymbol{x}), \quad U(\boldsymbol{x}) = \frac{1}{2} \sum_{i=1}^{n} U_i(\boldsymbol{x}) \tag{9}$$

with conventional multiplicative factors $1/4$ and $1/2$. The full cost function is then

$$W(x) = V(x) + U(x). \tag{10}$$

The consensual network dynamics acts on the individual opinion variables x_i through the iterative process

$$x_i \rightsquigarrow x_i' = x_i - \gamma \frac{\partial W}{\partial x_i} \qquad i = 1, \dots, n. \tag{11}$$

Analyzing the effect of the two dynamical components V and U separately we obtain

$$\frac{\partial V}{\partial x_i} = v_i(x_i - \bar{x}_i) \qquad i = 1, \dots, n \tag{12}$$

where the coefficients v_i were defined in (5) and the average preference $\bar{x}_i$ was defined in (6), and therefore

$$x_i' = (1 - \gamma v_i)x_i + \gamma v_i \bar{x}_i \qquad i = 1, \dots, n. \tag{13}$$

On the other hand, we obtain

$$\frac{\partial U}{\partial x_i} = u_i(x_i - s_i) \qquad i = 1, \dots, n \tag{14}$$

where the coefficients u_i were defined in (5), and therefore

$$x_i' = (1 - \gamma u_i)x_i + \gamma u_i s_i \qquad i = 1, \dots, n. \tag{15}$$

The full dynamics associated with the cost function $W = V + U$ acts iteratively according to

$$x_i' = (1 - \gamma (v_i + u_i))x_i + +\gamma v_i \bar{x}_i + \gamma u_i s_i \qquad i = 1, \dots, n \tag{16}$$

and the decision maker i is in dynamical equilibrium, in the sense that $x_i' = x_i$, if the following stability equation holds,

$$x_i = (v_i \bar{x}_i + u_i s_i)/(v_i + u_i) \qquad i = 1, \dots, n \tag{17}$$

that is, if the present opinion x_i coincides with an appropriate weighted average of the original opinion s_i and the average opinion value $\bar{x}_i$ for $i = 1, \dots, n$.

3 The Multidistance Framework

The definition of multidistance has been introduced by Martín and Mayor in [43, 44] as an extension of the classical notion of binary distance to the case of more than two points.

Consider a domain $X \subseteq \mathbb{R}$, with points in X^n being denoted as $x = (x_1, \ldots, x_n)$. The multidistance definition given in [44] is as follows.

Definition 1 Given a domain $X \subseteq \mathbb{R}$, a *multidistance* is a function

$$D : \bigcup_{n \geq 2} X^n \to \mathbb{R}$$

with the following properties

(P1) $D(x_1, \ldots, x_n) = 0$ if and only if $x_i = x_j$ for all $i, j = 1, \ldots, n$
(P2) $D(x_1, \ldots, x_n) = D(x_{\pi(1)}, \ldots, x_{\pi(n)})$ for any permutation π of $1, \ldots, n$
(P2) $D(x_1, \ldots, x_n) \leq D(x_1, y) + \cdots + D(x_n, y)$ for all $y \in X$

for all $x_1, \ldots, x_n \in X$ and $n \geq 2$. Note that (P1), (P2) and (P3) extend the usual distance axioms. In particular, (P3) generalizes the triangle inequality.

An important class of multidistances, the functionally expressible multidistances, are studied in [45, 48]. Applications of multidistances to the problem of consensus measuring can be found in [16, 22].

Starting from the results obtained in [13, 14] in [16] some connections between m-ary adjacency relations and multidistances were highlighted. It has been shown how m-ary adjacency relations can be modeled on the basis of OWA-based multidistances, and some consensus related optimization problems on m-ary adjacency relations are equivalent to corresponding multidistance minimization problems.

In this paper, a multidistance dissensus measure is introduced as an extension of the relationship between the dissensus measure in the traditional soft consensus model proposed in [27] and the multidistance approach to consensus introduced in [16]. This measure is based on a binary distance defined by means of a subadditive function whose effect is that of emphasizing small distances and attenuating large distances.

There are several methods to construct multidistances. As suggested in [44] given a binary distance $d(x_i, x_j)$, a multidistance may be defined on the basis of the pairwise binary distances, multiplying their sum by a sufficiently small value $\lambda(n)$ depending on n. This type of multidistance is called *sum–based multidistance*.

Proposition 1 (Martín and Mayor [44]) *A function* $D : \bigcup_{n \geq 2} X^n \to \mathbb{R}$ *defined as*

$$D(x_1, \ldots, x_n) = \lambda(n) \sum_{i,j=1}^{n} d(x_i, x_j) \qquad n \geq 2$$

is a multidistance if and only if the coefficient $\lambda(n)$ satisfies $\lambda(2) = 1/2$ and

$$0 < \lambda(n) \le \frac{1}{2(n-1)} \qquad n \ge 3 \tag{18}$$

where $x_1, \ldots, x_n \in X$.

In this paper we use the domain $X = [0, 1]$ equipped with the classical distance $d(x, y) = |x - y| \in [0, 1]$, for $x, y \in [0, 1]$, with the usual triangular inequalities $|x + y| \le |x| + |y|$ and $d(x, y) \le d(x, z) + d(y, z)$, for all $x, y, z \in [0, 1]$.

Moreover, we consider the particular coefficient choice

$$\lambda(n) = \frac{1}{n(n-1)} \tag{19}$$

which corresponds to constructing the sum-based multidistance by averaging pairwise binary distances.

Consider now an increasing and subadditive function $f : [0, 1] \to \mathbb{R}$. Due to the subadditivity of the function f, the composition of the distance d with the function f yields a new distance denoted $d_f(x, y) = f(d(x, y))$, which satisfies the triangle inequality $d_f(x, y) \le d_f(x, z) + d_f(y, z)$. This is obtained as follows,

$$d(x, y) \le d(x, z) + d(y, z) \tag{20}$$

$$f(d(x, y)) \le f(d(x, z) + d(y, z)) \le f(d(x, z)) + f(d(y, z)) \tag{21}$$

where the first inequality is due to the increasingness of f and the second inequality is due to the subadditivity of f. Finally, we obtain

$$d_f(x, y) \le d_f(x, z) + d_f(y, z). \tag{22}$$

We consider the construction of multidistances based on the binary distance d_f, in particular by averaging pairwise binary distances. In this way we define a multidistance $D_f : \bigcup_{n \ge 2}[0, 1]^n \to \mathbb{R}$ as

$$D_f(x_1, \ldots, x_n) = \frac{1}{n(n-1)} \sum_{i,j=1}^{n} d_f(x_i, x_j). \tag{23}$$

Consider for instance the case $n = 3$. The multidistance D_f is given by

$$D_f(x) = \frac{1}{6} \sum_{i,j=1}^{3} d_f(x_i, x_j) = \tag{24}$$

$$= \frac{1}{3}\left(d_f(x_1, x_2) + d_f(x_1, x_3) + d_f(x_2, x_3) \right) =$$

$$= \frac{1}{3}\Big(f(d(x_1, x_2)) + f(d(x_1, x_3)) + f(d(x_2, x_3))\Big).$$

Notice that each binary distance term is of the form

$$d_f(x_i, x_j) = c_f(x_i, x_j) \cdot d(x_i, x_j) \qquad (25)$$

where each classical binary distance $d(x_i, x_j)$ is multiplied by a coefficient

$$c_f(x_i, x_j) = f(d(x_i, x_j))/d(x_i, x_j) \qquad (26)$$

depending on the choice of the function f.

The multidistance D_f corresponds to a linear combination of the classical binary distances d, with non negative coefficients c_f. However, notice that these coefficients do not have unit sum and therefore the multidistance D_f does not correspond to a weighted mean of of the classical binary distances d.

4 The Soft Dissensus Measure in the Multidistance Framework

The traditional soft consensus model in group decision making [27, 28] is based on a non linear dissensus measure whose role is that of emphasizing small distances and attenuating large distances in the preference domain.

In this section we reformulate the soft dissensus measure as a sum-based multidistance, in the approach introduced in [12]. In the new multidistance framework, the usual binary distance is composed with a non linear subadditive function $f : [0, 1] \to \mathbb{R}$ defined as

$$f(u) = \frac{2}{\alpha} \ln\left(\frac{1 + e^{\alpha\beta}}{1 + e^{-\alpha(u-\beta)}}\right) \qquad \alpha \in (0, \infty) \quad \beta \in [0, 1] \qquad (27)$$

for all $u \in [0, 1]$. The two parameters are $\alpha \in (0, \infty)$ and $\beta \in [0, 1]$, but we can extend the domain of the former by defining $f(u) = u$ for $\alpha = 0$, which in fact corresponds to the asymptotic form of definition (27) at $\alpha = 0^+$.

In Fig. 1 we plot the function $f(u)$ with $u \in [0, 1]$ for various choices of the parameters α, β. In each plot the diagonal line $f(u) = u \in [0, 1]$ is associated with the case $\alpha = 0$.

In the following result we determine the values of the function f at the boundaries of its domain $[0, 1]$, for any choice of the parameters α, β.

Proposition 2 *In relation with the function f defined in (27), we obtain $f(0) = 0$ for any choice of the parameters $\alpha \in (0, \infty)$ and $\beta \in [0, 1]$, and*

- $f(1) < 1$ *for all $\beta \in [0, 1/2)$*

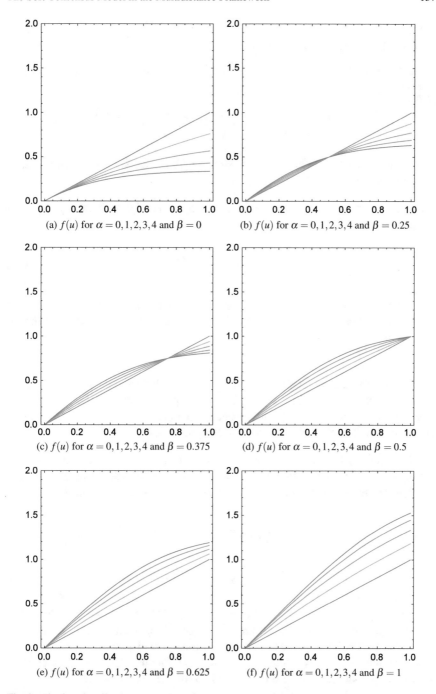

Fig. 1 The function $f(u)$ for $u \in [0, 1]$. Each plot is associated with a choice of the parameter β and shows the graph of $f(u)$ for various choices of the parameter α

- $f(1) = 1$ *for* $\beta = 1/2$
- $f(1) > 1$ *for all* $\beta \in (1/2, 1]$

for any choice of the parameter $\alpha \in (0, \infty)$. *In particular, the values of* $f(1)$ *for* $\beta = 0, 1$ *are*

$$f(1) = \frac{2}{\alpha} \ln \left(\frac{2}{1 + e^{-\alpha}} \right), \; \beta = 0 \qquad f(1) = \frac{2}{\alpha} \ln \left(\frac{1 + e^{\alpha}}{2} \right), \; \beta = 1 .$$

The limit at $\alpha = 0^+$ *is* $f(1) = 1$ *in both cases* $\beta = 0, 1$, *whereas the limit at* $\alpha = \infty$ *is* $f(1) = 0$ *for* $\beta = 0$ *and* $f(1) = 2$ *for* $\beta = 1$.

Proof From definition (27) we obtain immediately that $f(0) = 0$ for any choice of the parameters α, β, plus also

$$f(1) = 1 + \frac{2}{\alpha} \ln \left(\frac{1 + e^{\alpha\beta}}{e^{\alpha/2} + e^{\alpha\beta} e^{-\alpha/2}} \right) \tag{28}$$

which leads immediately to $f(1) = 1$ for $\beta = 1/2$. Otherwise, writing $N = 1 + e^{\alpha\beta}$ for the numerator and $D = e^{\alpha/2} + e^{\alpha\beta} e^{-\alpha/2}$ for the denominator of the logarithm, it follows that

$$N - D = (1 - e^{-\alpha/2})(e^{\alpha\beta} - e^{\alpha/2}) . \tag{29}$$

Considering the second factor in the product, we conclude that the logarithmic term in (29) is negative ($N < D$) for $\beta \in [0, 1/2)$ and is positive ($N > D$) for $\beta \in (1/2, 1]$. The asymptotic limits of $f(1)$ with respect to the parameter α for $\beta = 0, 1$ can be obtained straightforwardly by means of l'Hospital's rule. $\qquad\qquad\square$

The function f is continuous, strictly increasing and strictly concave in $u \in [0, 1]$ for any choice of the parameters α, β. Continuity is clear from definition (27) and the other properties follow directly from the first and second derivatives of f,

$$f'(u) = \frac{2}{1 + e^{\alpha(u-\beta)}} \qquad f'(u) \in (0, 2) \tag{30}$$

$$f''(u) = -\frac{2\alpha e^{\alpha(u+\beta)}}{(e^{\alpha\beta} + e^{\alpha u})^2} \qquad f''(u) \in (-\alpha/2, 0) \tag{31}$$

Notice that $f'(u = \beta) = 1$ and $f''(u = \beta) = -\alpha/2$. Moreover, we can show that $f''(u) = -\alpha f'(u)(2 - f'(u))/2$ for any choice of the parameters α, β. In the case $\alpha = 0$, for any choice of β, we have the linear form $f(u) = u$ for all $u \in [0, 1]$.

Given that f is (strictly) increasing and $f(0) = 0$ for any choice of the parameters α, β, we have that $f(u) \geq 0$ for all $u \in [0, 1]$. Moreover, we can write

$$f(u) = 2u + \frac{2}{\alpha} \ln \left(\frac{1 + e^{\alpha\beta}}{e^{\alpha u} + e^{\alpha\beta}} \right) \tag{32}$$

for all $u \in [0, 1]$, where the logarithmic term is always non positive. Therefore, we obtain $0 \leq f(u) \leq 2u$ for all $u \in [0, 1]$ and any choice of the parameters α, β.

The function f is subadditive, in the sense that $f(u + v) \leq f(u) + f(v)$. The proof is as follows: assuming $u, v \in [0, 1]$ and $u + v \neq 0$, concavity of f implies

$$f(u) \geq \frac{v}{u + v} f(0) + \frac{u}{u + v} f(u + v) = \frac{u}{u + v} f(u + v) \tag{33}$$

$$f(v) \geq \frac{u}{u + v} f(0) + \frac{v}{u + v} f(u + v) = \frac{v}{u + v} f(u + v) \tag{34}$$

and therefore we obtain $f(u) + f(v) \geq f(u + v)$ for $u, v \in [0, 1]$.

The composition of the distance d with the subadditive function f yields a new distance denoted

$$d_f(x, y) = f(d(x, y)) \tag{35}$$

satisfying the triangle inequality $d_f(x, y) \leq d_f(x, z) + d_f(y, z)$ as in (20)–(22).

We define the multidistance D_f by averaging pairwise binary distances d_f,

$$D_f(\boldsymbol{x}) = \frac{1}{n(n - 1)} \sum_{i,j=1}^{n} d_f(x_i, x_j) = \frac{1}{n(n - 1)} \sum_{i,j=1}^{n} f(d(x_i, x_j)) . \tag{36}$$

This sum–based multidistance is a natural nonlinear measure of dissensus, analogous but not equivalent to the traditional soft dissensus measure V in (9). The new soft dissensus measure, however, has a more appealing geometrical interpretation as a multidistance.

Finally, recall that each term in (36) is of the form $d_f(x_i, x_j) = c_f(x_i, x_j) \cdot d(x_i, x_j)$, where each binary distance $d(x_i, x_j)$ is multiplied by a coefficient $c_f(x_i, x_j) = f(d(x_i, x_j))/d(x_i, x_j)$ depending on the choice of the function f.

In other words, each term is of the form $f(u) = (f(u)/u) \cdot u$, where each single binary distance u is multiplied by a coefficient $f(u)/u$ which is decreasing with respect to the distance u.

In Fig. 2 we plot the function $f(u)/u$ with various choices of the parameters α and β. In each plot the horizontal line is associated with $\alpha = 0$ and the remaining lines are associated with $\alpha = 1, 2, 3, 4$. In the case $\alpha = 0$ all pairwise distances have the same weight $1/6$ and thus the multidistance corresponds to the weighted average. In the case $\alpha = 1, 2, 3, 4$ the function $f(u)/u$ is monotonically decreasing with respect to pairwise distances, in which a larger weight is assigned to a small distance and a smaller weight is given to a large distance.

The multidistance is defined as the weighted sum of pairwise distances. In this approach, the sum-based multidistance is closely related with the disensus measure in the soft consensus model. There is essentially a single difference: the basic pairwise distance $d_f(x, y)$ involves $|x - y|$ and not $(x - y)^2$, which is not a binary distance.

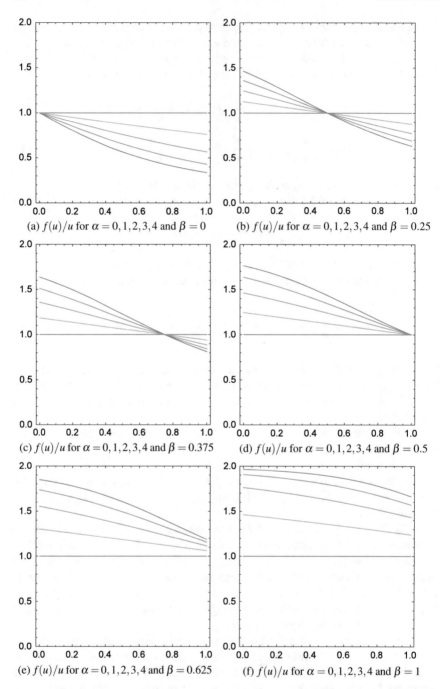

Fig. 2 The function $f(u)/u$ for $u \in [0, 1]$. Each plot is associated with a choice of the parameter β and shows the graph of $f(u)/u$ for various choices of the parameter α

5 Conclusions

We introduce a multidistance measure of dissensus within the framework of the soft consensus model of group decision making. The multidistance dissensus measure is based on a fundamental binary distance d_f associated with a subadditive function f over the domain $X = [0, 1]$, with $d_f(x, y) = f(|x - y|)$. This subadditive function has the effect of emphasizing small distances and attenuating large distances, in analogy with the subadditive scaling function g which plays a central role in the traditional soft consensus model [27, 28, 30].

References

1. Alonso, S., Pérez, I.J., Cabrerizo, F.J., Herrera-Viedma, E.: A linguistic consensus model for Web 2.0 communities. Appl. Soft Comput. **13**, 149–157 (2013)
2. Beliakov, G., Bustince Sola, H., Calvo, T.: A practical guide to averaging functions, studies in fuzziness and soft computing, vol. 329. Springer, Heidelberg (2016)
3. Beliakov, G., Calvo, T., James, S.: Consensus measures constructed from aggregation functions and fuzzy implications. Knowl.-Based Syst. **55**, 1–8 (2014)
4. Beliakov, G., Gagolewski, M., James, S.: Penalty-based and other representations of economic inequality. Int. J. Uncertainty Fuzziness Knowl.-Based Syst. **24**(1), 1–23 (2016)
5. Beliakov, G., James, S.: Unifying approaches to consensus across different preference representations. Appl. Soft Comput. **35**, 888–897 (2015)
6. Beliakov, G., James, S., Wilkin, T.: Aggregation and consensus for preference relations based on fuzzy partial orders. Fuzzy Opti. Decis. Making (online since 23 November 2016)
7. Beliakov, G., Pradera, A., Calvo, T.: Aggregation functions: a guide for practitioners, studies in fuzziness and soft computing, vol. 221. Springer, Heidelberg (2007)
8. Ben-Arieh, D., Chen, Z.: Linguistic group decision-making: opinion aggregation and measures of consensus. Fuzzy Optim. Decis. Making **5**(4), 371–386 (2006)
9. Ben-Arieh, D., Chen, Z.: Linguistic-labels aggregation and consensus measure for autocratic decision making using group recommendations. IEEE Trans. Syst. Man Cybern. **36**(3), 558–568 (2006)
10. Ben-Arieh, D., Easton, T.: Multi-criteria group consensus under linear cost opinion elasticity. Decis. Support Syst. **43**(3), 713–721 (2007)
11. Ben-Arieh, D., Easton, T., Evans, B.: Minimum cost consensus with quadratic cost functions. IEEE Trans. Syst. Man Cybern. **39**(1), 210–217 (2008)
12. Bortot, S., Fedrizzi, M., Fedrizzi, M., Marques Pereira, R.A.: A multidistance approach to consensus modeling. In: Collan, M., Fedrizzi, M., Kacprzyk, J. (eds.) Fuzzy Technol. Stud. Fuzziness Soft Comput. vol. 335, pp. 103–114. Springer, Heidelberg (2016)
13. Brunelli, M., Fedrizzi, M.: A fuzzy approach to social network analysis. In: 2009 International Conference on Advances in Social Network Analysis and Mining, pp. 225–230. IEEE Computer Society (2009)
14. Brunelli, M., Fedrizzi, M., Fedrizzi, M.: OWA-based fuzzy m-ary adjacency relations in social network analysis. In: Yager, R.R., Kacprzyk, J., Beliakov, G. (eds.) Recent Developments in the Ordered Weighted Averaging Operators: Theory and Practice, Studies in Fuzziness and Soft Computing, vol. 265, pp. 255–267. Springer, Heidelberg (2011)
15. Brunelli, M., Fedrizzi, M., Fedrizzi, M.: Fuzzy m-ary adjacency relations in social network analysis: optimization and consensus evaluation. Inf. Fusion **17**, 36–45 (2014)
16. Brunelli, M., Fedrizzi, M., Fedrizzi, M., Molinari, F.: On some connections between multidistances and valued m-ary adjacency relations. In: Greco, S., Bouchon-Meunier, B.,

Coletti, G., Fedrizzi, M., Matarazzo, B., Yager, R.R. (eds.) Advances in Computational Intelligence, Communications in Computer and Information Science, vol. 297, pp. 201–207. Springer, Heidelberg (2012)

17. Bustince, H., Barrenechea, E., Calvo, T., James, S., Beliakov, G.: Consensus in multi-expert decision making problems using penalty functions defined over a Cartesian product of lattices. Inf. Fusion **17**, 56–64 (2014)

18. Bustince, H., Beliakov, G., Pereira Dimuro, G., Bedregal, B., Mesiar, R.: On the definition of penalty functions in data aggregation. Fuzzy Sets and Syst. (online since 21 September 2016)

19. Cabrerizo, F.J., Moreno, J.M., Pérez, I.J., Herrera-Viedma, E.: Analyzing consensus approaches in fuzzy group decision making: advantages and drawbacks. Soft Comput. **14**(5), 451–463 (2010)

20. Calvo, T., Beliakov, G.: Aggregation functions based on penalties. Fuzzy sets and Syst. **161**(10), 1420–1436 (2010)

21. Calvo, T., Kolesárová, A., Komorníková, M., Mesiar, R.: Aggregation operators: properties, classes and construction methods. In: Calvo, T., Mayor, M., Mesiar, R. (eds.) Aggregation operators: new trends and applications, studies in fuzziness and soft computing, vol. 97, pp. 3–104. Springer, Heidelberg (2002)

22. Calvo, T., Martín, J., Mayor, G.: Measures of disagreement and aggregation of preferences based on multidistances. In: Greco, S., Bouchon-Meunier, B., Coletti, G., Fedrizzi, M., Matarazzo, B., Yager, R.R. (eds.) Advances in Computational Intelligence, Communications in Computer and Information Science, vol. 300, pp. 549–558. Springer, Heidelberg (2012)

23. Chen, Z., Ben-Arieh, D.: On the fusion of multi-granularity linguistic label sets in group decision making. Comput. Ind. Eng. **51**, 526–541 (2006)

24. Chiclana, F., Tapia García, J.M., del Moral, M., Herrera-Viedma, E.: A statistical comparative study of different similarity measures of consensus in group decision making. Inf. Sci. **221**, 110–123 (2013)

25. Contreras, I.: A distance-based consensus model with flexible choice of rank-position weights. Group Decis. Negot. **19**(5), 441–456 (2010)

26. Cook, W.D.: Distance-based and ad hoc consensus models in ordinal preference ranking. Eur. J. Oper. Res. **172**(2), 369–385 (2006)

27. Fedrizzi, M., Fedrizzi, M., Marques Pereira, R.A.: Soft consensus and network dynamics in group decision making. Int. J. Intel. Syst. **14**(1), 63–77 (1999)

28. Fedrizzi, M., Fedrizzi, M., Marques Pereira, R.A.: Consensus modelling in group decision making: dynamical approach based on fuzzy preferences. New Math. Nat. Comput. **3**(2), 219–237 (2007)

29. Fedrizzi, M., Fedrizzi, M., Marques Pereira, R.A., Brunelli, M.: Consensual dynamics in group decision making with triangular fuzzy numbers. In: 41st Annual Hawaii International Conference on System Sciences, pp. 70–78. IEEE Computer Society (2008)

30. Fedrizzi, M., Fedrizzi, M., Marques Pereira, R.A., Brunelli, M.: The dynamics of consensus in group decision making: investigating the pairwise interactions between fuzzy preferences. In: Greco, S., Marques Pereira, R.A., Squillante, M., Yager, R.R., Kacprzyk, J (eds.) Preferences and Decisions: Models and Applications, Studies in Fuzziness and Soft Computing, vol. 257, pp. 159–182. Springer, Heidelberg (2010)

31. Fedrizzi, M., Kacprzyk, J., Nurmi, H.: Consensus degrees under fuzzy majorities and fuzzy preferences using OWA (ordered weighted average) operators. Control Cybern. **22**(4), 71–80 (1993)

32. Fedrizzi, M., Pasi, G.: Fuzzy logic approaches to consensus modelling in group decision making. In: Ruan, D., Hardeman, F., van der Meer, K. (eds.) Intelligent Decision and Policy Making Support Systems, Studies in Computational Intelligence, vol. 117, pp. 19–37. Springer, Heidelberg (2008)

33. Fodor, J., Roubens, M.: Fuzzy Preference Modelling and Multicriteria Decision Support, Theory and Decision Library, series D, vol. 14. Kluwer Academic Publishers, Dordrecht (1994)

34. Grabisch, M., Marichal, J.L., Mesiar, R., Pap, E.: Aggregation functions, encyclopedia of mathematics and its applications, vol. 127. Cambridge University Press (2009)

35. Herrera-Viedma, E., Cabrerizo, F.J., Kacprzyk, J., Pedrycz, W.: A review of soft consensus models in a fuzzy environment. Inf. Fusion **17**, 4–13 (2014)
36. Herrera-Viedma, E., García-Lapresta, J.L., Kacprzyk, J., Fedrizzi, M., Nurmi, H., Zadrożny, S.: Consensual Processes, Studies in Fuzziness and Soft Computing, vol. 267. Springer, Heidelberg (2011)
37. Kacprzyk, J., Fedrizzi, M.: Soft consensus measures for monitoring real consensus reaching processes under fuzzy preferences. Control Cybern. **15**(3–4), 309–323 (1986)
38. Kacprzyk, J., Fedrizzi, M.: A soft measure of consensus in the setting of partial (fuzzy) preferences. Eur. J. Oper. Res. **34**(3), 316–325 (1988)
39. Kacprzyk, J., Fedrizzi, M.: A human-consistent degree of consensus based on fuzzy login with linguistic quantifiers. Math. Soc. Sci. **18**(3), 275–290 (1989)
40. Kacprzyk, J., Fedrizzi, M., Nurmi, H.: Group decision making and consensus under fuzzy preferences and fuzzy majority. Fuzzy Sets Syst. **49**(1), 21–31 (1992)
41. Kacprzyk, J., Nurmi, H., Fedrizzi, M.: Consensus Under Fuzziness, International Series in Intelligent Technologies, vol. 10. Kluwer Academic Publishers, Dordrecht (1997)
42. Martın, J., Mayor, G.: How separated Palma, Inca and Manacor are? In: Proceedings of the 5th International Summer School of Aggregation Operators AGOP 2009, pp. 195–200. Palma de Mallorca, Spain (2009)
43. Martín, J., Mayor, G.: Some properties of multi-argument distances and fermat multidistance. In: Hllermeier, E., Kruse, R., Hoffmann, F. (eds.) Information Processing and Management of Uncertainty in Knowledge-Based Systems. Theory and Methods, Communications in Computer and Information Science, vol. 80, pp. 703–711. Springer, Heidelberg (2010)
44. Martín, J., Mayor, G.: Multi-argument distances. Fuzzy Sets Syst. **167**(1), 92–100 (2011)
45. Martín, J., Mayor, G., Valero, O.: Functionally expressible multidistances. In: S. Galichet, J.M.G. Mauris (eds.) Proceedings of the 7th Conference of the European Society for Fuzzy Logic and Technology (EUSFLAT-2011) and LFA-2011, Advances in Intelligent Systems Research, pp. 41–46. Atlantis Press, Amsterdam (2011)
46. Merigó, J.M., Casanovas, M.: Decision-making with distance measures and induced aggregation operators. Comput. Ind. Eng. **60**, 66–76 (2011)
47. Meskanen, T., Nurmi, H.: Distance from consensus: a theme and variations. In: Simeone, B., Pukelsheim, F. (eds.) Mathematics and Democracy, Studies in Choice and Welfare, pp. 117–132. Springer, Heidelberg (2006)
48. Molinari, F.: About a new family of multidistances. Fuzzy Sets Syst. **195**, 118–122 (2012)
49. Palomares, I., Martínez, L., Herrera, F.: A consensus model to detect and manage noncooperative behaviors in large-scale group decision making. IEEE Trans. Fuzzy Syst. **22**(3), 516–530 (2014)
50. Parreiras, R.O., Ekel, P.Y., Martini, J.S.C., Palhares, R.M.: A flexible consensus scheme for multicriteria group decision making under linguistic assessments. Inf. Sci. **180**, 1075–1089 (2010)
51. Xu, Z.: Fuzzy ordered distance measures. Fuzzy Optim. Decis Making **11**(1), 73–97 (2012)
52. Xu, Z., Chen, J.: Ordered weighted distance measure. J. Syst. Sci. Syst. Eng. **17**(4), 432–445 (2008)
53. Yu, L., Lai, K.K.: A distance-based group decision-making methodology for multi-person multi-criteria emergency decision support. Decis. Support Syst. **51**, 307–315 (2011)

Fuzzy Multi-criteria Decision Making Methods Applied to Usability Software Assessment: An Annotated Bibliography

Yamilis Fernández-Pérez, Ailyn Febles-Estrada, Carlos Cruz and José Luis Verdegay

Abstract The software impacts every day in all aspects of our life. Therefore it is essential to assess the impact on the sustainability of any aspect of its application. A conscious and responsible use of resources is needed. One feature that is associated with the term sustainable software is the usability. It is very usual to define usability as a software ease of use, but this definition is ambiguous. For this reason, there are several definitions according to different approaches to measure it. In the last decades, a lot of papers have been published about software usability assessment models. These models have evolved from the use of algorithms, conventional statistics techniques, and Soft Computing methods. The goal of this work is to offer a collection of annotated bibliography about the characteristics, attributes and metrics used in usability assessment, focused in the Soft Computing techniques.

1 Introduction

Computer and software have become indispensable parts in our daily life. The intensive use of different types of software is increasing. Therefore the environmental impact of any aspect of its application is a current problem. Venters et. al. defined software sustainability as a composite, non-functional requirement which is a measure of a systems extensibility, interoperability, maintainability, portability, reusability, scalability, and usability. Because of the complexity of this feature and

Y. Fernández-Pérez · A. Febles-Estrada
Universidad de las Ciencias Informáticas, Havana, Cuba
e-mail: yamilisf@uci.cu

A. Febles-Estrada
e-mail: ailyn@uci.cu

C. Cruz · J. L. Verdegay (✉)
Universidad de Granada, Granada, Spain
e-mail: verdegay@decsai.ugr.es

C. Cruz
e-mail: carloscruz@decsai.ugr.es

© Springer International Publishing AG 2018
C. Berger-Vachon et al. (eds.), *Complex Systems: Solutions and Challenges in Economics, Management and Engineering*, Studies in Systems, Decision and Control 125, https://doi.org/10.1007/978-3-319-69989-9_11

its importance in the sustainability of the software, this chapter is devoted to the analysis of usability evaluation.

Usability is one of the most important attributes of Software Quality. It is very usual to define usability as a software ease of use, but this definition is ambiguous. For this reason, there are several definitions according to different approaches to measure it. Best known definitions appear in: ISO 9126, ISO 25000 and ISO 9241. Usability is defined in ISO 9241-11 as "the extent to which a product can be used by specified users to achieve specified goals with effectiveness, efficiency and satisfaction in a specified context of use." On the other hand, ISO 9126-1 defines usability as "the capability of the software product to be understood, learned, used and attractive to the user, when used under specified conditions." It also conceptualizes quality in use, as the result of the combined effects of the six categories of software quality when the product is used. Quality in use is defined as: "the capability of the software product to enable specified users to achieve specified goals with effectiveness, productivity, safety and satisfaction in a specified context of use". It is similar to the concept given by ISO 9241, except that safety category is added. In the last decades, a lot of papers have been published about software usability assessment. So it could be interesting and useful to research community to make an analysis about these assessment usability models taking into account many aspects which characterize the usability assessment process, such as:

- The goal of calculation of the usability is to isolate the specific problems and to obtain a global value.
- In this process different experts participate to determine criteria of preferences and those involved in carrying out usability testing.
- The collected information during the process is heterogeneous, ambiguous, and it comes from different sources such as experts, final users and clients.
- The criteria used for usability assessment are interdependent from each other.
- There is an inherent uncertainty in these assessment models.

Usability has two sides, one side relates to objective measures of interaction and the other side relates to a subjective perception of the product. Assigning numerical values for different usability subjective criteria will always be questionable and weak, due to the fact that it is necessary to use Fuzzy Logic. The use of Fuzzy Logic reduces the ambiguity in the assignation of values from parameters. The combination of different methods with Fuzzy Logic is the major trend in assessment usability models. These models have evolved from the use of algorithm and conventional statistics techniques to the use of Soft Computing ones. Among the Soft Computing techniques used to assess the usability the ones that have been developed are fuzzy inference system, fuzzy integral and in the last years, the use of integration of fuzzy logic and Multi Criteria Decision Methods (MCDM) has become the most outstanding one. The main research papers published about the topic are Fuzzy Analytic Hierarchy Process (AHP), Fuzzy Technique for Order of Preference by Similarity to Ideal Solution (TOPSIS) or their derivation. The search for a solution which will fulfill every aspect of usability assessment process is still a challenge. The goal of this work is to offer a collection of annotated bibliography about the characteristics, attributes and metrics used in usability assessment focused in the Soft Computing techniques mentioned before. For that reason there have been reviewed and commented more than 60 publications, which appear in magazines, books, papers of congress and others. The first step in constructing the review was to define what to search, where the search should be conducted and how it should be defined (search terms). The search was conducted using databases such as Springer, IEEExplore Digital Library, Science Direct, ACM digital library, Scopus and Google Scholar, using terms like usability evaluation, usability metrics and Soft Computing. Authors expect this work to be a guide for students and researchers of the subject. The contribution is organized as follows: Section 2 describes the used methodology to select the discussed literature. Section 3 describes the annotated bibliography emphasizing the criteria and the techniques used to measure the usability of software as well as detail how to add the information, if uncertainty is present and whether information which handled is heterogeneous. Finally, in Sect. 4 a brief summary, conclusions and some suggestions for future research are provided.

2 Bibliography Search and Repository Details

The first step in constructing the review was to define what to search, where the search should be conducted and how the search should be defined (search terms). The search was conducted using the databases described in Table 1, using terms like usability evaluation, usability metrics and this search was refined using terms like fuzzy and Soft Computing.

In this research, several papers were considered as well as book chapters and conference papers and the investigations that include the criteria and techniques to measure usability were selected. Finally 60 papers were thoroughly revised. With the purpose of studying the evolution of assessment usability models, the date of the works revised covers from the end of the last century up to nowadays. The distribution of the papers per years appears in Fig. 1.

The study include the following aspects: evolution of the concept of usability, criteria, metrics and techniques used to measure it, determining if it is possible to treat

Table 1 Search engines where the bibliography was collected

Search databases	URL
Springer	https://www.link.springer.com
IEEExplore digital library	https://www.ieeexplore.ieee.org
Science direct	https://www.sciencedirect.com
ACM digital library	https://www.dl.acm.org
Google scholar	https://www.scholar.google.com
Scopus	https://www.scopus.com

Fig. 1 Number of revised
works that measure usability
per year of publication

■ Decade 90   ■ 2000 - 2005   ■ 2006 - 2010   ■ 2011 - 2015

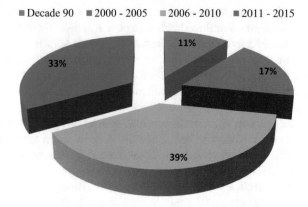

the uncertainty, if the information manipulated is heterogeneous or not and the way to add it. The criteria used to measure usability were the ones organized in hierarchized models, which have been enriched by many authors with several perspectives, using different criteria divided in sub criteria in a low level till reaching the leave node with the necessary measures that calculates an element of the top level. On the other hand, different techniques to assess Usability have been developed. In the analyzed bibliography different aspects to assess these techniques were considered such as:

- The information is heterogeneous, ambiguity and comes from different sources such as: experts in usability, final users and clients. The nature of the criteria involved in the assessment is both qualitative and quantitative.
- The criteria are interdependent, there are relationships among them.
- The experts participation to determine the preference of criteria and the staffs participation in the usability testing.
- The uncertainty inherent to these assessment models.
- The goal of calculating the usability is to isolate the specific problems and obtain a global usability value.

The different steps to carry out the assessment process were analyzed for covering:

- The definition of the problem and the determination of the assessment goal.
- The selection of the criteria, the hierarchical model, the measures and the tools to be used.
- The selection of the staff participating in the process (the usability expert and the testing staff).
- The assignation of weight for every criterion.
- The development of the usability test and other methods to generate a wide data set with the criteria values for every alternative.

- The application of the technique, adding the stored information to obtain the partial values of the criteria and a global usability value.
- The analysis of usability assessment results.

Starting from these aspects, the different techniques included in the bibliography consulted were analyzed.

The papers and their annotations are presented in chronological order in Sect. 3. In this way, the evolution of concepts and techniques used for usability assessment is shown. In addition, it shows the evolution of Soft Computing techniques used to assess the usability and also the use of integration of fuzzy logic and Multi Criteria Decision Methods (MCDM)

All the bibliography is organized using a free reference manager and academic social network known as Mendeley. In this same social network the published papers consulted may be found thus the correspondent notes and detailed notes in every paper.

3 Annotated Bibliography

N. Beyan. The MUSIC methodology for usability measurement. *Posters and Short Talks of the 1992 SIGCHI Conference on Human Factors in Computing Systems. ACM*, pages 123–124, 1992.

This paper presents the project Metrics for Usability Standards in Computing (MUSIC). This project is based on ISO 9241. It works with criteria that are crisp numbers, for that reason the value of some usability features will be inconsistent and questionable. It presents (Diagnostic Recorder for Usability Measurement) DRUM tool. This model can be classified among those that use conventional techniques.

J. Nielsen. *Usability Engineering.* Academic press, San Diego, CA., 1993.

It is a basic book for usability. It presents one of the most cited models in the area of usability, known as Nielsen, which enunciates five attributes: learnability; efficiency; memorability; low error rate, easy mistake recovery and Subjective satisfaction. It makes a detail analysis of the methods of usability and decides which of them will be used in each phase of the lifecycle. The book does not include any method to aggregate information and obtains an overall value for usability and possible comparison with other products.

B. D. Harper and K. L. Norman. Improving user satisfaction: the questionnaire for user interaction satisfaction version 5.5. In *Proceedings of the 1st Annual Mid-Atlantic Human Factors Conference.*, pages 224–228., 1993.

This paper discusses the Questionnaire for User Interaction Satisfaction (QUIS). Each area measures the overall satisfaction on a 9-point scale. The questionnaire data is aggregated by calculating the mean and standard deviation from the answers given by the users of the software. It gets the correlation matrix for all measures in the short version QUIS 5.5, aspect to be analyzed to determine the relationship between the criteria. This model can be classified among those using conventional techniques.

E. J. Chang, T. S. Dillon, and D. Cook. An Intelligent System Based Usability Evaluation Metric. In *Intelligent Information Systems, IASTED International Conference on. IEEE Computer Society*, pages 218–226, 1997.

This paper measures usability through criteria such as efficiency, preventing and recovering from errors, user like/dislike, consistency and feedback system. Each criterion is characterized with different aspects. It develops a model for each of the dimensions using fuzzy set theory. It employs the Generalised Bell (G-Bell) shapes for the membership functions. It then uses the Takagi Sugeno fuzzy inference approach in developing an overall measure of usability. The paper considers a system with m fuzzy rules of TS form. This method is intended for the second order to obtain an overall value of usability. It performs tests and generates data that will be used to help form the different control rules and aggregation of the results for a value of usability. The actual tuning algorithm used was the ANFIS algorithm since it tunes the parameters of both, the input membership function as well as the output coefficients. It does not expose the relationship between criteria.

A. R. Gray and S. G. MacDonell. Fuzzy Logic for Software Metric Models Throughout the Development Life-Cycle. *The Information Science Discussion Paper Series*, 99(20), 1999.

This paper discusses the ambiguity and uncertainty of different metrics to characterize the process and the product. For each stage of the life cycle of these measures may be more ambiguous and high level of uncertainty, so the software metrics at some point, becomes difficult to determine an accurate value. The use of fuzzy logic as a useful tool in this setting is highlighted in this paper. The use of fuzzy inference systems is proposed.

E. Folmer, J. van Gurp, and J. Bosch. A framework for capturing the relationship between usability and software architecture. *Software Process: Improvement and Practice*, 8(2):67–87, 2003.

This paper presents a framework that expresses the relationship between usability and software architecture. With this framework the software architect has a guide with the aim of reducing the usability problems that can present the final product, it integrates usability patterns and usability properties set. It associates usability attributes with properties and patterns. This is a guide for designing the software architecture considering usability. This paper does not include any technique to calculate the usability.

M. A. Sicilia, E. G. Barriocanal, and T. Calvo. An inquiry-based method for Choquet integral-based aggregation of interface usability parameters. *Kybernetika*, 39(5):601–614, 2003.

In this paper, some attributes to measure usability are used. Among them it can be mentioned: efficiency (E), memorability (M), satisfaction (S) and Learnability (I).

The work also analyzes the relationship between attributes and proposes the use of Choquet Integral as aggregation operator for usability. Finally author acknowledges that the applicability of such an inquiry-based method is restricted to problems with a small amount of input criteria, and for which criteria interactions can be assessed in some uncertain but definite ways. It is a paper that gives relevance to the relationship between attributes.

M. A. Sicilia and E. García. Modelling Interacting Web Usability Criteria through Fuzzy Measures. *Web Engineering. Springer Berlin Heidelberg*, pages 182–185, 2003.

This paper is similar to the previous one; this is applied to Web Applications. It analyzes usability as a multifaceted concept that is usually evaluated in terms of a number of aspects or attributes that are not independent, but interact among themselves. Considering this characteristic, it proposes the use of the aggregation operator Choquet integral which can be used as a generalization of the weighted arithmetic average that takes into account the interaction of the criteria.

L. Mich, M. Franch, and L. Gaio. Evaluating and Designing Web Site Quality. *MultiMedia, IEEE*, 10(1):34–43, 2003.

This paper proposes a model to calculate the quality of Web sites. It proposes a model of quality which includes 7 attributes: Identity, Content, Services, Location, Management, Usability, and Feasibility. Usability is measured from the accessibility, navigability, understandability. It proposes an evaluation process into three phases: Evaluation setup, Evaluation design (see Fig. 2). Using a scale that goes from 0 to 4 for the evaluation (0 = nonexistent, 1 = poor, 2 = adequate, 3 = good, 4 = excellent). It assigned to each sub-attribute a weight from 0 to 1 to factor in its relevance. The result of each feature shows as a radial diagram. It does not get a global value of usability. Only fulfills the first objective.

W. Siu Keung, T. Thi Nguyen, E. Chang, and N. Jayaratna. Usability Metrics for E-learning. In *On The Move to Meaningful Internet Systems 2003: OTM 2003 Workshops. Springer Berlin Heidelberg*, pages 235–252, 2003.

This paper assumes the approach adopted in [11] to measure usability in e-learning system. In this case, the criteria used are E-learning System Feedback, Consistency, Error Prevention, Perfor-mance/Efficiency, User like/dislike, Error Recovery, Cognitive load, Internationalization, Privacy and On-Line Help. It uses the first order Takagi Sugeno approach for fuzzy inference. The tuning algorithms that are available seek to minimize the value of this error function. There are many different tuning algorithms for doing these. They will only discuss the three main approaches for doing this and they are: Least Square Error Optimization Approach, Gradient Method and Neural Net Method.

T. S. Tullis and J. N. Stetson. A Comparison of Questionnaires for Assessing Website Usability. *Usability Professional Association Conference*, pages 1–12, 2004.

This paper discusses a comparison among the questionnaires used to measure usability of the websites. It analyzes Questionnaire for User Interface Satisfaction (QUIS) of 1988, Computer System Usability Questionnaire (CSUQ) of 1995, System Usability Scale (SUS) of Microsoft and the questionnaire proposed by the authors of this paper. It emphasizes the use of SUS as one of the simplest questionnaires studied and produced one of the most reliable results for all sample sizes. It is very important to analyze these questionnaires widely used in industry because they are more reliable results.

E. Folmer and J. Bosch. Architecting for usability: a survey. *Journal of Systems and Software*, 70(1-2):61–78, feb 2004.

This reference is a survey of usability from the perspective of a software architect; it emphasizes the great importance of the design of usability at the architectural level, which implies greater influence on the usability of a system. It shows that there are no design techniques and assessment tools that allow the design of usability at the architectural level. This research is essential to identify usability patterns that are sensitive to the architecture and the relationship of these patterns with the usability, through a study of the relationship between patterns of usability, usability properties and usability attributes. It analyzes various techniques and methods for evaluating usability. It does not take into account the calculation of usability, it is rather a guide to the designer.

A. Zenebe and A. F. Norcio. Evaluation Framework for Fuzzy Theoretic-Based Recommender System. In *11th International Conference on Human-Computer Interaction (To Appear), Las Vegas, Nevada, USA*, 2005.

This paper shows that the assessment of usability in user model driven recommender systems is still a challenge and explains aspects that are not covered.

E. Chang and T. Dillon. A usability-evaluation metric based on a soft-computing approach. *IEEE Transactions on Systems, Man, and Cybernetics - Part A: Systems and Humans*, 36(2):356–372, mar 2006.

This work calculates software usability using the criteria: System Feedback, Consistency, Error Prevention, Performance/Efficiency, User Like/Dislike or User Opinion, Error Recovery. It develops a solution similar to [11, 57]. It uses Singletons that represents a particular case of the TakagiSugeno model. The actual tuning algorithm used was the adaptive-network-based fuzzy-inference system (ANFIS) algorithm since it tunes the parameters of both the input membership function, as well as the output coefficients. It employs the generalized Bell (G-Bell) shapes for the membership functions. Here it works with linguistic labels of two and three terms, and concludes that if what is interesting is the overall value of usability then, the use of two terms is adequate one, but if what is needed are the details of usability criteria, then the linguistic label of three terms is the appropriate one. It is relevant to consider this criterion for calculating usability.

K. Hornbæk. Current practice in measuring usability: Challenges to usability studies and research. *International Journal of Human-Computer Studies*, 64(2):79–102, feb 2006.

This paper is a review on the measures of usability. It analyzes not only measures used for usability but also if they actually measure usability, if they cover usability broadly, if they meet recommendations on how to measure usability, etc. It details the metrics used to measure effectiveness, efficiency and satisfaction, based on ISO 9241. It analyzes the challenges of usability measures as involving objective and subjective measures, the need for a more complete understanding of the evolution of usability aspects throughout a period of time. It expands measures of satisfaction beyond post-use questionnaires, and focus more on the validation and standardization of measures of satisfaction, correlation studies between measures among others. It is significant to stand out that it determines the relationships between the measures.

M. F. Bertoa, J. M. Troya, and A. Vallecillo. Measuring the usability of software components. *Journal of Systems and Software*, 79(3):427–439, mar 2006.

This paper details the metrics for usability assessment of software components. It analyzes the validity of the measures taking into account its ease of calculation, its characteristics of objectiveness and reproduction, plus correspondence with quality features like Understandability, Learnability and/or Operability of a component. It is interesting in this paper, the specification of the usability measures for software components.

N. Juristo, A. M. Moreno, and M.-I. Sanchez-Segura. Analysing the impact of usability on software design. *Journal of Systems and Software*, 80(9):1506–1516, sep 2007.

This work analyzes the usability features that directly impact the design: Feedback, Undo, Cancel, Form/Field validation, Wizard, User Expertise, Multi-level Help, Use of Different Languages, Alert, User Input Errors Prevention/Correction. It quantifies the effect of incorporating features of usability on the design. It is important the results of this paper on the design phase of software. A value of usability as a result is not obtained.

N. J. Pizzi. Software quality prediction using fuzzy integration: a case study. *Soft Computing*, 12(1):67–76, jun 2007.

This paper uses fuzzy classifiers and demonstrates the effectiveness of fuzzy integration, combining classification results from multiple classifiers. The fuzzy integration method produced to improvement in predicting software complexity compared to the best single classifier using linear discriminant analysis. This usability assessment technique is presented in a study case.

R. Costa, M. Augusta, S. Machado, D. J. Santos, R. Tavares, and C. D. Campos. Uma Aplicação da Matemática Nebulosa na Usabilidade de Pacotes Estatísticos. *RESI Revista Eletrônica de Sistemas de Informação*, 10(1):1–8, 2007.

In this paper a set of indicators are used to measure the usability of statistical packages, these indicators are the following: learnability, efficiency, effectiveness, memorability, user satisfaction and error control. This model was developed in Brazil based on their interests, taking as a basement a questionnaire completed by 18 users. It describes the results using statistical analysis and gets fuzzy triangular numbers.

Then fuzzy triangular numbers for each resulting metric and its interpretation are presented. The required calculations for the fuzzy analysis of results were carried out using the mathematical MatLab software and the results for each metric study were generated graphically. The graph shown for each metric represents two sets. The first set represents the fuzzy number of the average of all the frequencies found for the questions of the metric. The second set represents the fuzzy number in the most similar triangular shape to the first set, which is the final result for the metric evaluated. An analysis of the different factors that affect usability is made but does not give an overall value of usability.

S. Winter, S. Wagner, and F. Deissenboeck. A Comprehensive Model of Usability. *Engineering interactive systems. Springer Berlin Heidelberg*, pages 106–122, 2008.

This paper measures the usability based on ISO 9241 by adding safety. It proposes a two-dimensional model of usability that associates system properties with the activities carried out by users. It is based on a metamodel, which promotes the accuracy and completeness. The entire model is composed of the activities with attributes, the facts with the corresponding attributes and the impacts between attributed facts and attributed activities. The model with all these details is too large to be described in detail; in this paper some interesting examples are presented: triplets of an attributed fact, an attributed activity, and a corresponding impact. It is a very useful tool in the design stage. It does not measure usability.

Y.-x. Li and Z.-x. Man. A Fuzzy Comprehensive Quality Evaluation for the Digitizing Software of Ethnic Antiquarian Resources. *2008 International Conference on Computer Science and Software Engineering*, pages 1271–1274, 2008.

In this paper, the usability is calculated by ISO 9126 considering only the understandability, learnability and operability. The values of the different factors are represented by linguistic labels (Excellent, good, usually unqualified). It is adopted fuzzy statistical methods to determine their membership. Then it identifies the weight of each factor in the evaluation through applying entropy, a concept in information theory. Then what you do is to calculate by levels, the weighted sum of the ratings of each factor. Finally, the final value is multiplied by a scaling factor for each linguistic label and then a crisp value is obtained. It uses MCDM elements.

G. A. Montazer and H. Q. Saremi. An Application of Type-2 Fuzzy Notions in Website Structures Selection: Utilizing Extended TOPSIS Method. *WSEAS TRANSACTIONS on COMPUTERS*, 7(1):8–15, 2008.

This paper determines the selection of the structure of Websites, to do this, it uses the following criteria: Easy to navigate, clear layout of info, up-to-date info, search tools and accuracy of info. The ultimate goal is to improve usability. It uses the fuzzy TOPSIS method with triangular fuzzy number of type 2. In this paper the three ways are compared, using TOPSIS, fuzzy TOPSIS and using TOPSIS to fuzzy triangular type number 2. It also measures the distance from the positive and negative ideal solutions using the euclidean distances for each alternative from the ideal positive and negative solutions respectively, it calculates the relative closeness to the ideal solution and obtains the ranking of alternatives. It is interesting, the detailed explanation of each method. It shows that fuzzy TOPSIS is a useful troubleshooting method for calculating usability.

D. Tamir, O. V. Komogortsev, and C. J. Mueller. An Effort and Time Based Measure of Usability. *Proceedings of the 6th international workshop on Software quality. ACM*, pages 47–52, 2008.

This work propose metrics for measuring reliability, learnability, and understandability through effort. An effort is a physical activity performed by a person trying to achieve a specific goal. A physical activity could include pressing a key on a keyboard, move the mouse, clicking a mouse button, or the movement of a person's eyes, so associated the need physical effort to perform tasks with ease of use of software. The definition of effort is using continuous functions. In practice, given the discrete nature of human-computer interaction, these measures are quantified through the convention of integrals sums. The way to get the measures is not trivial.

O. V. Komogortsev, C. J. Mueller, D. Tamir, and L. Feldman. An Effort Based Model of Software Usability. *Proceedings of the International Conference on Software Engineering Theory and Practice (SETP-08)*, (July):1–9, 2009.

This paper is similar to the one developed by [58]. Each of the indicators presented in the paper [58] are refined. Experiments with two travel reservation systems based on Web are made. To carry out the experiment, 20 persons were asked to participate. It shows that there is a correlation between the calculated effort as proposed and usability, moreover it shows that in the case of the eye there is a physical and mental strain on it. The experiments are thoroughly explained.

B. S. Jin and Y. G. Ji. Development of a Usability Evaluation Framework with Quality Function Deployment : From Customer Sensibility to Product Design. *Human Factors and Ergonomics in Manufacturing*, 19(2):177–194, 2009.

This paper defines usability to several factors selected from the study of the standards, and integration, this selection is refined by analyzing generality, involvement, relationship, correlation factors. Getting 23 factors including understandability, Simplicity, accessibility, ability to learn, familiarity/controllable memory consistency, feedback, performance, satisfaction, and other design as buttons, labels, special fonts, etc. 30 people participated in the experiment. Questionnaires and interviews were used to assess usability. Evaluation of usability data were analyzed by statistical methods and regression analysis. Sensation was evaluated. After that it integrates data using Quality Function Deployment (QFD). It used AHP for pairwise comparison. This paper shows that in the process of usability assessment if a precise and reliable value is desired when working with many criteria and experts then, a feasible solution for usability total value is the use of MCDM.

Y. Danfeng and Y. Fangchun. Fuzzy evaluation of sla-oriented QoSM (the quality of service management) in NGN. *Proceedings of IC-BNMT2009*, 2009.

In this work authors consider the multi criteria approach. The way of performing the pairwise comparison is underlined, where the experts perform it by using comparison signs and then the values of the criteria are determined following this property: aij + aji = 1 for all i different from j. It is important to detach that the solution avoids many questions to be asked to experts for pairwise comparison, due to the fact that the act of asking too many questions to the experts is considered as limitation of methods such as Analytic Hierarchy Process (AHP) and Analytic Network Process (ANP).

M. Hub and M. Zatloukal. Usability Evaluation of Selected Web Portals 2 Case study Usability evaluation of selected Web portals. *Proceedings of the 9th WSEAS International Conference on Applied Informatics and Communications (AIC '09)*, pages 259–264, 2009.

The authors of this paper propose a method for assessing the usability of web portals and demonstrate it by evaluating the usability of information systems in public administration using fuzzy logic based on factors of accessibility, immediate understanding, information retrieval, current information, simplicity of navigation, design preference, orientation, number of graphics, and speed of loading. Users assess the criteria using natural language, due to this; it is indispensable to work with linguistic labels. Each linguistic variable has triangular membership functions, 3 linguistic states (low, medium, high) and universe of discourse in range from 0 to 100. It puts into practice a fuzzy inference system. The following three defuzzification methods were used: center of gravity (COG), height method (HM) and weighted center of area (WCA). The authors consider the COG method as the most preferable. It is one of the most detailed and reasonable solutions.

L. Liang, X. Deng, and Y. Wang. Usability Measurement Using a Fuzzy Simulation Approach. *2009 International Conference on Computer Modeling and Simulation*, pages 88–92, feb 2009.

This paper assesses the usability using ISO 9241. It uses Finite State Machine (FSM) to represent the abstract structure of the user interface. The FSM can be described by a state transition diagram (STD). The method first represents the FSM through a STD. After redundant states are reduced and extended FSM is obtained. Finally, the state transition probability can be calculated by a fuzzy mathematical formula proposed by the author based on a fuzzy metric. After that an algorithm for measuring the efficiency and effectiveness in the simulated data is applied and thus it is possible to analyze the usability. This method is appropriate to the design stage of small or medium embedded software. In a complex system is difficult to represent the STD.

A. Parvinder S. Sandhu and S. S. Priyanka Kakkar. A Survey on Software Reusability. *2010 International Conference on Mechanical and Electrical Technology (ICMET 2010)*, pages 769–773, 2010.

This paper consists of a survey of different models or measurement proposed in literature for predicting the software reutilization. Here Neuro-fuzzy inference engine and Taguchi's approach are revised for analyzing the significance of different structural attributes or factors in deciding the level of reusability component owners.

There are metrics for identifying the quality of reusable components, but the function that uses these metrics to find reuse of software components is still unclear.

M. Hub and M. Zatloukal. Model of Usability Evaluation of Web Portals Based on the Fuzzy Logic. *WSEAS Transactions on Information Science and Applications*, 7(4):522–531, 2010.

This paper is similar to [25]. It compares its results with SUS for validating them. The results evaluated by both methods show very good level of consistency. The differences might be caused by different complexity of criteria and lower precision of the SUS method.

R. Lamichhane and P. Meesad. A Usability Evaluation for Government Websites of Nepal Using Fuzzy AHP. *The 7th International Conference on Computing and Information Technology IC2IT2011*, pages 99–104, 2011.

The work considers usability as Adequacy and Quality of Information Services, Updates and Interaction, Appearance and Outline, Navigation and Accessibility. Use Fuzzy AHP to determine the value of usability. It does a fuzzification of data and builds the fuzzy judgment matrix, continued to defuzzification of the matrix based in bisect the area and get matrix with crisp numbers. It uses trapezoidal fuzzy number (TrFN) for diffuse aggregation claiming that the trapezoidal fuzzy number covers a larger fuzzy area than triangular number. The same is done for the values of the criteria and the process continues with crisp values. Once the experimental results are obtainined and compared with triangular fuzzy AHP, the results have not significant differences.

J. S. Challa, A. Paul, Y. Dada, and V. Nerella. Integrated Software Quality Evaluation : A Fuzzy Multi-Criteria Approach. *Journal of Information Processing Systems*, 7(3):473–518, 2011.

The paper determines the quality of software in general, based on ISO 9126, in the case of usability it adds reusability. It uses the fuzzy multi-criteria approach. Firstly it assigns fuzzy ratings to each and every metric that exists in the software model. After that it assigns fuzzy weights to the sub characteristics, characteristics and perspectives. Then it takes the weighted average of the metrics in different levels. Finally it takes the weighted average of the quality perspective in Level 1 (using their weights and ratings) under the corresponding perspectives to evaluate the fuzzy rating of the different perspectives in Level 0. This value has to be defuzzified by using the centroid formula to get the crisp value of the software quality.

J. S. Challa, A. Paul, Y. Dada, V. Nerella, and P. R. Srivastava. Quantification of Software Quality Parameters Using Fuzzy Multi Criteria Approach. *2011 International Conference on Process Automation, Control and Computing*, pages 1–6, jul 2011.

In this paper, as it happens in the previous one, not only the usability but the quality of software in its entirety is determined, based on the ISO 9126. It uses fuzzy multi criteria approach. The result is given as a linguistic term. It works with heterogeneous data; each value is led to a linguistic label modeled as a triangular fuzzy number. It assigns a triangular fuzzy number to the weights of the metric. At every level perspective from characteristic to sub characteristic to metric, every parameter is associated with a corresponding rating and weight. So first the fuzzy weighted average is taken of the metrics to evaluate the rating of the sub characteristic. Then the fuzzy weighted average of the sub characteristics is taken to get the rating of the characteristics. Then the fuzzy weighted average is taken of the characteristics to get the rating of the perspective. Fuzzy weighted average of different perspectives is taken to get the ultimate software quality in terms of a fuzzy set. Centroid formula is employed then a fuzzy set on this triangular end to calculate the software quality.

M. L. Etaati, S. Sadi-Nezhad, A. Using Fuzzy Analytical Network Process and ISO 9126 Quality Model in Software Selection: A case study in E-learnig Systems. *Journal of Applied Sciences*, 11(1):96–103, 2011.

This paper proposes a model for calculating the quality of software using Fuzzy Group Analytical Network Process and uses the ISO 9126. It takes into account the relationship between attributes obtained from some literature review and interview. Linguistic variables are used. It is interesting the use of ANP for calculating the quality of software in general.

H. Mittal, M. Sharma, and J. Mittal. Analysis and Modelling of Websites Quality Using Fuzzy Technique. *2012 Second International Conference on Advanced Computing & Communication Technologies*, pages 10–15, jan 2012.

This paper evaluates the quality of the Websites and it is carried out through attributes such as: load time, response time, mark-up validation, broken links, accessibility error size, page rank, frequency of update, traffic and design optimization. Each attribute is associated with linguistic variables (Low, Medium, High) these are taken as triangular fuzzy number. When the defuzzification is made, it obtains each quality factor using a function of 5 intervals linking complexity and coefficient. Finally it aggregates the factors multiplying the values that are proportional to the quality and dividing those that are inversely proportional, and multiplying the result by a scaling factor.

S. K. Dubey, A. Mittal, and A. Rana. Measurement of Object Oriented Software Usability using Fuzzy AHP. *International Journal of Computer Science and Telecommunications*, 3(5):98–104, 2012.

This paper uses two models; the first proposed by ISO 9241 and the other is adding learnability to ISO 9241. To calculate the weights and the overall value of usability Fuzzy AHP is used. Establishing Triangular Fuzzy Numbers (Lower; Middle; Upper). Where Lower is minimum of a judgment of k experts, Upper is maximum and Middle is the geometric mean of a judgment of k experts. After, it establishes fuzzy pairwise comparison matrix and its defuzzification. The weight vector and the global value of usability are given. The model which adds learnability is more usable than the model which does not add it.

H. A. Al-Jamimi and M. Ahmed. Prediction of Software Maintainability Using Fuzzy Logic. *Software Engineering and Service Science (ICSESS), IEEE 3rd International Conference*, pages 702–705, 2012.

This paper evaluates the performance of a fuzzy inference model to predict a quality attribute: maintenance. Fuzzy logic is used for the treatment of uncertainty and imprecision. They used the Object Oriented metrics as well as the QUES and UMIS datasets. This experiment shows that an approach based on Mamdani inference engine provides high performance to predict maintenance. Each input variable has three Gaussian membership functions with three different means based on the data range available for each input variable.

V. Kumar, A. Sharma, R. Kumar, and P. S. Grover. Quality aspects for component-based systems : A metrics based approach. *Software Practice and Experience*, 42(January):1531–1548, 2012.

The paper analyzes different quality metrics of component-based systems. This analysis is focused through different aspects such as the formal definition, the techniques used, if they are validated with experiments, if they are easy to use, data source, whether they have comparative analysis, applicability and extensibility. It is concluded that while algorithmic and statistical methods are widely used and tested, due to the lack of adaptability and power prediction, the research community is moving to the following non-conventional techniques: fuzzy logic, neuronal networks, neuro-fuzzy, probabilistic reasoning and combination of the above methods.

S. Bhatnagar, S. K. Dubey, and A. Rana. Quantifying Website Usability using Fuzzy Approach. *International Journal of Soft Computing and Engineering (IJSCE) ISSN: 2231-2307*, 2(2):424–428, 2012.

This paper uses ISO 9241 where learnability is added. Usability is calculated in a similar way to the multi criteria approach. The information is fuzzified. Then, the factors in fuzzy relation matrix could be calculated using the following formula: $R_{ij} =$ (*Number of corresponding average rank*)/(*Number of the Participants*). The weight vectors are made by AHP. The usability evaluation result ($Ub = W \times R$) is obtained and can be defuzzified to a global score. Here the classification of evaluation is defined as 95, 82, 67, 50 and 31 for excellent, good, average, poor, very poor, thus vector B can be defuzzified.

A. Dasso and A. Funes. Software Quality Metrics Aggregation. *13th Argentine Symposium on Software Engineering, ASSE 2012 ISSN: 1850-2792*, pages 312–323, 2012.

This paper calculates the quality from ISO 9126. It is interesting the way of adding information to calculate the quality of software. It uses a way to aggregate different metrics into a single value using a Continuous Logic, the one proposed in Logic Score of Preference (LSP) method. It uses operators from a Continuous Logic, specifically the Logic employed by the LSP method that proposes the aggregation of preferences by using a group of logic functions called Generalized Conjunction

Disjunction (GCD) operators. The aggregation of software quality metrics using the LSP method allows not only a complete aggregation of the desired characteristics but also allows partial aggregations, thus only one or more metrics software quality are added according to the method chosen.

P. Jain, S. K. Dubey, and A. Rana. Software Usability Evaluation Method. *International Journal of Advanced Research in Computer Engineering & Technology*, 1(2):28–33, 2012.

This paper uses the metric for measuring usability based on ISO 9126, the technique recommended is: consider the problem and design the fuzzy set on the basis of linguistic terms. Define the fuzzy rules. Calculate the fuzzy weighted indicators and defuzzify the solution. Make a study of the different methods of assessment.

S. K. Dubey, A. Gulati, and A. Rana. Usability Evaluation of Software Systems using Fuzzy Multi- Criteria Approach. *IJCSI International Journal of Computer Science Issues ISSN (Online): 1694-0814*, 9(3):404–409, 2012.

This paper calculates the usability based on ISO 9241 adding understandability and security. It uses the multi criteria approach. It assigns Fuzzy value to all leaf nodes of the hierarchical structure. In the process of fuzzification, fuzzy sets were assigned to real time values. The triangular fuzzy numbers are assigned to the fuzzy ratings and weights obtained by the users. It assigns fuzzy weights to all the nodes (sub-characteristics, characteristics, sub-attributes, attributes) in the hierarchy structure. To aggregate the values of the experts the arithmetic mean is used.

First, the fuzzy weighted average of the sub Characteristics (level 4) is taken to evaluate the rating of the characteristic. Then the fuzzy weighted average of the Characteristics (level 3) is taken to evaluate- the rating of the Corresponding sub-attribute. Then the fuzzy weighted average of the sub-attributes (level 2) is taken to evaluate-the rating of the Corresponding attribute. Lastly, the fuzzy weighted average of the fuzzy attributes gives the rating for the usability. From the fuzzy rating of usability Obtained in the previous step, a crisp value is calculated by the defuzzification process using the Centroid Method and usability value between 0 and 1 is determined.

R. Nagpal, D. Mehrotra, A. Sharma, and P. Bhatia. ANFIS Method for Usability Assessment of Website of an Educational Institute. *World Applied Sciences Journal*, 23(11):1489–1498, 2013.

To calculate the usability of an educational website the following parameters are used: ease of use (it can be defined as the ease or how quickly the visitor is able to access a particular website without much effort), Informative Quality of content, Response Time (how quickly a website Responds to a request) and Ease of navigation. It proposes a model for usability assessment using the Adaptive Neuro Fuzzy Inference System Approach (ANFIS). ANFIS integrates ANN and FL, That is, it combines the benefits of both methods in a single framework. In this model Sugeno's approach is used to assess usability. The fuzzy logic toolbox using the Fuzzy MATLAB software is used to create the ANFIS model.

B. Furer, F. Ruggeri, S. M. Voci, C. A. Borges, and B. Slater. Avaliação da usabilidade de um sistema computadorizado de epidemiologia nutricional. *Bras Epidemiol*, 16(4):966–975, 2013.

This paper uses ease of learning, efficiency, effectiveness, ease of recall, satisfaction, error control to measure usability. The method is similar to [3, 12, 50]. One of the limitations of this method is that it does not obtain a final value of usability.

A. Singh and S. K. Dubey. Evaluation of Usability Using Soft Computing Technique. *International Journal of Scientific & Engineering Research*, 4(12):162–166, 2013.

The ISO 9126 is used to measure usability, which subdivides each usability sub-characteristic in various factors affecting it. It uses the multi-criteria approach similar to [16].

L. J. Basto Cordero, L. F. Ribeiro Parente Filho, R. Costa dos Santos, W. Gassenferth, and M. A. Soares Machado. Ipod System's Usability: An Application of the Fuzzy Logic. *Global Journal of Computer Science and Technology*, 13(6), 2013.

This work is similar to all previous researches done in Brazil [12, 20]. For each constructor grouped in usability attributes: ease of learning, efficiency, effectiveness, ease of recall, satisfaction, error control; reports statistics on the responses from the questionnaires filled by a number of users. Then fuzzy triangular numbers for each metric result and its interpretation are presented. The calculations required for the fuzzy analysis of results were carried out using the mathematical MatLab software, which graphically generated the results for each metric studied. An analysis of the various factors affecting the usability is done but a global value of usability is not obtained.

V. Sharma and A. Sharma. Software Usability Assessment Models and Metrics: A Survey. *International KIET Journal of Software and Communication Technologies (IKJSCT)*, 1(1):7–15, 2013.

This paper is a survey of usability assessment standards and models, and also details other usability assessment techniques by pointing out their findings and limitations. It describes different models to evaluate the usability and groups them into three categories: those that use the approach of neural networks, which uses the fuzzy approach and other models. This evaluates the coexistence of different factors and parameters for assessing usability. Most of these models are static in nature. Some of them are just proposed but they have not been applied or validated. Very few of them have to do with the experimental calculation of the usability on a particular application.

R. Madhavan and K. Alagarsamy. Usability Evaluation of Object Oriented Applications Using Learning Curves. *International Journal of Engineering and Technical Research (IJETR)*, 2(2 ISSN: 2321-0869):35–39, 2014.

This paper uses the ISO 9241 for calculating usability adding error tolerant and easy to learn. Attributes are considered to have no interaction between them and work with learning curves relating experience and productivity. The time required to produce a unit decreases as more units occur. Learning curves with the parameters time of task and completion rates are developed. It does not give a value of usability.

P. Kortum and S. C. Peres. The relationship between system effectiveness and subjective usability scores using the System Usability Scale. *International Journal of Human-Computer Interaction*, 30(November 2015):575–584, 2014.

This paper assesses the usability using System Usability Scale (SUS) based on ISO 9241. This study examined if the average usability score for a series of tasks was the same as the usability score for the product. The result of the usability measured at task level was significantly higher than at the product level. This result is important for evaluators, who must analyze and interpret it because leads to different measures of usability and therefore the decision to take may be different.

S. Hedegaard and J. Simonsen. Mining until it hurts: automatic extraction of usability issues from online reviews compared to traditional usability evaluation. *Proceedings of the 8th Nordic Conference on Human-Computer Interaction Fun, Fast, Foundational—NordiCHI '14*, pages 157–166, 2014.

A new method for evaluating usability through finding data in reviews, tweets, forum postings, and contain user narratives on interaction with products is developed in this work. The data are extracted manually to identify usability problems, and these are clasified after training several machine learning classifiers. Results obtained from this method are compared quantitatively and qualitatively against usability problems identified by traditional methods. This work demonstrates that traditional techniques involving quality experts give better results than this method, but it is not conclusive and much more research is needed.

S. Roy, P. K. Pattnaik, and R. Mall. A quantitative approach to evaluate usability of academic websites based on human perception. *Egyptian Informatics Journal*, 15(3):159–167, 2014.

Two types of evaluation techniques of usability were used in this study. The first is based on a questionnaire ("Website Analysis and Measurement Inventory" was used). The second technique is based on performance evaluation. The results of the questionnaire-based assessment were observed to be consistent with the results of the assessment based on performance. The usability evaluation was performed analyzing the results of the success rates of task completion times of tasks and satisfaction. A statistical analysis was performed and a negative relationship between the time of task completion and satisfaction was obtained. This implies that a decrease in the time of task completion improves the satisfaction levels of participants. This work demonstrates the relationship between quality attributes and the importance of take this into account in the software evaluation.

S. Patalano, A. Lanzotti, D. M. Del Giudice, F. Vitolo, and S. Gerbino. On the usability assessment of the graphical user interface related to a digital pattern software tool. *International Journal on Interactive Design and Manufacturing (IJIDeM)*, (Dii), 2015.

In this paper the authors evaluate the usability of a tool that belongs to knowledge-based engineering (KBE) system. The Analytic Hierarchy Process (AHP) as multiple-criteria decision analysis and the ISO 9241 as model of usability are used in the evaluation process. The authors determine efficiency through the two metrics: the number of errors and the completion of the task. The efficiency is measured using the number of operations and time. Finally the satisfaction is established with levels of satisfaction.

M. Kurosu. Human-computer interaction users and contexts: 17th international conference, HCI international 2015 Los Angeles, CA, USA, August 2–7, 2015 proceedings, Part III. *Lecture Notes in Computer Science (including subseries Lecture Notes in Artificial Intelligence and Lecture Notes in Bioinformatics)*, 9171:35–42, 2015.

This paper shows the importance of the participation of people with disabilities in the evaluation the usability of applications that aim to serve to the majority of the population, such as websites of the administration and public services. The results of the experiments in this paper suggest that can reliably measure the level of satisfaction of these users by using fast and short questionnaires as SUS, UMUX and UMUX-LITE.

R. Ackerman, A. Parush, F. Nassar, and A. Shtub. Metacognition and system usability: Incorporating metacognitive research paradigm into usability testing. *Computers in Human Behavior*, 54:101–113, 2016.

This paper proposes a methodology to enhance in-depth testing of users' performance and the perceived usability at the task level. Also, a Metacognitive Usability Profile which includes a comprehensive set of measures based on collecting confidence in the success of each particular task is proposed, and triangulating it with objective measures. Thus, this methodology combines objective and subjective measures. The work emphasizes that the real success and response time for a collection of tasks allow understand the cognitive processing that involves the user interface. Users can assess their chances of success more reliably improving the user interface using this proposal.

4 Summary and Conclusion

The assessment usability software process is too expensive because it implies the use of material recourses and a team of well-trained specialists, for that reason it is necessary to achieve correlation among the software assessment results and the usability that this product actually shows. This paper exposes the commented bibliography of the related works on models to measure software usability emphasizing what has been done in the last decades. Such models can be analyzed through two points of view: The criteria for assessing the usability and the information aggregation techniques. The first one, refers to the attributes and metrics used to measure the usability and the latter, refers to the techniques to add the information obtained from the evaluation of different metrics.

Regarding the criteria analyzed, the tendency in last years is to use one of the following standards: ISO 9241, ISO 9126 or extensions or adaptations of these norms with some inner attributes of the software to be tested or a complementation due to unified criteria of both. ISO 9241 expresses usability based on three major attributes: effectiveness, efficiency and satisfaction. This standard is used in researches such as: [5, 6, 24, 36], different researches developed in 2012 such as [7, 16] add learnability. Winter et al. [60] adds Safety meanwhile [15] adds comprehensibility and safety. In [53] efficiency, memorability, satisfaction and learnability are used. In [37] error tolerant and easy to learn are added.

ISO 9126-1 describes usability as a combination of understandability, learnability, operability, attractiveness and usability compliance. 9126 is applied in [14, 55]. In [35, 58] does it considering only understandability, learnability and operability. [8, 9, 27] use metrics to measure usability based in 9126, adding reusability. These authors detail different metrics associated to each characteristic and when metrics are associated to more than one characteristic it is a way of showing the relation among characteristics.

Other criteria for different models were defined [3, 6, 10–12, 20, 28, 29, 42, 54], it is nothing more than the joint of previous models as well as the description in details of some criteria.

Sometimes it is difficult to measure usability if suitable measures, considering the type of product, are not taken. For example:

1. A website [25, 26, 34, 41, 42, 51, 56].
2. E-learning system [44, 57]
3. Virtual community [48].
4. Software component [4].

These models largely overlap; the attributes in different models are superimposed. Different name for the same attribute are used and in some case, there are equal names for different attributes, which is determined when the actually measured for this attribute in a low level is examined. Different ways to mix attributes are used and they are located in different places in hierarchy. The authors have different opinions about which attributes are indispensable in the usability calculation. Starting from this analysis, it is confirmed what Folmer and J. Bosch stated, the term of usability is still ambiguous and many researchers spend a lot of enforce trying to find the best manner to define usability through the definition of attributes which can be measured [18].

Regarding the techniques which are used to obtain a usability value, The models that handle crisp numbers are not completely reliable because the numerical value assigned to the different usability characteristics will always be questionable and weak; for that reason, it is necessary the use of fuzzy logic. Fuzzy logic also helps to reduce ambiguity in the assignation of values to the parameters. The combination of different methods with fuzzy logic is the principal trend of the methods for usability assessing. For the study of these techniques it is recommended to classify it in: algorithm and conventional statistics techniques, Soft Computing techniques and others. As examples of algorithms and conventional techniques it can be mentioned the model presented in [5, 22, 38]. These techniques are widely used and exactly tested but their limitation is the lack of adaptation capacity and power of prediction [32]. For studying the Soft Computing techniques, they were organized in four groups: the methods which use: fuzzy inference system, multi criteria decision methods, fuzzy integral (Choquet integral [53, 54]) and others.

Among the models analyzed, the ones that use fuzzy inference systems are [10, 11, 25–27, 42, 57]. The majority of those models use Takagi Sugeno Inference method.

There are many methods which combine conventional techniques of MCDM with fuzzy concepts. Some use fuzzy TOPSIS [41], AHP [7, 16, 34], and others that use fuzzy multi criteria approach [8, 9, 15, 55].

There are other techniques of soft computing, such as: [12, 20, 35, 40]. And, finally, [47] which uses fuzzy classification and shows the effectiveness of fuzzy integration for combining the results of multiple classifiers.

There are other techniques, such as the one that the finite-state machine (FSM) [36] uses. It uses a way to aggregate different metrics into a single value using a Continuous Logic: the one proposed in the Logic Score of Preference (LSP) method [14].

Among the most used Soft Computing techniques since 2008, is the one that uses a fuzzy multi criteria approach, highlighting the Fuzzy TOPSIS and Fuzzy AHP methods, or their derivatives. But it has the limitation that it does not take into account the relation between the criteria. This entails an analysis with ANP.

Some techniques have only been proposed. They have not been validated yet and others have been applied experimentally in the calculation of usability in a particular system.

In the case of fuzzy modeling, it has been proved that the use of triangular fuzzy numbers as valid for usability assessment models. In the case of using linguistic variables, they must have at least three terms.

After this review, the authors consider that the following aspects should deserve further consideration:

On criteria:

- The tendency in the last years is using one of the standards ISO 9241 and 9126 or extensions or adaptations of these norms, but it is essential to analyze ISO 25000. This does not have a wide use but it is new and it emerged from the review of 9126.
- The values of the criteria are heterogeneous because they come from objective and subjective criteria.
- In the majority of the bibliography, it is not reflected the interde-pendency among the criteria. There are models that reflect this relation because they repeat measures for the related attributes.
- Proving that the use of usability attributes and their metrics fulfill with completeness, decomposability, no redundancy and minimality is nowadays a challenge.

On techniques

- The majority of these models have a static nature.
- The use of any Soft Computing techniques allows obtaining better results. The conclusion about one specific Soft Computing technique being better than another is not appropriate.
- In the case of the aggregation to obtain a global value of usability, the complete aggregation as well as the partial aggregation must be permitted.

After reviewing the bibliography, it is possible to conclude that the usability assessment is still a challenge for Software Engineering.

To conclude, the authors would like to recognize that this review has been a hard task. Any review process is also a filtering process, so we would like to apologize to the readers and researchers for any possible omissions. The authors will be happy to receive their feedback in order to improve it. If this study motivates students and researchers to use and investigate on these problems, then the aim of the paper will be fulfilled.

Acknowledgements This research has been supported by projects TIN2014-55024-P from the Spanish Ministry of Economy and Competitiveness, and P11-TIC-8001 from the Andalusian Government (both including FEDER funds).

References

1. Ackerman, R., Parush, A., Nassar, F., Shtub, A.: Metacognition and system usability: Incorporating metacognitive research paradigm into usability testing. Comput. Hum. Behav. **54**, 101–113 (2016)
2. Al-Jamimi, H.A., Ahmed, M.: Prediction of software maintainability using fuzzy logic. In: IEEE 3rd International Conference Software Engineering and Service Science (ICSESS), pp. 702–705 (2012)
3. Basto Cordero, L.J., Ribeiro Parente Filho, L.F., Costa dos Santos, R., Gassenferth, W., Soares Machado, M.A.: Ipod system's usability: an application of the fuzzy logic. Glob. J. Comput. Sci. Technol. **13**(6) (2013)
4. Bertoa, M.F., Troya, J.M., Vallecillo, A.: Measuring the usability of software components. J. Syst. Softw. **79**(3), 427–439 (2006)
5. Bevan, N., Macleod, M.: Usability measurement in context. Behav. nf. Technol. **13**(1–2), 132–145 (1994)
6. Beyan, N.: The MUSIC methodology for usability measurement. Posters and Short Talks of the 1992 SIGCHI Conference on Human Factors in Computing Systems, pp. 123–124. ACM (1992)
7. Bhatnagar, S., Dubey, S.K., Rana, A.: Quantifying website usability using fuzzy approach. Int. J. Soft Comput. Eng. (IJSCE) **2**(2), 424–428 (2012). ISSN: 2231-2307
8. Challa, J.S., Paul, A., Dada, Y., Nerella, V.: Integrated software quality evaluation: a fuzzy multi-criteria approach. J. Inf. Process. Syst. **7**(3), 473–518 (2011)
9. Challa, J.S., Paul, A., Dada, Y., Nerella, V., Srivastava, P.R.: Quantification of software quality parameters using fuzzy multi criteria approach. In: 2011 International Conference on Process Automation, Control and Computing, pp. 1–6, July 2011
10. Chang, E., Dillon, T.: A usability-evaluation metric based on a soft-computing approach. IEEE Trans. Syst. Man Cybern.— Part A: Syst. Hum. **36**(2), 356–372 (2006)

11. Chang, E.J., Dillon, T.S., Cook, D.: An intelligent system based usability evaluation metric. In: IASTED International Conference on Intelligent Information Systems, pp. 218–226. IEEE Computer Society (1997)
12. Costa, R., Augusta, M., Machado, S., Santos, D.J., Tavares, R., Campos, C.D.: Uma Aplicação da Matemática Nebulosa na Usabilidade de Pacotes Estatísticos. RESI Revista Eletrônica de Sistemas de Informação **10**(1), 1–8 (2007)
13. Danfeng, Y., Fangchun, Y.: Fuzzy evaluation of sla-oriented QoSM (the quality of service management) in NGN. In: Proceedings of IC-BNMT2009 (2009)
14. Dasso, A., Funes, A.: Software quality metrics aggregation. In: 13th Argentine Symposium on Software Engineering, ASSE 2012, pp. 312–323 (2012). ISSN: 1850-2792
15. Dubey, S.K., Gulati, A., Rana, A.: Usability evaluation of software systems using fuzzy multi-criteria approach. IJCSI Int. J. Comput. Sci. Issues ISSN (Online) 1694–0814, **9**(3), 404–409 (2012)
16. Dubey, S.K., Mittal, A., Rana, A.: Measurement of object oriented software usability using fuzzy AHP. Int. J. Comput. Sci. Telecommun. **3**(5), 98–104 (2012)
17. Etaati, M.L., Sadi-Nezhad, S., Using, A.: Fuzzy analytical network process and ISO 9126 quality model in software selection: a case study in E-learnig systems. J. Appl. Sci. **11**(1), 96–103 (2011)
18. Folmer, E., Bosch, J.: Architecting for usability: a survey. J. Syst. Softw. **70**(1–2), 61–78 (2004)
19. Folmer, E., van Gurp, J., Bosch, J.: A framework for capturing the relationship between usability and software architecture. Softw. Process Improv. Pract. **8**(2), 67–87 (2003)
20. Furer, B., Ruggeri, F., Voci, S.M., Borges, C.A., Slater, B.: Avaliação da usabilidade de um sistema computadorizado de epidemiologia nutricional. Bras Epidemiol **16**(4), 966–975 (2013)
21. Gray, A.R., MacDonell, S.G.: Fuzzy logic for software metric models throughout the development life-cycle. Inf. Sci. Discuss. Pap. Ser. **99**(20) (1999)
22. Harper, B.D., Norman, K.L.: Improving user satisfaction: the questionnaire for user interaction satisfaction version 5.5. In: Proceedings of the 1st Annual Mid-Atlantic Human Factors Conference, pp. 224–228 (1993)
23. Hedegaard, S., Simonsen, J.: Mining until it hurts: automatic extraction of usability issues from online reviews compared to traditional usability evaluation. In: Proceedings of the 8th Nordic Conference on Human-Computer Interaction Fun, Fast, Foundational—NordiCHI '14, pp. 157–166 (2014)
24. Hornbæk, K.: Current practice in measuring usability: challenges to usability studies and research. Int. J. Hum.-Comput. Stud. **64**(2), 79–102 (2006)
25. Hub, M., Zatloukal, M.: Usability evaluation of selected web portals 2 case study usability evaluation of selected web portals. In: Proceedings of the 9th WSEAS International Conference on Applied Informatics and Communications (AIC '09), pp. 259–264 (2009)
26. Hub, M., Zatloukal, M.: Model of usability evaluation of web portals based on the fuzzy logic. WSEAS Trans. Inf. Sci. Appl. **7**(4), 522–531 (2010)
27. Jain, P., Dubey, S.K., Rana, A.: Software usability evaluation method. Int. J. Adv. Res. Comput. Eng. Technol. **1**(2), 28–33 (2012)
28. Jin, B.S., Ji, Y.G.: Development of a usability evaluation framework with quality function deployment : from customer sensibility to product design. Hum. Factors Ergon. Manuf. **19**(2), 177–194 (2009)
29. Juristo, N., Moreno, A.M., Sanchez-Segura, M.-I.: Analysing the impact of usability on software design. J. Syst. Softw. **80**(9), 1506–1516, Sept 2007
30. Komogortsev, O.V., Mueller, C.J., Tamir, D., Feldman, L.: An effort based model of software usability. In: Proceedings of the International Conference on Software Engineering Theory and Practice (SETP-08), vol. (July), pp. 1–9 (2009)
31. Kortum, P., Peres, S.C.: The relationship between system effectiveness and subjective usability scores using the system usability scale. Int. J. Hum.-Comput. Interact. **30**(Nov 2015), 575–584 (2014)
32. Kumar, V., Sharma, A., Kumar, R., Grover, P.S.: Quality aspects for component-based systems: a metrics based approach. Softw. Pract. Exp. **42**(January), 1531–1548 (2012)

33. Kurosu, M.: Human-computer interaction users and contexts. In: proceedings of the 17th international conference, HCI international 2015 Los Angeles, CA, USA, August 2–7, 2015, Part III. Lecture Notes in Computer Science (including subseries Lecture Notes in Artificial Intelligence and Lecture Notes in Bioinformatics), vol. 9171, pp. 35–42 (2015)
34. Lamichhane, R., Meesad, P.: A usability evaluation for government websites of nepal using fuzzy AHP. In: The 7th International Conference on Computing and Information Technology IC2IT2011, pp. 99–104 (2011)
35. Li, Y.-X., Man, Z.-X.: A fuzzy comprehensive quality evaluation for the digitizing software of ethnic antiquarian resources. In: 2008 International Conference on Computer Science and Software Engineering, pp. 1271–1274 (2008)
36. Liang, L., Deng, X., Wang, Y.: Usability measurement using a fuzzy simulation approach. In: 2009 International Conference on Computer Modeling and Simulation, pp. 88–92, Feb 2009
37. Madhavan, R., Alagarsamy, K.: Usability evaluation of object oriented applications using learning curves. Int. J. Eng. Tech. Res. (IJETR) 2(2 ISSN: 2321-0869), 35–39 (2014)
38. Mansor, Z., Kasirun, Z.M., Yahya, S., Arshad, N.H.: The evaluation of webcost using software usability measurement inventory (SUMI). Int. J. Digit. Inf. Wirel. Commun. (IJDIWC) 2(2), 197–201 (2012)
39. Mich, L., Franch, M., Gaio, L.: Evaluating and designing web site quality. Multimed. IEEE 10(1), 34–43 (2003)
40. Mittal, H., Sharma, M., Mittal, J.: Analysis and modelling of websites quality using fuzzy technique. In: 2012 Second International Conference on Advanced Computing & Communication Technologies, pp. 10–15, Jan 2012
41. Montazer, G.A., Saremi, H.Q.: An application of Type-2 fuzzy notions in website structures selection: utilizing extended TOPSIS method. Wseas Trans. Comput. 7(1), 8–15 (2008)
42. Nagpal, R., Mehrotra, D., Sharma, A., Bhatia, P.: ANFIS method for usability assessment of website of an educational institute. World Appl. Sci. J. 23(11), 1489–1498 (2013)
43. Nielsen, J.: Usability Engineering. Academic Press, San Diego, CA (1993)
44. Oztekin, A., Kong, Z.J., Uysal, O.: UseLearn: a novel checklist and usability evaluation method for eLearning systems by criticality metric analysis. Int. J. Ind. Ergon. 40(4), 455–469 (2010)
45. Parvinder, A., Sandhu, S., Priyanka Kakkar, S.S.: A survey on software reusability. In: 2010 International Conference on Mechanical and Electrical Technology (ICMET 2010), pp. 769–773 (2010)
46. Patalano, S., Lanzotti, A., Del Giudice, D.M., Vitolo, F., Gerbino, S.: On the usability assessment of the graphical user interface related to a digital pattern software tool. Int. J. Interact. Des. Manuf. (IJIDeM) (Dii) (2015)
47. Pizzi, N.J.: Software quality prediction using fuzzy integration: a case study. Soft Comput. 12(1), 67–76, June 2007
48. Preece, J.: Sociability and usability in online communities: determining and measuring success. Behav. Inf. Technol. 2001 20(5), 347–356 (2001)
49. Roy, S., Pattnaik, P.K., Mall, R.: A quantitative approach to evaluate usability of academic websites based on human perception. Egypt. Inf. J. 15(3), 159–167 (2014)
50. Santos, D.J., Tavares Carneiro de Campos, R., Parente Filho, L.F., Costa dos Santos, R., Soares Machado, M.A.: Usabilidade de sistemas: uma aplicação da lógica fuzzy na avaliação do PHSTAT (2007)
51. Schenkman, B.N., Jönsson, F.U.: Aesthetics and preferences of web pages. Behav. Inf. Technol. 19(5), 367–377 (2000)
52. Sharma, V., Sharma, A.: Software usability assessment models and metrics : a survey. Int. KIET J. Softw. Commun. Technol. (IKJSCT) 1(1), 7–15 (2013)
53. Sicilia, M.A., Barriocanal, E.G., Calvo, T.: An inquiry-based method for Choquet integral-based aggregation of interface usability parameters. Kybernetika 39(5), 601–614 (2003)
54. Sicilia, M.A., García, E.: Modelling Interacting Web Usability Criteria through Fuzzy Measures. Web Engineering. Springer, Berlin (2003)
55. Singh, A., Dubey, S.K.: Evaluation of usability using soft computing technique. Int. J. Sci. Eng. Res. 4(12), 162–166 (2013)

56. Sitar-Taut, D.-A., Stanca, L.-M., Buchmann, R., Lacurezeanu, R.: A case study on usability metrics applied in romanian E-Commerce environment. Wseas Trans. Inf. Sci. Appl. 6(10), 1697–1706 (2009)
57. Siu Keung, W., Thi Nguyen, T., Chang, E., Jayaratna, N.: Usability metrics for E-learning. In: On The Move to Meaningful Internet Systems 2003: OTM 2003 Workshops, pp. 235–252. Springer, Berlin (2003)
58. Tamir, D., Komogortsev, O.V., Mueller, C.J.: An effort and time based measure of usability. In: Proceedings of the 6th International Workshop on Software Quality, pp. 47–52. ACM (2008)
59. Tullis, T.S., Stetson, J.N.: A comparison of questionnaires for assessing website usability. In: Usability Professional Association Conference, pp. 1–12 (2004)
60. Winter, S., Wagner, S., Deissenboeck, F.: A comprehensive model of usability. Engineering Interactive Systems, pp. 106–122. Springer, Berlin (2008)
61. Zenebe, A., Norcio, A.F.: Evaluation framework for fuzzy theoretic-based recommender system. In: 11th International Conference on Human-Computer Interaction (To Appear), Las Vegas, Nevada, USA (2005)

Production Systems Optimization Using Hierarchical Planning

F. González Santoyo, B. Flores Romero, A. M. Gil Lafuente, J. Flores
Juan and R. Chávez Rivera

Abstract This chapter proposes a new and efficient heuristic algorithm to solve the
problem of hierarchical production of lumber in sawmills. The proposed solution is
based on Mixed-Integer Linear Programming (MILP), using Benders' decomposition and Lagrangian relaxation techniques. The proposed methodology achieves a
higher computational efficiency than the state of the art for the solution of this kind
of problems.

1 Introduction

Mexican companies are immerse in a world-wide competition. González Santoyo
[5, 6, 8, 9], states that in a globalized market, Mexican companies will find difficult
to remain in the market to distribute its products, unless they modernize to a
competent infrastructure. This modernization includes material and human
resources, and a control system that allow them to operate with a high degree of
efficacy and efficiency: the requirements in our current reality.

Under those circumstances, the company requires an optimal production planning and scheduling to gain competitive advantage. This can be done assuming that
goods and/or services provided by the company are certified and competitive in
price and quality.

F. González Santoyo (✉) · B. Flores Romero · J. Flores Juan · R. Chávez Rivera
Universidad Michoacana de San Nicolás de Hidalgo México, Morelia, Mexico
e-mail: fsantoyo@umich.mx

B. Flores Romero
e-mail: betyf@umich.mx

J. Flores Juan
e-mail: juanf@umich.mx

A. M. Gil Lafuente
Universidad de Barcelona, Barcelona, Spain
e-mail: amgil@ub.edu

© Springer International Publishing AG 2018   191
C. Berger-Vachon et al. (eds.), *Complex Systems: Solutions and Challenges
in Economics, Management and Engineering*, Studies in Systems,
Decision and Control 125, https://doi.org/10.1007/978-3-319-69989-9_12

The production-planning problem [1] addressed in this contribution is classified as NP-Complete. The goal of this problem is to plan and schedule production over time—[2, 7, 10]. Production planning has fundamental objectives: the planning function determines the requirements sources, the current point in time determines the planning horizon and the order of demand satisfaction. The scheduling function determines how the available production sources act, locating the individual products provided to consumers at a minimum production cost. The problem is to achieve a production volume for the analysis period at high efficiency levels. Achieving this goal guarantees the company's permanence in the global markets. This contribution addresses this problem in the framework of a lumber production sawmill; we present its mathematical model, a solution algorithm and the results of the production plan.

2 Problem Statement

On of the most important economic activities of the state of Michoacan, Mexico is the industrialization of the forestry resources. The first industrialization stage from the mechanical point of view of this resource is sawmills. The stages that form this process [4] are: Reception and classification of raw material, edging, multiple head remover, classification table and storage. The problem's representative variables are: raw material, machine time, processing time availability, supplies, labor, electricity, maintenance and sawn products market.

One problem found in the process is that logs present different diameters, lengths and uniformity ranges, so they need to be classified according to those variables. This enables a more efficient operation and the choice of an optimal cut schedule for the main saw and other stages in the process.

In the state of Michoacan, Mexico, the main problem is the log length. Logs are supplied in lengths of 8, 10, 12, 14, 16, 18, 20, and 22 ft., with diameters from 12 to 30 in. Commercial measures are ½, ¾, 1½, 3, 3½ in. thick, 8, 10, and 12 in. wide. These conditions allow us to deal with this problem using Mixed-Integer Linear Programming (MILP). We have 8 kinds with 1, 2, and 4 families, which means to have 160 elements of sawn lumber in the company.

Using MILP, production is aggregated in families and families in types of products [3]. This aggregation structure [5], lies in the research line known as hierarchical production planning. Planning of the original production is divided in a hierarchy of sub problems, where a structuring production plan considers that individual parts and final products are aggregated.

The grouping criterion follows [3], using the description provided by [5]: **Items** are final products, required by the market in a time unit, **types of products** are groups of items that have similar production costs, present the same demand model and the same production range. Types are characterized according to log length. Families of products are represented by a set of items that share a common characteristic. In this case this characteristic is mean width.

The general MILP problem is characterized as a large-scale problem. The hierarchical planning problem will be solved using MILP and a heuristic algorithm, based on Bender's decomposition theory [1, 7].

3   Mathematical Model

The hierarchical planning problem (PS) can be formulated as follows:

$$
Z_{ps}Min. \sum_t \left(C_t O_t + \sum_i h_{it} I_{it} \right) + \sum_j \sum_t S_{jt} X_{jt}
$$

$$
\sum_{j \in T(i)} FP - P = 0, \forall i, t
$$

$$
\sum_l K_i P_{it} - O_t \le r_t, \forall t,
$$

$$
\sum_{j \in T(i)} FI_{jt} - I_{it} = 0, \forall i, t \tag{1}
$$

$$
FP_{jt} + FI_{j,t-1} - FI_{jt} = d_{jt}, \forall j, t
$$

$$
FP_{jt} - m_{jt} X_{jt} \le 0, \forall j, t
$$

$$
O_t, P_{lt}, I_{it}, FP_{jt}, FI_{jt} \ge 0, \forall i, j, t
$$

$$
X_{jt} \in \{0, 1\}, \forall j, t
$$

where t is current the time period, C_t the cost of 1 h of extra time, h_{it} is the inventory cost for items of type i, S_{jt} is the preparation cost per family j, d_{it} (d_{jt}) is the demand of item i (family j), K_i is the required production time for i, $T(i)$ is the set of families that belong to type i, m_{jt} is the production amount for family j, and r_t is the available production time. The model's decision variables re: C is the number of extra time hours for production at time t, I_{it} (FI_{jt}) is the inventory of type i (family j), P_{it} (FP_{jt}) is the production amount of type I (family j), X_{jt} is a 0–1 variable that indicates the update of family j.

The solution process is based on Bender's decomposition techniques. Complex variables in PS are X_{jt}, FI_{jt}, and FP_{jt}; these variables allow us to structure the problem on a type and family levels. For this variable partition, the Bender's sub problem $PSUB_T$ can be stated as:

$$
Z_{PSUB}(t) = Min.C_t O_t + \sum_i h_{it} I_{it}
$$

$$
P = \sum_{j \in T(i)} FPjt, \forall i
$$

$$
\sum_i K_i P_{it} - O_t \le r_t \tag{2}
$$

$$
I_{it} = \sum_{j \in T(i)} FI_{it}, \forall i
$$

$$
O_t, I_{it}, P_{it} \ge 0, \forall i
$$

An upper bound on the optimal value is provided by the sub problem, when the constant $\sum_j \sum_t S_{jt} X_{jt}$, is added to $\sum_t Z_{PSUB}(t)$, becomes a PS constraint. The sub problem has unique feasible primal solution, which is found by inspection as follows:

$$P_{it} = \sum_{j \in T(i)} FP_{jt}, \forall i, t$$

$$I = \sum_{j \in T(i)} FI_{jt}, \forall i, t \tag{3}$$

$$O_t = Max. \left\{ 0, \sum_i K_i \left(\sum_{j \in T(i)} FP_{jt} \right) - r_t \right\}, \forall t$$

Let U_{it}, V_t, W_{it}, be the dual variables associated to the primal constraints, given in the formulation $PSUB_t$. The $PSUB_t$ dual problem $DPSUB_t$ can be stated as:

$$Z_{DPSUB}(t) = Max. \sum_i U_{it} \left(\sum_{j \in T(i)} FP_{jt} \right) + V_t r_t + \sum_i W_{it} \left(\sum_{j \in T(i)} FI_{it} \right)$$
$$U_{it} + K_i V_t \leq 0, \forall i$$
$$-V_t \leq C_t$$
$$W_{it} \leq h_{it}, \forall i \tag{4}$$
$$V_t \leq 0$$
$$U_{it}, W_{it}, \forall i$$

The solution of the dual problem, obtained by inspection, is:

$$W_{it} = h_{it, \forall i, t}$$
$$V_t = \begin{cases} -C_t & si \quad O_t > 0, \forall t \\ 0 & si \quad O_t = 0, \forall t \end{cases} \tag{5}$$
$$U_{it} = -K_i V_t, \forall i, t$$

DPSUB has multiple solutions, since PSUB is a degenerate solution. An alternative is to make U_{it} y W_{it} zero, if $\sum_{j \in T(i)} FP_{jt}$ y $\sum_{j \in T(i)} FI_{jt}$ are zero, respectively.

In the case where $O_t = 0$, we can obtain the Bender's cut $Z_t \geq \sum_i h_{it} (\sum_{j \in T(i)} FI_{jt})$, where Z_t, is an auxiliary variable from Bender's master problem.

These kinds of cuts are favorable and are included in the master problem. Let U_{it}^* y V_t^* be the dual solutions corresponding to the case $O_t > 0$. Let us assume PSUB has been solved for a sequence of given variables and that the $O_t > 0$ occurs at least once for the periods $T^* \subseteq \{1, \ldots, T\}$. The master problem $PM1$ can be stated as:

$$Z_{PM} = Min. \sum_j \sum_t S_{jt} X_{jt} + \sum_t z_t$$

$$z_t \geq \sum_i U_{it}^* (\sum_{j \in T(i)} FP_{jt}) + V_t^* r_t$$

$$+ \sum_i h_{it} (\sum_{j \in T(i)} FI_{jt}), t \in T^*$$

$$z \geq \sum_i h_{it} (\sum_{j \in T(i)} FI_{jt}), \forall t \tag{6}$$

$$FP_{jt} + FI_{j,t-1} - FI_{jt} = d_{jt}, \forall j, t$$

$$FP_{jt} - m_{jt} X_{jt} \leq 0, \forall j, t$$

$$O_t, P_{it}, FP_{jt}, FI_{jt}, I_{it} \geq 0, \forall i, j, t$$

$$X_{jt} \in \{0, 1\}$$

A variable substitution allows us to reformulate the master problem *PM2* as follows:

$$Z_{PM} = Min. \sum_j \sum_t S_{jt} X_{jt} + \sum_i \sum_t h_{it} (\sum_{j \in T(i)} FI_{jt}) + q_t$$

$$q_t \geq \sum_i U_{it}^* (\sum_{j \in T(i)} FP_{jt}) + V_t^* r_t, t \in T^*$$

$$q_t \geq 0, \forall t \tag{7}$$

$$FP_{jt} + FI_{j,t-1} - FI_{jt} = d_{jt}, \forall j, t$$

$$FP_{jt} - m_{jt} X_{jt} \leq 0, \forall j, t$$

$$O_t, P_{it}, FP_{jt}, FI_{jt}, I_{it} \geq 0, \forall i, j, t$$

$$X_{jt} \in \{0, 1\}, \forall j, t$$

The master problem is a relaxation of PS, where Z_{PM} provides a lower bound to Z_{PS}. González Santoyo [5], suggests the use of a Lagrangian relaxation as a strategy to solve problems of PM2. By relaxing Bender's cuts with Lagrangian multipliers λ_t, the objective function gets the form:

Min.

$$\sum_j \sum_t S_{jt} X_{jt} + \sum_i \sum_t h_{it} (\sum_{j \in T(i)} FI_{jt}) + \sum_t q_t + \sum_{t \in T^*} I_t (g_t(FP, FI) - q_t$$

where

$$g(FP, FI) = \sum_i U_{it}^* (\sum_{j \in T(i)} FP_{jt}) + V_t^* r_t \tag{8}$$

Each q_t is a non-negative continuous variable that does not appear in the rest of the constraints. Assuming $\lambda_t \leq 1$, $t \in T^*$, q_t's coefficient becomes non-negative, and q_t's optimal value approaches zero as the objective function approaches its minimum.

PM2's Lagrangian relaxation *LRPM* can be stated as:

$$Z_{LRPM}(\lambda) = Min. \sum_j \sum_t S_{jt}X_{jt} + \sum_i \sum_t h_{it}\left(\sum_{j \in T(i)} FI_{jt}\right)$$
$$+ \sum_{j \in T(i)} \lambda_t g_t(FP, FI)$$
$$FP_{jt} + FI_{j,t-1} - FI_{jt} = d_{jt}, \forall j, t \qquad (9)$$
$$FP_{jt} - m_{jt}X_{jt} \leq 0, \forall j, t$$
$$O_t, P_{it}, FP_{jt}, FI_{jt}, I_{it} \geq 0, \forall i, j, t$$
$$X_{jt} \in \{0, 1\}, \forall j, t$$

By relaxing PM, the resulting problem LRPM, is separated by families in a set of problems of Economic Lot Size without capacity constraint. This kind of problems can be solved efficiently by dynamic programming [11]. From the Lagrangian Duality theory for Integer Programming, Z_{LRMP} is a lower bound to Z_{PM}. The largest available lower bound, D, is obtained by the solution to:

$$Z_D = Max. Z_{LRPM}(\lambda)$$
$$0 \leq \lambda \leq 1, t \in T^* \qquad (10)$$

Which is the Lagrangian Dual with respect to the relaxed Bender's cuts. This terminology is used since $\lambda = \{\lambda_{tt}\}$ plays a similar roll LRPM with Lagrangian Multipliers normally used in the continuous problem. The dual objective function $Z_{LRPM}(\lambda)$ is continuous, concave, piecewise linear, and sub-differentiable.

A standard procedure to solve the dual problem is the sub-gradient optimization algorithm, which generates dual solutions according to the following rule.

$$\lambda_t^{l+1} = \lambda_t^l + \theta\gamma_t^l, \quad t \in T^*, \quad l = 0, 1, \ldots \qquad (11)$$

where $\gamma = \{\gamma_t\}$ is a sub-gradient of $Z_{LRPM}(\lambda)$ for a particular value of λ and Θ_l is the step size.

4 Solution Algorithm

1. Assume I T*I Bender's cuts have been generated and that PM2 is solved using Lagrangian Relaxation and Subgradient Optimization.
2. Vector λ_t is updated a number of times determined by the equation.

$$\lambda^{l+1} = \lambda_t^l + \theta_1\gamma_t^l, \quad t \in T^*, \quad l = 0, 1, \ldots$$

For a particular value of λ, $\gamma = \{\gamma_t\}$ is a subgradient of $\mathbf{Z_{lrpm}}$ (λ) and Θ_1 is the step size. At every point in time LRPM is solved and lower bound to $\mathbf{Z_{ps}}$ is determined.

3. The values of variables Xjt, FIjt, FPjt, obtained from LRPM solution, are used to execute PSUB, while the set of updated variables is improved for the heuristic exchange.
4. The optimal dual variables sent to PSUB are based on new Bender's cuts. Nonetheless, a cut may already be included in PM2. A previously generated cut cannot be included again in PM2.
5. An upper bound for $\mathbf{Z_{ps}}$ is obtained by a heuristic exchange for PSUB's solution; this value depends on the lower bound.

In the next iteration, the subgradient procedure for PM2 continues from the last solution to LRPM. The procedure is initialized by initial values $\{\mathbf{X_{jt}}\}$, which indicate the amount of product for family $\mathbf{j}$ at time $\mathbf{t}$. The algorithm continues until a given number of Bender's subproblems have been solved or the difference between the upper and lower bounds is small enough, according to a previously established parameter. Dual variables $\mathbf{U_{it}}$, $\mathbf{W_{it}}$ are part of the input to LRPM's objective function. These variables can be interpreted as production and storage costs for families that belong to a certain type of product.

5 Study Case

The proposed algorithm was used to solve problems that included 1, 2, and 4 families from the 8 different types in the problem. 160 different products were included (commercial sawn wood products), which are being produced in the sawmill under the current supply conditions. A family's model includes 5 different commercial sizes of sawn products; 4 families include 20 different sizes for each one of the types they produce.

Each production plan specifies the commercial size, its location within the family, and its type. The size contributes to the objective function to minimize cost; to satisfy demand, the production amount is specified and the inventory level at the planning horizon (assumed to be 1 year). Since this is a dynamic analysis process, it

Table 1 Commercial sawn wood sizes

Size	Size	Size	Size
$\frac{1}{2}" \times 6" \times 8$	$\frac{1}{2}" \times 8" \times 8$	$\frac{1}{2}" \times 10" \times 8$	$\frac{1}{2}" \times 12" \times 8$
$\frac{3}{4}" \times 6" \times 8$	$\frac{3}{4}" \times 8" \times 8$	$\frac{3}{4}" \times 10" \times 8$	$\frac{3}{4}" \times 12" \times 8$
$11/2" \times 6" \times 8$	$11/2" \times 8" \times 8$	$11/2" \times 10" \times 8$	$11/2" \times 12" \times 8$
$2" \times 6" \times 8$	$2" \times 8" \times 8$	$2" \times 10" \times 8$	$2" \times 12" \times 8$
$31/2" \times 6" \times 8$	$31/2" \times 8" \times 8$	$31/2" \times 10" \times 8$	$31/2" \times 12" \times 8$

allows the manager to define the plant's operation form. This section presents an example with the results of the production plan for type 1, characterized by 8′ length, 4 families, and different commercial diameters (indicated in the following tables). The plan information is expressed in P.U. (per unit) (Table 1).

6 Results

The product type and family for Plan I, characterized by 8′ (feet) long in different width and thick sizes are shown in the Tables 2 and 3.

These results allow the manager to make efficient and effective decisions, associated to the plan's operation cost, the product volume and type of commercial sizes the market demands. It is also important an appropriate management of an minimal inventory system capable of satisfying the market's demand in contingency conditions with the least financial resources invested in it.

Table 2 Production plan type I

	Family 1			Family 2			Family 3			Family 4		
T	D	P	I	D	P	I	D	P	I	D	P	I
1	40	125	0	40	90	0	65	65	0	70	120	0
2	60	0	85	50	0	50	120	120	0	50	0	50
3	25	0	25	50	90	0	120	145	0	100	190	0
4	65	65	0	40	0	40	25	0	25	90	0	90
5	120	120	0	70	120	0	60	120	0	80	120	0
6	120	145	0	50	0	50	60	0	60	40	0	40
7	25	0	25	100	190	0	85	125	0	60	85	0
8	60	120	0	90	0	90	40	0	40	25	0	25
9	60	0	60	80	120	0	40	90	0	65	65	0
10	85	165	0	40	0	40	50	0	50	120	120	0
11	40	0	80	60	85	0	50	90	0	120	145	0
12	40	0	40	25	0	25	40	0	40	25	0	25

T = time (months), D = Demand, P = Production Volume (units), I = Inventory Level

Table 3 Product type and family for Plan I

Families	Product	T (min)	Cost ($)
1	(3/4″ × 6″ × 8″)	0.826	3,801.00
2	(1/2″ × 8″ × 8″)		
3	(2″ × 10″ × 8″)		
4	(3/4″ × 12″ × 8″)		

7 Conclusions

From the resulting production plan, we conclude that for large-scale problems, the proposed algorithm is computationally efficient. The solution is obtained in 4 iterations. The proposed methodology presents a practical flexibility for efficient decision making in the company. Thus, the algorithm can be easily deployed on a personal computer, presenting a greater flexibility than the currently available commercial software to solve this kind of problem.

References

1. Aardal, K., Larsson, T.: A benders decomposition based heuristic for the hierarchical production problem. Eur. J. Oper. Res. (1990)
2. Bitran, G.R., Hax, A.C.: On the design of hierarchical production planning systems. Dec. Sci. **8** (1977)
3. Bitran, G.R., Haas, E.A., Hax, A.C.: Hierarchical production planning: a single stage system. Oper. Res. (1981)
4. González, S.F.: Modelado Matemático de un Aserradero Estándar en el Estado de Michoacán. Bol. 10, CIC- UMSNH. México (1987)
5. González, S.F.: Planeación de la Producción Jerárquica empleando Técnicas de Descomposición. Tesis PhD (Doctor en Ingeniería). Universidad Nacional Autónoma de México (1995)
6. González Santoyo, F.: Estrategias para la toma de decisions empresariales en un entorno de incertidumbre. Tesis de Doctor en Ciencias. Centro de Investigación y Desarrollo del Estado de Michoacán. México (2008)
7. González, S.F., Pérez Morelos, G.: Planeación de la Producción en la Pequeña Empresa Mexicana. Economía y Sociedad. Facultad de Economía UMSNH-México (1996)
8. González, S.F., Brunet, I.I, Flores, B., Chagolla, M.: Diseño de Empresas de Orden Mundial. URV (España)-UMSNH (México), FeGoSa-Ingeniería Administrativa (México) (2004)
9. González, S.F., Terceño, G.A., Flores, B., Diaz, R.: Decisiones Empresariales en la Incertidumbre. URV (España)-UMSNH (México), FeGoSa-Ingeniería Administrativa (México) (2005)
10. González Santoyo, F., Flores Romero, B., Gil Lafuente, A.M.: Procesos para la toma de decisions en un entorno globalizado. Editorial Universitaria Ramón Areces. España (2011)
11. Wagner, M.H., Whithin, M.: Dynamic version of the economic lot size model. Manag. Sci. (1958)

Mathematical Model and Parametrical Identification of Ecopyrogenesis Plant Based on Soft Computing Techniques

Yuriy P. Kondratenko, Oleksiy V. Kozlov, Galyna V. Kondratenko
and Igor P. Atamanyuk

Abstract This paper presents the development of the mathematical model and
system of parametrical identification for the ecopyrogenesis (EPG) plant as a
complex multi-coordinate control object on the basis of soft computing techniques.
The synthesis procedure of the main parts of the EPG plant's mathematical model,
including its fuzzy parametrical identification system, adaptive-network-based
fuzzy inference system for calculating of multiloop circulatory system
(MCS) temperature and Mamdani type fuzzy inference system for calculating of
reactor load level, is presented. The analysis of computer simulation results in the
form of static and dynamic characteristics graphs of the EPG plant confirms the
high adequacy of the developed complex neuro-fuzzy model to the real processes.
The developed mathematical model with parametrical identification based on
neuro-fuzzy technologies gives the opportunity to investigate the behavior of the
given complex control object in steady and transient modes, in particular, to synthesize and adjust the intelligent controllers of the multi-coordinate automatic
control system of the EPG plant.

Y. P. Kondratenko (✉) · G. V. Kondratenko
Petro Mohyla Black Sea State University, 68-Th Desantnykiv str. 10,
54003 Mykolaiv, Ukraine
e-mail: y_kondrat2002@yahoo.com

Y. P. Kondratenko · O. V. Kozlov · G. V. Kondratenko
Admiral Makarov National University of Shipbuilding, Heroes of Ukraine ave. 9,
54025 Mykolaiv, Ukraine
e-mail: oleksiy.kozlov@nuos.edu.ua

I. P. Atamanyuk
Mykolaiv National Agrarian University, Georgiy Gongadze str. 9,
54010 Mykolaiv, Ukraine
e-mail: atamanyuk_igor@mail.ru

© Springer International Publishing AG 2018
C. Berger-Vachon et al. (eds.), *Complex Systems: Solutions and Challenges
in Economics, Management and Engineering*, Studies in Systems,
Decision and Control 125, https://doi.org/10.1007/978-3-319-69989-9_13

1   Introduction

The environmental problem of organic waste (OW) accumulation is one of the major problems of our time caused by the development of industry and growth of urbanization in many countries of the world. Quite a prospective method of this problem solution is the use of the ecopyrogenesis technology, which allows complete utilization of the whole scope of the organic solid waste in the environmentally-friendly and energy-saving modes [1]. The EPG technology provides full utilization of all kinds of polymer waste, including polyethylene, polypropylene, polystyrene, polyvinyl chloride (but not more than 2%), worn tires, rubber, oil sludge, etc. by means of the multi-loop circulatory pyrolysis (MCP). It allows to obtain from the mass of raw materials up to 60–85% of the alternative liquid fuel (LF) of light fractions with characteristics comparable to diesel fuel, that can be used in combustion engines without any extra filtration, and up to 12–28% of pyrolysis gas (PG) with characteristics comparable to natural gas [1].

For realization of the EPG technology, the specific technological plants are used, which are, in turn, complicated multi-component technical objects. Complex automation of such technological plants allows to significantly increasing their operation efficiency and economic indicators [2].

Stabilization of the set values of the EPG plant's technological parameters is one of the important tasks of automatic control of the EPG process [2]. The possibility of automatic control of pyrolysis reactor temperature and load level as well as MCS temperature with high quality indicators allows controlling of the thermal destruction process of waste at various depth of hydrocarbon decay. This gives the opportunity to obtain the high-quality liquid fractions of alternative fuel on the outlet with the set molecular mass and, in turn, requires a special multi-coordinate automatic control system (MACS) of the EPG plant [2].

To study the MACS effectiveness at the stage of its design it is reasonable to use mathematical and computer modeling methods that are quite effective and low-cost, comparing with experimental and other approaches, especially while studying the behavior of thermal power and chemical-technological objects as well as their control systems [3–8]. In particular, the processes of development and adjustment of MACS intelligent controllers requires an availability of an adequate mathematical model. Also quality indicators of the EPG plant MACS significantly depend on the accuracy of the synthesized model and its adequacy to the real processes.

Analysis of physical properties and technical characteristics of the EPG plant as a complex multi-coordinate control object shows the reasonability of developing of mathematical models of its certain components with the use of combination of optimization approaches [9–15], various artificial intelligence principles [16–26] and soft computing techniques [27–32]. Fuzzy, neural and hybrid neuro-fuzzy mathematical models are developed and successfully implemented in such areas as technological processes control [24, 25, 33–35], robotics [24, 36], management of enterprises and human resources management [37–43], transport logistics [44–48], financial management [49–51], sport [52, 53] and political management [54], medical and technical diagnostics, stock forecasting, pattern recognition, and others [55–61].

Therefore, the aim of this work is development and research of the mathematical model of the EPG plant as a complex multi-coordinate control object on the basis of soft computing techniques.

2 Functional Structure of the EPG Plant as a Complex Multi-coordinate Control Object

The functional structure of the EPG plant as a complex multi-coordinate control object is shown in Fig. 1 [1], where the following notations are used: LFR—linear flow regulator of gas; ECV—electrically controlled valve; GB—gas burner which heats the reactor; HD—hydraulic drive; CA—cooling air; LG—liquefied gas; OC —output condenser; 1C, 2C, 3C—first, second and third MCS cooling circuits; u_{LFR}, u_{ECV}—control signals of LFR and ECV; Q_{CA}, Q_{LG}—flow values of cooling air and liquefied gas; X_V, Y_P—linear movement of ECV and HD piston; P_{GB}— heating power of GB; T_{MCS}—MCS outlet temperature; T_R—heating temperature of the reactor bottom wall; T_{WR}—heating temperature of waste in the reactor; L_R— waste load level in the reactor.

Temperature values T_R, T_{WR} and T_{MCS} are important parameters of the EPG plant, which essentially influence on speed and depth of the waste thermal destruction and subsequently affect on the quality of the output liquid alternative fuel [1, 2]. Also, to provide the continuous passage of the MCP process it is necessary to realize the automatic reloading of waste and control of it current level value in the reactor L_R [2].

Fig. 1 Functional structure of the EPG plant as a complex multi-coordinate control object

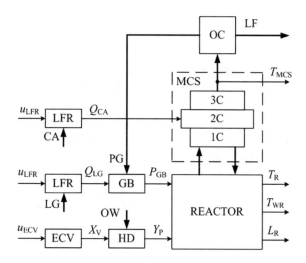

Control of the temperature values of the reactor bottom wall T_R, waste T_{WR} and MCS outlet T_{MCS} is implemented by changing of the gas burner power P_{GB} and cooling intensity of MCS circuits, respectively. In turn, control of the waste loading process and it level value L_R in the reactor is implemented by controlling of the EPG plant hydraulic drive.

The EPG plant's MCS (Fig. 1) includes three sequentially connected circuits with various cooling types: the 1st—with non-regulated air cooling; the 2nd—with regulated air cooling; the 3rd—with non-regulated water cooling. Thus, its cooling intensity control can be performed due to the cooling air flow Q_{CA} change of the 2nd circuit. The unregulated air fan delivers the constant value of cooling air through the 2nd circuit, which then can be changed using a linear flow regulator of air. The LFR, in turn, is a valve with a servodrive and has the linear characteristic of the dependence of the gas (air) consumption from the input voltage. LFR of liquefied gas is used for the control of gas flow Q_{LG}, that is fed to the GB. ECV is a spool valve that is moved by a servodrive and is intended for the control of HD, in particular, its piston linear movement Y_P.

Let's consider the development of the mathematical models of the EPG plant's main components in details.

3 Mathematical Model of the EPG Plant's Reactor with Fuzzy Parametrical Identification as a Temperature Control Object

Previous research, conducted in [63] shows that it is advisable to implement the development of the mathematical model of the EPG plant's reactor on the basis of an approach that involves the use of the transfer function of the generalized heat power control object and its parameters identification based on experimental data [63].

3.1 Structure of the Reactor's Mathematical Model with Fuzzy Parametrical Identification

The transfer function of the generalized heat power control object [63], that has heating device power as an input and heating temperature as an output, has the following form

$$W_R(p) = \frac{U_{out}(p)}{U_{in}(p)} = \frac{ke^{-\tau p}}{(T_1 p + 1)(T_2 p + 1)^n}, \tag{1}$$

where $W_R(p)$—transfer function of the control object; $U_{out}(p)$—image of the controlled coordinate $U_{out}(t)$ (reactor heating temperature); $U_{in}(p)$—image of the control action $U_{in}(t)$ (gas burner heat power); k—gain; τ—time delay; T_1, T_2—time constants of the aperiodic and inertial links respectively; n—order of the inertial link.

Transient characteristic $h(t)$ and other dynamic characteristics of the generalized heat power control object [63], that has transfer function (1), are given in Fig. 2, where $h(t)$—control object transient characteristic, that is obtained experimentally; $h''(t)$—second derivative nature of the transient characteristic $h(t)$; $h(t_e)$—value of the transient characteristic at the extreme point; h_s—steady value of the transient characteristic; T_0 i τ_0—time constant and time delay that can be found from the experimental transient characteristic.

The problem of the EPG plant reactor's mathematical model is reduced to the identification of the transfer function (1) parameters and inertial link order n [63]. In this case, the approximation criterion has the following form

$$\left.\begin{aligned} h_a(0) &= h(0) \\ h_{a,\,steady} &= h_{steady} \\ h_a(t_i) &= h(t_i) \\ h'_a(t_i) &= h'(t_i) \end{aligned}\right\} \quad (2)$$

It requires the coincidence of the experimental transient characteristic of a real object $h(t)$ with the approximate transient characteristic $h_a(t)$ of the mathematical model at the points $t = 0$, $t = \infty$ and inflection point t_i, which is determined from the requirement $h''(t) = 0$ [63].

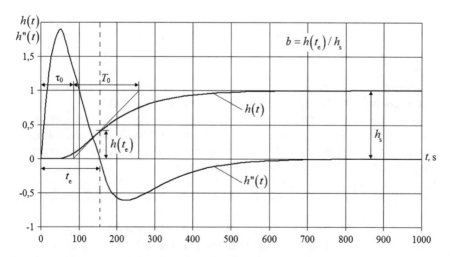

Fig. 2 Transient characteristics of the generalized heat power control object

Based on the analytical solution of the differential equation corresponding to the transfer function (1) at $n = 1$, we can create a parameter identification algorithm for approximating the transfer function [63]. The approximation criterion (2) after substituting the results of the analytical solution of the differential equation becomes

$$\left.\begin{array}{c} xe^{-y} = e^{-y/x} \\ (1+x)e^{-y} = 1 - b \\ T_1/T_0 = e^{-y} \end{array}\right\}, \tag{3}$$

where $x = T_1/T_2$ and $y = t_{i,\,a}/T_1$—dimensionless coefficients.

On the basis of known experimental transfer function values b and T_0 and the use of numerical methods for the solution of transcendental Eq. (3) the calculation of the variables x and y, as well as time constants T_1, T_2 and moment of inflection $t_{i,a}$ of the transient characteristic of the mathematical model (1) at $\tau = 0$ is possible. To bring the point of inflection of the approximating function to the real value of t_a it is necessary to enter a delay $\tau = t_{i,\,a} - t_a$ [63].

The reserve of further increasing the degree of adequacy of the reactor mathematical model is the optimization of its structure [62], which is done by varying the exponent n of Eq. (1) using the methods of nonlinear programming, in particular optimization gradient methods [9].

The problem formulation of the nonlinear programming concedes the objective function choice, the definition of the optimized parameters set, the set of constraints, as well as the formation of the primary hypothesis of the optimal parameters values.

The quadratic integral functional (4) of experimental transfer function $h(t)$ deviation from the identified approximating transfer function $h_a(t)$ is proposed to be used as the objective function [9].

$$I[h_a(t), h(t)] = \int_0^{T_{\max}} (h_a(t) - h(t))^2 dt. \tag{4}$$

Herewith reactor model transfer function $h_a(t)$ is uniquely determined by the parameters and structure of the approximating transfer function (1).

As the exponent n of the Eq. (1) can take only integer nonnegative values, so it's inserting to the optimized parameters set of the nonlinear optimization algorithm is not advisable. In particular, the definition of the n value is possible by means of complete listing of the set of its admissible values $n \in \{1, 2, 3, \ldots, n_{\max}\}$, where $n_{\max}$ is the limit of the aperiodic link order, $n_{\max} = 8$. Thus, the optimized parameters set for the nonlinear optimization algorithm is reduced to the form: $\mathbf{P} = \{T_1, T_2, \tau\}$.

Let's accomplish the identification of the transfer function (1) parameters on the basis of the given approach and the heating transient experimental characteristics of the EPG plant reactor's bottom wall (Fig. 3) and waste in the reactor (Fig. 4).

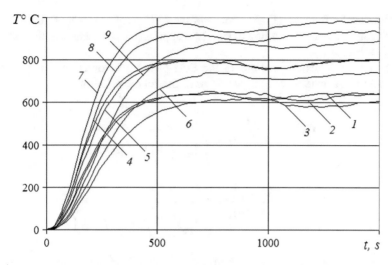

Fig. 3 Experimental transient characteristics of the reactor's bottom wall heating $T_R = f(t)$

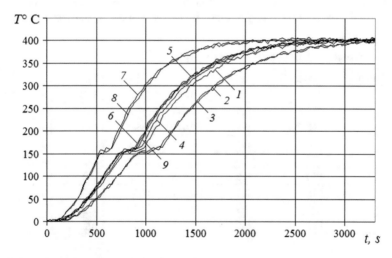

Fig. 4 Experimental transient characteristics of waste heating in the reactor $T_{WR} = f(t)$

The experimental transient characteristics of heating are obtained for the reactor's capacity of 14 L and following values of gas burner power P_{GB} and reactor load level L_R: (1) $P_{GB} = 16$ kW, $L_R = 0,2$ m; (2) $P_{GB} = 16$ kW, $L_R = 0,3$ m; (3) $P_{GB} = 16$ kW, $L_R = 0,5$ m; (4) $P_{GB} = 20$ kW, $L_R = 0,2$ m; (5) $P_{GB} = 20$ kW, $L_R = 0,3$ m; (6) $P_{GB} = 20$ kW, $L_R = 0,5$ m; (7) $P_{GB} = 25$ kW, $L_R = 0,2$ m; (8) $P_{GB} = 25$ kW, $L_R = 0,3$ m; (9) $P_{GB} = 25$ kW, $L_R = 0,5$ m.

The transfer function (1) parameters of the EPG plant's reactor are found in the identification process according to the given approach and presented in the Tables 1 and 2, respectively.

In turn, the transfer functions of the EPG plant's reactor as control objects of temperature of bottom wall and waste have the following forms

$$W_R(p) = \frac{T_R(p)}{P_{GB}(p)} = \frac{k_{\sim R1}}{\left(T_{\sim R1}p+1\right)\left(T_{\sim R2}p+1\right)^2}; \tag{5}$$

$$W_{WR}(p) = \frac{T_{WR}(p)}{P_{GB}(p)} = \frac{k_{\sim WR1}}{\left(T_{\sim WR}p+1\right)\left(T_{\sim WR2}p+1\right)^3}. \tag{6}$$

As Tables 1 and 2 show the parameters k, T_1 and T_2 of reactor transfer function (5) and (6) change at different values of the gas burner power P_{GB} and reactor's load level L_R. Therefore, for the synthesis of a universal mathematical model of the reactor on the basis of the data presented in Tables 1 and 2 it is advisable to use the

Table 1 Parameters of the transfer function W_R (p)

Curve number	P_{GB}, kW	L_R, m	k_R	T_{R1}	T_{R2}	τ_R	n_R
1	15	0,1	0,043	85,2	34,3	0	2
2	15	0,3	0,042	92,8	56,8	0	2
3	15	0,5	0,04	99,3	103,2	0	2
4	20	0,1	0,041	84,9	33,6	0	2
5	20	0,3	0,04	92,5	56,4	0	2
6	20	0,5	0,038	99,1	102,5	0	2
7	25	0,1	0,041	84,5	32,8	0	2
8	25	0,3	0,037	92,4	55,3	0	2
9	25	0,5	0,035	99,1	102	0	2

Table 2 Parameters of the transfer function W_{WR} (p)

Curve number	P_{GB}, kW	L_R, m	k_{WR}	T_{WR1}	T_{WR2}	τ_{WR}	n_{WR}
1	15	0,1	0,0251	426,5	343,7	0	3
2	15	0,3	0,0247	464,2	356,2	0	3
3	15	0,5	0,0256	482,3	383,6	0	3
4	20	0,1	0,021	377,7	303,4	0	3
5	20	0,3	0,0197	398,6	312,5	0	3
6	20	0,5	0,0207	418,4	324,1	0	3
7	25	0,1	0,0162	306,1	216,8	0	3
8	25	0,3	0,0164	309,7	220,2	0	3
9	25	0,5	0,0159	368,2	295,6	0	3

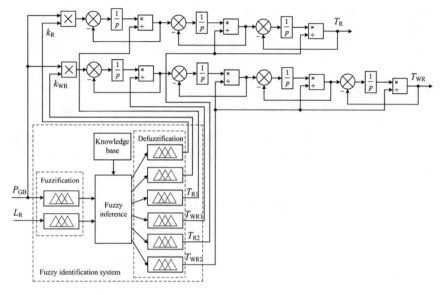

Fig. 5 Functional structure of the mathematical model of the EPG plant's reactor with fuzzy parametrical identification system

specific identification system that determine the coefficients k_R, T_{R1}, T_{R2}, k_{WR}, T_{WR1} and T_{WR2} of reactor transfer functions (5) and (6), depending on the parameters P_{GB} and L_R [64].

The analysis of dependences $k_{\sim R} = \text{var} = f(P_{GB}, L_R)$; $T_{\sim R1} = \text{var} = f(P_{GB}, L_R)$; $T_{\sim R2} = \text{var} = f(P_{GB}, L_R)$; $k_{\sim WR1} = \text{var} = f(P_{GB}, L_R)$; $T_{\sim WR1} = \text{var} = f(P_{GB}, L_R)$; $T_{\sim WR2} = \text{var} = f(P_{GB}, L_R)$, presented in the Tables 1 and 2 shows the reasonability of the parametrical identification system development on the basis of fuzzy logic principles and algorithms [64], that are widely used in the synthesis of mathematical models and control devices of objects with significant uncertainties and nonlinearities [26, 55–61].

The functional structure of the mathematical model of the EPG plant's reactor with fuzzy parametrical identification system is presented in Fig. 5 [64].

Let us consider the synthesis procedure particularities of fuzzy identification system of Mamdani type in detail.

3.2 Synthesis Procedure of the Mamdani Type Identification System of the Reactor's Mathematical Model

The main stages of the Mamdani type fuzzy logic inference are: fuzzification, aggregation, activation, accumulation and defuzzification [20]. The according linguistic meaning and degree of fuzzy set membership are determined for each input variable on the fuzzification stage [21]. In this case it is advisable to choose the following linguistic terms for the input and output variables, whose parameters are presented in Fig. 6.

For input variables P_{GB} and L_R range of values are determined in relative units. In Fig. 6 the following notation are accepted: VS—very small; S—small; LM—less than middle; M—middle; MM—more than middle; B—big; VB—very big; L—low; H—high.

The knowledge base is formed to implement the fuzzy inference. The rules of the developed knowledge base according to the Mamdani algorithm are the linguistic statements in the form:

$$\text{IF}\,''P_{GB} = a''\,\text{AND}\,''L_R = b''\,\text{THEN}\,''k_R = c''\,\text{AND}\,''T_{R1} = d''\,\text{AND}\,''T_{R2} = e''$$
$$\text{AND}\,''k_{WR} = f''\,\text{AND}\,''T_{WR1} = g''\,\text{AND}\,''T_{WR2} = h'',$$

where a, b, c, d, e, f, g, h—according linguistic terms values.

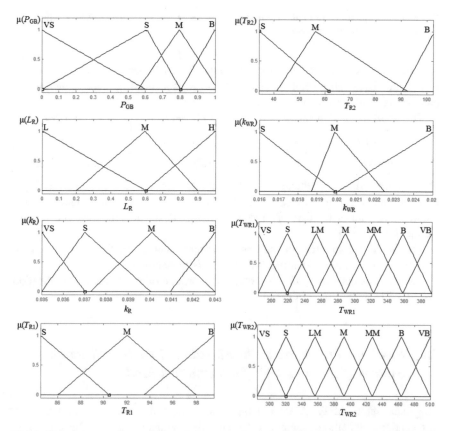

Fig. 6 Linguistic terms parameters of the reactor's mathematical model fuzzy identification system

Table 3 Knowledge base of the reactor's mathematical model fuzzy identification system

Rule number	Input variables		Output variable					
	P_{GB}	L_R	k_R	T_{R1}	T_{R2}	k_{WR}	T_{WR1}	T_{WR2}
1	VS	L	B	S	S	B	MM	MM
2	VS	M	B	M	M	M	B	B
3	VS	H	B	B	B	S	VB	VB
4	S	L	B	S	S	B	M	M
5	S	M	B	M	M	M	B	B
6	S	H	M	B	B	S	VB	VB
7	M	L	M	S	S	B	M	M
8	M	M	M	M	M	M	MM	MM
9	M	H	S	B	B	S	B	B
10	B	L	B	S	S	B	VS	VS
11	B	M	M	M	M	M	S	S
12	B	H	S	B	B	S	LM	LM

The developed knowledge base consists of 12 rules, which correspond to all possible combinations of two input fuzzy variables. The reactor's mathematical model identification system knowledge base is presented in Table 3.

The truth degree is determined for every rule of the fuzzy inference system at the next stage (aggregation), and truth degree finding procedure for each fuzzy output rule sub-conclusion is implemented on the activation stage [21].

The further stage of the fuzzy logic inference is accumulation that is the membership function finding procedure for every output linguistic variable [24]. The aim of accumulation is to combine all output linguistic terms with according truth degrees of each rule for the obtaining of output variable membership function.

Thus, the resultant membership function for the fuzzy decision is formed at the accumulation stage, which is necessary to be converted to precise output signal value.

The procedure of finding of output variables $k_{\sim R}$, $T_{\sim R1}$, $T_{\sim R2}$, $k_{\sim WR}$, $T_{\sim WR1}$, $T_{\sim WR2}$ precise numerical values is the defuzzification stage.

There are several methods of defuzzification: the gravity center method, the square center method, the left modal value method, the right modal value method and the other [20, 25].

In this case the gravity center method is chosen, according to which the output signal value is calculated by the formula

$$u_{\text{out}} = \frac{\sum_{i=1}^{n} u_i \cdot \mu(u_i)}{\sum_{i=1}^{n} \mu(u_i)}, \tag{7}$$

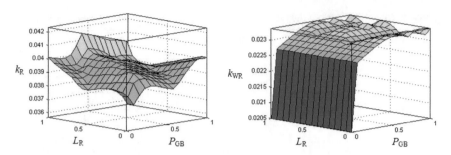

Fig. 7 Characteristic surfaces of the reactor's mathematical model fuzzy identification system: **a** $k_R = f(P_{GB}, L_R)$; **b** $k_{WR} = f(P_{GB}, L_R)$

where u_{out}—one of the identification system output signals ($k_{\sim R}$, $T_{\sim R1}$, $T_{\sim R2}$, $k_{\sim WR}$, $T_{\sim WR1}$, $T_{\sim WR2}$), n—the number of output linguistic variable values; u_i—ith value of the according output linguistic variable; $\mu(u_i)$—the number of the resultant membership function for the according value u_i.

The characteristic surfaces of the developed fuzzy identification system are presented in Fig. 7.

3.3 Comparative Analysis and Adequacy Evaluation of the Reactor's Mathematical Models Based on Heat Exchange Equations and Fuzzy Parametrical Identification

The computer simulation of the EPG plant's reactor heating transients (Figs. 8 and 9) was carried out for the experimental EPG plant with the help of the developed in this paper mathematical model with fuzzy parametrical identification system and well tested mathematical model based on heat transfer processes equations developed and presented in [65].

The simulation was carried out for the experimental EPG plant's reactor with a volume of 14 L at gas burner power $P_{GB} = 18$ kW and reactor's load level $L_R = 0,4$ m, constant temperature of ambient $T_A = 0°C$.

In Figs. 8 and 9 the following notations are accepted: *1*—transition process of the real object, *2*—transition process of the model based on the heat exchange processes equations, *3*—transitional process of the model with fuzzy parametrical identification system.

Adequacy analysis of the developed in this paper mathematical model with fuzzy parametrical identification system and mathematical model based on heat transfer processes equations developed in [65] is carried out using hypotheses assessing methods of mathematical statistics [64], namely:

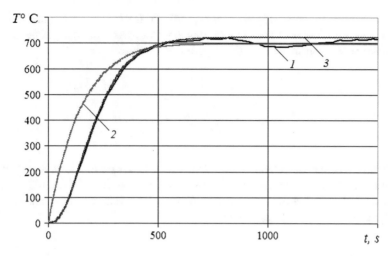

Fig. 8 Transient characteristics of the reactor's bottom wall heating $T_R = f(t)$

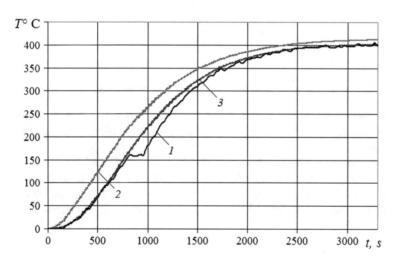

Fig. 9 Transient characteristics of waste heating in the reactor $T_{WR} = f(t)$

(1) The sum of squared errors (SSE), which shows the total deviation of values of the mathematical model $T_m(t)$ from the corresponding values of the experimental data $T_e(t)$

$$SSE = \sum_{i=1}^{k} [T_{ei} - T_{mi}]^2 \to 0; \qquad (8)$$

(2) The root of the mean for the squared error for (*RMSE*) is an estimation of the standard deviation of the random component between the data of the synthesized regression model and the experimental values

$$RMSE = \sqrt{\frac{1}{k} \sum_{i=1}^{k} [T_{ei} - T_{mi}]^2} \to 0; \tag{9}$$

(3) The coefficient of determination (R^2), which is a part of the variance of the variable deviation dependent from its average value. In other words, R^2 is the square of mixed correlation between experimental values and the values of the synthesized mathematical model

$$R^2 = 1 - \frac{\sum_{i=1}^{k} [T_{ei} - T_{mi}]^2}{\sum_{i=1}^{k} [T_{ei} - \overline{T_{ei}}]^2} \to 1, \tag{10}$$

where $\overline{T_{ei}} = \frac{1}{k} \sum_{i=1}^{k} T_{ei}$ is the arithmetic mean of the experimental sample.

The calculation results of statistical evaluations of the adequacy of the EPG plant's reactor mathematical models as a temperature control object are shown in Table 4.

Based on the transient characteristics (Figs. 8 and 9) and the statistical data (Table 4) we can conclude that the reactor's mathematical model as a temperature control object developed in this work has better results of experimental samples compliance than the model based on heat transfer processes equations that is presented in [65].

Table 4 Comparative analysis of the adequacy of the EPG plant's reactor mathematical models

Calculation of transients of the reactor bottom wall heating $T_R = f(t)$			
Mathematical model type	SSE	R^2	RMSE
Based on the heat exchange processes equations	319726,7	0,854	79,177
With fuzzy parametrical identification	13891,6	0,993	16,504
Calculation of transients of waste heating in the reactor $T_{WR} = f(t)$			
Mathematical model type	SSE	R^2	RMSE
Based on the heat exchange processes equations	172230,9	0,921	39,39
With fuzzy parametrical identification	18162,77	0,991	12,79

4 Neuro-Fuzzy Mathematical Model of the EPG Plant's MCS

At the synthesis of the MCS mathematical model it is reasonable also to use the approach based on the experimental data, that is used for the development of the reactor model as a temperature control object [62–64]. But, in this case the value of the heat flow, that is given to the MSC, is almost impossible to define. So, it is also impossible to determine the proportional dependence of the MSC outlet temperature T_{MCS} from the heat flow in the form of the gain k of the transfer function (1) [62, 63]. In turn, the total heat flow, that is given to the MSC, basically depends on such known parameters, as: heating temperature of waste T_{WR}, cooling air flow Q_{CA} and heating temperature of the reactor bottom wall T_R. Thus, to develop the MSC mathematical model on the basis of the given above approach and to consider the special features of the MCS multicircuit structure it is reasonable to use the adaptive-network-based fuzzy inference system for temperature calculating (ANFISTC) [66], according to the dependence $T_{MCS} = f_{ANFISTC} (T_{WR}, Q_{CA}, T_R)$, that can be synthesized on the basis of experimental data.

4.1 Structure of the MCS Neuro-Fuzzy Mathematical Model

The functional structure of the MCS neuro-fuzzy mathematical model is presented in Fig. 10 [66].

The following values are given to the input of this model: heating temperature of waste T_{WR}, heating temperature of the reactor bottom wall T_R and the flow of the cooling air of the 2nd circuit of the MCS Q_{CA}. The MCS outlet temperature T_{MCS} is formed on the output. Time constants T_{1MCS}, T_{2MCS}, τ_{MCS} and n order are determined from the experimental characteristic of the MCS heating transient using the approach [63] given in the item 3.1. The mathematical model with fuzzy parametrical identification system, developed in the item 3, is used as the EPG plant reactor model in this case. In turn, the LFR mathematical model is represented as an

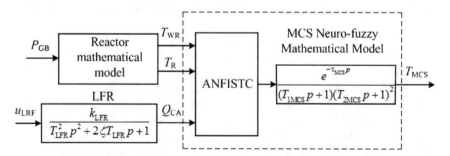

Fig. 10 Functional structure of the MCS Neuro-fuzzy mathematical model

oscillatory dynamic link [66], where k_{LFR}, T_{LFR}, ζ—gain constant, time constant and damping coefficient, which are determined by the parameters of the servodrive and gas valve, which are the parts of the linear flow regulator.

Let us consider the synthesis procedure particularities of the adaptive-network-based fuzzy inference system for temperature calculating of the MCS mathematical model in detail.

4.2 Synthesis Procedure of the ANFISTC

Adaptive-network-based fuzzy inference system (ANFIS) is a variant of hybrid neuro-fuzzy network—neural network of direct signal propagation of particular type [67]. ANFIS implements the Sugeno type fuzzy inference system in the form of a five-layer neural network of the signal forward propagation. The neuro-fuzzy network architecture is isomorphic to the fuzzy knowledge base. In the neuro-fuzzy networks the differentiated implementation of triangular norms (multiplication and probabilistic OR) are used, as well as the smooth membership functions [67–69]. This allows to apply for adjustment of neuro-fuzzy networks fast algorithms for neural networks training, based on the method of back-propagation.

Fuzzy rule with serial number r has the following form [67]

$$\text{IF } x_1 = a_{1,r} \text{ AND} \ldots \text{AND } x_n = a_{n,r} \text{ THEN } y = b_{0,r} + b_{1,r} x_1 + \ldots + b_{n,r} x_n, \quad (11)$$

where $r = 1,\ldots, m$—the number of rules; $a_{i,r}$—fuzzy term with membership function $\mu_r(x_i)$, that is used for the linguistic evaluation of variable x_i in the r-th rule ($r = 1,\ldots, m$; $i = 1,\ldots, n$); $b_{q,r}$—real numbers in conclusion of r-th rule ($r = 1,\ldots, m$; $q = 1,\ldots, n$).

The functional structure of typical ANFIS with two inputs x_1, x_2 and one output y is presented in Fig. 11 [67].

ANFIS-network functions as follows:

Layer 1. Each node of the first layer is one term with certain membership function. Network inputs x_1, $x_2,\ldots, x_n$ are connected only with their terms [67]. Nodes amount is equal to the sum of the terms of all variables. The node output is a membership degree of the input variable value to the corresponding fuzzy term

$$\mu_r(x_i) = \frac{1}{1 + \left| \frac{x_i - g}{c} \right|^{2d}}, \quad (12)$$

where c, d, g—adjustable parameters of the membership function.

Layer 2. The number of nodes of the second layer is m. Each node of this layer corresponds to one fuzzy rule [68]. A node of the second layer is connected with the nodes of the first layer, which form the antecedents of the corresponding rule. Therefore, each node of the second layer can receive from 1 to n input signals. The node output is the degree of fulfillment of the rules, which is calculated as the

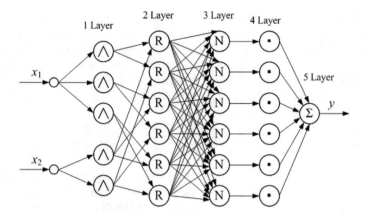

Fig. 11 Functional structure of typical ANFIS

product of the input signals. Let's denote the nodes outputs of this layer as $\tau_r, r = \overline{1, m}$.

Layer 3. The number of nodes of the third layer is also m. Each node in this layer calculates the relative degree of fulfillment of the fuzzy rule [67]

$$\tau_r^* = \frac{\tau_r}{\sum\limits_{j=\overline{1,m}} \tau_j}. \tag{13}$$

Layer 4. The number of nodes of the fourth layer is also m [69]. Each node is connected to one node of the third layer and also with all network inputs (in Fig. 11 the connection with the inputs are not shown). The node of the fourth layer calculates the contribution of a fuzzy rule in the output of the network

$$y_r = \tau_r^* \cdot (b_{0,r} + b_{1,r}x_1 + \ldots + b_{n,r}x_n). \tag{14}$$

Layer 5. The only node of this layer sums the contributions of all the rules [67]

$$y = y_1 + \ldots y_r \ldots + y_m. \tag{15}$$

The authors' studies [66] have shown, that the shape of the input variables linguistic terms membership functions significantly affects the training process of the ANFISTC and MCS model accuracy on the whole. So, to achieve the highest accuracy of the MCS neuro-fuzzy mathematical model, at the stage of its design the ANFISTC synthesis for the different types of linguistic terms membership functions (triangular, trapezoidal, Gaussian 2) of the input variables T_{WR}, T_R and Q_{CA} is considered. For the variables T_{WR} and T_R 3 linguistic terms are chosen for each one: L—Low, M—middle and H—High. In turn, for the variable Q_{CA}—5 terms: Z— Zero, S—small, M—middle, B—big and VB—Very Big. The proposed ANFISTC knowledge base consists of 45 rules, which correspond to all combinations of three input fuzzy variables. In this case the amount of rules coefficients $b_{q,r}$ is 135.

ANFISTC training was conducted with the help of training sample with the length of 246 points and the hybrid training method, which combines the back-propagation method and the least square method. The training sample is based on the experimental characteristics of the MCS heating transients of the real EPG plant in different operation modes.

Training results of the proposed ANFISTC for different types of membership functions (triangular, trapezoidal, Gaussian 2) of the input variables linguistic terms are shown in Table 5.

Parameters of the input variables linguistic terms as well as the rules coefficients $b_{q,r}$, found at the training process, are shown in Fig. 12 and Table 6, respectively.

The characteristic surfaces of the developed ANFISTC for the input variables linguistic terms membership functions of different types at $T_R = 1000$ °C are presented in Fig. 13.

After analyzing the above research we can conclude that the highest accuracy at the ANFISTC training process can be achieved at the trapezoidal type of membership function of input variables linguistic terms (training error is 0,0934).

4.3 Comparative Analysis and Adequacy Evaluation of the MCS Mathematical Models Based on ANFISTC with Input Variables Linguistic Terms Membership Functions of Different Types

The computer simulation of the EPG plant's MCS heating transient processes (Figs. 14 and 15) was carried out for the experimental EPG plant at $T_R = 1000$°C, $Q_{CA} = 0$.

In Figs. 14 and 15 the following notations are accepted: 1—transient process of the real object, 2—transient process of the model based on the ANFISTC with input variables linguistic terms membership functions of Gaussian 2 type, 3—transient process of the model based on the ANFISTC with input variables linguistic terms membership functions of triangular type; 4—transition process of the model based on the ANFISTC with input variables linguistic terms membership functions of trapezoidal type.

Table 5 Training results of the ANFISTC

Input variables linguistic terms membership function type	Number of training epochs	Minimum training error
Triangular	3147	0,879
Trapezoidal	4725	0,0934
Gaussian 2	3012	1,443

Fig. 12 Parameters of the ANFISTC input variables linguistic terms at different types of membership functions: **a** triangular; **b** trapezoidal; **c** Gaussian 2

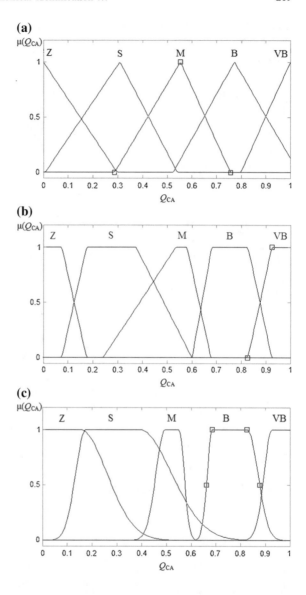

After analyzing the simulation results we can confirm that the mathematical model has rather high level of adequacy as the nature of its transient processes with high accuracy repeat the character of real processes of MCS heating. The graphics of transient processes (Figs. 14 and 15) represent the processes of MCS heating, that proceed very slowly before waste start to boil and is carried mainly out at the cost of light evaporation when waste heating and melting. Then, after the waste boiling, at the cost of it intensive evaporation, the MCS heating rate significantly increases.

Table 6 ANFISTC knowledge base fragment for the input variables linguistic terms membership functions of different types

Rule number	Input variables and rules coefficients					
	T_{WR}	T_R	Q_{CA}	$b_{1,r}$	$b_{1,r}$	$b_{3,r}$
Membership functions of triangular type						
1	L	L	Z	0	0	0
3	L	L	M	0,193	0,0014	0,001
6	L	M	Z	0,192	0,0167	0,0087
9	L	M	B	0,195	0,0014	1,89
12	L	H	S	0,257	0,096	2,056
15	L	H	VB	0,324	0,09	2,53
20	M	L	VB	0,256	0,097	2,39
25	M	M	VB	0,138	0,043	1,58
30	M	H	VB	0,194	0,056	4,578
41	H	H	Z	0,313	0,076	0,0096
45	H	H	VB	0,245	0,023	7,89
Membership functions of trapezoidal type						
1	L	L	Z	0	0	0
3	L	L	M	0,0087	0	0,003
6	L	M	Z	0,184	0,0107	0,0044
9	L	M	B	1,498	0,035	0,914
12	L	H	S	0,8676	0,089	2,0798
15	L	H	VB	0,3647	0,094	2,52
20	M	L	VB	0,259	0,082	2,396
25	M	M	VB	0,298	0,3986	2,004
30	M	H	VB	0,198	0,0136	4,523
41	H	H	Z	0,368	0,072	0,0091
45	H	H	VB	0,213	0,032	7,12
Membership functions of Gaussian 2 type						
1	L	L	Z	0	0	0
3	L	L	M	0	0,009	0
6	L	M	Z	0,202	0,0171	0,0037
9	L	M	B	0,159	0,0054	1,95
12	L	H	S	0,495	0,089	0,984
15	L	H	VB	0,356	0,089	2,38
20	M	L	VB	0,251	0,094	2,14
25	M	M	VB	0,596	0,746	0,948
30	M	H	VB	0,209	0,061	5,67
41	H	H	Z	0,421	0,079	0,0096
45	H	H	VB	0,241	0,026	7,81

Adequacy analysis of the developed MCS mathematical models based on ANFISTC with input variables linguistic terms membership functions of different

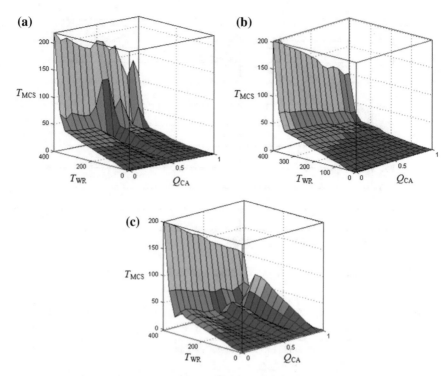

Fig. 13 ANFISTC characteristic surfaces for the input variables linguistic terms at different types of membership functions: **a** triangular; **b** trapezoidal; c Gaussian 2

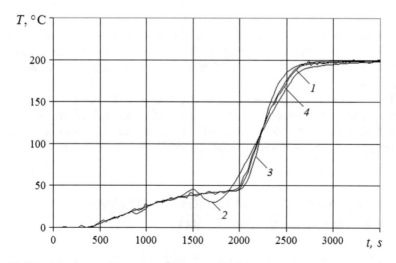

Fig. 14 Transient characteristics of the MCS $T_R = 1000°C$, $Q_{CA} = 0$

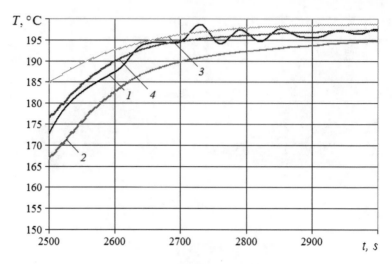

Fig. 15 Detailed graphics of transient characteristics of the MCS $T_R = 1000°C$, $Q_{CA} = 0$

Table 7 Comparative analysis of the adequacy of the EPG plant MCS mathematical models

Mathematical model type	SSE	R^2	RMSE
Based on the ANFISTC with input variables linguistic terms membership functions of triangular type	2307,2	0,997	4,366
Based on the ANFISTC with input variables linguistic terms membership functions of trapezoidal type	323,3	0,999	1,634
Based on the ANFISTC with input variables linguistic terms membership functions of Gaussian 2 type	2940,8	0,996	4,929

types is carried out using hypotheses assessing methods of mathematical statistics [64], that are presented in the paragraph 3.3. The calculation results of statistical evaluations of the adequacy of the EPG plant MCS mathematical models are shown in Table 7.

Based on the transient characteristics (Figs. 14 and 15) and the statistical data (Table 7) we can conclude that the MCS model developed on the basis of the ANFISTC with input variables linguistic terms membership functions of trapezoidal type has the best result of experimental samples compliance.

5   Mathematical Model of the EPG Plant's Reactor as a Load Level Control Object

At developing the mathematical model of the EPG plant's reactor as a load level control object the following processes should be considered: waste loading to the reactor with the help of hydraulic drive as well as waste evaporation, destruction and conversion into pyrolysis gas and liquid fuel according to the MCP process at a certain temperature value achievement [70]. The waste level in the reactor increases stepwisely at downloading due to the reciprocating movement of the HD piston [1, 70]. In turn, at heating to the certain temperature the processes of waste evaporation, destruction and conversion into pyrolysis gas and liquid fuel are going and it level in the reactor gradually decreases. The level value of waste in the reactor at loading process with the help of the HD can be determined on the basis of the equation of volume calculating, considering the cross-sectional areas of the HD piston S_P and the reactor S_R as well as the HD piston linear movement and waste compression ratio k_C at it passing through the cone of the loading system [70]. In turn, the compression ratio k_C depends on the physical properties of solid waste and is defined for each type of raw materials experimentally. Also, in the evaporation and MCP processes the speed of waste level decreasing depends on the temperature mode in the reactor and can be determined from experimental data obtained for the EPG plant reactor with a volume of 14 L (Fig. 16).

Level measurement (Fig. 16) was carried out for three values of the reactor heating temperature: *1*, *2*, *3*—$T_R = 900°C$; *4*, *5*, *6*—$T_R = 700°C$; *7*, *8*, *9*—$T_R = 600°C$; three values of the MCS outlet temperature: *3*, *6*, *9*—$T_{MCS} = 100°C$; *2*, *5*, *8*—$T_{MCS} = 150°C$; *1*, *4*, *7*—$T_{MCS} = 200°C$; at three points: $t = 0$ s; $t = 3000$ s; $t = 6000$ s after the establishment of these values of the reactor and MCS temperature.

So the point $t = 0$ corresponds to the beginning of the processes of evaporation and destruction of waste.

From the graphs of level change, shown in Fig. 16, the values of the average speed of waste level decreasing can be determined, which are: (1) $dL_R/dt = 0,000083$ m/s; (2) $dL_R/dt = 0,00008$ m/s; (3) $dL_R/dt = 0,000077$ m/s; (4) $dL_R/dt = 0,000069$ m/s; (5) $dL_R/dt = 0,000066$ m/s; (6) $dL_R/dt = 0,000063$ m/s; (7) $dL_R/dt = 0,000061$ m/s; (8) $dL_R/dt = 0,000059$ m/s; (9) $dL_R/dt = 0,000056$ m/s.

To calculate the approximated values of the speed of the level change dL_R/dt at different values of the reactor heating temperature T_R and MCS outlet temperature

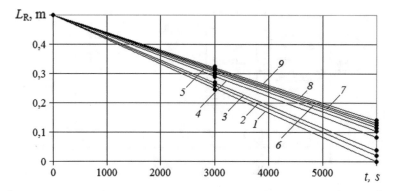

Fig. 16 Transients of waste level change in the reactor at the passage of the processes of evaporation and destruction

T_{MCS} it is advisable to use the fuzzy system for calculating of the speed of the level change (FSCSL) in the reactor.

5.1 Structure of the Mathematical Model of Reactor as a Load Level Control Object

The functional structure of the synthesized mathematical model, whose inputs are the linear movement of the HD piston Y_{P}, the reactor's bottom wall heating temperature T_{R} as well as the MCS outlet temperature T_{MCS} and output is the reactor load level L_{R}, is shown in Fig. 17.

In order to receive the value of the HD piston movement speed dY_{P}/dt from the value of it linear movement Y_{P} the time derivative is calculated by the differentiator D. The cut-off element CE allows only positive values of the piston speed dY_p^+/dt, which is then integrated by integrator I. To calculate the volume of waste, loaded into the reactor by the HD, and then the load level value L_{LR} the received positive movement of the piston is multiplied by a ratio k_{R} [70]

$$k_{\mathrm{R}} = \frac{S_{\mathrm{P}}}{k_{\mathrm{C}} S_{\mathrm{R}}}. \tag{16}$$

5.2 Synthesis Procedure of the Mamdani Type FSCSL

At the synthesis of Mamdani type FSCSL on the fuzzification stage 3 linguistic terms are chosen for the first and for the second input variables T_{R} and T_{MCS},

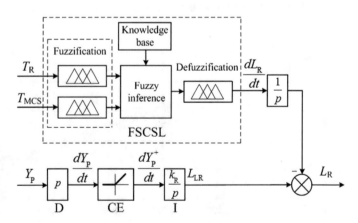

Fig. 17 Functional structure of the mathematical model of the EPG plant's reactor as a load level control object

respectively, and 9 linguistic terms are chosen for the output variable dL_R/dt. Parameters of the selected linguistic terms are presented in Table 8.

The rules of the developed FSCSL knowledge base according to the Mamdani algorithm are the linguistic statements in the form:

$$\text{IF } ''T_R = x'' \text{ AND } ''T_{MCS} = y'' \text{ THEN } ''dL_R/dt = z'',$$

where x, y, z—according FSCSL linguistic terms values.

The developed FSCSL knowledge base (Table 9) consists of 9 rules, which correspond to all possible combinations of two input fuzzy variables.

The gravity center method is also chosen for the defuzzification procedure in case.

The characteristic surface of the developed FSCSL is presented in Fig. 18.

Table 8 Linguistic terms parameters of the developed FSCSL

Term	Membership function type	Range of values
For input variable T_R		
L—Low	Triangular	[500 600 700]
M—Middle	Triangular	[600 700 900]
H—High	Triangular	[700 900 1000]
For input variable T_{MCS}		
L—Low	Triangular	[80 100 120]
M—Middle	Triangular	[100 150 180]
H—High	Triangular	[150 200 250]
For output variable dL_R/dt		
1	Triangular	$[5,3 \cdot 10^{-5} \ 5,6 \cdot 10^{-5} \ 5,9 \cdot 10^{-5}]$
2	Triangular	$[5,6 \cdot 10^{-5} \ 5,9 \cdot 10^{-5} \ 6,1 \cdot 10^{-5}]$
3	Triangular	$[5,9 \cdot 10^{-5} \ 6,1 \cdot 10^{-5} \ 6,3 \cdot 10^{-5}]$
4	Triangular	$[6,1 \cdot 10^{-5} \ 6,3 \cdot 10^{-5} \ 6,5 \cdot 10^{-5}]$
5	Triangular	$[6,4 \cdot 10^{-5} \ 6,6 \cdot 10^{-5} \ 6,8 \cdot 10^{-5}]$
6	Triangular	$[6,7 \cdot 10^{-5} \ 6,9 \cdot 10^{-5} \ 7,1 \cdot 10^{-5}]$
7	Triangular	$[6,8 \cdot 10^{-5} \ 7,7 \cdot 10^{-5} \ 8 \cdot 10^{-5}]$
8	Triangular	$[7,8 \cdot 10^{-5} \ 8 \cdot 10^{-5} \ 8,2 \cdot 10^{-5}]$
9	Triangular	$[8,1 \cdot 10^{-5} \ 8,3 \cdot 10^{-5} \ 8,5 \cdot 10^{-5}]$

Table 9 Knowledge base of the developed FSCSL

		T_{MCS}		
		L	M	H
T_R	L	1	2	3
	M	4	5	6
	H	7	8	9

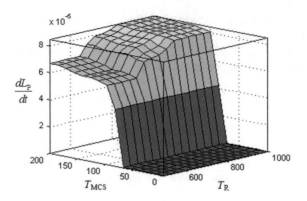

Fig. 18 Characteristic surface of the developed FSCSL

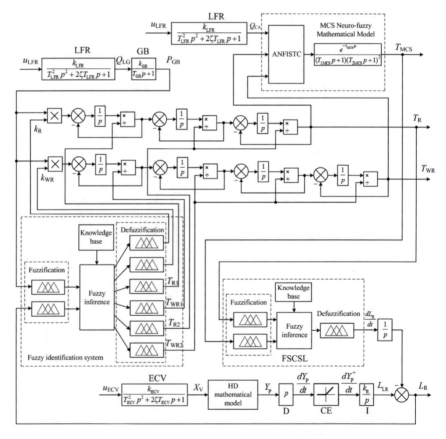

Fig. 19 Functional structure of the EPG plant as a complex multi-coordinate control object

6 Mathematical Model of the EPG Plant as a Multi-coordinate Control Object

Functional structure of the generalized mathematical model of the EPG plant as a multi-coordinate control object, that is synthesized on the basis of the models presented and described in paragraphs 3–5, is shown in Fig. 19.

The ECV mathematical model is represented as an oscillatory dynamic link [70], where k_{ECV}, T_{ECV}, ζ—gain constant, time constant and damping coefficient, which are determined by the parameters of the servodrive and spool valve, which are the parts of the electrically controlled valve. In turn, the GB, that heats the reactor, has the transfer function of the first-order inertial link [62], where k_{GB}—the gain of the gas burner, that is determined by the lower specific heat of combustion Q_1^s of gas, that is used as a fuel; T_{GB}—gas burner time constant, that is determined by the rate of gas inflammation.

Also the well tested model developed and presented in [70] was chosen as the mathematical model of the HD.

The computer simulation of the EPG plant as a multi-coordinate control object is carried out with the help of the synthesized model for the experimental EPG plant with the reactor volume of 14 L at $P_{GB} = 20$ kW; $Q_{CA} = 0{,}3$; $L_{Rmax} = 0{,}5$ m.

The simulation results in the form of graphs of transients of reactor heating and loading as well as MCS heating are shown in Fig. 20, where the following notations are accepted: t_1—time of the waste initial loading to the reactor; t_2—time of waste heating in the reactor up to the boiling temperature T_b; t_3—time of the MCS heating; t_4—time at which the EPG plant achieves the steady operating mode; t_5—time of the waste evaporation and destruction according to the MCP process.

After analyzing the simulation results (Fig. 20) we can conclude that the nature of the transients of the developed mathematical model on the basis of soft computing techniques accurately replicates the real nature of the processes which take place in the EPG plant.

The initial loading of waste to the reactor using HD is going during time t_1. Then begins the reactor and loaded in it waste heating, respectively. This process takes place during the time t_2. At reaching the boiling temperature ($T_b \approx 400$ °C) the process of intensive evaporation begins, and waste temperature stays almost unchanged.

As for the MCS heating (during the time t_3), before the beginning of waste boiling in the reactor this process is very slow and is carried mainly out at the cost of light evaporation at waste heating and melting. Then, after the beginning of waste boiling, at the cost of it intensive evaporation, the MCS heating rate significantly increases. At this point the EPG plant begins to operate in the steady mode, and the value of waste L_R in the reactor, in turn, gradually decreases.

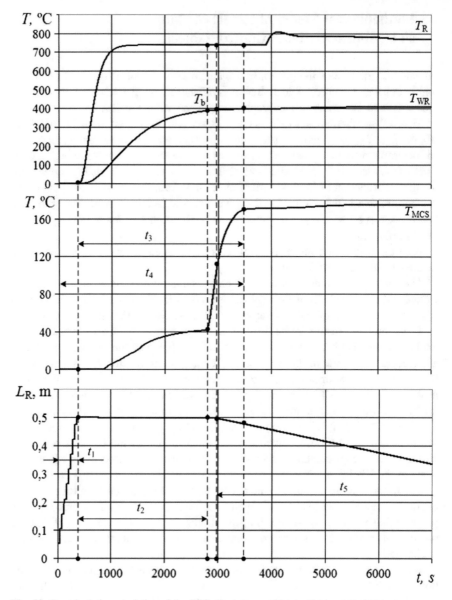

Fig. 20 Transient characteristics of the EPG plant as a multi-coordinate control object

7   Conclusions

In this work the development of the mathematical model and system of parametrical identification for the EPG plant as a complex multi-coordinate control object on the basis of soft computing techniques is presented.

The mathematical model of the EPG plant's reactor as a temperature control object based on experimental data with the fuzzy parametrical identification system is synthesized. The adequacy analysis using hypotheses assessing methods of mathematical statistics revealed that the obtained model based on experimental data with fuzzy parametrical identification system has higher accuracy and adequacy to the real processes ($R^2 = 0,993$), than the model based on the equations of heat transfer processes ($R^2 = 0,854$).

The mathematical models of the EPG plant MCS as a temperature control object based on ANFISTC with input variables linguistic terms membership functions of different types (triangular, trapezoidal, Gaussian 2) are presented. The simulation results and adequacy analysis using hypotheses assessing methods of mathematical statistics revealed that the MCS model developed on the basis of the ANFISTC with input variables linguistic terms membership functions of trapezoidal type has the highest adequacy to the real processes ($R^2 = 0,999$).

The mathematical model of the EPG plant's reactor as a load level control object based on the fuzzy system for calculating of the speed of the level change in the reactor is synthesized. The simulation results in the form of transient graphs of reactor loading and level reducing at waste evaporation and destruction show a rather high adequacy of the developed model to the real processes, that take place in the reactor.

The mathematical model of the EPG plant as a complex multi-coordinate control object is synthesized on the basis of the models of it main parts with the help of soft computing techniques. The analysis of computer simulation results in the form of static and dynamic characteristics graphs of the EPG plant confirms the high adequacy of the developed complex neuro-fuzzy model to the real processes. The developed mathematical model with parametrical identification system based on soft computing techniques gives the opportunity to investigate the behavior of the given complex control object in steady and transient modes, in particular, to synthesize and adjust the intelligent controllers of the multi-coordinate automatic control system of the EPG plant.

The application of the mathematical apparatus of soft computing, in particular, fuzzy logic and artificial neural networks at the development of this model allows to take into account the specific features of the EPG plant's multi-circuit structure and to display with quite high accuracy its basic features as the control object with essentially non-linear and undefined parameters.

References

1. Markina, L.M.: Development of new energy-saving and environmental safety technology at the organic waste disposal by ecopyrogenesis. J. Collected Works NUS **4**, 8 (2011). (in Ukrainian)
2. Kondratenko, Y.P., Kozlov, O.V.: Computerized Monitoring and Control System for Ecopyrogenesis Technological Complex. Tribuna Plural: La revista cientifica 1, Barcelona: Reial Academia de Doctors, pp. 223–255 (2014)

3. Fiss, D., Wagenknecht, M., Hampel, R.: Modeling a Boiling Process Under Uncertainties. 19th Zittau Fuzzy Colloquium, Proceedings of East-West Fuzzy Colloquium 2012, Zittau, Hochschule Zittau/Goerlitz, Germany, pp. 141–146 (2012)

4. Han, Z.X., Yan, C.H., Zhang, Z.: Study on robust control system of boiler steam temperature and analysis on its stability. J. Zhongguo Dianji Gongcheng Xuebao Proc. Chinese Soc. Elect. Eng. 30(8), 101–109 (2010)

5. Xiao, Z., Guo, J., Zeng, H., Zhou, P., Wang, S.: Application of fuzzy neural network controller in hydropower generator unit. J. Kybernetes 38(10), 1709–1717 (2009)

6. Chaikin, B.S., Mar'yanchik, G.E., Panov, E.M., Shaposhnikov, P.T., Vladimirov, V.A., Volovik, I.S., Makarevich, B.A.: State-of-the-art plants for drying and high-temperature heating of ladles. Int. J. Refract. Indust. Ceramics 47(5), 283–287 (2006)

7. Štemberk, P., Lanska, N.: Heating System for Curing Concrete Specimens under Prescribed Temperature. In: 13th Zittau Fuzzy Collquium, Proceedings of East-West Fuzzy Colloquium 2006, Zittau, Hochschule Zittau/Goerlitz, Germany, pp. 82–88 (2006)

8. Skrjanc, I.: Design of fuzzy model-based predictive control for a continuous stirred-tank reactor. 12th Zittau Fuzzy Colloquium, Proceedings of East-West Fuzzy Colloquium 2005, Zittau, Hochschule Zittau/Goerlitz, Germany, pp. 126–139 (2005)

9. Himmelblau, D.: Applied Nonlinear Programming. Translated from English. a. ed. by of Bihovckiy, M.L., Mir, M. 534 (1974) (in Russian)

10. Atamanyuk, I.P., Kondratenko, V.Y., Kozlov, O.V., Kondratenko, Y.P.: The Algorithm of optimal polynomial extrapolation of random processes. In: Modeling and Simulation in Engineering, Economics and Management. Engemann, K.J., Gil-Lafuente, A.M., Merigo, J L. (Eds.), International Conference MS 2012, New Rochelle, NY, USA, Proceedings. Lecture Notes in Business Information Processing 115, Springer, pp. 78–87 (2012)

11. Drozd, J., Drozd, A.: Models, methods and means as resources for solving challenges in co-design and testing of computer systems and their components. In: Proceedings of 9th International Conference on Digital Technologies 2013, Zhilina, Slovak Republic, pp. 225–230 (2013)

12. Tkachenko, A.N., Brovinskaya, N.M., Kondratenko, Y.P.: Evolutionary adaptation of control processes in robots operating in non-stationary environments. J. Mech. Mach. Theor. Printed in Great Britain, 18 (4), 275–278 (1983)

13. Gerdt, V.P., Prokopenya, A.N.: Simulation of quantum error correction with Mathematica. In: Computer Algebra in Scientific Computing/ CASC'2013, Gerdt, V.P., Koepf, W., Mayr, E. W., Vorozhtsov, E. V. (Eds.), Lecture Notes in Computer Science 8136, Springer-Verlag, Berlin, pp. 116—129 (2013)

14. Drozd, J., Drozd, A., Zashcholkin, K., Antonyuk, V., Kuznetsov, N., Kalinichenko, V., A: Concept of computing based on resources development analysis. In: Proceedings of IEEE East-West Design & Test Symposium, Rostov-on-Don, Russia, pp. 102–107 (2013)

15. Palagin, A.V., Opanasenko, V.N.: Reconfigurable-computing technology. Cybernet. Syst. Anal. 43(5), 675–686 (2007)

16. Trunov, A.N.: An adequacy criterion in evaluating the effectiveness of a model design process. East. Eur. J. Enterprise Technol. 1, 4 (73), 36–41 (2015)

17. Zadeh, L.A.: Fuzzy sets. Inf. Control 8, 338–353 (1965)

18. Gil-Aluja, J.: Investment in Uncertainty. Kluwer Academic Publishers, Dordrecht, Boston, London (1999)

19. Takagi, T., Sugeno, M.: Fuzzy identification of systems and its applications to modeling and control. IEEE Trans. Syst. Man Cybernet. 15 (1) (1985)

20. Zimmermann, H.-J.: Fuzzy Set Theory—and Its Applications. Kluwer Academic Publishers, Boston/Dordrecht/London (1992)

21. Jamshidi, M., Vadiee, N., Ross, T.J. (eds.): Fuzzy Logic and Control: Software and Hardware Application. Prentice Hall Series on Environmental and Intelligent Manufacturing Systems 2, M. Jamshidi (ed.), Prentice Hall, Englewood Cliffs, NJ (1993)

22. Zadeh, L.A.: The role of fuzzy logic in modeling, identification and control. Model. Identific. Control 15(3), 191–203 (1994)

23. Yager, R.R., Filev, D.P.: Essentials of Fuzzy Modeling and Control. Wiley, New York, NY (1994)
24. Hampel, R., Wagenknecht, M., Chaker, N.: Fuzzy Control: Theory and Practice. Physika-Verlag, Heidelberg, New York (2000)
25. Piegat, A.: Fuzzy Modeling and Control. Physica-Verlag, Heidelberg, New York (2001)
26. Yager, R.R., Filev, D. P.: Unified structure and parameter identification of fuzzy models. Syst. Man Cybernet. 23 (4) (1993)
27. Lodwick, W.A., Kacprzhyk, J. (Eds): Fuzzy Optimization. Studies in Fuzziness and Soft Computing 254, Springer-Verlag, Berlin, Heidelberg (2010)
28. Kondratenko, Y.P., Kondratenko, N.Y.: Reduced library of the soft computing analytic models for arithmetic perations with fuzzy numbers. In: Soft Computing: Developments, Methods and Applications, Alan Case (Ed), Series: Computer Science, technology and applications, NOVA Science Publishers, Hauppauge, New York, pp. 1–38 (2016)
29. Kondratenko, Y., Kondratenko, V.: Soft computing algorithm for arithmetic multiplication of fuzzy sets based on universal analytic models. In: Information and Communication Technologies in Education, Research, and Industrial Application. Communications in Computer and Information Science 469, Ermolayev, V. et al. (Eds.): ICTERI'2014, Springer International Publishing Switzerland, pp. 49–77 (2014)
30. Simon, D.: Training fuzzy systems with the extended Kalman filter. Fuzzy Sets Syst. 132, 189–199 (2002)
31. Gil-Aluja, J., Gil-Lafuente, A.M., Klimova, A.: The optimization of the economic segmentation by means of fuzzy algorithms. J. Comput. Optim. Econom. Finance 1(3), 169–186 (2011)
32. Shebanin, V., Atamanyuk, I., Kondratenko, Y., Volosyuk Y.: Application of fuzzy predicates and quantifiers by matrix presentation in informational resources modeling. Perspective Technologies and Methods in MEMS Design: Proceedings of the International Conference MEMSTECH-2016 Lviv-Poljana, Ukraine, pp. 146–149 (2016)
33. Kondratenko, Y.P., Klymenko, L.P., Al Zu'bi, E.Y.M.: Structural optimization of fuzzy systems' rules base and aggregation models. Kybernetes 42(5), 831–843 (2013)
34. Kondratenko, Y.P., Altameem, T.A., Al Zubi, E.Y.M.: The optimisation of digital controllers for fuzzy systems design. Advanc. Model. Anal. AMSE Period. A 47(1–2), 19–29 (2010)
35. Kondratenko, Y.P., Al Zubi, E.Y.M.: The optimisation approach for increasing efficiency of digital fuzzy controllers. Annals of DAAAM for 2009 & Proceeding of the 20th International DAAAM Symposium. Intelligent Manufacturing and Automation, Published by DAAAM International, Vienna, Austria, pp. 1589–1591 (2009)
36. Kondratenko, Y.P.: Robotics, automation and information systems: future perspectives and correlation with culture, sport and life science. In book: Decision Making and Knowledge Decision Support Systems. Lecture Notes in Economics and Mathematical Systems 675, Gil-Lafuente, A.M., Zopounidis, C. (Eds), Springer International Publishing Switzerland, pp. 43–56 (2015)
37. Kauffman, A., Gil-Aluja, J.: Introduction of fuzzy sets theory to management of enterprises. Minsk, Higher School (1992). (in Russian)
38. Gil-Aluja, J.: Elements for a theory of decision in uncertainty. Springer Science & Business Media vol. 32 (1999)
39. Gil-Aluja, J.: Fuzzy sets in the management of uncertainty. Springer Science & Business Media vol. 145 (2004)
40. Gil-Aluja, J.: The interactive management of human resources in uncertainty. Springer Science & Business Media vol. 11 (2013)
41. Merigo, J.M., Gil-Lafuente, A.M., Gil-Aluja, J.: Decision making with the induced generalized adequacy coefficient. Appl. Comput. Math. 2(2), 321–339 (2011)
42. Gil-Aluja, J.: Handbook of management under uncertainty. Springer Science & Business Media vol. 55 (2013)

43. Gil-Aluja, J., Gil-Lafuente, A.M., Klimova, A.: The optimization of the economic segmentation by means of fuzzy algorithms. J. Comput. Optim. Econom. Finance **1** (3), Nova Science Publishers, 169–186 (2008)
44. Werners, B., Kondratenko, Y.P.: Tanker routing problem with fuzzy demands of served ships. Int. J. Syst. Res. Informat. Technol. **1**, 47–64 (2009)
45. Kondratenko, Y.P., Sidenko, Ie. V.: Decision-making based on fuzzy estimation of quality level for cargo delivery. In book: Recent Developments and New Directions in Soft Computing. Studies in Fuzziness and Soft Computing 317, Zadeh, L.A. et al. (Eds), Springer International Publishing Switzerland, pp. 331–344 (2014)
46. Kondratenko, G.V., Kondratenko, Y.P., Romanov, D.O.: Fuzzy Models for Capacitive Vehicle Routing Problem in Uncertainty. In: Proceedings of 17th International DAAAM Symposium Intelligent Manufacturing and Automation: Focus on Mechatronics & Robotics, Vienna, Austria, pp. 205–206 (2006)
47. Werners, B., Kondratenko, Y.P.: Fuzzy multi-criteria optimization for vehicle routing with capacity constraints and uncertain demands. In: Proceedings of the International Congress on Cost Control, Publ. by ACCID/ASEPUC, Barcelona, Spain, pp. 145–159 (2011)
48. Kondratenko, Y.P., Encheva, S.B., Sidenko, E.V.: Synthesis of intelligent decision support systems for transport logistic. In: Proceedings of the 6th IEEE International Conference on Intelligent Data Acquisition and Advanced Computing Systems: Technology and Applications 2, IDAACS'2011, Prague, Czech Republic, pp. 642–646 (2011)
49. Gil-Lafuente, A.M.: Fuzzy Logic in Financial Analysis. Studies in Fuzziness and Soft Computing vol. 175, Springer, Berlin (2005)
50. Merigó, J.M., Gil-Lafuente, A.M., Gil-Aluja, J.: A new aggregation method for strategic decision making and its application in assignment theory. Afr. J. Bus. Manag. **5**(11), 4033–4043 (2011)
51. Gil-Aluja, J., Gil-Lafuente, A.M., Merigó, J.M.: Using homogeneous groupings in portfolio management. Expert Syst. Appl. **38**(9), 10950–10958 (2011)
52. Butenko, S., Gil-Lafuente, J., Pardalos, P.M. (eds.): Optimal Strategies in Sports Economics and Management. Springer-Vertag, Heidelberg, Dordrecht, London, New York (2010)
53. Gil Aluja, J. (ed.): Les Universitats En El Centenari Del Futbol Club Barcelona. Estudis En L'Ambit De L'Esport, Proleg, Josef Lluis Nunez (1999)
54. Gil-Lafuente, A.M., Merigo, J.M.: Decision making techniques in political management. In book: Fuzzy Optimization. In: Studies in Fuzziness and Soft Computing vol. 254, Lodwick, W.A., Kacprzhyk, J. (Eds), Springer, Berlin, Heidelberg, pp. 389–405 (2010)
55. Kacprzyk, J., Yager, R.R., Zadrożny, S.: A fuzzy logic based approach to linguistic summaries of databases. Int. J. Appl. Math. Comput. Sci. **10**(4), 813–834 (2000)
56. Oh, S.K., Pedrycz, W.: The design of hybrid fuzzy controllers based on genetic algorithms and estimation techniques. J. Kybernetes **31**(6), 909–917 (2002)
57. Suna, Q., Li, R., Zhang, P.: Stable and optimal adaptive fuzzy control of complex systems using fuzzy dynamic model. J. Fuzzy Sets Syst. **133**, 1–17 (2003)
58. Vachkov, G., Kiyota, Y., Komatsu, K.: Identification of dynamic cause-effect relations for systems performance evaluation. Appl. Sci. Soft Comput. Advan. Soft Comput. **24**, 187–194 (2004)
59. Hayajneh, M.T., Radaideh, S.M., Smadi, I.A.: Fuzzy logic controller for overhead cranes. Eng. Comput. **23**(1), 84–98 (2006)
60. Wang, L., Kazmierski, T.J.: VHDL-AMS based genetic optimization of fuzzy logic controllers. Int. J. Comput. Math. Elect. Electron. Eng. **26**(2), 447–460 (2007)
61. Ho, G.T.S., Lau, H.C.W., Chung, S.H., Fung, R.Y.K., Chan, T.M., Lee, C.K.M.: Fuzzy rule sets for enhancing performance in a supply chain network. Indust. Manag. Data Syst. **108**(7), 947–972 (2008)
62. Kondratenko, Y.P., Kozlov, O. V.: Fuzzy Controllers in Reactors Control Systems of Multiloop Pyrolysis Plants. 19th Zittau Fuzzy Colloquium, Proceedings of East-West Fuzzy Colloquium 2012, Zittau, Hochschule Zittau/Goerlitz, Germany, 15–22 (2012)

63. Rotach, V. Y.: Automatic Control Theory of Heat and Power Processes: M, Energoatomizdat, 296 (1985). (in Russian)
64. Kondratenko, Y. P., Kozlov, O. V.: Mathematical Model of Ecopyrogenesis Reactor with Fuzzy Parametrical Identification. In: Recent Developments and New Direction in Soft-Computing Foundations and Applications. Studies in Fuzziness and Soft Computing 342. Lotfi A. Zadeh et al. (Eds.). Berlin, Heidelberg: Springer, pp. 439–451 (2016)
65. Kondratenko, Y. P., Kozlov, O. V.: Mathematic Modeling of Reactor's Temperature Mode of Multiloop Pyrolysis Plant. Lecture Notes in Business Information Processing: Modeling and Simulation in Engineering, Economics and Management 115, K. J. Engemann, A. M. Gil-Lafuente, J. M. Merigo (Eds.), Berlin, Heidelberg: Springer, pp. 178–187 (2012)
66. Kondratenko, Y. P., Kozlov, O. V., Klymenko, L. P., Kondratenko, G.V.: Synthesis and Research of Neuro-Fuzzy Model of Ecopyrogenesis Multi-circuit Circulatory System. Advance Trends in Soft Computing, Studies in Fuzziness and Soft Computing 312, Berlin, Heidelberg: Springer, pp. 1–14 (2014)
67. Jang, J.-S.R.: ANFIS: Adaptive-Network-based Fuzzy Inference Systems. IEEE Transactions on Systems, Man, and Cybernetics 23(3), 665–685 (1993)
68. Jang, J.-S. R., Sun, C.-T., Mizutani, E.: Neuro-Fuzzy and Soft Computing: A Computational Approach to Learning and Machine Intelligence. Prentice Hall (1996)
69. Dimirovski, G. M., Lokevenc, I. I., Tanevska, D. J.: Applied Adaptive Fuzzy-neural Inference Models: Complexity and Integrity Problems. Intelligent Systems, Proceedings of 2nd International IEEE Conference 1 (22–24), 45–52 (2004)
70. Kondratenko, Y. P., Kozlov, O. V., Atamanyuk, I. P., Korobko, O. V.: Computerized Control System for the Pyrolysis Reactor Load Level Based on the Neural Network Controllers. Computing in Science and Technology 2012/2013, T. Kwater, B. Twarog (Eds.), Monographs in Applied Informatics, Wydawnictwo Uniwersytety Rzeszowskiego, Rzeszow, Poland, 97–120 (2013)

Part III
Intelligent Data Analysis and Processing

Fuzzy Data Processing Beyond Min t-Norm

Andrzej Pownuk, Vladik Kreinovich and Songsak Sriboonchitta

Abstract Usual algorithms for fuzzy data processing—based on the usual form of Zadeh's extension principle—implicitly assume that we use the min "and"-operation (t-norm). It is known, however, that in many practical situations, other t-norms more adequately describe human reasoning. It is therefore desirable to extend the usual algorithms to situations when we use t-norms different from min. Such an extension is provided in this chapter.

1 Need for Fuzzy Data Processing

Need for data processing. In many real-life situations, we are interested in the value of a quantity y which are difficult (or even impossible) to measure directly. For example, we may be interested in the distance to a faraway star, in the amount of oil in a given well, or in tomorrow's weather.

Since we cannot measure the quantity y directly, the way to estimate this value is to measure related easier-to-measure quantities $x_1, \ldots, x_n$, and then to use the known relation $y = f(x_1, \ldots, x_n)$ between the desired quantity y and the quantities x_i to estimate y; see, e.g., [8].

For example, since we cannot directly measure the distance to a faraway star, we can measure the directions x_i to this star in two different seasons, when the Earth is on the opposite sides of its Solar orbit, and then use trigonometry to find the desired distance. Since we cannot directly measure the amount y of oil in the well, we can

A. Pownuk (✉) · V. Kreinovich
Computational Science Program, University of Texas, El Paso 500 W University,
El Paso, TX 79968, USA
e-mail: ampownuk@utep.edu

V. Kreinovich
e-mail: vladik@utep.edu

S. Sriboonchitta
Faculty of Economics, Chiang Mai University, Chiang Mai, Thailand
e-mail: sognsakecon@gmail.com

© Springer International Publishing AG 2018
C. Berger-Vachon et al. (eds.), *Complex Systems: Solutions and Challenges in Economics, Management and Engineering*, Studies in Systems,
Decision and Control 125, https://doi.org/10.1007/978-3-319-69989-9_14

measure the results of artificially set seismic waves propagating through the corresponding region, and then use the known equations of seismic wave propagation to compute y. Since we cannot directly measure tomorrow's temperature y, to estimate y, we measure temperature, moisture, wind speed, and other meteorological characteristics in the vicinity of our area, and use the known equations of aerodynamics and thermal physics to predict tomorrow's weather.

The computation of $y = f(x_1, \ldots, x_n)$ based on the results of measuring x_i is known as *data processing*.

In some cases, the data processing algorithm $f(x_1, \ldots, x_n)$ consists of direct application of a formula, but in most cases, we have a rather complex algorithm—such as solving a system of partial differential equations.

Need for fuzzy data processing. In some cases, instead of the measurement results $x_1, \ldots, x_n$, we have expert estimates for the corresponding quantities. Experts can rarely describe their estimates by exact numbers, they usually describe their estimates by using imprecise ("fuzzy") words from natural language. For example, an expert can say that the temperature is *warm*, or that the temperature is *around 20*.

To process such imprecise natural-language estimates, L. Zadeh designed the technique of *fuzzy logic*; see, e.g., [3, 7, 10]. In this technique, each imprecise word like "warm" is described by assigning, to each possible value of the temperature x, the degree to which the expert believes that this particular temperature is warm.

This degree can be obtained, e.g., by asking an expert to mark, on a scale from 0 to 10, where 0 means no warm and 10 means warn, how much x degrees means warm. Then, if for some temperature x, the expert marks, say, 7 on a scale from 0 to 10, we assign the degree $\mu(x) = 7/10$.

The function that assigns, to each possible value of a quantity x, the degree $\mu(x)$ to which the quantity satisfies the expert's estimate, is known as the *membership function*.

If for each of the inputs $x_1, \ldots, x_n$, we only have expert estimates, then what can we say about $y = f(x_1, \ldots, x_n)$? Computing appropriate estimates for y is known as *fuzzy data processing*.

To perform fuzzy data processing, we need to deal with propositional connectives such as "and" and "or". In fuzzy data processing, we know that $y = f(x_1, \ldots, x_n)$, and we have fuzzy information about each of the inputs x_i. Based on this information, we need to estimate the corresponding degrees of confidence in different values y.

A value y is possible if for some tuple $(x_1, \ldots, x_n)$ for which $y = f(x_1, \ldots, x_n)$, x_1 is a possible value of the first variable, *and* x_2 is a possible value of the second variable, $\ldots$, *and* x_n is a possible value of the n-th variable. So, to describe the degree to which each value y is possible, we need to be able to describe our degrees of confidence in statements containing propositional connectives such as "and" and "or".

Dealing with "and"- and "or" in fuzzy logic is not as easy as in the 2-valued case. In the traditional 2-valued logic, when each statement is either true or false, dealing with "and" and "or" connectives is straightforward: if we know the truth values of two statements A and B, then these truth values uniquely determine the truth values of the propositional combinations $A \& B$ and $A \vee B$.

In contrast, when we deal with degrees of confidence, the situation is not as straightforward. For example, for a fair coin, we have no reason to be more confident that it will fall head or tail. Thus, it makes sense to assume that the expert's degrees of confidence $a = d(A)$ and $b = d(B)$ in statement $A =$ "coin falls head" and $B =$ "coin falls tail" are the same: $a = b$.

Since a coin cannot at the same time fall head and tail, the expert's degree of belief that both A and B happen is clearly 0: $d(A \& B) = 0$. On the other hand, if we take $A' = B' = A$, then clearly $A' \& B'$ is equivalent to A and thus, $d(A' \& B') = d(A) > 0$. This is a simple example where for two different pairs of statements, we have the same values of $d(A)$ and $d(B)$ but different values of $d(A \& B)$:

- in the first case, $d(A) = d(B) = a$ and $A(A \& B) = 0$, while
- in the second case, $d(A') = d(B') = a$, but $d(A' \& B') = a > 0$.

This simple example shows that, in general, the expert's degree of confidence in a propositional combination like $A \& B$ or $A \vee B$ is not uniquely determined by his/her degrees of confidence in the statements A and B. Thus, ideally, in addition to eliciting, from the experts, the degrees of confidence in all basic statements $A_1, \ldots, A_n$, we should also elicit from them degrees of confidence in all possible propositional combinations. For example, for each subset

$$I \subseteq \{1, \ldots, n\},$$

we should elicit, from the expert, his/her degree of confidence in a statement $\&_{i \in I} A_i$.

The problem is that there are 2^n such statement (and even more if we consider different propositional combinations). Even of a small size knowledge base, with $n = 30$ statements, we thus need to ask the expert about $2^{30} \approx 10^9$ different propositional combinations—this is clearly not practically possible.

Need for "and"- and "or"-operations. Since we cannot elicit the degrees of confidence for all propositional combinations from an expert, it is therefore desirable to estimate the degree of confidence in a propositional combination like $A \& B$ or $A \vee B$ from the expert's degrees of confidence $a = d(A)$ and $b = d(B)$ in the original statements A and B—knowing very well that these are *estimates*, not exactly the original expert's degrees of confidence in the composite statements.

The corresponding estimate for $d(A \& B)$ will be denoted by $f_\&(a, b)$. It is known as an *"and"-operation*, or, alternatively, as a *t-norm*. Similarly, the estimate for $d(A \vee B)$ will be denoted by $f_\vee(a, b)$. This estimate is called an *"or"-operation*, or a *t-conorm*.

There are reasonable conditions that these operations should satisfy. For example, since "A and B" means the same as "B and A", it makes sense to require that the estimates for $A \& B$ and $B \& A$ are the same, i.e., that $f_\&(a, b) = f_\&(b, a)$ for all a and b. Similarly, we should have $f_\vee(a, b) = f_\vee(b, a)$ for all a and b. The fact that $A \& (B \& C)$ means the same as $(A \& B) \& C$ implies that the corresponding estimates should be the same, i.e., that we should have $f_\&(a, f_\&(b, c)) = f_\&(f_\&(a, b), c)$ for all a, b, and c. There are also reasonable monotonicity requirements.

There are many different "and"- and "or"-operations that satisfy all these require-ments. For example, for "and", we have the min-operation $f_\&(a, b) = \min(a, b)$. We also have an *algebraic product* operation $f_\&(a, b) = a \cdot b$. In addition, we can con-sider general *Archimedean operations*

$$f_\&(a, b) = f^{-1}(f(a) \cdot f(b)),$$

for some strictly increasing continuous function $f(x)$.

Similarly, for "or", we have the max-operation $f_\vee ee(a, b) = \max(a, b)$, we have *algebraic sum*

$$f_\vee(a, b) = a + b - a \cdot b = 1 - (1 - a) \cdot (1 - b),$$

and we have general Archimedean operations

$$f_\vee(a, b) = f^{-1}(f(a) + f(b) - f(a) \cdot f(b)).$$

Which "and" and "or"-operations should we select? Our objective is to describe the expert's knowledge. So, it is reasonable to select "and"- and "or"-operations which most adequately describe the reasoning of this particular expert (or this par-ticular group of experts).

Such a selection was first made by Stanford researchers who designed the world's first expert system MYCIN—for curing rare blood diseases; see, e.g., [1]. It is in-teresting to mention that when the researchers found the "and"- and "or"-operations that best describe the reasoning of medical experts, they thought that they have found general laws of human reasoning. However, when they tried the same operations on another area of knowledge—geophysics—it turned out that different "and'- and "or'-operations are needed.

In hindsight, this difference make perfect sense: in medicine, one has to be very cautious, to avoid harming a patient. You do not start a surgery unless you are ab-solutely sure that the diagnosis is right. If a doctor is not absolutely sure, he or she recommends additional tests. In contrast, in geophysics, if you have a reasonable de-gree of confidence that there is oil in a given location, you dig a well—and you do not wait for a 100% certainty: a failed well is an acceptable risk.

What this experience taught us is that in different situations, different "and"- and "or"-operations are more adequate.

Let us apply "and"- and "or"-operation to fuzzy data processing. Let us go back to fuzzy data processing. To find the degree to which y is a possible value, we need to consider all possible tuples $(x_1, \ldots, x_n)$ for which $y = f(x_1, \ldots, x_n)$. The value y is possible if one of these tuples is possible, i.e., either the first tuple is possible *or* the second tuple is possible, etc.

Once we know the degree $d(x)$ to which each tuple $x = (x_1, \dots, x_n)$ is possible, the degree to which y is possible can then be obtained by applying "or"-operation $f_\vee(a, b)$ to all these degrees:

$$d(y) = f_\vee\{d(x_1, \dots, x_n) : f(x_1, \dots, x_n)\}.$$

Usually, there are infinitely many such tuples x for which $f(x) = y$. For most "or"-operations, e.g., for

$$f_\&(a, b) = a + b - a \cdot b = 1 - (1 - a) \cdot (1 - b),$$

if we combine infinitely many terms, we get the same meaningless value 1. In effect, the only widely used operation for which this is not happening is $f_\vee(a, b) = \max(a, b)$. Thus, it is makes sense to require that

$$d(y) = \max\{d(x_1, \dots, x_n) : f(x_1, \dots, x_n) = y\}.$$

To fully describe this degree, we thus need to describe the degree $d(x_1, \dots, x_n)$ to which a tuple $x = (x_1, \dots, x_n)$ is possible. A tuple $(x_1, \dots, x_n)$ is possible if x_1 is a possible values of the first variable, *and* x_2 is a possible of the second variable, etc.

We know the expert's degree of confidence that x_1 is a possible value of the first variable: this described by the corresponding membership function $\mu_1(x_1)$. Similarly, the expert's degree of confidence that x_2 is a possible value of the second variable is equal to $\mu_2(x_2)$, etc. Thus, to get the degree to which the tuple $(x_1, \dots, x_n)$ is possible, we need to use an appropriate "and"-operation $f_\&(a, b)$ to combine these degrees $\mu_i(x_i)$:

$$d(x_1, \dots, x_n) = f_\&(\mu_1(x_1), \dots, \mu_n(x_n)).$$

Substituting this expression into the above formula for $\mu(y) = d(y)$, we arrive at the following formula.

General form of Zadeh's extension principle. If we know the the quantity y is related to the quantities $x_1, \dots, x_n$ by a relation $y = f(x_1, \dots, x_n)$, and we know the membership functions $\mu_i(x_i)$ that describe each inputs x_i, then the resulting membership function for y takes the form

$$\mu(y) = \max\{f_\&(\mu_1(x_1), \dots, \mu_n(x_n)) : f(x_1, \dots, x_n) = y\}.$$

This formula was originally proposed by Zadeh himself and is thus known as *Zadeh's extension principle*.

Algorithms for fuzzy data processing: what is already known and what we do in this paper. Usually, Zadeh's extension principle is applied to the situations in which we use the min "and"-operation $f_\&(a, b) = \min(a, b)$. In this case, Zadeh's extension principle takes a simplified form

$$\mu(y) = \max\{\min(\mu_1(x_1), \ldots, \mu_n(x_n)) : f(x_1, \ldots, x_n) = y\}.$$

For this case, there are efficient algorithms for computing $\mu(y)$. (We will mention these algorithms in the nest section.)

The problem is that, as it is well known, in many practical situations, other "and"-operations more adequately describe expert reasoning [3, 7]. It is therefore desirable to generalize the existing efficient algorithms for fuzzy data processing from the case of min to the general case of an arbitrary "and"-operation. This is what we do in this paper.

2 Possibility of Linearization

In most practical cases, measurement and estimation errors are relatively small. Before we get deeper into algorithms, let us notice that in the above text, we consider the generic data processing functions $f(x_1, \ldots, x_n)$.

In the general case, this is what we have to do. However, in most practical situations, the expert uncertainty is relatively small—just like measurement errors are usually relatively small, small in the sense that the squares of the measurement or estimation errors are much smaller than the errors themselves; see, e.g., [8].

Indeed, if we have a measurement or an expert estimate with accuracy 10% (i.e., 0.1), then the square of this inaccuracy is 0.01, which is much smaller than the original measurement or estimation error. If we measure or estimate with an even higher accuracy, e.g., 5% or 1%, then the quadratic term is even much more small than the measurement or estimation error.

Possibility of linearization. In such situations, instead of considering general functions $f(x_1, \ldots, x_n)$, we can expand the corresponding data processing function in Taylor series around the estimates $\widetilde{x}_1, \ldots, \widetilde{x}_n$, and keep only linear terms in this expansion—thus, ignoring the quadratic terms. As a result, the original expression

$$f(x_1, \ldots, x_n) = f(\widetilde{x}_1 + \Delta x_1, \ldots, \widetilde{x}_n + \Delta x_n),$$

where we denoted $\Delta x_i \overset{\text{def}}{=} x_i - \widetilde{x}_i$, is replace by its linear approximation:

$$f(x_1, \ldots, x_n) = f(\widetilde{x}_1 + \Delta x_1, \ldots, \widetilde{x}_n + \Delta x_n) \approx c_0 + \sum_{i=1}^{n} c_i \cdot \Delta x_i,$$

where $c_0 \overset{\text{def}}{=} f(\widetilde{x}_1, \ldots, \widetilde{x}_n)$ and $c_i \overset{\text{def}}{=} \dfrac{\partial f}{\partial x_i}_{|x_i = \widetilde{x}_i}$. Substituting $\Delta x_i = x_i - \widetilde{x}_i$ into the above formula, we get a linear dependence

$$f(x_1, \ldots, x_n) = a_0 + \sum_{i=1}^{n} c_i \cdot x_i,$$

where $a_0 \overset{\text{def}}{=} c_0 - \sum_{i=1}^{n} c_i \cdot \tilde{x}_i$.

Zadeh's extension principle in case of linearization. For linear functions, the general Zadeh's extension principle takes the form

$$\mu(y) = \max \left\{ f_\&(\mu_1(x_1), \ldots, \mu_n(x_n)) : a_0 + \sum_{i=1}^{n} c_i \cdot x_i = y \right\}.$$

In particular, the min-based extension principle takes the form

$$\mu(y) = \max \left\{ \min(\mu_1(x_1), \ldots, \mu_n(x_n)) : a_0 + \sum_{i=1}^{n} c_i \cdot x_i = y \right\}.$$

In the linearized case, we can reduce a general problem to several sums of two variables. From the computational viewpoint, linearization has an additional advantage: it enables us to reduce a general problem, with many inputs, to a sequence of problems with only two inputs. To be more precise, we can reduce it to the problem of computing the sum of two variables.

Indeed, our goal is to compute the membership degrees corresponding to the sum

$$y = a_0 + c_1 \cdot x_1 + c_2 \cdot x_2 + \ldots + c_n \cdot x_n.$$

To simplify this problem, let us recall how this expression would be computed on a computer.

- First, we will compute $y_1 \overset{\text{def}}{=} c_1 \cdot x_1$, then the first intermediate sum $s_1 = a_0 + y_1$.
- After that, we will compute the second product $y_2 = c_2 \cdot x_2$, and the second intermediate sum $s_2 = s_1 + y_2$.
- Then, we compute the third product $y_3 = c_3 \cdot x_3$, and the third intermediate sum $s_3 = s_2 + y_3$.
- ...
- At the end, we compute $y_n = c_n \cdot x_n$ and $y = s_{n-1} + y_n$.

Let us follow the same pattern when computing the membership function $\mu(y)$. First, let us find the membership functions $v_i(y_i)$ corresponding to $y_i = c_i \cdot x_i$. For each y_i, there is only one value x_i for which $y_i = c_i \cdot x_i$ – namely, the value $x_i = y_i/c_i$ – so in Zadeh's formula, there is no need to apply "or" or "and", we just have $v_i(y_i) = \mu_i(y_i/c_i)$.

Similarly, for $s_1 = a_0 + y_1$, we have $\eta_1(s_1) = v_1(s_1 - a_0) = \mu_1((s_1 - a_0)/c_1)$. Now, we can do the following:

- first, based on the membership functions $\eta_1(s_1)$ and $v_2(y_2)$, we find the membership function $\eta_2(s_2)$ corresponding to the sum $s_2 = s_1 + y_2$,
- then, based on the membership functions $\eta_2(s_2)$ and $v_3(y_3)$, we find the membership function $\eta_3(s_3)$ corresponding to the sum $s_3 = s_2 + y_3$,
- etc.
- until, based on the membership functions $\eta_{n-1}(s_{n-1})$ and $v_n(y_n)$, we find the desired membership function $\mu(y)$ corresponding to the sum $y = s_{n-1} + y_n$.

On each step, we need to several times apply Zadeh's extension principle to the sum of two variables.

For the sum of two variables, these formulas have the following simplified form.

Zadeh's extension principle: case of the sum of two variables. For a general "and"-operation $f_\&(a, b)$, when we know the membership functions $\mu_1(x_1)$ and $\mu_2(x_2)$ describing two quantities x_1 and x_2, then, once we have selected x_1, the value x_2 for which $x_1 + x_2 = y$ can be easily described as $x_2 = y - x_1$. Thus, the membership function $\mu(y)$ for the sum $y = x_1 + x_2$ takes the form

$$\mu(y) = \max_{x_1} f_\&(\mu_1(x_1), \mu_2(y - x_1)).$$

In particular, for $f_\&(a, b) = \min(a, b)$, we have

$$\mu(y) = \max_{x_1} \min(\mu_1(x_1), \mu_2(y - x_1)).$$

In the following text, this is what we will consider: algorithms for computing these two formulas.

3   Efficient Fuzzy Data Processing for the min t-Norm: Reminder

Our objective is to come up with efficient algorithm for fuzzy data processing for a general "and"-operation.

To come up with such an algorithm, let us recall how fuzzy data processing is performed in the case of the min t-norm.

What is the input for the algorithm. To describe each of the two input membership functions $\mu_1(x_1)$ and $\mu_2(x_2)$, we need to describe their values at certain number of points $v_0 < v_1 < \cdots < v_{N-1}$, where N is the total number of these points. Usually, these points are equally spaced, i.e., $v_i = v_0 + i \cdot \Delta$ for some $\Delta > 0$. Thus, the inputs consist of $2N$ values $\mu_i(v_j)$ corresponding to $i = 1, 2$ and to $j = 0, 1, \ldots, N - 1$.

First-naive-algorithm for fuzzy data processing. When the values of each of the two variables x_1 and x_2 go from v_0 to $v_{N-1} = v_0 + (N - 1) \cdot \Delta$, their sum

$y = x_1 + x_2$ takes values $w_0 \stackrel{\text{def}}{=} 2v_0$, $w_1 = w_0 + \Delta = 2v_0 + \Delta$, ..., all the way to $w_{2(N-1)} = 2v_{N-1} = 2v_0 + 2 \cdot (N-1) \cdot \Delta$.

Each value w_k can be represented as the sum $v_i + v_{k-i}$ of possible values of x_1 and x_2 when $0 \leq i \leq N - 1$ and $0 \leq k - i \leq N - 1$, i.e., when

$$\max(k + 1 - N, 0) \leq i \leq \min(k, N - 1).$$

Thus, we have

$$\mu(w_k) = \max(\min(\mu_1(v_i), \mu_2(v_{k-i})) : \max(k + 1 - N, 0) \leq i \leq \min(k, N - 1)\}.$$

What is the computational complexity of this naive algorithm. The computational complexity of an algorithm is usually gauged by its number of computational steps—which is roughly proportional to the overall computation time.

In the above case, for each k, we have up to N different values i:

- exactly N for $k = N - 1$,
- $N - 1$ for $k = N - 2$ and $k = N$,
- $N - 2$ values i for $k = N - 3$ and $k = N + 1$,
- etc.

For each k and i, we need one min operation, to compute the minimum

$$\min(\mu_1(v_i), \mu_2(v_{k-i})).$$

Then, we need to find the maximum of all these numbers. So, we need N steps to compute the value w_{N-1}, $N - 1$ steps to compute the values w_{N-2} and w_N, etc. Overall, we therefore need

$$N = 2 \cdot (N - 1) + 2 \cdot (N - 2) + \cdots + 2 \cdot 1 = N + 2 \cdot ((N - 1) + (N - 2) + \cdots + 1) =$$

$$N + 2 \cdot \frac{N \cdot (N - 1)}{2} = N + N \cdot (N - 1) = N^2$$

computational steps.

It is possible to compute $\mu(y)$ faster, by using α-cuts. It is known that for the case of the min t-norm, it is possible to compute $\mu(y)$ faster. This possibility comes from considering α-cuts

$$\mathbf{x}_i(\alpha) \stackrel{\text{def}}{=} \{x_i : \mu_i(x_i) \geq \alpha\}.$$

Indeed, according to Zadeh's formula, $\mu(y) \geq \alpha$ if and only if there exists values x_1 and x_2 for which $x_1 + x_2 = y$ and $\min(\mu_1(x_1), \mu_2(x_2)) \geq \alpha$, i.e., equivalently, for which $\mu_1(x_1) \geq \alpha$ and $\mu_2(x_2) \geq \alpha$.

Thus, the α-cut for y is equal to the set of all possible values $y = x_1 + x_2$ when x_1 belongs to the α-cut for x_1 and x_2 belongs to the α-cut for x_2:

$$\mathbf{y}(\alpha) = \{x_1 + x_2 : x_1 \in \mathbf{x}_1(\alpha) \,\&\, x_2 \in \mathbf{x}_2(\alpha)\}.$$

In many practical situations, each of the membership functions $\mu_i(x_i)$ increases up to a certain value, then decreases. In such situations, each α-cut is an interval: $\mathbf{x}_i(\alpha) = [\underline{x}_i(\alpha), \overline{x}_i(\alpha)]$. If we know that x_1 belongs to the interval $[\underline{x}_1(\alpha), \overline{x}_1(\alpha)]$ and that x_2 belongs to the interval $[\underline{x}_2(\alpha), \overline{x}_2(\alpha)]$, then possible values of $y = x_1 + x_2$ form the interval

$$[\underline{y}(\alpha), \overline{y}(\alpha)] = [\underline{x}_1(\alpha) + \underline{x}_2(\alpha), \overline{x}_1(\alpha) + \overline{x}_2(\alpha)].$$

(It is worth mentioning that this formula is a particular case of *interval arithmetic*; see, e.g., [2, 5].)

Thus, instead of N^2 elementary operations, we only need to perform twice as many operations as there are possible levels α. When we use $2N$ values $\mu_i(x_i)$, we can have no more than $2N$ different values α, so the computation time is $O(N) \ll N^2$.

This is the reason why α-cuts—and interval computations—are mostly used in fuzzy data processing.

4   Efficient Fuzzy Data Processing Beyond min t-Norm: The Main Result of This Paper

Now that we recalled how fuzzy data processing can be efficiently computed for the min t-norm, let us find out how we can efficiently compute it for the case of a general t-norm. Reminder: we know the membership functions $\mu_1(x_1)$ and $\mu_2(x_2)$, we want to estimate the values

$$\mu(y) = \max_{x_1} f_\&(\mu_1(x_1), \mu_2(y - x_1)).$$

Naive algorithm. Similar to the case of the min t-norm, the naive (straightforward) application of this formula leads to an algorithm that requires N^2 steps: the only difference is that now, we use $f_\&$ instead of min.

Let us show that for t-norms different from min, it is also possible to compute $\mu(y)$ faster.

Reducing to the case of a product t-norm. It is know that Archimedean t-norms, i.e., t-norms of the type $f^{-1}(f(a) \cdot f(b))$ for monotonic $f(x)$, are *universal approximators*, in the sense that for every t-norm and for every $\varepsilon > 0$, there exists an Archimedean t-norm whose value is ε-close to the given one for all possible a and b; see, e.g., [6].

Thus, from the practical viewpoint, we can safely assume that the given t-norm is Archimedean. For the Archimedean t-norm, the above formula takes the form

$$\mu(y) = \max_{x_1} f^{-1}(f(\mu_1(x_1)) \cdot f(\mu_2(y - x_1))).$$

The inverse function f^{-1} is monotonic, thus, the largest values of $f^{-1}(z)$ is attained when z is the largest. So, we have

$$\mu(y) = f^{-1}(v(y)),$$

where we denoted

$$v(y) = \max_{x_1} f(\mu_1(x_1)) \cdot f(\mu_2(y - x_1)).$$

So, if we denote $v_1(x_1) \overset{\text{def}}{=} f(\mu_1(x_1))$ and $v_2(x_2) \overset{\text{def}}{=} f(\mu_2(x_2))$, we arrive at the following problem: given functions $v_1(x_1)$ and $v_2(x_2)$, compute the function

$$v(y) = \max_{x_1}(v_1(x_1) \cdot v_2(y - x_1)).$$

This is exactly the original problem for the product t-norm. Thus, fuzzy data processing problem for a general t-norm can indeed be reduced to the problem of fuzzy data processing for the product t-norm.

Let us simplify our problem. Multiplication can be simplified if we take logarithm of both sides; then it is reduced to addition. This is why logarithms were invented in the first place—and this is why they were successfully used in the slide rule, the main computational tool for the 19 century and for first half of the 20 century—to make multiplication faster.

The values $v_i(x_i)$ are smaller than 1, so their logarithms are negative. To make formulas easier, let us use positive numbers, i.e., let us consider the values $\ell_1(x_1) \overset{\text{def}}{=} -\ln(v_1(x_1))$, $\ell_2(x_2) \overset{\text{def}}{=} -\ln(v_2(x_2))$, and $\ell(y) \overset{\text{def}}{=} -\ln(v(y))$. Since we changed signs, max becomes min, so we get the following formula:

$$\ell(y) = \min_{x_1}(\ell_1(x_1) + \ell_2(y - x_1)).$$

We have thus reduced our problem to a known problem in convex analysis. The above formula is well-known in *convex analysis* [9]. It is known as the *infimal convolution*, or an *epi-sum*, and usually denoted by

$$\ell = \ell_1 \square \ell_2.$$

There are efficient algorithms for solving this convex analysis problem. At least for situations when the functions $\ell_i(x_i)$ are convex (and continuous), there is an efficient algorithm for computing the infimal convolution. This algorithm is based on the use of *Legendre-Fechnel transform*

$$\ell^*(s) = \sup_x(s \cdot x - \ell(x)).$$

Specifically, it turns out that the Legender transform of the infimal convolution is equal to the sum of the Legendre transforms:

$$(\ell_1 \square \ell_2)^* = \ell_1^* + \ell_2^*,$$

and that to reconstruct a function form its Legendre transform, it is sufficient to apply the Legendre transform once again:

$$\ell = (\ell^*)^*.$$

Thus, we can compute the infimal convolution as follows:

$$\ell_1 \square \ell_2 = (\ell_1^* + \ell_2^*)^*.$$

Similarly, we can compute the infimal composition of several functions

$$\ell_1 \square \cdots \square \ell_n = \min\{\ell_1(y_1) + \cdots + \ell_n(y_n) : y_1 + \cdots + y_n = y\},$$

as

$$\ell_1 \square \cdots \square \ell_n = (\ell_1^* + \cdots + \ell_n^*)^*.$$

There exists a fast (linear-time $O(N)$) algorithm for computing the Legendre transform; see, e.g., [4]. So, by using this algorithm, we can compute the results of fuzzy data processing in linear time.

Let us summarize the resulting algorithm.

5 Resulting Linear Time Algorithm for Fuzzy Data Processing Beyond min t-Norm

What is given and what we want to compute: reminder. We are given:

- a function $f(x_1, \ldots, x_n)$;
- n membership functions $\mu_1(x_1), \ldots, \mu_n(x_n)$; and
- an "and"-operation $f_\&(a, b)$.

We want to compute a new membership function

$$\mu(y) = \max\{f_\&(\mu_1(x_1), \ldots, \mu_n(x_n)) : f(x_1, \ldots, x_n) = y\}.$$

Algorithm.

- First, we represent the given "and"-operation in the Acrhiemdean form $f_\&(a, b) = f^{-1}(f(a) \cdot f(b))$ for an appropriate monotonic function $f(x)$.
  In the following text, we will assume that we have algorithms for computing both the function $f(x)$ and the inverse function $f^{-1}(x)$.
- Then, for each i, we find the value $\tilde{x}_i$ for which $\mu_i(x_i)$ attains its largest possible value.

  For normalized membership functions, this value is $\mu_i(\tilde{x}_i) = 1$.
- We then compute the values $c_0 = f(\tilde{x}_1, \ldots, \tilde{x}_n)$ and $c_i = \dfrac{\partial f}{\partial x_i}\Big|_{x_i = \tilde{x}_i}$, and

$$a_0 = c_0 - \sum_{i=1}^{n} c_i \cdot \tilde{x}_i.$$

In the following text, we will then use a linear approximation

$$f(x_1, \ldots, x_n) = a_0 + \sum_{i=1}^{n} c_i \cdot x_i.$$

- After that, we compute the membership functions

$$v_1(s_1) = u_1((s_1 - a_0)/c_1)$$

and $v_i(y_i) = \mu_i(y_i/c_i)$ for $i > 2$.

In terms of the variables $s_1 = a_0 + c_1 \cdot x_1$ and $y_i = c_i \cdot x_i$, the desired quantity y has the form $y = s_1 + y_2 + \cdots + y_n$.
- We compute the minus logarithms of the resulting functions:

$$\ell_i(y_i) = -\ln(v_i(y_i)).$$

- For each i, we then use the Fast Legendre Transform algorithm from [4] to compute ℓ_i^*.
- Then, we add all these Legendre transforms and apply the Fast Legendre Transform once again to compute the function

$$\ell = (\ell_1^* + \ldots + \ell_n^*)^*.$$

- This function $\ell(y)$ is equal to $\ell(y) = -\ln(v(y))$, so we can reconstruct $v(y)$ as

$$v(y) = \exp(-\ell(y)).$$

- Finally, we can compute the desired membership function $\mu(u)$ as

$$\mu(y) = f^{-1}(v(y)).$$

6 Conclusions

To process fuzzy data, we need to use Zadeh's extension principle. In principle, this principle can be used for any t-norm. However, usually, it is only used for the min t-norm, since only for this t-norm, an efficient (linear-time) algorithm for fuzzy data processing was known.

Restricting oneselves to min t-norm is not always a good idea, since it is known that in many practical situations, other t-norms are more adequate in describing expert's reasoning. In this paper, we show that similar efficient linear-time algorithms can be designed for an arbitrary t-norm. Thus, it become possible to use a t-norm that most adequately describes expert reasoning in this particular domain—and still keep fuzzy data processing algorithm efficient.

Acknowledgements This work was supported in part by the National Science Foundation grants HRD-0734825 and HRD-1242122 (Cyber-ShARE Center of Excellence) and DUE-0926721, and by an award "UTEP and Prudential Actuarial Science Academy and Pipeline Initiative" from Prudential Foundation. We also acknowledge the partial support of the Center of Excellence in Econometrics, Faculty of Economics, Chiang Mai University, Thailand.

References

1. Buchanan, B.G., Shortliffe, E.H.: Rule Based Expert Systems: The MYCIN Experiments of the Stanford Heuristic Programming Project. Addison-Wesley, Reading, Massachusetts (1984)
2. Jaulin, L., Kiefer, M., Dicrit, O., Walter, E.: Applied Interval Analysis. Springer, London (2001)
3. Klir, G., Yuan, B.: Fuzzy Sets and Fuzzy Logic. Prentice Hall, Upper Saddle River, New Jersey (1995)
4. Lucet, Y.: Faster than the fast Legendre transform, the linear-time Legendre transform. Numer. Alg. **16**, 171–185 (1997)
5. Moore, R.E., Kearfott, R.B., Cloud, M.J.: Introduction to Interval Analysis. SIAM, Philadelphia (2009)
6. Nguyen, H.T., Kreinovich, V., Wojciechowski, P.: Strict Archimedean t-Norms and t-Conorms as universal approximators. Int. J. Approx. Reason. **18**(3–4), 239–249 (1998)
7. Nguyen, H.T., Walker, E.A.: A First Course in Fuzzy Logic. Chapman and Hall/CRC, Boca Raton, Florida (2006)
8. Rabinovich, S.G.: Measurement Errors and Uncertainty: Theory and Practice. Springer, Berlin (2005)
9. Rockafeller, R.T.: Convex Snslysis. Princeton University Press, Princeton, New Jersey (1997)
10. Zadeh, L.A.: Fuzzy sets. Inf. Control **8**, 338–353 (1965)

Detecting Changes in Time Sequences with the Competitive Detector

Leszek J. Chmielewski and Arkadiusz Orłowski

Abstract The concept of the competitive edge detector is revisited and extended. In the case of application to 1D signals it can be denoted as the detector of changes. In the detector two approximators are used working one at the 'past' and one at the 'future' side of the considered data point. The difference of their outputs makes it possible to find the change of the value and the derivative of the signal. The new features introduced consist in performing robust analysis and in adding the option to use a quadratic function as an approximator. Weighted voting of elemental subsets is used with weights related to the significance of a subset for the result. Weak fuzzification is used to increase the robustness. Results of change detection on test data as well as some real-life economic data are encouraging.

Keywords Change detector · Competitive · Robust · Fuzzy · Weighted

1 Introduction

The most interesting phenomena which manifest themselves in the data collected about the world are related to the changes. The change in the existing signal or the emergence of a new one is something that invokes an accelerated cognitive process of deciding whether it is necessary to react.

The method to be presented here has its origin primarily in the domain of image processing; therefore, we shall first review the approaches to change detection in image processing setting, where the problem of change detection is incessantly one of the main topics of research. If the changes in space are considered, then the problem

L. J. Chmielewski (✉) · A. Orłowski
Faculty of Applied Informatics and Mathematics – WZIM, Warsaw University of Life Sciences – SGGW, Nowoursynowska 159, bldg 34, 02-776 Warsaw, Poland
e-mail: leszek_chmielewski@sggw.pl
URL: http://www.wzim.sggw.pl

A. Orłowski
e-mail: arkadiusz_orlowski@sggw.pl

© Springer International Publishing AG 2018
C. Berger-Vachon et al. (eds.), *Complex Systems: Solutions and Challenges in Economics, Management and Engineering*, Studies in Systems, Decision and Control 125, https://doi.org/10.1007/978-3-319-69989-9_15

251

is related to the detection of edges. This comprises the detection of changes in image intensity or color, but also finding the precise localization, directionality, continuity, and finally the significance of the edges. The changes in time relate to the detection of movement or lighting. The research in edge detection has been summarized in the reviews [2, 3, 19, 21] and older surveys [12, 25, 36]. The research in motion detection was recapitulated among others in the surveys [16, 28]. The motion is related to the movement of the objects viewed (cf. e.g. [18]) and is the necessary step towards solving the problem of surveillance (e.g. [13]). A more general problem is the detection of changes between multiple images, reviewed for example in [27]. One of the well-established applications is the detection of differences in medical images (e.g. [34]).

The change detection in general is broadly treated in the statistical meaning [30] (see [26] for a state-of-the-art survey; see also e.g. [1, 15, 32] and an extensive list of references in [33]). One of its main applications is the detection of faults and damages. In simple words it can be formulated as testing the hypothesis that a new sample comes from a different statistical distribution than the previous samples [14, 17]. The sample can consist of more than one measurement.

In this study we shall come back to the concept of fitting a model of the signal to the data, present in the image processing domain from its early years. This concept was used by so many authors that we can name just some of them. Blake and Zisserman in [4] proposed to approximate the image intensity function with models based on mechanical analogies, which remove noise and simultaneously detect edges in the simulated process of stretching and breaking a membrane. The filter performed very well on two-dimensional noisy data but its speed was small [8]. A roof edge detector proposed by Pajdla and Hlaváč [24] consisted in fitting the edge model to the image with the edge direction found beforehand to speed up the process. Niedźwiecki, Sethares and Suchomski in [22, 23] proposed a filter denoted as the *competitive filter*. It consisted of two filters working simultaneously from the two sides of an edge. The filter in which the approximation error was smaller won the competition and the output from that filter was used as the filtering result. Each filter approximated the image intensity in 2D with the simple constant function which was the reason why the algorithm was extremely effective.

In 1996 one of us proposed to use the concept of competitive filtering to detect edges [6] (the idea went back to 1994 [5]). The concept was successfully developed for one-dimensional data in the form of two linear filters. The extension to two dimensions failed due to the due to the problematic definition of the two sides of an edge in face of the complexity of shapes of the image intensity function near the edge junctions, and was abandoned [7].

We have revisited the concept of the competitive edge detection in the application to one-dimensional data in a pilot study [11] where we have added the fuzzy weighted robust analysis mechanism to the fitting of the two filters. We have made preliminary analyzes of some real-life data The detector performed in a promising way. In this paper we have implemented a second degree polynomial function as an additional option of the approximators and we have tried to process economical data.

The robustness is understood as immunity to outlying data and as such it will not always be applicable to economic data. An outlying data point can appear as a result of an erroneous measurement or observation and then it is a gross error and as such it should be rejected form the analysis. However, if it appears as a result of a proper observation it is the evidence of an unknown or unexpected phenomenon, and hence it is a valuable piece of information which should be paid special attention in the analysis.

The method presented here will not deal with the change of the statistical nature of the signal but rather with the detection of steps or jumps in the value and the first derivative, or slope, of the signal. This could serve as a source of information on the signal as well as a hint for a human observer to pay attention to the details related to the signal. There exist a large number of stock exchange state indicators, like the moving averages or the Relative Strength Index. Our detector can be viewed in a similar way, although it is neither directly related to the market analysis nor has emerged or was derived with such analysis in view. However, we deem it useful in looking at time series of economy-related data in a more insightful way.

In fact, what we had in mind, was the detection of early signals of important events to come in the near future [29]. This far-reaching goal by no means can be attained with simple methods, but we hope our proposition can be an incentive in the search.

The detector originated from the domain of image processing. Therefore, some image processing terminology will still appear in places, although the relation of the detector to images is only historical. In particular, the notions of *change*, *step*, *jump* will be interchanged according to the context.

The software and the graphs were produced within the Matlab® environment.

The remaining part of this chapter will be organized as follows. In the next section the detector in its previous form will be described and its new features will be explained. The description will be illustrated with examples of detection in synthetic images. Then, the results of the detection of changes in some real-life data will be presented. The propositions of further development of the method and some concluding remarks will come at the end.

2   Method

2.1   General Concept

According to [23] let us take a sequence of measurements $z(t) = y(t) + n(t)$, where $n(t)$ is noise. Time t is discrete. The measurements are known up to the time $t_0 + \Delta t$. Two predictors, or approximators, will be used to find $y(t_0)$, one running from the past towards the future, using $z(t), t \in [t_0 - s - \delta, t_0 - \delta]$ to find $\hat{y}_-(t_0)$, and one running towards the past, using $z(t), t \in [t_0 + \delta, t_0 + s + \delta]$ to find $\hat{y}_+(t_0)$. The approximators will be referred to as *past* and *future*, or *left* and *right*. The data used in one approximation will form its support. For each prediction its error is estimated yielding $e_-(t_0)$

and e_+t_0. As the estimate of the result at the point of interest, $y(t_0)$, the output of the filter which has smaller error is used. The filter can work providing enough measurements are known in advance. The parameter s can be understood as the scale of the filter. Parameter δ is the gap between the point of interest and each of the estimators.

In [6] linear least square approximators were used as filters and their mean square errors were used as their approximation errors. The idea of using least median of squares as a robust approximator was mentioned, but not developed. The concept of using the difference of values and their derivatives as the estimates of the step and roof edge at point t_0 was introduced. The conditions for the existence of the step was that the graphs of the approximation errors crossed in such a way that for increasing t the error from the past increased and that for the future decreased. These conditions were expressed in [6] in a complicated way but they can be simply written down, respectively, as

$$e_+(t_0 - \varepsilon) > e_-(t_0 - \varepsilon) \ \land \ e_+(t_0 + \varepsilon) < e_-(t_0 + \varepsilon), \tag{1}$$
$$e_+(t_0 - \varepsilon) > e_+(t_0 + \varepsilon) \ \land \ e_-(t_0 - \varepsilon) < e_-(t_0 + \varepsilon). \tag{2}$$

If the steps should be found not *at*, but *between* the points, these conditions could be reformulated accordingly. Here we shall not do so; instead, we can point out that it seems reasonable that ε should be as small as possible and $\varepsilon \le \delta$, and that both can be equal to one. This was assumed in [6] and so it will be in the present paper. Because the past error should be known for $t_0 + \varepsilon$ then the measurements for $t_0 + \Delta t = t_0 + \delta + s + 2\varepsilon = t_0 + s + 3$ should be known for the detector to operate.

The process of error graphs crossing is illustrated in Fig. 1. Let us describe it in a figurative rather than rigorous way. Let us imagine that both approximators together with the analyzed point are moved along the data from left to right. When a step is encountered, first the right approximator moves over it so the step enters the right approximator's support. Therefore, the error of the right approximator goes up. As the analyzed point is moved forward, the step leaves the support of the right approximator, so its error goes down, and enters that of the left one (this particular moment is shown in the figure). Now, the error of the left approximator increases.

The result of formulating the condition for a step using the values in two points distant by 2 is that a step can be detected in two points, like this at $x = 29, 3$ in Fig. 1. This is reasonable, because in fact the step is formed by data in both points.

There is no separate step existence condition for steps of the function and of its derivative. These two appear together, except the points where the value of one of them is zero or small.

At present, as it was in [6], as the measure of error the mean square error is used. This is not the best solution if the approximators are found with the robust analysis. In the application of our present interest not the filtered value but the detection of changes is important. The error measure do not influence the value of the step, but only its location. The question of error measure in relation to detectability and location of change points will be commented on further in the end of the next section, as soon as some more images will be shown.

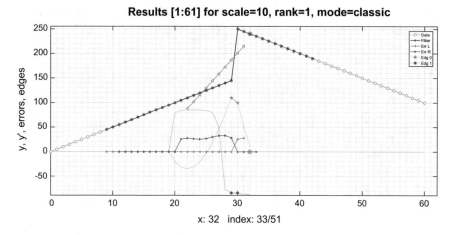

Fig. 1 Intermediate results for the two approximators up to point indexed 32; see the legend (Edg 0 and 1 mean zeroth and first derivative, respectively). Graphs of errors cross between points indexed 29, 30. See text for more details

2.2 Extensions and Changes

The new extension is the quadratic approximator, so it will be described first. The extensions are applied to the detector in pairs, triples etc. In the following, the detector without the robustness feature will be denoted as the *classic* one. The independent variable will be denoted by x instead of t.

Quadratic Approximator

The least square approximation can be easily applied to the quadratic function. The formulas are so well known that we shall refrain from showing them here. However, this approximation was not used before in the context of the competitive filter or detector. The result of applying the quadratic detector to the data representing the combined step and roof jump is shown in Fig. 2. In Fig. 3 the location of the approximators at a selected point can be seen.

Robustness

Let us recall the Hough transform for lines with two point voting subsets (e.g. [20, 31]). Such a subset defines the two parameters of a straight line univocally so it votes for one point in the 2D parameter space. This space is represented approximately by the accumulator array. In the present implementation the votes from each subset formed from the points of the approximator are calculated and collected. The ranges of the parameters serve to calculate the dimensions of the accumulator so that a change in a parameter by a unit is represented by 100 elements, but the dimensions

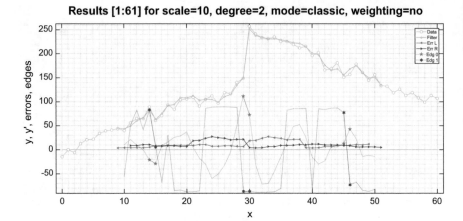

Fig. 2 Step and roof edge like in Fig. 4a with some noise added, analyzed with the classic quadratic detector

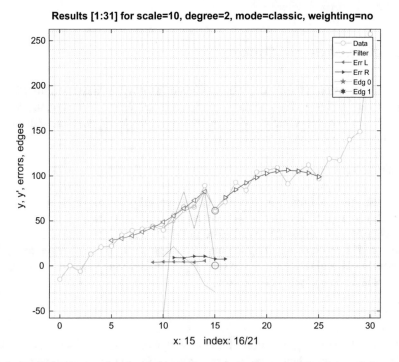

Fig. 3 Result for Fig. 2 analyzed with the classic, quadratic detector—a detailed result for one of the points with a step, $x = 15$. Quadratic functions can fit to the data more closely than the linear ones (some colors desaturated to enhance the view of approximators)

of the accumulator are limited to 1001×1001. The votes are stored in the accumulator which is then fuzzified by convolution with the array containing the inverted quadratic function, clipped to nonnegative values. Each dimension of the window containing the fuzzifying function is chosen as 0.1 of the respective dimension of the accumulator, so the conditions of *weak fuzzification* is fulfilled [9, 10]. Accumulator for each approximator is searched for maxima which indicate the solutions for them.

The results of using the robust method can be seen in Fig. 4 Approximators for a point near to the step are displayed to show how some point were not taken into account in the analysis due to that they could be treated as outlying from the major part of the data. In the left accumulator shown in subfigure b some local maxima corresponding to the voting pairs which contain outliers were postponed, and the global

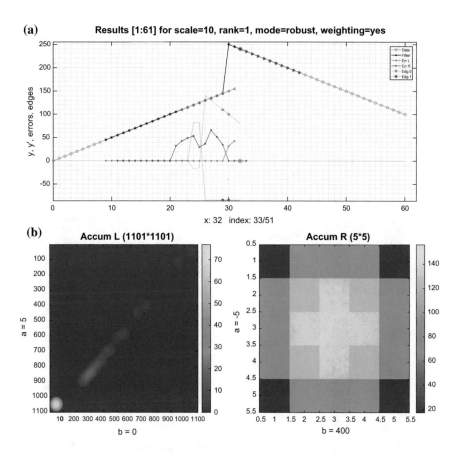

Fig. 4 Results for the image of Fig. 1 found with the robust version: **a** illustration of graphs of errors cross between points 29, 30 (approximators for $x = 32$ shown with larger triangles, empty in general but filled-in if approximation error is zero); **b** accumulators for robust filters at $x = 30$, left and right, respectively. See text for more details

maximum formed by votes coming from inliers was chosen. Quadratic shape of the fuzzifying function can be seen. The right accumulator is degenerated to a single value (fuzzified) due to all points lie on a common line.

The similar image but with some noise added is analyzed in Fig. 5. The robust approximators omit some extraneous data points.

The robust quadratic approximator is designed in an analogous way as the linear one, with the following changes. The voting subsets now contain three points, so triples are formed from the data points within an approximator. The parameters of the approximating quadratic function are calculated by solving the set of three equations, the structure of which does not have to be explained. The accumulator is now three dimensional, and so is the fuzzifying function.

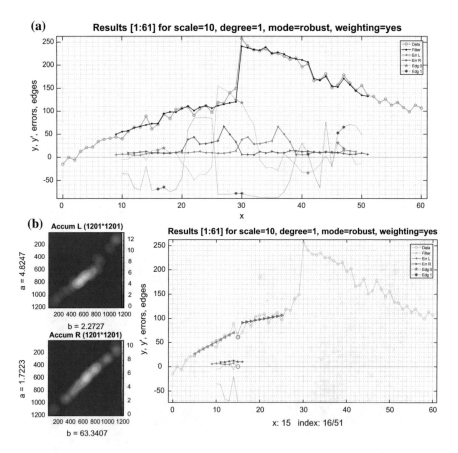

Fig. 5 Step and roof edge like in Fig. 4a with some noise added, analyzed with the robust detector. **a** result for all the data; **b** details for the step at $x = 15$ (some colors desaturated to enhance the view of approximators)

The ability to omit some data points can sometimes have a negative influence on the result. This frequently appears with the robust quadratic approximator. An example can be seen in Fig. 6. The approximated line passes with precision through some points and omits the points which are the nearest to the central point of the detector located at $x = 18$. In the case shown, this leads to an un acceptable result. This phenomenon can be treated by proper weighting the influence of the data points. The weighting will be the subject of the next paragraph.

Weighting

It is not reasonable to treat votes equally important. Note that the noise in pairs with data points which are close to each other have larger impact on the parameters of the approximation than those which contain distant points. Therefore, the pairs with distance equal to one were dismissed. Attention should be paid to weighting the remaining votes. Let us concentrate upon the linear approximator. If the votes were weighted with the distance of points in their pairs, then it is very probable that the

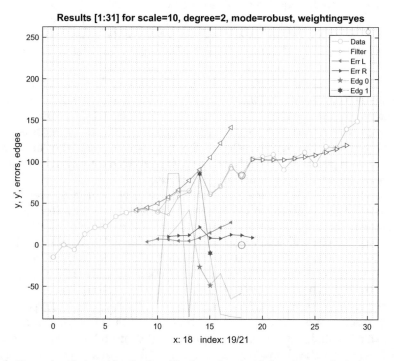

Fig. 6 Step and roof edge with noise from Fig. 5a, analyzed with the robust quadratic detector. An extremal example of the problematic local solution as the result of robustness. The approximated line passes with precision through points $x = 8, 9, 11, 12, 14$ and omits the points $10, 13$ which is positive, and points $15, 16, 17$ which are the nearest to the central point $x = 18$

voting pair consisting in the endpoints of the support would always win. Therefore, not to promote the most elongated pairs excessively, at present the votes are weighted with the support length to the power of 0.25. The pairs which contain data points for which the x coordinates differ by one are dismissed.

The problem of weighting is complex and at present it was only partly solved. As shown in the previous paragraph, it is necessary to promote those data which come from points closer to the central point of the detector. This means that there are at least two criteria for finding the weights, which makes it necessary to find some optimum between them. This will be the subject of further studies. At present the robust quadratic detector is too vulnerable to omitting important data and will not be used in the examples which follow.

Output without extrapolation

In the present implementation, as the filtered value the output at $t_0 - \delta$ from the left approximator, and for $t_0 + \delta$ for the right one, is used, to avoid using extrapolated values. This additionally stabilizes the results, especially when robust approximators are used. Due to the requirements of the step detector, the analysis of a point t_0 is finished when the approximators are placed around $t_0 + 2$, so the necessary outputs and their errors are already known. As said before, in the present application the filtering functionality of the algorithm is of secondary importance, however it is always reasonable to use the best approximation available.

In Fig. 7 the results for data containing all the detectable changes can be seen. Some point noise or outlying values are added, mainly to illustrate the advantageous functioning of the robust detector. It should be pointed out that if the outlying point at $x = 42$ were moved towards the slope between $x = 36$ and 37, for example to point 39, it would be doubtful whether this is an outlier or a part of the slope at 36, 37, interrupted at 38 and continued at 39. The data element is an outlier if it does not follow the trend, which is not always univocal.

The graph in Fig. 7 was drawn so that the scale s of the detector were smaller than the distance between the changes to be detected. Let us now look closely at subfigure b, point $x = 37$. At this point the error of the right approximator is zero (dark magenta line with small triangles pointing to the right). What would happen if the already mentioned outlier at $x = 42$ were moved left by one, so that this error went up instead of falling down to zero? graphs would not cross and the jump would not be detected. In this way it can be seen that the jumps which are closer to each other than the scale, do interfere, and this can make them undetectable. It is expected that the formulation of the error measure in a better way than as the mean square error, for example with the use of information coming from the robust approximators on which points are treated as extraneous, would limit the detrimental mutual influence of the close jumps. This will be studied in the future research.

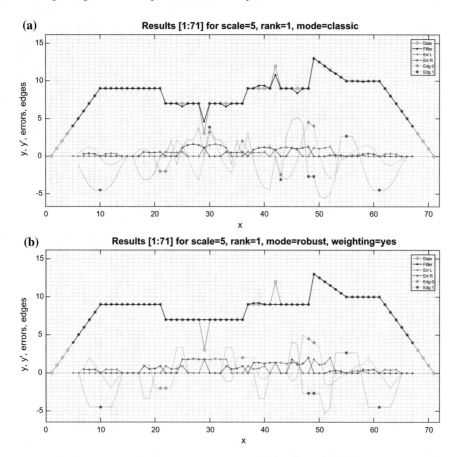

Fig. 7 Results for the classic and robust version of the detector for data with all the detectable changes and some outlying data represented. (**a**) Classic; (**b**) robust. Angles shown in tens of degrees

3 Real-Life Examples

Let us consider daily closing values of WIG, the main index of the Warsaw Stock Exchange ([35]→Market Data→Market Indices→WIG), between the days February 1 and October 30, 2007. Near the middle of this period WIG at closing attained its global maximum of 67735.30, on July 6, 2007. The lowest value in this period was 50782.08 on March 5, day 23 of the period. The results of processing these values with the change detector is shown in Fig. 8. Scale was set to $s = 10$ to observe the changes in 2 week trends. It can be seen that all the versions of the detector had problems with precisely pointing at the maximum. The majority of important jumps up and down and the deep minima were indicated well.

Let us concentrate at the day of the nearest maximum at day $x = 18$ before the lowest value at day 23. After this day the fall started. The locations of the predictors

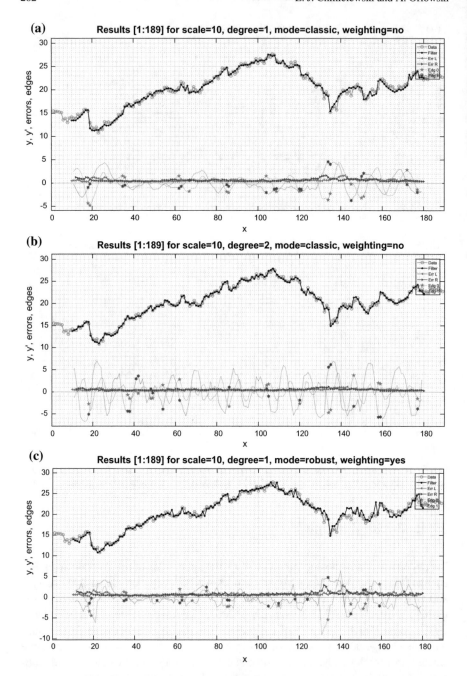

Fig. 8 Main index of the Warsaw Stock market WIG from February 1 to October 30, 2007, and the results of its processing with the detectors: **a** classic, linear; **b** classic, quadratic; **c** robust. Scale $s = 10$. For better visibility of error and edge graphs it is $y = \mathrm{WIG}/1000 - 40$

around this day are shown in Fig. 9. It can be seen that each pair of approximators has a different location and underlines different aspects of the values analyzed. The classic linear approximators indicate the typical trends within their support. The robust linear approximators are positioned according to the trends established by the majority of data points, but the most extraneous points, for example, days 13 and 23 are omitted. The quadratic approximators bend to show the more detailed changes, so the estimated values of the jump can be more precisely fitted to the data. Each detector finds the jump, but its values differ according to the characteristics of the detectors.

The scale of the images and lack of access to the intermediate data in the paper edition of this study makes it impossible to notice, but as previously, it can be said also now that the difficulties with detecting the locations of the change points is related to the existence of many small jumps in the data which interfere within the supports of the approximators and make it impossible for the approximation errors to go down to the extent which would make it possible for the error graphs to cross. This problem will be studied further.

4   Discussion

At present we see two problems needing further studies. As already mentioned, these are: the interference between nearby changes and the question of optimal weighting the votes in the robust detector. The question of weighted voting and the question of choosing the right error measure needs further work.

It should be reminded that it is easy to introduce weighting also in the classic least squares detector so the experience gained will be used also in it.

The calculations in the detection for each point are performed for small sets of data so they are effective. The only operation which need longer time is the processing of the 3D accumulator in the quadratic robust version. It should be considered to use another method to eliminate outstanding sets of parameters, for example some of the unsupervised clustering methods. This should be easy due to that the number of objects (set of parameters) is small and distance measures between the sets seem to be easy to design.

The features of the proposed detector can be summarized as follows.

Advantages

- Detection is performed together with the filtering process.
- The classic version in the linear as well as quadratic form has only one parameter: the scale.
- Higher order derivatives can be estimated according to the order of the approximating function.
- The processes in some versions of the detector are relatively effective.

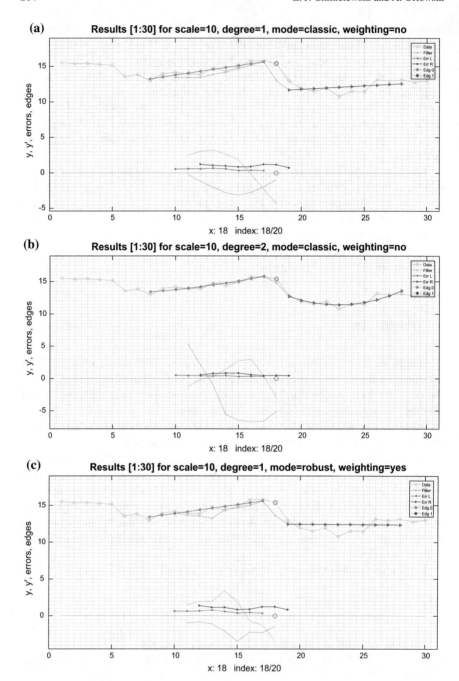

Fig. 9 Close view of the closest maximum at $x = 18$ before the minimum at day 23 of Fig. 8. Two approximators are located at the day after which the fall started (some colors desaturated to enhance the view of approximators). Graphs of edges (dotted lines) are visible but edge points could not ba marked at this time point due to the structure of (1), (2). Detectors: **a** classic, linear; **b** classic, quadratic; **c** robust, linear

Drawbacks

- A considerable part of data about the future must be known before the detection can be made. It can be too late for reaction.
- The interference between changes which are closer to each other than the scale is the source of problems with jump localization.
- The detector in its weighted version has many parameters and their tuning needs optimization.
- The robust procedure by fuzzy voting is time-consuming. This can be overcome by using a different data clustering technique than the accumulation.

5 Summary and Prospects

The concept of the competitive edge detector was recalled, extended and used to analyze some sets of data. In the proposed detector of changes two approximators are used. One of them works at the 'past' of 'left' and one at the 'future' or 'right' side of the considered data point. The outputs from the approximators are used to calculate the change of the value and the derivative of the data. The detector can perform robust analysis with weak fuzzification which make it possible to postpone extraneous data points. In the present study the option to use a quadratic function as an approximator was added. Weighted voting of elemental subsets is used with weights related to the significance of a subset for the result. Results of change detection on test data as well as some real-life economic data are encouraging.

As the directions of future work the optimization of the weighting process, the question of reducing the interference of neighboring jumps in data, and the choice of an effective clustering method for votes in the robust algorithm were indicated.

References

1. Basseville, M.: Statistical methods for change detection. In: H. Unbehauen (ed.) Control Systems, Robotics and Automation. Encyclopedia of Life Support Systems (EOLSS), Developed under the Auspices of the UNESCO, vol. XIV. Eolss Publishers, Paris, France (2002). http://www.eolss.net
2. Basu, M.: Gaussian-based edge-detection methods-a survey. IEEE Trans. Syst. Man Cybern. Part C (Applications and Reviews) **32**(3), 252–260 (2002). https://doi.org/10.1109/TSMCC.2002.804448
3. Bhardwaj, S., Mittal, A.: A survey on various edge detector techniques. Procedia Technol. **4**, 220–226 (2012). https://doi.org/10.1016/j.protcy.2012.05.033
4. Blake, A., Zisserman, A.: Visual Reconstruction. MIT Press, Cambridge, MA, London (1987)
5. Chmielewski, L.: The concept of step and roof edge detector derived from a competitive filter. EPSILON ARG Report (1994)
6. Chmielewski, L.: The concept of a competitive step and roof edge detector. Mach. Graphics Vis. **5**(1–2), 147–156 (1996)

7. Chmielewski, L.: Failure of the 2D version of the step and roof edge detector derived from a competitive filter. Report of the Division of Optical and Computer Methods in Mechanics, IFTR PAS (1997)
8. Chmielewski, L., Skłodowski, M., Cudny, W., Nieniewski, M., Kuriański, A., Michalski, B.: Fringe image enhancing in the light wavelength stepping method. Mach. Graphics Vis. **3**(3), 543–578 (1994)
9. Chmielewski, L.J.: Fuzzy histograms, weak fuzzification and accumulation of periodic quantities. Application in two accumulation-based image processing methods. Pattern Anal. Appl. **9**(2–3), 189–210 (2006). https://doi.org/10.1007/s10044-006-0037-7
10. Chmielewski, L.J.: Evidence accumulation methods in digital image analysis. Corrected edition (in Polish). Akademicka Oficyna Wydawnicza EXIT Andrzej Lang, Warsaw (2015). http://www.lchmiel.pl/akum06
11. Chmielewski, L.J., Orłowski, A.: Detecting changes with the robust competitive detector. In: Proceedings 8th Iberian Conference on Pattern Recognition and Image Analysis IbPRIA 2016. Lecture Notes in Computer Science. Springer, Faro, Portugal (2017) (Submitted for review)
12. Davis, L.: A survey of edge detection techniques. Comp. Graph. Image Proc. **4**, 248–270 (1975). https://doi.org/10.1016/0146-664X(75)90012-X
13. Frejlichowski, D., Forczmański, P., et al.: SmartMonitor: An approach to simple, intelligent and affordable visual surveillance system. In: Bolc, L., et al. (ed.) Computer Vision and Graphics: Proceedings International Conference ICCVG 2012. Lecture Notes in Computer Science, vol. 7594, pp. 726–734. Springer, Heidelberg, Warsaw, Poland (2012). https://doi.org/10.1007/978-3-642-33564-8_87
14. Furmańczyk, K., Jaworski, S.: Large parametric change-point detection by a v-box control chart. Sequ. Anal. **35**(2), 254–264 (2016). https://doi.org/10.1080/07474946.2016.1165548
15. Gordon, L., Pollak, M.: An efficient sequential nonparametric scheme for detecting a change of distribution. Ann. Statist. **22**(2), 763–804 (1994). https://doi.org/10.1214/aos/1176325495
16. Hu, W., Tan, T., Wang, L., Maybank, S.: A survey on visual surveillance of object motion and behaviors. IEEE Trans. Syst. Man Cybern. Part C (Applications and Reviews) **34**(3), 334–352 (2004). https://doi.org/10.1109/TSMCC.2004.829274
17. Jaworski, S., Furmańczyk, K.: On the choice of parameters of change-point detection with application to stock exchange data. Quant. Methods Econ. **XII**(1), 87–96 (2011)
18. Kuriański, A., Nieniewski, M.: A model of the MRF with three observation sources for obtaining the masks of moving objects. In: Borgefors, G. (ed.) Proceedings 9th Scandinavian Conference on Image Analysis, vol. 2, pp. 931–940. IAPR, Uppsala (1995)
19. Maini, R., Aggarwal, H.: Study and comparison of various image edge detection techniques. Int. J. Image Process. (IJIP) **3**(1), 1–11 (2009)
20. Maître, H.: Un panorama de la transformation de Hough. Traitement du Signal **2**(4), 305–317 (1985)
21. Narendra, V.G., Hareesh, K.S.: Study and comparison of various image edge detection techniques used in quality inspection and evaluation of agricultural and food products by computer vision. Int. J. Agric. Biol. Eng. **4**(2), 83 (2011). https://doi.org/10.3965/j.issn.1934-6344.2011.02.083-090
22. Niedźwiecki, M., Sethares, W.: New filtering algorithms based on the concept of competitive smoothing. In: Proceedings 23rd International Symposium on Stochastic Systems and Their Applications, pp. 129–132. Osaka (1991)
23. Niedźwiecki, M., Suchomski, P.: On a new class of edge-preserving filters for noise rejection from images. Mach. Graphics Vis. **1–2**(3), 385–392 (1994)
24. Pajdla, T., Hlaváč, V.: Surface discontinuities in range images. In: 5th International Conference Computer Vision, pp. 524–528. IEEE Computer Society Press, Berlin (1993). https://doi.org/10.1109/ICCV.1993.378168
25. Peli, T., Malah, D.: A study of edge detection algorithms. Comput. Graphics Image Process. **20**(1), 1–21 (1982). https://doi.org/10.1016/0146-664X(82)90070-3
26. Polunchenko, A.S., Tartakovsky, A.G.: State-of-the-art in sequential change-point detection. Method. Comput. Appl. Probab. **14**(3), 649–684 (2012). https://doi.org/10.1007/s11009-011-9256-5

27. Radke, R.J., Andra, S., Al-Kofahi, O., Roysam, B.: Image change detection algorithms: a systematic survey. IEEE Trans. Image Process. **14**(3), 294–307 (2005). https://doi.org/10.1109/TIP.2004.838698
28. Räty, T.D.: Survey on contemporary remote surveillance systems for public safety. IEEE Trans. Syst. Man Cybern. Part C (Applications and Reviews) **40**(5), 493–515 (2010). https://doi.org/10.1109/TSMCC.2010.2042446
29. Scheffer, M., Bascompte, J., Brock, W.A., et al.: Early-warning signals for critical transitions. Nature **461**, 53–59 (2009). https://doi.org/10.1038/nature08227
30. Shiryaev, A.N.: On optimum methods in quickest detection problems. Theory Probab. Appl. **8**(1), 22–46 (1963). https://doi.org/10.1137/1108002
31. Strauss, O.: Use the fuzzy Hough transform towards reduction of the precision-uncertainty duality. Pattern Recognit. **32**, 1911–1922 (1999)
32. Tartakovsky, A., Nikiforov, I., Basseville, M.: Sequential analysis: hypothesis testing and changepoint detection. In: Monographs on Statistics and Applied Probability. CRC Press, Chapman & Hall/CRC (2012)
33. Veeravalli, V.V., Banerjee, T.: Quickest change detection. (2012). arXiv:1210.5552
34. Venot, A., Golmard, J., Lebruchec, J., et al.: Digital methods for change detection in medical images. In: Deconink, F. (ed.) Information Processing in Medical Imaging. Martinus Nijhoff Publishers, Dodrecht, The Netherlands (1984)
35. WGPW—Warszwska Giełda Papierów Wartościowych (Warsaw Stock Exchange) (2016). http://www.gpw.pl. Accessed 15 Nov 2016
36. Ziou, D., Tabbone, S.: Edge detection techniques—An overview. Pattern Recognit. Image Anal. **8**(4), 537–559 (1998)

Deep Learning Architecture for High-Level Feature Generation Using Stacked Auto Encoder for Business Intelligence

Vikas Singh and Nishchal K. Verma

Abstract In the era of modern world, faster development and wider use of digital technology generates large amount of data in digital space. Handling such large amount of data by conventional machine learning algorithms is difficult because of heterogeneous nature and large size of data. Deep learning strategy, is an advancement in machine learning research to deal with such heterogeneous nature and large size of data and extract high-level representations of data through a hierarchical learning process. This paper proposes novel multi-layer feature selection with conjunction of Stacked Auto-Encoder (SAE) to extract high level features or representations and eliminate the lower level features or representations from data. The proposed approach is validated on the Farm Ads dataset and the result is compared with various conventional machine learning algorithms. The proposed approach has outperformed as compared to conventional machine learning algorithms for the given dataset.

Keywords Business data · Stacked auto-encoder · Deep learning
Feature selection

1 Introduction

In the modern era, due to advancement of digital technology, the marketers are moving more and more into the digital space, the amount of data generated is increasing at exponential rate. These data carry a very high potential for many industries like social media, health-care, entertainment, commerce, banking, online retail markets and government agencies who can analyze these data for trends, patterns, fraud detection, relationships, and other useful information to provide a better experience to their users. Several relevant examples are as follow;

V. Singh · N. K. Verma (✉)
Department of Electrical Engineering, Indian Institute of Technology Kanpur, Kanpur
208016, India
e-mail: nishchal.iitk@gmail.com

© Springer International Publishing AG 2018　　　　　　　　　　269
C. Berger-Vachon et al. (eds.), *Complex Systems: Solutions and Challenges
in Economics, Management and Engineering*, Studies in Systems,
Decision and Control 125, https://doi.org/10.1007/978-3-319-69989-9_16

- Purchase and Click Data records are pool of information for marketers especially for e-commerce. It is widely used by direct marketers to know buying patterns of customers i.e. their liking for different products. It helps to enrich their experience by presenting similar kind of ads matching to their interests. Then marketers target different types of customers in various ways to maximize profit and provide what they need.
- Tourists database collected from hotels, travel agencies etc. is very helpful to build customize tour packages for users. By using this information clever pricing strategies are planned, where different customers pay different prices for the same accommodation with some added perks.
- Financial planning services are other market giants where large amount of data about security prices are kept. They have large customers with differing preferences for different types of investments. Trying to find which investments match to particular customer is a very challenging problem. Therefore, financial planners are under competitive pressure to analyze such data and provide best package to their customers. It should be also flexible enough to accommodate on the spot need of customers.

To deal such type of problems, machine learning enables a system to learn from past or present and use that knowledge to make decisions on future unseen events. But from past couple of decades due to advancement of technologies large amount of data is generated in all sectors, which is not only large in volume but it is also heterogeneous or complex in nature. Dealing such type of data by conventional machine learning algorithms is very difficult and time consuming. To deal with these challenges lots of advanced algorithms have been proposed. Deep Learning, which is a major advancement in machine learning research was proposed by Hinton et al. [1] in 2006. It is a promising area of research where automatic extraction of complex data representation is done at high level without human intervention. This algorithm develops a layer by layer, hierarchical framework of learning and representation of data, where high-level features or representations are characterized in terms of low-level features or representations. Hierarchical framework of deep learning is inspired by human brain which follows a layer by layer learning phenomena same as primary sensorial areas of the neocortex inside human brain. In the algorithm features are extracted automatically from data is inspired by artificial intelligence (AI) emulating the deep. Deep learning approach is beneficial for handling large amount of unsupervised data but it is also used for supervised and semi-supervised data. This paper proposes novel multi-layer feature selection with conjunction of stacked auto-encoder (SAE) to extract high-level features or representations and eliminate the lower level features or representations from data. In this approach we reduce the number of neurons in each hidden layers. While the error is propagated in the backward direction we do not consider the neurons which are drooped during forward propagation. This is similar to the drop out condition. The proposed approach is validated on the Farm Ads dataset [2] and is also compared with various conventional machine learning algorithms and multi-layer feature selection with conjunction of stacked auto-encoder algorithm when we consider the drooped hidden neurons when error is propagated in backward direction. The proposed approach has outperformed in the term of accuracy and shows substantiates effectiveness of our approach in term Area Under ROC Curve (AUC) values for the given Farm Ads dataset.

The rest of the paper is organized as follows. Section 2 deal, with reviews of conventional machine learning and traditional machine learning for business data applications. Brief description of stacked auto-encoder is discussed in Sect. 3. Proposed multi-layer feature selection with conjunction of stacked auto-encoder (SAE) is described in Sect. 4. Results and discussions on selected financial dataset with different algorithms are given in Sect. 5. Section 6 finally concludes the paper.

2   Literature Survey

Machine learning enables a system to learn from the past or present and use that knowledge to make decisions on unseen or unknown future data. The goodness of system is that it learns relevant features from training data and gives good performance on testing data. If system is not good, then it shows poor representation of data that reduces the performance of the system. Thus, feature selection is an important step in machine learning, which focuses on constructing features from raw data [3]. Breiman [4] discussed two cultures of statistical modeling. In first he assumes that data are generated from stochastic model. In the other he uses algorithmic models and assumes that property of data is unknown and he also described the diverse set of tools to deal such problems.

Apte [5] describes how machine learning concepts are being utilized in various business applications developed by IBM for its internal use. A new machine learning approach is proposed in [6] for predicting users churn in telecommunication industry by using Genetic Programming. Fantazzini et al. [7] proposes a non-parametric approach for classification which is based on Random Survival Forests and compared its performance with a standard logit model. Shi et al. [8] proposes a weighted feature support vector machine for credit scoring model and applied for credit risk assessment, in which an F-score is adopted for relevant feature ranking. Peng et al. [9] presents the concept of maximal statistical dependency criterion for the relevant selection which is based on mutual information. Taffler et al. [10] described the discriminant analysis model for the identification of British company risk failure and also discussed the weakness of multivariate statistical techniques in the application of financial sector. In all the cases performing feature selection is more automated and general trends. This becomes an advancement in machine learning research and allows automatic extraction of such features without direct human intervention. Hence the concept of deep learning algorithms is automatic extraction of features or representation from the data [11–13]. Deep learning algorithm uses large amount of unsupervised data for automatic extraction of heterogeneous features or representations form data. These algorithms are basically inspired by the field of artificial intelligence (AI), where main objective is to emulate the human brain's abilities i.e. observe, analysis, learn, and take decisions, mainly for complex problems. Models which is based on shallow learning framework such as random forest [14], decision trees [15] and support vector machines [16–18] etc. may not perform good when trying to extract desired information from data. In contrast to these methods, Deep Learning framework has the capability to generalize globally and generating learning patterns and relationships from heterogeneous data [11]. Ding et al. [19] proposed a deep learning approach in which deep convolutional neural network model was used for stock market prediction in which both short-term and long-term effects of events on stock price variations where analyzed. Sirignano [20] proposes a new neural network framework for modeling spatial distributions which is computationally efficient to take advantage of the spatial structure of limit order books. Takeuchi et al. [21] use an auto-encoder which is stacked with restricted Boltzmann machines to extract high-level features from the history of individual stock prices. Hence deep learning becomes an important factor in artificial intelligence. It not only allows to analyze heterogeneous representation of data but it also makes the system independent of human interference which is an essential characteristic of artificial intelligence. Deep learning model generally trying to extracts high-level features or representations directly from supervised or unsupervised data without human interference.

3   Stacked Auto-Encoder (SAE)

A stacked auto-encoder is a neural network architecture comprising of multiple hidden layers of auto-encoders (AE) in which outputs of every hidden layer is connected to inputs of the successive layer [22]. It is an unsupervised learning algorithm that applies backpropogation to learn high-level features or representation from unlabeled data. The network architecture of AE is shown in Fig. 1. In the proposed approach we have considered three layers of AE as three hidden layers. The SAE which represents the three layers of AE is used to learn high-level feature from the data as shown in Figs. 2 and 3.

In 2006, Hinton et al. [1] drew additional attention by showing how multi-layered feedforward neural network could be effectively pre-trained one layer at a time. In the same year Bengio et al. [23] and LeCun et al. [24] also presented the same approach. This was the emergence of deep learning era. The idea beneath it is common to all, uses unlabeled data to initialize the weights using unsupervised blocks in greedy layer wise manner, and then uses labelled data to fine tune the network. Bengio et al. [23] concluded that this approach improves performance and weight initialization of the network.

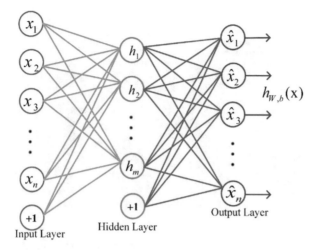

Fig. 1 Basic structure of auto-encoder (AE)

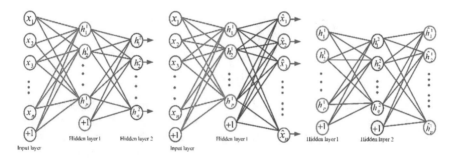

Fig. 2 High level feature generation using simple SAE, where $q < p$ is hidden units

3.1 Auto-Encoder (AE)

Auto-encoder is basically an artificial neural network, which is used to learn a representation form data (input). Simplest form of an auto-encoder comprises of an input, an output and one or more hidden layers connected to them. Since, it does not use labels so this process is an unsupervised learning. An auto-encoder learns the hypothesis (prediction) function $h_{\mathbf{W},\mathbf{b}}(x) = x$ with unlabeled data, where $\mathbf{W}$ and $\mathbf{b}$ are weight matrices and bias vector of the network respectively. Given unlabeled training samples, the cost (error) function for an auto-encoder is formulated as below,

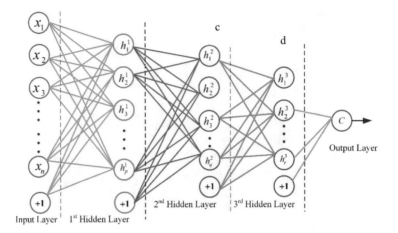

Fig. 3 Proposed approach for high -level feature generation using SAE, where $r < q < p$ is hidden units

$$J(\mathbf{W}, \mathbf{b}) = \frac{1}{m} \sum_{i=1}^{N} \frac{1}{2} \left\| \left(h_{\mathbf{W}, \mathbf{b}}(x^l) - \hat{x}^l \right)^2 \right\|$$
$$+ \frac{\lambda}{2} \sum_{i=1}^{s_l} \sum_{j=1}^{s_{l+1}} \mathbf{W}_{ji}^2 + \beta \sum_{j=1}^{s_l} KL(\rho \| \hat{\rho}_j) \tag{1}$$

In Eq. (1), KL is the Kullback-Leibler divergence function [25], λ is regularization parameter are used to minimize the problem of overfitting, ρ is sparsity parameter are used to put restriction on hidden units, which minimize the dependency between features and β is a parameter that controls sparsity penalty term. The parameter $\hat{\rho}_j$ is the average activation of hidden unit j and the activation function of a hidden unit implicitly depends on W and b. Our aim is to minimize the cost function J (W, b) with respect to W and b to train our network. The back propagation (BP) algorithm is mainly used for computing the gradients by using batch gradient descent optimization for learning the weights of network. The intuition for using back propagation is that for given training samples (x, y), we first run the forward propagation to compute the activation function of every node in all the layer of the complete network. Then, for every node i in layer l, compute the error $\delta_i^{(l)}$ that shows how much corresponding units (nodes) is responsible for errors at output of every layer in the network. The training of stacked auto-encoder (deep neural network) using back propagation is described in Algorithm 1.

Algorithm 1: Training of Stacked Auto-Encoder

Step 1. Initialize bias vector **b** as a zero vector and weight matrix **W** between $[-\upsilon, \upsilon]$, where υ is a constant dependent on number of neurons in hidden layers.

Step 2. Set, $\Delta \mathbf{W}^l = 0$ and $\Delta \mathbf{b}^l = 0$ for all l

Step 3. Apply back-propagation algorithm to compute

$$\nabla_{\mathbf{w}^l} J(\mathbf{W}, \mathbf{b}; x, y) = \delta^{l+1} (a^l)^T$$

$$\nabla_{\mathbf{b}^l} J(\mathbf{W}, \mathbf{b}; x, y) = \delta^{l+1}$$

and   a. Set $\Delta \mathbf{W}^l := \Delta \mathbf{W}^l + \nabla_{\mathbf{w}^l} J(\mathbf{W}, \mathbf{b}; x, y)$

       b. Set $\Delta \mathbf{b}^l := \Delta \mathbf{b}^l + \nabla_{\mathbf{b}^l} J(\mathbf{W}, \mathbf{b}; x, y)$

for all training examples. where a^l is activation of lth hidden layer.

Step 4. Update the weight matrix **W** and bias vector **b**

$$\mathbf{W}^l = \mathbf{W}^l - \eta [\frac{1}{m} \Delta(\mathbf{W}^l) + \lambda(\mathbf{W}^l)]$$

$$\mathbf{b}^l = \mathbf{b}^l - \eta [\frac{1}{m} \Delta(\mathbf{b}^l)]$$

where, η is learning rate.

Some necessary constraints are also put for useful learning and to minimize the problem of overfitting and under fitting in machine learning. The following constraints are as follows:

- **Number of hidden nodes (neurons)**: Putting constraints on the network such that number of neurons in hidden layers are lesser than input features. This forces the auto-encoder to learn heterogeneous feature representation from data, which also has the ability to reconstruct the output from data. This is similar to that of a nonlinear principal component analysis (PCA) [26].
- **Sparsity**: Sparsity is defined as the average activation value of hidden layer units in an auto-encoder. Sparsity constraint ensures that at any particular time, the feature representation(s) generated at hidden layer is sparse in size, i.e. the new feature representation must have features that are not active most of the time. Such feature representation not only forces data compression to happen, but it also improves robustness of the network [24, 25]. Auto-encoders with sparsity constraint are known as sparse auto-encoders.
- **Regularization**: This constraint is used to avoid the problem of over-fitting during learning of the network, and ensures that individual weight values do not take large values during fine-tuning of the network parameters.

3.2 Training Deep Neural Network with Stacked Auto-Encoders

The Algorithm 1 describes the learning of weights for a single AE. The weights of deep neural network (DNN) are initialized by training AE in greedy layer-wise manner. It shows that weights between input land first hidden layer are initialized by learning an AE which has an architecture that matches the input and first hidden layer. Once the weights are learned, the input training data is forward propagated till first hidden layer, to get first level feature representation form of the data. The data which is obtained in first level of feature representation is further used to initialize weights between first hidden layer and second hidden layer, by learning an AE of same architecture. The process of initializing weights, layer by layer using the AE is kept on till the last hidden layer [27]. Figures 2 and 3 illustrate the complete process of simple feature selection using multi-layer SAE and proposed approach. Once the weights of DNN are initialized by stacking of an AE, they are finally fine-tuned based on the objective. The tuning is done by defining cost function for the objective namely classification, and then learned the complete architecture is done using back-propagation (BP) algorithm.

3.3 The Output Layer

The high-level features which are learned from the SAE are subsequently applied at output layer of classifier to classify data in two different classes.

- **Support Vector Machine**: It is used to classify the high-level features learned from the output of SAE, with a proper non-linear mapping to a sufficiently large dimension [16]. The data from different classes are always separated by a hyperplane.
- **Random Forest Classifier**: It is an ensemble based learning algorithm used for classification and regression [28]. In this algorithm every tree is constructed by selecting a high level feature which is learned by SAE from training example. At every internal node of the tree some variables are selected at random and used to find the best split. By using combination of features, trees are constructed on new data set to the maximum depth without pruning.

4 Proposed Methodology

Farm Ads dataset has specific properties like high dimensionality and large sample size. Handling such type of data by classical machine learning is time consuming and does not give good performance. Additionally, some of the features in this dataset are irrelevant for classification and their inclusion not only adds noise but also they perturb the classifier and reduce the performance of system. So the feature selection is a useful approach that reduces dimensions by dropping the irrelevant features and improves the classifier performance. In most of the literature some of the feature selection techniques have been proposed [29, 30]. But we believe that a fair observation should be paid for feature selection. Herein, we proposed a novel multi-layer feature selection algorithm using stacked auto-encoder to learn relevant features from data.

The flow chart in Fig. 4 depicts the complete process proposed for feature selection and classification. The complete process is illustrated below.

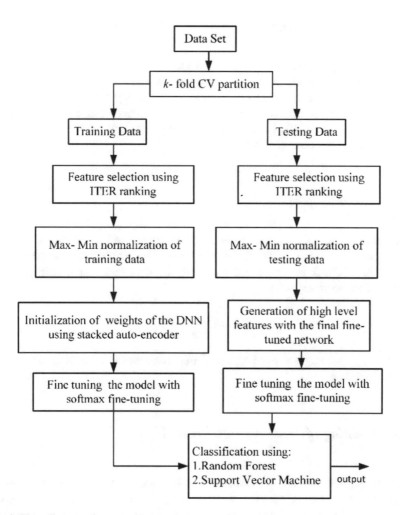

Fig. 4 Flow diagram of proposed approach

4.1   Cross Validation Partition (CV)

In this proposed approach we use CV partition to construct an object c in Matlab using equation

$$c = \text{cvpartition}\left(n,\, 'k\text{Fold}',\, k\right)$$

where n is class label, that defines a random partition for k-fold cross validation. We obtained k random disjoint sub-samples or folds of dataset that are in general equal in size.

4.2   Feature Selection

The second step is to select relevant features from data by using DNN. Dealing with large number of features is very inconvenient as described above. Here we have used a very simple and effective feature selection algorithm called Individual Training Error Reduction (ITER) ranking. We first ranked features of trained data and then selected top 1024 features from training data and used the same features for test data at the time of validation.

4.3   Normalization

The selected features using feature selection algorithm are in different scales which create problem of training the DNN. Before sending training data of the selected feature to DNN, it is normalized by using the max and min normalization of training data as given below.

$$x_{norm} = \frac{x - x_{min}}{x -\,_{max}}$$

where, x_{max} and x_{min} are maximum and minimum values of training data of the selected features.

4.4   Training of Deep Neural Network

- **Initialization**: Firstly, we initialized weight and bias vector of the network randomly and then brought to best initial values by training the stacked auto-encoder as described in Sect. 2. Then sparsity parameter ρ was varied from

0.1, 0.2, … to 0.9 and for each ρ we trained our model on dataset. The proposed approach we used three hidden layered architectures. At every hidden layer ITER ranking was used to rank the nodes using the label (stored these ranks for future use to validate on test dataset). We dropped some of the nodes with least rank and trained the next layer with remaining nodes as shown in Algorithm 2. Training of DNN is done by dropping number of features at successive layers and not considering these features for fine tuning of the model.

Algorithm 2: Training of DNN

Step 1. Obtain **b** and **W** using AE on input layer (data) containing 1024 features.
Step 2. Find features '*a*' of size 900 of first hidden layer using $x=$**W***data+**b** and $a=f(x)$, where f is sigmoidal function
Step 3. Using labels, rank the features '*a*' and drop 50 least ranked features to get '*a'* ' with size of 850 features.
Step 4. Repeat step 2 and 3 for the second hidden layer using '*a'* ' as input to obtain '*c*' and '*c'* ' with features 800 and 750 respectively.
Step 5. Repeat step 2 and 3 for the third hidden layer using '*c'* 'as input to obtain '*d*' and '*d'* ' with features 700 and 650 respectively.

Algorithm 3 used to compare the results of proposed approach describes training of DNN by dropping number of same number of features in Algorithm 2 at successive layers but considers these features for fine tuning of the model.

Algorithm 3: Training of DNN

Step 1. Obtain **b** and **W** using AE on input layer (data) containing 1024 features.
Step 2. Find features '**a**' of size 900 of first hidden layer using $x=$**W***data+**b** and $a=f(x)$, where f is sigmoidal function.
Step 3. Using output of first hidden layer as input for second hidden layer of size of 750 neurons.
Step 4. Using output of second hidden layer as input for third hidden layer of size of 650 neurons.

- **Fine Tuning**: Fine tuning is a strategy that is used to increase the performance of stacked auto-encoder. In fine tuning all layers of SAE are treated as single network so that in a single iteration, we improve all the weights in SAE. The DNN is tuned with softmax based finetuning [31] and high-level features or representations are generated at third hidden layer of network. This method uses softmax classification at the last layer of DNN which predicts the output. Error gradients of classifier propagated in backward direction are used to improve each hidden layer weights as well as the performance of deep network.

4.5 Classification

The high-level features generated after fine tuning at third layer is fed as input to the classifiers. Herein, we have used linear support vector machine, kernal support vector machine and random forest classifier. While testing, we converted the test data into new form by propagating forward through the same fine-tuned network. Then this new test data was tested with the trained classifier. For SVM, we used LibSVM and its optimal parameters were obtained using grid search method [2]. For Random Forest, we used TreeBagger class of Statics and Machine Learning toolbox of MATLAB and it was designed to have 1000 decision trees.

5   Results and Discussions

This section deals with results and discussions was obtained through proposed algorithm and different conventional machine learning algorithms on Farm Ads data set. Detailed description of data set is discussed below:

- **Farm Ads Data Set**: The data set comprises of 4143 number of Instances (samples) and 54877 number of Attributes (features). It was collected from twelve websites that deal with different farm animal related topics and are in text format. The class labels are binary values based on whether owner approves or disapproves the ad [2]

The datasets are processed using the proposed algorithm and different classical machine learning algorithms. We have used 10-fold cross validation in our experimentation. In the proposed approach top 1024 performing features from the 54877 features from the Farm Ads data set are selected by using Individual Training Error Reduction (ITER) rank. The DNN was tuned separately with softmax based fine-tuning and all optimization was performed using the minFunc toolbox [27]. Result obtained is compared with classical machine learning approach in which selected 1024 features after ITER ranking is also fed to different conventional machine learning algorithms like principle component analysis (PCA), mutual information feature selection (MIFS), linear discriminant analysis (LDA) and chi-squared (Chi2). In which we selected 650 features same as last hidden layer of proposed approach. These selected features are fed to different classifiers but the result obtained by proposed approach is much better in comparison to classical machine learning and feature selection using multi-layer stacked auto-encoder in

Table 1 Accuracy of dataset using conventional learning with different classifier

Data set	ITER initial-reduced	Feature reduction techniques	High level features initial-reduced	Classifiers	Accuracy (%)
Farm ads [2]	54877-1024	PCA, Jolliffe [26]	1024-650	RF	55.86
				Linear SVM	50.46
				RBF SVM	53.34
		MIFS, Peng et al. [9]	1024-650	RF	77.51
				Linear SVM	49.23
				RBF SVM	72.38
		LDA, Zenman [32]	1024-650	RF	46.66
				Linear SVM	49.50
				RBF SVM	53.34
		Chi 2, nlp.staford. edu [33]	1024-650	RF	64.65
				Linear SVM	50.32
				RBF SVM	53.58

Table 2 Accuracy and AUC of dataset using deep learning with different classifiers

ITER initial-reduced	Feature reduction techniques	High level features initial-reduced	Classifiers	Accuracy (%)	AUC (%)
54877-1024	Conventional	1024-650	RF	76.32	78.39
			Linear SVM	53.39	54.21
			RBF SVM	53.65	55.43
	Proposed	1024-650	RF	**77.51**	**79.21**
			Linear SVM	**94.62**	**95.49**
			RBF SVM	**96.65**	**97.32**

which dropped features are not taken into account at the time of fine tuning. This is shown in Tables 1 and 2. The substantiates effectiveness of our proposed approach in term Area Under ROC Curve (AUC) values for the given Farm Ads dataset is also listed in Table 2 (Fig. 5).

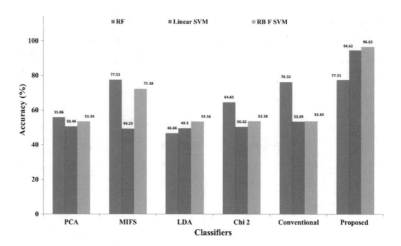

Fig. 5 Comparison of accuracy with proposed approach

6   Conclusion

Business strategies need to be dynamic to face competitive market which requires to take advantage of information gathering from market data. It helps to make better, smarter, real time and fact based decisions. In the modern era large amount of data is generated in all sectors which carries useful information. The presented approach is validated on the Farm Ads data which carry the advertisement on different website related to farm animal topics. The proposed approach auto-encoders are stacked to build DNN which help in leveraging information from data. This approach appears to work very effectively. Future researches in this area can address other feature ranking techniques.

References

1. Hinton Geoffrey, E., et al.: A fast learning algorithm for deep belief nets. Neural Comput. **18** (7), 1527–1554 (2006)
2. Lichman, M.: UCI Machine Learning Repository. Irvine, CA University of California, School of Information and Computer Science. http://archive.ics.uci.edu/ml (2013)
3. Domingos, P.: A few useful things to know about machine learning. Commun. ACM **55**(10) (2012)
4. Breiman, L.: Statistical modeling: The two cultures (with comments and a rejoinder by the author). Stat. Sci. **16**(3), 199–231 (2001)
5. Apte, C.: The role of machine learning in business optimization. In: Proceedings of the 27th International Conference on Machine Learning (ICML-10), pp. 1–2 (2010)
6. Faris, H., et al.: A genetic programming based framework for churn prediction in telecommunication industry. In: International Conference on Computational Collective Intelligence September 24, pp. 353–362. Springer (2014)

7. Dean, F., Silvia, F.: Random survival forests models for SME credit risk measurement. Methodol. Comput. Appl. Probab. **11**(1), 29–45 (2009)
8. Jian, S., et al.: Credit scoring by feature-weighted support vector machines. J. Zhejiang Univ. Sci. C **14**(3), 197–204 (2013)
9. Peng, H., et al.: Feature selection based on mutual information criteria of max-dependency, max-relevance, and min-redundancy. IEEE Trans. Pattern Anal. Mach. Intell. **27**(8), 1226–1238 (2005)
10. Taffler, R.J., et al.: Forecasting company failure in the UK using discriminant analysis and financial ratio data. J. R. Stat. Soc. Ser. A (General) 342–358 (1982)
11. Bengio, Y., et al.: Representation learning: a review and new perspectives. IEEE Trans. Pattern Anal. Mach. Intell. **35**(8), 1798–1828 (2013)
12. Bengio, Y.: Learning Deep Architectures for AI. Now Publishers Inc., Hanover, MA, USA (2009)
13. Bengio, Y.: Deep learning of representations: looking forward. In: Proceedings of the 1st International Conference on Statistical Language and Speech Processing. SLSP'13, pp. 1–37. Springer, Tarragona, Spain (2013)
14. Breiman, L.: Random forests. Mach. Learn. **45**(1), 5–32 (2001)
15. Safavian, S.R., Landgrebe, D.: A Survey of Decision Tree Classifier Methodology (1990)
16. Corinna, C., Vapnik, V.: Support-vector networks. Mach. Learn. **20**(3), 273–297 (1995)
17. Sevakula, R.K., et al.: Fast data sampling for large scale support vector machines. In: IEEE Workshop on Computational Intelligence: Theories, Applications and Future Directions (IEEE WCI 2015), India (2015)
18. Sevakula, R.K., et al.: Data preprocessing methods for sparse auto-encoder based fuzzy rule classifier. In: IEEE Workshop on Computational Intelligence: Theories, Applications and Future Directions (IEEE WCI 2015), India (2015)
19. Ding, X., et al.: Deep learning for event-driven stock prediction. In: Proceedings of the 24th International Joint Conference on Artificial Intelligence (ICJAI 15), pp. 2327–2333 (2015)
20. Sirignano, J.A.: Deep Learning for Limit Order Books. arXiv:1601.01987 (2016)
21. Takeuchi, L., Lee, Y.-Y.: Applying Deep Learning to Enhance Momentum Trading Strategies in Stocks. http://cs229.stanford.edu/proj2013/
22. Ng, A.: Sparse autoencoder. In: CS294A Lecture Notes, vol. 72, pp. 1–19 (2011)
23. Bengio, Y., et al.: Greedy layer-wise training of deep networks. Adv. Neural. Inf. Process. Syst. **19**, 153 (2007)
24. Poultney, C., et al.: Efficient learning of sparse representations with an energy-based model. In: Advances in Neural Information Processing Systems, pp. 1137–1144 (2006)
25. Thirukovalluru, R., et al.: Generating feature sets for fault diagnosis using denoising stacked auto-encoder. In: 2016 IEEE International Conference on Prognostics and Health Management (ICPHM), pp. 1–7. IEEE (2016)
26. Jolliffe, I.: Principal Component Analysis. Wiley (2002)
27. Chih-Chung, C., Lin, C.-J.: LIBSVM: a library for support vector machines. ACM Trans. Intell. Syst. Technol. (TIST) **2**(3), 27 (2011)
28. Biau, G.: Analysis of a random forests model. J. Mach. Learn. Res. **13**, 1063–1095 (2012)
29. Yu, L., Liu, H.: Feature selection for high-dimensional data: a fast correlation-based filter solution. In: ICML, vol. 3, pp. 856–863 (2003)
30. Teng, C.M.: Combining noise correction with feature selection. In: International Conference on Data Warehousing and Knowledge Discovery, pp. 340–349. Springer (2003)
31. Ng, A., et al.: UFLDL Tutorial (2016)
32. Izenman, A.J.: Linear discriminant analysis. Modern Multivariate Statistical Techniques, pp. 237–280. Springer, New York (2013)
33. http://nlp.stanford.edu/IR-book/html/htmledition/feature-selectionchi2-feature-selection-1.html

Comparison of Type-2 Fuzzy Integration for Optimized Modular Neural Networks Applied to Human Recognition

Daniela Sánchez, Patricia Melin and Oscar Castillo

Abstract In this paper optimization techniques for Modular Neural Networks (MNNs) and their combination with a granular approach is presented. A Firefly Algorithm (FA) and a Grey Wolf Optimizer (GWO) are developed to perform modular neural networks (MNN) optimization. These algorithms perform the optimization of some parameters of MNN such as; number of sub modules, percentage of information for the training phase and number of hidden layers (with their respective number of neurons) for each sub module and learning algorithm. The modular neural networks are applied to human recognition based on face, iris, ear and voice. The minimization of the error of recognition is the objective function. To combine the responses of the modular neural networks different type-2 fuzzy inference system are proposed and a comparison of results is performed.

1 Introduction

Combining different intelligent techniques allow emerging a new kind of more powerful systems called hybrid intelligent systems. This systems allows to overcome the limitations of each intelligent technique has individually [14, 18, 34]. This kind of systems have demonstrated to allow better results than individual techniques, such as in [1, 12, 22]. Some of these techniques are fuzzy logic, neural networks and genetic algorithms, among these techniques shine a lot of computing techniques dedicated to performing the optimization problems such as: Genetic Algorithm [15], Particle Swarm Optimization [16], Ant Colony Optimization [10], Bat Algorithm [29], Harmony Search [13], and Gravitational Search Algorithm [24] among others. In this paper different intelligent techniques are combined such as neural networks, fuzzy logic, firefly algorithm and grey wolf optimizer. The proposed method is

D. Sánchez (✉) · P. Melin · O. Castillo
Tijuana Institute of Technology, Tijuana, Mexico
e-mail: danielasanchez.itt@hotmail.com

P. Melin
e-mail: pmelin@tectijuana.mx

© Springer International Publishing AG 2018
C. Berger-Vachon et al. (eds.), *Complex Systems: Solutions and Challenges in Economics, Management and Engineering*, Studies in Systems, Decision and Control 125, https://doi.org/10.1007/978-3-319-69989-9_17

applied to human recognition based on face, iris, ear and voice biometrics. Optimizing some parameters of MNN such as; number of sub modules, percentage of information for the training phase and number of hidden layers (with their respective number of neurons) for each sub module and learning algorithm, having as objective function the minimization of the error of recognition. To combine the modular neural networks responses type-2 fuzzy integrators are used to compare which of them allows having better results. This paper is organized as follows: The basic concepts used in this work are presented in Sect. 2. Section 3 contains the general architecture of the proposed method. Section 4 presents experimental results and in Sect. 5, the conclusions of this work are presented.

2 Basic Concepts

In this section the basic concepts used in this research work are presented.

2.1 Modular Neural Networks

A neural network (NN) is integrated by many artificial neurons and its objective is to convert the inputs into significant outputs. A NN has a large number of features similar to the brain due to its constitution and its foundations, as it can be to learn from the experience [11]. The learning mode can be supervised or unsupervised [25]. A neural network is said to be modular if the computation performed by the network can be decomposed into two or more modules (sub-systems) that operate on distinct inputs without communicating with each other. The modular neural networks are comprised of modules which can be categorized on the basis of both distinct structure and functionality which are integrated together via an integrating unit. With functional categorization, each module is a neural network which carries out a distinct identifiable subtask. Also, using this approach different types of learning algorithms can be combined in a smooth fashion [2, 17].

2.2 Fuzzy Logic

Fuzzy logic is an area of soft computing that enables a computer system to reason with uncertainty [4]. Fuzzy logic is a useful tool for modeling complex systems and deriving useful fuzzy relations or rules [23]. However, it is often difficult for human experts to define the fuzzy sets and fuzzy rules used by these systems [6]. The basic structure of a fuzzy inference system consists of three conceptual components: a rule base, which contains a selection of fuzzy rules, a database (or dictionary) which defines the membership functions used in the rules, and a reasoning mechanism that performs the inference procedure [6, 33].

The concept of a type-2 fuzzy set, was introduced by Zadeh (1975) as an extension of the concept of an ordinary fuzzy set (henceforth called a "type-1 fuzzy set"). A type-2 fuzzy set is characterized by a fuzzy membership function, i.e., the membership grade for each element of this set is a fuzzy set in [0, 1], unlike a type-1 set where the membership grade is a crisp number in [0, 1]. Fuzzy logic is a useful tool for modeling complex systems and deriving useful fuzzy relations or rules However, it is often difficult for human experts to define the fuzzy sets and fuzzy rules used by these systems, for that reason different techniques to perform the optimization are used to set the fuzzy system architecture [5].

2.3 Granular Computing

Granular computing is a term for the problem solving paradigm and may be viewed more on a philosophical rather than technical level [30]. In granular computing, computing with words, and the computational theory of perceptions, the information granulation plays a central role. Information granules are the generic conceptual and computing objects of granular computing [32]. It is clear that an underlying idea of granular computing is the use of groups, classes, or clusters of elements called granules [31].

There are different areas where granular computing has begun to play important roles such as bioinformatics, e-Business, security, machine learning, data mining, high-performance computing and wireless mobile computing due to its efficiency, effectiveness, robustness and uncertainty [3].

2.4 Firefly Algorithm

Firefly Algorithm (FA) was introduced in [26, 27]. This optimization algorithm is based on fireflies (their flashing patterns and behavior). FA is based on three rules: (1) Fireflies are unisex. A firefly can attract to other fireflies independently of their sex. (2) The attractiveness is proportional to the brightness, and they both decrease as their distance increases, where for any two flashing fireflies; the firefly with less brightness will move toward the brighter one. If there is no brighter one than a particular firefly, it will move randomly. (3) The brightness of a firefly is determined by objective function [28]. It is important to say, in this work the attractiveness of a firefly is represented by the objective function, and it is also associated with its brightness.

2.5 Grey Wolf Optimizer

Grey Wolf Optimizer was introduced in [20]. Grey wolves mostly prefer to live in a pack; the group size is approximately between 5 and 12 grey wolves. Each grey wolf has an important role. They are divided into 4 categories: alpha, beta, delta and omega. The alpha wolf is the dominant wolf and his/her orders should be followed by the pack. The second level is beta [19]. The betas are subordinate wolves that help the alpha in decision-making or other pack activities. Delta wolves have to submit to alphas and betas, but they dominate the omega. Scouts, sentinels, elders, hunters, and caretakers belong to this category. The omega plays the role of scapegoat. Omega wolves always have to submit to all the other dominant wolves. They are the last wolves that are allowed to eat. In addition to the social hierarchy of wolves, group hunting is another interesting social behavior of grey wolves. The main phases of grey wolf hunting are as follows: Tracking, chasing, and approaching the prey, pursuing, encircling, and harassing the prey until it stops moving and attack towards the prey [21].

3 Proposed Method

The proposed method combines responses of modular granular neural networks. The modular granular neural networks architectures were optimized. A firefly algorithm and a grey wolf optimizer were proposed to improve the performance of the MGNN. The optimization of MGNN consist into find an optimal granulation of the information. This granulation divides the number of persons for each sub module up to "m" sub modules and also the number of data for the training phase.

Other parameters must be find by the optimization technique, these parameters are the number of hidden layers (with their respective number of neurons) for each sub module and learning algorithm. For each database a modular granular neural network is used, and its architecture can be different each other. The different responses provided by the modular granular neural network are combined using a type-2 fuzzy inference system. Figure 1 shows the architecture of proposed method for MGNNs optimization and integration.

The minimum and maximum values used for the optimization techniques for the MGNNs are shown in Table 1. The search space of each optimization techniques is established with those parameters.

To combine the responses of the modular granular neural networks, in this work five type-2 fuzzy inference systems are proposed. Each of these fuzzy integrators has 81 fuzzy rules. In Fig. 2, the fuzzy inference system #1 has in all its variables Gaussian membership functions.

In Fig. 3, the fuzzy inference system #2 has in all its variables trapezoidal membership functions.

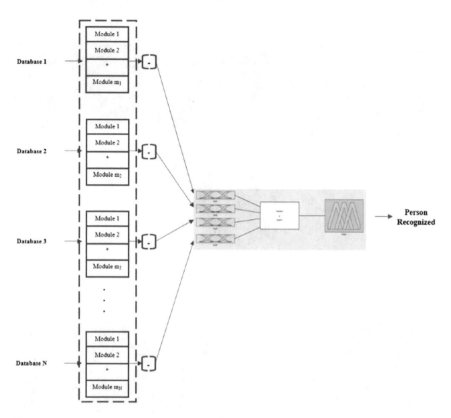

Fig. 1 Architecture of proposed method for combination integration of MGNNs

Table 1 Values for MGNNs

Parameters of MNNs	Minimum	Maximum
Modules (m)	1	10
Percentage of data for training	20	80
Error goal	0.000001	0.001
Learning algorithm	1	3
Hidden layers (h)	1	5
Neurons for each hidden layers	20	400

In Fig. 4, the fuzzy inference system #3 has in all its variables combinations of membership functions (Gaussian and trapezoidal).

In Figs. 5 and 6, the fuzzy inference systems have in all its variables combinations of membership functions (Gaussian and trapezoidal), in these fuzzy integrator, the footprint of uncertainty is different among each membership function.

Entradas

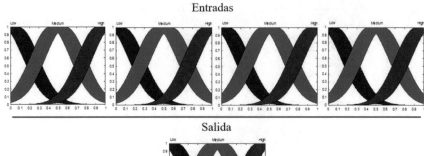

Fig. 2 Fuzzy integrator #1

Entradas

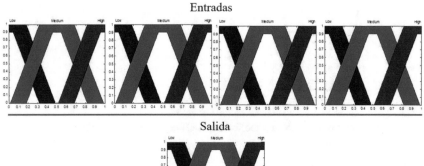

Fig. 3 Fuzzy integrator #2

3.1 Databases

The databases used in this work are presented in this section. The human recognition is performed to 77 persons, only the first 77 persons of each database are used.

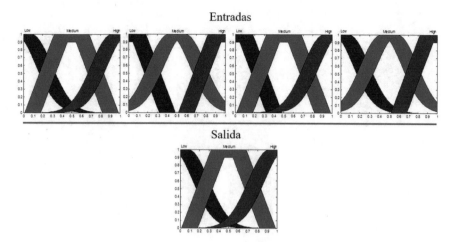

Fig. 4 Fuzzy integrator #3

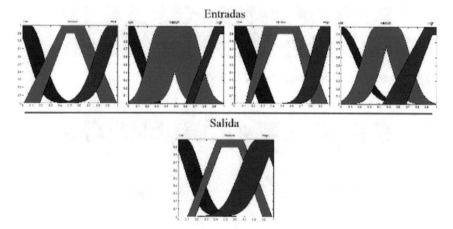

Fig. 5 Fuzzy integrator #4

3.1.1 Face Database

The database of human iris from the Institute of Automation of the Chinese Academy of Sciences was used [8]. Each person has 5 images. The image dimensions are 640 × 480, BMP format. Figure 7 shows examples of the face images.

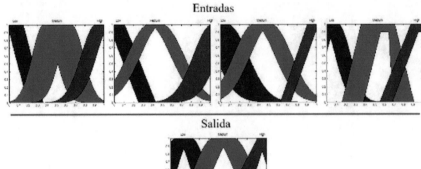

Fig. 6 Fuzzy integrator #5

Fig. 7 Examples of the face images from CASIA database

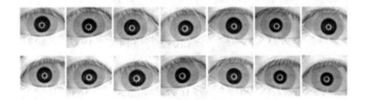

Fig. 8 Examples of the human iris images from CASIA database

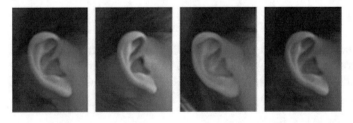

Fig. 9 Examples of Ear Recognition Laboratory from the University of Science and Technology Beijing (USTB)

Table 2 The best result of each biometric measure (FA, 80%)

Evolution	Images for training	Number of neurons	Persons per module	Recognition rate
Face FA80F1	71% (1, 3, 4 and 5)	131, 195, 106, 176 109, 184 123, 33, 236 180, 226, 88, 78, 83 134, 164, 59, 119 129, 23, 42	Module #1 (1–17) Module #2 (18–39) Module #3 (40–50) Module #4 (51–71) Module #5 (72–77)	100% 0 (77/77)
Iris FA80I9	59% (2, 3, 4, 5, 6, 7, 8 and 11)	101, 47, 67, 139, 42 93, 164, 147, 44, 25 143, 139, 67, 156, 90 110, 52, 77 74, 21, 92, 96 85, 89, 63, 158, 179 143, 168, 58, 143, 146 117, 123, 181 119, 62, 86, 107	Module #1 (1–6) Module #2 (7–21) Module #3 (22–41) Module #4 (42–44) Module #5 (45–48) Module #6 (49–58) Module #7 (59–65) Module #8 (66–68) Module #9 (69–77)	99.13% 0.0087 (458/462)
Ear FA80O1	64% (1, 2 and 3)	150, 137, 100, 158, 112 106, 108, 97, 178, 155 153, 201, 104, 100, 71 125, 104, 61, 125, 74 185, 60, 174, 89, 199 93, 216, 46, 31 121, 209, 133, 168, 106 97, 148, 181, 87	Module #1 (1–8) Module #2 (9–12) Module #3 (13–21) Module #4 (22–26) Module #5 (27–33) Module #6 (34–39) Module #7 (40–58) Module #8 (59–77)	100% 0 (77/77)
Voice FA80V1	66% (1, 2, 5, 6, 8, 9 and 10)	114, 158, 101, 162 100, 115, 108, 174, 78 117, 166, 96, 174 113, 111, 76, 38 127, 85, 112, 161, 58 139, 128, 132, 113	Module #1 (1–4) Module #2 (5–9) Module #3 (10–17) Module #4 (18–24) Module #5 (25–60) Module #6 (61–77)	100% 0 (231/231)

3.1.2 Iris Databases

The database of human iris from the Institute of Automation of the Chinese Academy of Sciences was used [9]. Each person has 14 images (7 for each eye). The image dimensions are 320 × 280, JPEG format. Figure 8 shows examples of the human iris images.

3.1.3 Ear Database

The database of the University of Science and Technology of Beijing was used [7]. Each person has 4 images, the image dimensions are 300 × 400 pixels and the format is BMP. Figure 9 shows examples of the human ear images.

3.1.4 Voice Database

In the case of voice, the database was made from students of Tijuana Institute of Technology. Each person has 10 voice samples, WAV format. The word that they said in Spanish was "ACCESAR". To preprocess the voice the Mel Frequency Cepstral Coefficients were used.

4 Experimental Results

The optimized results using the different optimization techniques are shown in this section. Two tests were performed for each optimization technique; in the first test, each algorithm used up to 80% of data for the training phase and, for the second test up to 50. 20 evolutions for each test were performed.

Table 3 Summary of result of each biometric measure (FA, 80%)

	Face	Iris	Ear	Voice
Best	100%	99.13%	100%	100%
	0	0.0087	0	0
Average	98.80%	98.22%	99.84%	99.84%
	0.0120	0.0178	0.0016	0.0016
Worst	96.10%	96.59%	98.05%	99.35%
	0.0390	0.0341	0.0195	0.0065

Table 4 The best result of each biometric measure (FA, 50%)

Evolution	Images for training	Number of neurons	Persons per module	Recognition rate
Face FA50R18	50% (2, 4 and 5)	20, 39, 46, 64, 104 191, 164, 156, 69, 240 163, 250, 212, 221, 224 170, 196, 28, 74, 221 98, 121, 132 74, 206, 111, 154, 151	Module #1 (1–4) Module #2 (5–20) Module #3 (21–34) Module #4 (35–42) Module #5 (43–57) Module #6 (58–77)	90.26% 0.0974 (139/154)
Iris FA50I9	41% (2, 3, 4, 5, 6 and 13)	89, 97, 92, 75 128, 127, 72, 171, 40 143, 168, 103, 94, 156 155, 101, 49 77, 62, 122, 90 123, 127, 58, 30	Module #1 (1–8) Module #2 (9–17) Module #3 (18–29) Module #4 (30–49) Module #5 (50–61) Module #6 (62–77)	96.92% 0.0308 (597/616)
Ear FA50O3	39% (2 and 3)	124, 132, 126, 83 129, 101, 209, 74 105, 161, 96, 129, 170 147, 79, 125, 150, 205 171, 143, 108, 164, 139 135, 103, 89, 212 118, 139, 110, 162	Module #1 (1–15) Module #2 (16–22) Module #3 (23–38) Module #4 (39–53) Module #5 (54–64) Module #6 (65–75) Module #7 (76–77)	97.40% 0.0260 (150/154)
Voice FA50V13	48% (1, 2, 6, 7 and 10)	133, 146, 145, 174, 146 183, 130, 236 156, 97, 129, 218 176, 158, 153 141 108, 127, 87, 115, 51 135, 83, 60	Module #1 (1–7) Module #2 (8–9) Module #3 (10–24) Module #4 (25–29) Module #5 (30–47) Module #6 (48–61) Module #7 (62–77)	98.96% 0.0104 (381/385)

4.1 Firefly Algorithm

The best results achieved with the proposed firefly algorithm are shown in this section.

4.1.1 Using up to 80% of Data

The best result of each biometric measure is shown in Table 2. For face, ear and voice a 100 of recognition rate is obtained in the evolution #1 of each biometric measure using 71, 64 and 66 of the percentage of data for the training phase.

In Table 3, a summary of result of each biometric measure is shown, where the best averages were obtained with the ear and voice.

4.1.2 Using up to 50% of Data

The best result of each biometric measure is shown in Table 4. The best results is obtained when the proposed method is applied to voice, with a 98.96 of recognition rate using 48 of the percentage of data for the training phase.

In Table 5, a summary of result of each biometric measure is shown, where the best average were obtained with voice.

4.2 Grey Wolf Optimizer

The best results achieved with the grey wolf optimizer are shown in this section.

4.2.1 Using up to 80% of Data

The best result of each biometric measure is shown in Table 6. Also with this optimization technique for face, ear and voice a 100 of recognition rate is obtained using 74, 80 and 75 of the percentage of data for the training phase.

Table 5 Summary of result of each biometric measure (FA, 50%)		Face	Iris	Ear	Voice
	Best	90.26% 0.0974	96.92% 0.0308	97.40% 0.0260	98.96% 0.0104
	Average	86.94% 0.1306	95.62% 0.0438	96.82% 0.0318	97.85% 0.0215
	Worst	85.28% 0.1472	94.08% 0.0592	96.10% 0.0390	96.97% 0.0303

Table 6 The best result of each biometric measure (GWO, 80%)

Evolution	Images for training	Number of neurons	Persons per module	Recognition rate
Face GWO80R3	74% (1, 3, 4 and 5)	47, 187, 138, 144, 99 100, 63, 190, 44, 158 176, 186, 48, 158, 93 154, 106, 191, 48, 116 63, 55, 168, 126, 117 71, 179, 150, 90, 218 117, 48, 66, 92, 76 100, 77, 29, 222, 60	Module #1 (1–2) Module #2 (3–27) Module #3 (28–30) Module #4 (31–34) Module #5 (35–42) Module #6 (43–56) Module #7 (57–71) Module #8 (72–77)	100% 0 (77/77)
Iris GWO80I4	78% (1, 2, 3, 4, 5, 6, 8, 9, 10, 11 and 14)	83, 150, 122, 88, 240 116, 143, 213, 83, 220 80, 68, 136, 130, 97 78, 162, 84, 108, 111 64, 137, 202, 70, 235 108, 116, 47, 163, 138	Module #1 (1–19) Module #2 (20–37) Module #3 (38–42) Module #4 (43–45) Module #5 (46–58) Module #6 (59–77)	99.57% 0.0043 (230/231)
Ear GWO80O1	80% (1, 2 and 3)	126, 96, 179, 239, 37 188, 196, 93, 171 109, 107, 110, 168, 29	Module #1 (1–12) Module #2 (13–40) Module #3 (41–77)	100% 0 (77/77)
Voice GWO80V1	75% (1, 2, 3, 6, 7, 8, 9 and 10)	248, 35, 142, 199, 68 112, 87, 106, 58, 162 62, 160, 117, 27, 136 119, 112, 126, 123, 111 134, 67, 142, 72, 126 132, 126, 90, 126, 92 137, 159, 63, 201, 220 179, 195, 225, 231, 57	Module #1 (1–9) Module #2 (10–16) Module #3 (17–20) Module #4 (21–32) Module #5 (33–40) Module #6 (41–49) Module #7 (50–69) Module #8 (70–77)	100% 0 (154/154)

Table 7 Summary of result of each biometric measure (GWO, 80%)

	Face	Iris	Ear	Voice
Best	100%	99.57%	100%	100%
	0	0.0043	0	0
Average	99.29%	99.05%	100%	99.95%
	0.0071	0.0095	0	0.005
Worst	98.70%	98.18%	100%	99.35%
	0.0130	0.0182	0	0.0065

In Table 7, a summary of result of each biometric measure is shown, where the best average were obtained with ear.

4.2.2 Using up to 50% of Data

The best result of each biometric measure is shown in Table 8. The best result is obtained when the proposed algorithm is applied to voice with a 99.22 of recognition rate is obtained using 47 of the percentage of data for the training phase.

In Table 9, a summary of result of each biometric measure for this test is shown, where the best average were obtained when the proposed method is applied to voice.

4.3 Fuzzy Integration

The different evolutions obtained by the different tests using the proposed algorithms of optimization presented above were combined and 5 cases were considered to be presented and test the proposed integration using type-2 fuzzy logic. In Table 10, the cases are shown.

For each case, the 5 fuzzy integrators were used to perform the combination of responses. In Table 11, the average and the results achieved by each fuzzy integrator are shown. For this work, the best average was obtained with the type-2 fuzzy integrator #4. This fuzzy integrator has different type of membership functions and footprint of uncertainty.

5   Conclusions

In this paper, optimization and integration of responses of modular neural networks were proposed. The results achieved by each database were shown. For these optimization, 20 evolutions were performed for each technique. For almost all the biometric measure (except ear biometric), the grey wolf optimizer achieved better results than the firefly algorithm.

Table 8 The best result of each biometric measure (GWO, 50%)

Evolution	Images for training	Number of neurons	Persons per module	Recognition rate
Face GWO50R1	50% (3, 4 and 5)	135, 91, 87, 126, 132 66, 78, 77, 172, 39 70, 148, 107, 139, 135 131, 113, 127, 91, 63 163, 232, 75, 131, 241 113, 179, 107, 100	Module #1 (1–15) Module #2 (16–26) Module #3 (27–47) Module #4 (48–58) Module #5 (59–70) Module #6 (71–77)	95.45% 0.0455 (147/154)
Iris GWO50I11	47% (2, 3, 4, 5, 7, 8 and 14)	143, 181, 129, 195, 168 102, 77, 137, 87, 177 94, 100, 155, 75, 187 172, 158, 168, 113, 178 100, 169, 249, 224, 225 98, 100, 38, 85, 80 74, 157, 31, 130, 104 121, 167, 65, 172, 93 111, 186, 130, 51	Module #1 (1 a 18) Module #2 (19–31) Module #3 (32 a 33) Module #4 (34 a 39) Module #5 (40 a 46) Module #6 (47 a 56) Module #7 (57 a 58) Module #8 (59 a 71) Module #9 (72 a 77)	97.59% 0.0241 (526/539)
Ear GWO50O2	43% (2 and 3)	115, 49, 187, 122, 194 182, 139, 50, 217, 54 132, 182, 56, 187, 159 167, 132, 121, 123, 219 116, 195, 54, 174 157, 108, 166, 95, 88 116, 119, 76, 121, 94 102, 58, 69, 111, 42	Module #1 (1 a 9) Module #2 (10 a 22) Module #3 (23 a 33) Module #4 (34 a 36) Module #5 (37 a 51) Module #6 (52 a 63) Module #7 (64 a 75) Module #8 (76 a 77)	96.75% 0.0325 (149/154)
Voice GWO50V4	47% (2, 5, 6, 9 and 10)	60, 208, 159, 70, 142 49, 78, 71, 42, 45 71, 132, 158, 93, 111 52, 87, 92, 147, 38 103, 161, 150, 169, 61 224, 92, 145, 112, 41 61, 181, 158, 145, 140 83, 176, 86, 63, 79	Module #1 (1–8) Module #2 (9–14) Module #3 (15–21) Module #4 (22–33) Module #5 (34–43) Module #6 (44–48) Module #7 (49–64) Module #8 (65–77)	99.22% 0.0078 (382/385)

To combine the modular neural networks responses, 5 fuzzy inference systems were proposed (using type-2 fuzzy logic). The best results were obtained when the fuzzy integrator used different kind of membership functions in the same variable and when the footprint of uncertainty also was different. As future work the same optimization techniques will be used to perform the fuzzy inference system optimization to perform a comparison between them.

Table 9 Summary of result of each biometric measure (GWO, 50%)

	Face	Iris	Ear	Voice
Best	95.45%	97.59%	96.75%	99.22%
	0.0455	0.0241	0.0325	0.0078
Average	91.71%	96.75%	96.20%	98.27%
	0.0829	0.0325	0.0380	0.0173
Worst	85.28%	95.82%	95.45%	97.40%
	0.1472	0.0418	0.0455	0.0260

Table 10 The 5 cases–perform the fuzzy integration

Evolution	Face	Iris	Ear	Voice
1	FA80R2	GWO50I6	FA50O12	GWO50V18
	100%	96.29%	96.75%	97.92%
2	FA80R12	GWO80I7	FA80O19	GWO50V2
	98.70%	99.13%	98.70%	98.18%
3	GWO80R8	FA80I19	GWO80O12	FA80V5
	98.70%	97.40%	100%	99.56%
4	GWO50R1	FA80I2	FA80O7	FA50V9
	95.45%	97.22%	100%	97.62%
5	FA80R4	GWO50I8	FA50O2	GWO80V17
	97.40%	96.85%	96.75%	100%

Table 11 Comparison of fuzzy integration results

Case	Fuzzy integrator #1 (%)	Fuzzy integrator #2 (%)	Fuzzy integrator #3 (%)	Fuzzy integrator #4 (%)	Fuzzy integrator #5 (%)
1	96.48	96.85	96.47	98.33	98.14
2	99.48	99.48	99.48	99.48	99.22
3	97.40	97.66	97.40	99.48	99.48
4	97.40	97.59	97.22	99.44	100
5	97.03	97.22	96.85	97.59	97.22
Average	97.56	97.76	97.48	98.86	98.81

References

1. Amezcua, J., Melin, P.: Optimization of the LVQ network architecture with a modular approach for arrhythmia classification using PSO. Design of Intelligent Systems Based on Fuzzy Logic, Neural Networks and Nature-Inspired Optimization 2015, pp. 119–126 (2015)
2. Azamm, F.: Biologically inspired modular neural networks. Ph.D. Thesis, Virginia Polytechnic Institute and State University, Blacksburg, Virginia, May 2000
3. Bargiela, A., Pedrycz, W.: The roots of granular computing. In: IEEE International Conference on Granular Computing (GrC), 2006, pp 806–809
4. Castillo, O., Melin, P., Alanis, Garza A., Montiel, O., Sepúlveda, R.: Optimization of interval type-2 fuzzy logic controllers using evolutionary algorithms. Soft. Comput. 15(6), 1145–1160 (2011)
5. Castillo, O., Melin, P.: Type-2 Fuzzy Logic Theory and Applications, pp. 29–43. Springer, Berlin (2008)
6. Castillo, O., Melin, P.: Soft Computing for Control of Non-linear Dynamical Systems. Springer, Heidelberg, Germany (2001)
7. Database Ear Recognition Laboratory from the University of Science & Technology Beijing (USTB). http://www.ustb.edu.cn/resb/en/index.htm. Accessed 21 Sep 2009
8. Database of Face. Institute of Automation of Chinese Academy of Sciences (CASIA). http://biometrics.idealtest.org/dbDetailForUser.do?id=9. Accessed 11 Nov 2012
9. Database of Human Iris. Institute of Automation of Chinese Academy of Sciences (CASIA). http://www.cbsr.ia.ac.cn/english/IrisDatabase.asp. Accessed 21 Sep 2009
10. Dorigo, M., Gambardella, L.M.: Ant colony system: a cooperative learning approach to the traveling salesman problem. IEEE Trans. Evol. Comput. 1, 53–66 (1997)
11. Funes, E., Allouche, Y., Beltrán, G., Jiménez, A.: A review: artificial neural networks as tool for control food industry process. J. Sens. Technol. 5, 28–43 (2015)
12. Gaxiola, F., Melin, P., Valdez, F., Castro, J.R., Castillo, O.: Optimization of type-2 fuzzy weights in backpropagation learning for neural networks using GAs and PSO. Appl. Soft Comput. 38, 860–871 (2016)
13. Geem, Z., Yang, X., Tseng, C.: Harmony search and nature-inspired algorithms for engineering optimization. J. Appl. Math. 438158:1–438158:2 (2013)
14. Hidalgo, D., Castillo, O., Melin, P.: Type-1 and type-2 fuzzy inference systems as integration methods in modular neural networks for multimodal biometry and its optimization with genetic algorithms. Inf. Sci. 179(13), 2123–2145 (2009)
15. Holland, J.H.: Adaptation in Natural and Artificial Systems. University of Michigan Press, MI (1975)
16. Kennedy, J., Eberhart, R.C.: Particle swarm optimization. In: Proceedings of the IEEE international Joint Conference on Neuronal Networks, pp. 1942–1948. IEEE Press (1995)
17. Khan, A., Bandopadhyaya, T., Sharma, S.: Classification of stocks using self organizing map. Int. J. Soft Comput. Appl. 4, 19–24 (2009)
18. Martínez-Soto, R., Castillo, O., Aguilar, L., Rodríguez Díaz, A.: A hybrid optimization method with PSO and GA to automatically design type-1 and type-2 fuzzy logic controllers. Int. J. Mach. Learn. Cybernet. 6(2), 175–196 (2015)
19. Mech, L.D.: Alpha status, dominance, and division of labor in wolf packs. Can. J. Zool. 77, 1196–1203 (1999)
20. Mirjalili, S., Mirjalili, S.M., Lewis, A.: Grey wolf optimizer. Adv. Eng. Softw. 69, 46–61 (2014)
21. Muro, C., Escobedo, R., Spector, L., Coppinger, R.: Wolf-pack (Canis lupus) hunting strategies emerge from simple rules in computational simulations. Behav. Proc. 88, 192–197 (2011)
22. Nawi, N.M., Khan, A., Rehman, M.Z.: A new back-propagation neural network optimized with cuckoo search algorithm. In: ICCSA 2013, Part I, LNCS 7971, pp. 413–426 (2013)

23. Okamura, M., Kikuchi, H., Yager, R., Nakanishi, S.: Character diagnosis of fuzzy systems by genetic algorithm and fuzzy inference. In: Proceedings of the Vietnam-Japan Bilateral Symposium on Fuzzy Systems and Applications, Halong Bay, Vietnam 1998, pp. 468–473
24. Rashedi, E., Nezamabadi-Pour, H., Saryazdi, S.: GSA: a gravitational search algorithm. Inf. Sci. **179**, 2232–2248 (2009)
25. Saravanan, K., Sasithra, S.: Review on classification based on artificial neural networks. Int. J. Amb. Syst. Appl. (IJASA) **2**(4), 11–18 (2014)
26. Yang, X.S.: Firefly algorithms for multimodal optimization. In: Watanabe, O., Zeugmann, T. (Eds.) Proceedings of 5th Symposium on Stochastic Algorithms, foundations and applications, Lecture Notes in Computer Science, vol. 5792, pp. 169–178 (2009)
27. Yang, X.S.: Nature-Inspired Metaheuristic Algorithms. Luniver Press, UK (2008)
28. Yang, X.S., He, X.: Firefly algorithm: recent advances and applications. Int. J. Swarm Intell. **1** (1), 36–50 (2013)
29. Yang, X.: Bat algorithm: literature review and applications. Int. J. Bio-Insp. Comput. **5**(3), 141–149 (2013)
30. Yao, J.T.: A ten-year review of granular computing. In: Proceedings of the 3rd IEEE International Conference on Granular Computing (GrC), pp. 734–739 (2007)
31. Yao, Y.Y.: Granular computing: basic issues and possible solutions. In: Proceedings of the 5th Joint Conferences on Information Sciences, pp. 186–189 (2000)
32. Yu, F., Pedrycz, W.: The design of fuzzy information granules: tradeoffs between specificity and experimental evidence. Appl. Soft Comput. **9**(1), 264–273 (2009)
33. Zadeh, L.A.: Fuzzy sets. J. Inf. Control **8**, 338–353 (1965)
34. Zhang, Z., Zhang, C.: An agent-based hybrid intelligent system for financial investment planning. In: 7th Pacific Rim International Conference on Artificial Intelligence (PRICAI), pp. 355–364 (2002)

Part IV
Sustainability

Fuzzy Measure of National Sustainable Development Aggregate Index

Gorkhmaz Imanov

Abstract In this paper is proposed fuzzy measure of national sustainable development aggregate index (NSDAI) taking into account the sub-indices of economic, social and environmental sustainability at the example of Azerbaijan Republic. In order to measure sub-indices we use instruments of intuitionistic fuzzy set, generalized entropy measure of intuitionistic fuzzy set, L. Zadeh's composite rules of inferences, algorithm of the weighted rules. Next, using the fuzzy method of forgotten effects proposed by A. Kaufman and J. Gil-Aluja, relations between socioeconomic indicators have been analyzed.

Keywords Economical sustainable index · Diversification level index · Financial stability index · Social sustainability index · Social quality index · Human capital index · Ecological quality index · Green economy index · National sustainable development aggregate index

1 Introduction

In the literatures sustainable development has several definitions. However common aspects of their concepts are equilibrium economic development, social justice and ecological security [1].

In September 25, 2015 the 70-th Session UN General Assembly adopted sustainable development goals. New targets are significantly different from the Millennium Development Goals of 2000–2015-th years [2]. A key feature of adopted goals is that they are related both to the third world countries and the developed countries. Achieving these goals will require the developed countries not only to assist developing countries, but also the political steps to change their own institutions. Among

G. Imanov (✉)
Laboratory of Fuzzy Modelling, Institute of Control Systems,
Azerbaijan National Academy of Science, Bakhtiyar Vahabzadeh street, 9,
Az 1141 Baku City, Azerbaijan
e-mail: korkmazi2000@gmail.com

© Springer International Publishing AG 2018
C. Berger-Vachon et al. (eds.), *Complex Systems: Solutions and Challenges in Economics, Management and Engineering*, Studies in Systems, Decision and Control 125, https://doi.org/10.1007/978-3-319-69989-9_18

305

the 17 goals called for developed countries marked a steady growth and employment for all, sustainable consumption, climate change mitigation and income equality.

The national strategy for sustainable development is defined as a strategic and participatory process of analysis, debate, capacity-building and planning activities aimed at sustainable development [3].

Azerbaijan signed The Millennium Declaration on Sustainable Development in 2000. The state policy of Azerbaijan in the field of sustainable development reflects the "State Program on Poverty Reduction and Sustainable Development". The activities on implementation of the principles of sustainable development continued on the concept of "Azerbaijan 2020: Look into the Future" [4]. In this concept, an export-oriented economic model is taken as a basis, and it is planned that increasing the competitiveness of the economy and improvement in the structure will boost non-oil exports. Along with the speedy development of the non-oil industry, the promotion and expansion of innovative activity will create favorable grounds for the formation of economy based on knowledge in the country.

The purpose of any economic growth is to increase social welfare. From this point of view, one of the main priorities of the concept is the development of social spheres and human capital. Within these priorities the main objectives are to increase the quality of education and services, to strengthen social security, to ensure gender equality and develop the family, youth potential and sports.

One of the main targets of the concept is to achieve sustainable socio-economic development from ecological point of view. The necessary measures will be continued in the future in order to protect biodiversity, neutralize the negative impact of the fuel-energy complex on the environment, eliminate the pollution of the sea basin and protect them, restore green areas and effectively protect the existing resources. In order to measure of the meaning of sub indices and aggregate index, we use instruments of fuzzy sets and fuzzy logic theory. This paper is systemized as follows: Sect. 2 presents the measurement of Economic Sustainability Index (ESI) and it includes fuzzy composite index of diversification level of economy and fuzzy approach to measuring financial stability. Estimation of Social Sustainability Index is carried out in Sect. 3. The section includes fuzzy model estimation of social economy and fuzzy model estimation of national human capital quality. Section 4 presents the measurement of Environmental Sustainability Index including fuzzy ecological quality index and the model of green economy. On the basis of defined indices, Sect. 5 presents Fuzzy Assessment of National Sustainable Development Aggregate Index. After that we have Estimation Degree of Factors' Impact to Indicators of Socioeconomic System in the Sect. 6 and finally, conclusion about this paper in Sect. 7.

2   Economic Sustainability Index

Sustainability economics like all other scientific endeavors and fields of human action is based on a specific basic understanding of the world. The basic structure of reality, that is the systematic of basic types of entities (objects, properties, processes) and their structural relationships is not universal and a priori given, but it derives from the specific perspective of sustainability economics on the world. It therefore differs from the basic understanding of the world of other scientific approaches and fields of human action [5]. In order to compute Economic Sustainability Index (ESI) have been used indices of diversification level of economy and financial stability.

2.1 Fuzzy Entropy Composite Index of Diversification Level of Economy

Diversification of the economic structure is the main state in achieving sustainable development, especially for oil exporting country such as Azerbaijan. Normally diversified economy provides optimal growth and relation among industries of national economy. There are various methods for determining the level of economic diversification. In literature, among which I would like to highlight Ogive Index [6], Entropy Index [7], and Herfindahl–Hirschman Index [8].

In order to calculate fuzzy entropy composite index of diversification level of Azerbaijan economy we have used structural indicators of GDP for 2013. Using the formula of equiproportional distribution ($1/N = 1/13 = 0.077$, where N—number of sectors), intervals and corresponding terms are determined:

Lowestnorm	**(VLN)**	$(0.010, 0.030, 0.050)$
Belowthenorm	**(LON)**	$(0.040, 0.053, 0.065)$
Norm	**(NOR)**	$(0.060, 0.080, 0.100)$
Abovethenorm	**(HAN)**	$(0.090, 0.295, 0.500)$

Further, membership degree of structural parameters to corresponding terms $\mu_A(x)$ is defined. Based on this information and fuzzy entropy of economic sectors $-E_i(A_i)$ [9], parameters of the model are computed.

The obtained results of parameters calculation are shown in Table 1.

On the basis of the following formula Fuzzy Entropy Composite Index of diversification—E(A) is defined:

$$E(A) = \sum_{i=1}^{n} E_i(A_i) = \sum_{i=1}^{n} \frac{|A_i \cap A_i^c|}{|A_i \cup A_i^c|}$$

$$= \frac{1.9 + 4.226 + 5.1 + 2.988}{19.1 + 15.774 + 15.9 + 17.012} = 0.209689, i = 1, ..., 13$$

Obtained results of calculations of Fuzzy Entropy Composite Index demonstrate a low level of diversification of Azerbaijani economy in 2013.

Table 1 Parameters of the entropy models

Economic sectors	2013	$\mu_A(x)$	Terms	$E_i(A_i)$
Agriculture	0.057	0.667	LON	0.7395
Mining industry	0.420	0.39	HAN	0.4535
Manufacturing	0.045	0.25	VLN	0.2698
Construction	0.124	0.166	HAN	0.1756
Trade	0.076	0.8	NOR	0.8627
Transport and communication	0.066	0.3	NOR	0.3208
Tourism	0.020	0.5	VLN	0.5833
Real estate	0.022	0.6	VLN	0.6964
State governance and social insurance	0.027	0.85	VLN	0.8779
Education	0.050	0.769	LON	0.8188
Health care	0.019	0.45	VLN	0.5094
Finance, insurance	0.062	0.25	LON	0.2679
Other services	0.012	0.1	VLN	0.0995

2.2 Fuzzy Approach to Measuring Financial Stability

The measurement performs two quite distinct roles. One is to help to ensure the accountability of the authorities responsible for performing the task. The other is to support the implementation of the chosen strategy in order to achieve the goal in the real time. The former calls for ex post measurement of financial instability, i.e. for assessments of whether financial instability prevailed or not at some point in the past. The latter relies on ex ante measurement, i.e. on assessment of whether the financial system is fragile or not today. While both ex ante and ex post measurements are fuzzy, the challenges in supporting strategy implementation are tougher [10].

In a fuzzy approach to aggregate index of financial stability the obtained values of sub-indices for 2005–2015 years are classified into the following terms:

- very low stability VLS = (−2.43, −2.43, −1.20);
- low stability LS = (−1.23, 0.00, 0.00);
- stable S = (0.00, 0.00, 0.65);
- high stability HS = (0.63; 1.28; 1.28).

The matrix of linguistic variables for the years of 2005–2015 is given in the following Table 2.

In determining the weights of the sub-indices are mainly used expert assessments. However, it should be noted that the values of these weights depend not only on time but also on situation existing in the various financial markets and global events. In

Table 2 Matrix of linguistic values of sub-indices in the period 2005–2015

Sub-indices	2005	2006	2007	2008	2009	2010	2011	2012	2013	2014	2015
FMI	VS	VL	LS	ST	ST	LS	LS	LS	LS	ST	HS
FVI	LS	LS	LS	ST	LS	LS	ST	ST	HS	HS	LS
FSI	HS	ST	ST	ST	ST	LS	LS	LS	LS	ST	LS
WEI	ST	ST	ST	LS	LS	ST	ST	ST	ST	ST	LS

order to define the weights of individual sub-indices of aggregated index of financial stability, we have used instruments of intuitionistic fuzzy set. The intuitionistic fuzzy set, proposed by K. Atanassov [11], is a generalization of L. Zadeh's fuzzy set. The intuitionistic fuzzy set is defined as:

$$A = \{< x, \mu_A(x), v_A(x) >| x \in X\}$$

where,

$$\mu_A : X \to [0, 1] v_A : X \to [0, 1]$$

if
$$0 \le \mu_A(x) + v_A(x) \le 1 \qquad \forall x \in X$$

$\mu_A(x), v_A(x) \in [0, 1]$ numbers indicate the degree of membership and non-membership of x to A respectively.

For each intuitionistic fuzzy set X, there is intuitionistic index x in A.

$$\pi_A(x) = 1 - \mu_A(x) - v_A(x)$$

Indicators of financial stability sub-indices of Azerbaijan for 2005–2015 years, corresponding to indicators of intuitionistic fuzzy set, are given in following Table 3.

In this study in order to define weights of financial stability sub-indices, we use generalized entropy measure of intuitionistic fuzzy set F, composed of n elements, proposed by E. Szmidt and J. Kacprzyk [12]:

$$E(A_i) = \frac{maxCount(A_i \cap A_i^c)}{maxCount(A_i \cup A_i^c)}, (i = 1, ..., n).$$

where,

$$F_i \cap F_i^c = \langle min(\mu_{F_i}, \mu_{F_i}^c), max(v_{F_i}, v_{F_i}^c)\rangle,$$

$$F_i \cup F_i^c = \langle min(\mu_{F_i}, \mu_{F_i}^c), max(v_{F_i}, v_{F_i}^c)\rangle.$$

The calculation of entropy for each individual sub-indices during the year of 2005 is given below:

Table 3 Indicators of intuitionistic fuzzy set

Sub indices years	FMI			FVI			FSI			WEI		
	$\mu_1 t$	$v_1 t$	$\pi_1 t$	$\mu_2 t$	$v_2 t$	$\pi_2 t$	$\mu_3 t$	$v_3 t$	$\pi_3 t$	$\mu_4 t$	$v_4 t$	$\pi_4 t$
2005	0.7	0.3	0	0.08	0.92	0	0.85	0.15	0	0.94	0.06	0
2006	0.92	0.08	0	0.69	0.31	0	0.12	0.88	0	0.28	0.72	0
2007	0.96	0.04	0	0.99	0.01	0	0.16	0.84	0	0.36	0.64	0
2008	0.18	0.82	0	0.85	0.15	0	0.78	0.22	0	0.50	0.50	0
2009	0.20	0.80	0	0.73	0.27	0	0.64	0.36	0	0.08	0.92	0
2010	0.99	0.01	0	0.95	0.05	0	0.70	0.30	0	0.27	0.73	0
2011	0.33	0.67	0	0.66	0.34	0	0.32	0.68	0	0.85	0.15	0
2012	0.53	0.47	0	0.70	0.30	0	0.57	0.43	0	0.89	0.11	0
2013	0.53	0.47	0	0.04	0.96	0	0.86	0.14	0	0.24	0.76	0
2014	0.89	0.11	0	0.31	0.69	0	0.85	0.15	0	0.54	0.46	0
2015	0.35	0.65	0	0.93	0.07	0	0.30	0.70	0	0.52	0.48	0

$$E(A_1) = \frac{(0.35, 0.65, 0) \cap (0.65, 0.35, 0)}{(0.35, 0.65, 0) \cup (0.65, 0.35, 0)} = \frac{0.35}{0.65} = 0.54$$

$$E(A_2) = \frac{(0.93, 0.07, 0) \cap (0.93, 0.07, 0)}{(0.07, 0.93, 0) \cup (0.07, 0.93, 0)} = \frac{0.07}{0.93} = 0.08$$

$$E(A_3) = \frac{(0.3, 0.7, 0) \cap (0.7, 0.3, 0)}{(0.3, 0.7, 0) \cup (0.7, 0.3, 0)} = \frac{0.3}{0.7} = 0.43$$

$$E(A_4) = \frac{(0.52, 0.48, 0) \cap (0.48, 0.52, 0)}{(0.52, 0.48, 0) \cup (0.48, 0.52, 0)} = \frac{0.48}{0.52} = 0.92$$

The entropy for each individual sub-index in 2005–2014 is as:

$$2005 - E(A_1) = 0.43; E(A_2) = 0.09; E(A_3) = 0.18; E(A_4) = 0.06$$
$$2006 - E(A_1) = 0.09; E(A_2) = 0.45; E(A_3) = 0.14; E(A_4) = 0.39$$
$$2007 - E(A_1) = 0.04; E(A_2) = 0.01; E(A_3) = 0.19; E(A_4) = 0.56$$
$$2008 - E(A_1) = 0.22; E(A_2) = 0.18; E(A_3) = 0.28; E(A_4) = 1$$
$$2009 - E(A_1) = 0.25; E(A_2) = 0.37; E(A_3) = 0.56; E(A_4) = 0.09$$
$$2010 - E(A_1) = 0.01; E(A_2) = 0.05; E(A_3) = 0.43; E(A_4) = 0.37$$
$$2011 - E(A_1) = 0.49; E(A_2) = 0.52; E(A_3) = 0.47; E(A_4) = 0.18$$
$$2012 - E(A_1) = 0.89; E(A_2) = 0.43; E(A_3) = 0.75; E(A_4) = 0.12$$
$$2013 - E(A_1) = 0.89; E(A_2) = 0.04; E(A_3) = 0.16; E(A_4) = 0.32$$
$$2014 - E(A_1) = 0.12; E(A_2) = 0.45; E(A_3) = 0.18; E(A_4) = 0.85$$

Then on the basis of the following formula weights of each individual index are defined:

$$w_i = \frac{1 - E(A_i)}{n - \sum_{i=1}^{n} E(A_i)}$$

The weights of individual sub-indices for the year of 2015 are calculated as follows:

$$w_1(2015) = \frac{1 - 0.54}{4 - 1.97} = \frac{0.46}{2.03} = 0.23$$

$$w_2(2015) = \frac{1 - 0.08}{4 - 1.97} = \frac{0.92}{2.03} = 0.45$$

$$w_3(2015) = \frac{1 - 0.43}{4 - 1.97} = \frac{0.57}{2.03} = 0.28$$

$$w_4(2015) = \frac{1 - 0.92}{4 - 1.97} = \frac{0.08}{2.03} = 0.04$$

Using weights of individual sub-indices and their linguistic values (Table 3), aggregate index of financial stability is calculated for the year of 2015:

$$AFSI(2015) = 0.23 * HS + 0.45 * LS + 0.28 * LS + 0.04 * LS$$
$$= 0.23 * (0.63, 1.28, 1.28) + 0.45 * (-1.23, 0, 0) + 0.28 * (-1.23, 0, 0) + 0.04$$
$$* (-1.23, 0, 0)$$
$$= (0.15, 0.29, 0.29) + (-0.55, 0, 0) + (-0.34, 0, 0) + (-0.05, 0, 0)$$
$$= (-0.79, 0.29, 0.29) = LS - ST$$

The weights of sub-indices and aggregate index of financial stability for the period 2005–2015 are given in Table 4.

The results of calculations for weights of sub-indices and aggregate indices of financial stability during the period 2005–2015 are given in Table 4.

As it is seen from the table, aggregate index level of financial stability of Azerbaijan in 2005, 2007, 2009–2013, 2015 years was between high and stable level, it was very low in 2006, stable in 2008, and stable and high stability level in 2014. Of course this value of aggregate index depended on the global economic crisis and oil prices in the world market.

3   Social Sustainability Index

In the literature [13] there are several definitions of social stability. These definitions are generally accepted concepts of social quality, human capital and well-being. Taking this into account we can say that social sustainability is the quality of the society, which is a measure of social quality. The terms of the definition of social quality are the socio-economic security, social cohesion, social inclusion and social powers.

Table 4 Weights of sub-indices and aggregate indices

Indicators years	w_1	w_2	w_3	w_4	AFSI
2005	0.18	0.25	0.28	0.29	(−0.57, −0.08, 0.33) **LS-ST**
2006	0.31	0.19	0.29	0.21	(−0.98, −0.75, −0.04) **LS**
2007	0.3	0.31	0.25	0.14	(−0.75, 0, 0251) **LS-ST**
2008	0.34	0.35	0.31	0	(0, 0, 0.65) **ST**
2009	0.28	0.23	0.16	0.33	(−0.69, 0, 0.228) **LS-ST**
2010	0.32	0.3	0.18	0.2	(−1, 0, 0.13) **LS-ST**
2011	0.22	0.21	0.23	0.35	(−0.55, 0, 0.37) **LS-ST**
2012	0.06	0.32	0.14	0.49	(−0.24, 0, 0.53) **LS-ST**
2013	0.04	0.37	0.32	0.26	(−0.2, 0.47, 0.64) **LS-ST**
2014	0.37	0.23	0.34	0.06	(0.15, 0.29, 0.79) **LS-ST**
2015	0.23	0.45	0.28	0.04	(−0.79, 0.29, 0.29) **LS-ST**

Social quality and human capital plays significant role in economic development. Social quality and human capital influences economic growth not only directly, but also through interaction with each-other. Social quality and human capital enables individuals, communities and firms to cope with the demands of rapid changes. Estimation of social and human capital quality is almost impossible by using traditional methods. In this paper we attempt to estimate social and human capital quality by using fuzzy approach. The article mainly consists of two parts—in the first part are investigated fuzzy model of social quality factors, in the second fuzzy model for estimation of national human capital quality. Problems corresponding to the models are solved by applying fuzzy method for defining quality indices. World Development and Azerbaijan Indicators are used in the process of solution of the problems.

3.1 Fuzzy Model Estimation of Social Quality

The theory of social quality has been offered by Beck et al. [13, 14], Van der Maeson et al. [15]. Social quality represents degree of participation of citizens in the social and economic life of a society in which their well-being and individual potential raises. Social quality defined on the base of indicators of the four conditional factors—social-economic security, social cohesion, social inclusion, social empowerment.

Social-economic security indicators are following: number of square meters per household member (NSM); proportion of population living in houses with lack of functioning basic amenities (PPL); proportion of people covered by compulsory/voluntary health insurance (PHI); number of medical doctors per 10 000 inhabitants (MED); length of notice before termination of labor contract (LNT); proportion of employed workforce with temporary, non-permanent, job contract (PET); proportion of workface that is illegal (PWI); number of fatal cases (NFC); number of nonfatal cases (NNC); number of hours a full-time employee typically works a week (NHE); proportion of pupils leaving education without finishing compulsory education (PLE); study fees in school as proportion of national mean net wage (SFS); study fees in high school as proportion of national mean net wage (SFH); proportion of students who, within a year of leaving school, are able to find employment (PSE); people affected by criminal offences per 10 000 inhabitants (CRI); eco-civilization index (ECC). Quality of socioeconomic security index (SESI) is the output indicator.

Social cohesion indicators are indicated as follows: Extent to which most people can be trusted (TRU); Trust to authorities (TRA); Trust to religion (TRR); Number of cases being referred to European Court of Law (ECO); Respect for parents (IFA); Blood donation (%) (BLO); Multiculturalism (tolerance) (TOL); Willingness to pay more taxes if you were sure that it would improve the situation of the poor (WMT); Help elders (VOL); Membership (active or inactive) of political, voluntary, charitable organizations or sport clubs (MVO); Frequency of contact with friends and colleagues (CWF); Sense of national pride (NAP). Quality of social cohesion index (SCOI) is the output indicator.

Social inclusion indicators are following: proportion having right to vote in local elections (POV) and proportions exercising it (PPV); proportion with right to a public pension (PEN); proportion of ethnic minority groups elected or appointed to parliament, boards of private companies and foundations (ETH); proportion of women elected or appointed to parliament, boards of private companies and foundations (WPA); long-term unemployment (12+ month) (LTU); proportion of population with entitlement to and using public primary health care (PPH); proportion homeless, sleeping rough (HLP); average waiting time for social housing (WAI); school participation rates and higher education participation rates (HED); proportion of people in need received care services (PPN); density of public transport system and road density (TRD); number of public sport facilities per 10 000 inhabitants

(NPS); number of public and private civic and cultural facilities (e.g. cinema, theatre, concerts) per 10 000 inhabitants (NPC); duration of contact with relatives (cohabitating and non-cohabitating (PRC). Quality of social inclusion index (SIQI) is the output indicator.

Social empowerment indicators are following: extend to which social mobility is knowledge-based (SOM); percentage of population literate and numerate (PLN); availability of free media (FME); percentage of labor force that is member of a trade union (TRU); percentage of labor force covered by a collective agreement (LCA); percentage of employed labor force receiving work-based training (TRA); index of democracy (DEM); percentage of organizations/institutions with work councils (WCC); percentage of the national and local public budget that is reserved for voluntary, non-for-profit citizenship initiatives (CIL); proportion of national budget allocated to all cultural activities (CUL); percent expenses of national and local budgets devoted to disabled people (DIL). Quality of social empowerment index (SEQI) is the output indicator. Indicators of conditional factors of social quality have been adopted from [15].

In order to estimate indices of social quality factors has been proposed method based on L. Zadehs composite rules of inference [16] consisting of the following steps:

1. development of the Table 5 describing parameters of the model on the basis of information obtained from international organizations and experts. In the first column of the table shows the input parameters of the model, and in the following columns—terms and their intervals. The last column specifies crisp meaning of input parameters for a fixed period;
2. definition of membership degrees of the crisp meaning of input parameters to the relevant terms;
3. determination of the minimum degree of membership to the corresponding term of input parameters, i.e. $\max_i(\min_j \mu_{ij})$;
4. determination of the maximum among minimum values of the degrees of membership to the corresponding term, i.e. $\max_i(\min_j \mu_{ij})$;

The obtained value will reflect the quality of the social factor.

The proposed methodology is tested on the basis of information on quality parameters of the model of socio-economic security (Table 5). The source materials are obtained from international socio-economic organizations and the expert opinion data. Information on the socio-economic security of Azerbaijan in 2010 is given in the last column of Table 5.

To calculate the quality of the social factors the following terms are used: low (L), medium (M), high (H) and very high (VH), which are scaled in the interval [0, 1].

In order to define linguistic variables intervals following calculations have been used, as given in [17]. In the second stage we have determined the degree of membership of national indicators of socio-economic security with appropriate term. In

Table 5 Parameters of the model of social-economic security

Input variable	Terms and its intervals				Azerbaijan
NSM	L 0 15	M 14–20	H 18–30	VH 28–70	L 12.6
PPL	L 0.5 0.25	M 0.24 0.15	H 0.14 0.1	VH 0.09–0	M 0.15
PHI	L 0 10	M 9–21	H 20–60	VH 59–100	L 0.2
MED	L 0 300	M 299–350	H 300–400	VH 370–600	H 36.8
LNT	L 1 31	M 30–51	H 50–61	VH 60–90	H 60
PET	L 100 50	M 49–20	H 19–10	VH 9–1	L 68
PWI	L 0.5 0.2	M 0.19 0.14	H 0.13 0.09	VH 0.18 0	VH 0.002
NFC	L 10 8	M 7–5	H 4–2	VH 1–0	VH 0.00128
NNC	L 10 8	M 7–5	H 4–2	VH 1–0	VH 0.00172
NHE	L 50 44	M 43–39	H 38–36	VH 35–20	M 42
PLE	L 50 20	M 18–9	H 8–7	VH 6–0	M 10
SFS	L 6 3	M 2.9–2	H 1 0.5	VH 0.4–0	M 2.8
SFH	L 7 3	M 2.9–2	H 1 0.5	VH 0.4–0	L 6
PSE	L 0 5	M 4–10	H 9–20	VH 19–100	VH 30
CRI	L 180 80	M 79–50	H 49–20	VH 19–0	VH 13.5
ECC	L 0–0.2	M 0.19 0.5	H 0.49 0.7	VH 0.7–1	H 0.632

determining the degrees of membership, we have used triangular membership functions.

At the task level membership of 16 indicators of the terms is as follows:

Among the minimum values the maximum value is determined, which is equal to 0.32. This value corresponds to the term-"low", thus quality index of socio-economic security is defined: SESI as low. Likewise, the indices of the quality of social inclusion—SIQI is 0.86 (high), the index of the quality of social empowerment—SEQI is 0.9 (high), the index of the quality of social cohesion—SCQI is 0.91 (moderate). Estimated meanings of the conditional factors give us possibility to estimate

Low (L)	Mean (M)	High (H)	Very high (VH)
$\mu_{NSM} = 0.32$	$\mu_{PPL} = 0.25$	$\mu_{MED} = 0.64$	$\mu_{PWI} = 0.05$
$\mu_{PHI} = 0.4$	$\mu_{NHE} = 0.66$	$\mu_{LNT} = 0.18$	$\mu_{NFC} = 0.003$
$\mu_{PET} = 0.72$	$\mu_{PLE} = 0.22$	$\mu_{ECC} = 0.65$	$\mu_{NNC} = 0.003$
$\mu_{SFH} = 0.5$	$\mu_{sfs} = 0$		$\mu_{PSE} = 0.27$
			$\mu_{CRI} = 0,58$
min:0.32	min:0	min:0.18	min:0.003

social quality index. In order to estimate social capital quality index (SCQ), we use fuzzy union operation, i.e. (moderate).

$$\mu_{SCQ} = max(\mu_{SESI}, \mu_{SIQI}, \mu_{SEQI}, \mu_{SCQI}) = max(0.32; 0.86; 0.91) = 0.91 \text{ (moderate)}$$

The proposed model for estimation of social quality gives us possibility to define human capital quality.

3.2 Fuzzy Model Estimation of Quality of National Human Capital

Human capital is the one of the main factors, which provides information on the development level of the socioeconomic system. Fundamental concepts of the human capital theory were founded by the American economists, Nobel Prize Laureates, T. Shultz [18] and G. Becker [19]. According to the definition of the Organization Economic Cooperation and Development (OECD) experts human capital is the knowledge, skills, competencies and attributes embodied in individuals that facilitate the creation of personal, social and economic well-being [20].

The conventional standard for measuring human capital stock has been largely categorized into three parts [20]:

- school enrollment rates, scholastic attainments, adult literacy, and average years of schooling are the examples of the output-based approach;
- cost based approach is based on calculating costs paid for obtaining knowledge;
- income-based approach is closely linked to each individuals benefits obtained by education and training investment.

Professor D. Kwon [21] considers that human capital is difficult to identify and measure directly. So, many researchers have used indirect measurements. Concept of the human capital needs to be considered both from monetary and non-monetary characteristics. Human capital is closely linked to the social quality. It is necessary to analyze the results of human capital measurement within the socio-cultural framework of a society and all levels such as individual, organizational, national.

Human capital on the national level expresses intellectual potential of the society.

In this section we study problem of quality estimation of the national level human capital—NHCQI. With this purpose, following elements are taken as recourses of human capital: education quality—QUE; level of the health care development—

QHC; cultural level of society—QSC; innovation index—INO; quantity of patent—PAT; quantity of articles—ART; labor productivity PRO; socioeconomic security index—SEQ; social cohesion index SCI; social inclusion index—SII; social empowerment index SEI.

As factices meaning results of expert opinion of the specialist, working in this sphere have been used.

In order to define quality of science following indices have been used: resident patent applications index, scientific and technical journal articles, patent application per million of population, innovation index. For this purpose, corresponding world development indicators [22] and component of knowledge economy are applied [23]. Labor productivity indicators are taken from [24]. Indicators of socioeconomic security quality index, social cohesion quality index, social inclusion quality index and social empowerment quality index are the results of the solution of the problem of social quality estimation. Along with this in order to define NHCQI have been used L. Zadehs composite rules of inference. As a result of National Human Capital Quality Index (NHCQI) is "low".

4 Environmental Sustainability Index

Environmental sustainability defines as a condition of balance, resilience, and interconnectedness that allows human society to satisfy its needs while neither exceeding the capacity of its supporting ecosystems to continue to regenerate the services necessary to meet those needs nor by our actions diminishing biological diversity [25]. In order to assess Environmental Sustainability Index (ENI) is computed fuzzy Ecological quality index (EQI) and Green economy index (GEI).

4.1 Fuzzy Ecological Quality Index

In order to build fuzzy model for estimation of ecological quality index we have used ecological data from various international organizations and data available for Azerbaijan. Table referring to fuzzy model is given in the Table 6.

In order to resolve the stated problem, which correspond to the model, algorithm of the weighted rules [26] has been used. Steps of the algorithm are as follows.

Fuzzification is carried out as the first step, and then a Gaussian function of accessories is applied. Further, on the basis of quantity of terms, initial fuzzy rules are defined (for example, if quantity of terms is 3, quantity of initial rules is equal to three). On the following step, other possible rules are defined by Cartesian product of terms in initial rules.

Then the peak point of each corresponding interval c_{ij} is defined on the basis of the matrix $C = (c_{ij})$, where the i-index corrected under construction $C = (c_{ij})$, j an index of terms, which is defined as c_{ij}. Initial rules are expressed on the basis of c_{ij}.

Table 6 Fuzzy model of ecological quality

Parameter	Definition	Terms and its values					Azerbaijan
I Air Quality Index (AQI)		Very low	Low	Moderate	Good	Very good	
		>50	35–50	30–40	15–30	Eki.20	
1. Annual Average SO2(SO2)	mgr/m³	Very high	High	Moderate	Low	Very low	Low
		0–20	19–40	39–60	59 80	79–100	15
2. Annual Average NO2 (NO2)	mgr/m³	Very high	High	Moderate	Low	Very low	High
		>40	30–45	20–35	Eki.25	0–15	50
3. Annual Average TSP (TSP)	mgr/m³	Very high	High	Moderate	Low	Very low	Very high
		>60	50–60	40–50	30–45	20–35	300
II Water Quality Index (WQI)		Very bad	Bad	Moderate	Good	Very good	Bad
		0–20	20–40	40–60	60–80	80–100	21.8
4. Dissolved oxygen concentrations (milliliters of dissolved oxygen per liter of water) (DOC)	(ml/l)	Very bad	Bad	Moderate	Good	Very good	Good
		>14	Kas.14	09.Ara	07.Eki	<7	8.27

(continued)

Table 6 (continued)

Parameter	Definition	Terms and its values					Azerbaijan
5. Fresh water resources (FWR)	m³/percapita	Very bad	Bad	Moderate	Good	Very good	Very bad
		<3500	3000–6000	5500–9000	8500–12000	11500–15000	948
6. Fresh water withdrawal 40% of available water (FWW)	% of internal resources	Very low	Low	Moderate	High	Very high	Very low
		>79	80–59	60–39	40–19	20–0	150
III Land Quality Index (LQI)		Very bad	Bad	Moderate	Good	Very good	Moderate
		0–20	19.05.1940	39.5–60	59.5–80	79.5–100	49.5
7. Percentage of agricultural land (AGL)	% of land area	Very low	Low	Moderate	High	Very high	High
		0–15	14.05.2025	24.05.1950	49.5–70	>69.5	58
8. Annual average forest area (AAF)	% of land area	Very bad	Bad	Moderate	Good	Very good	Bad
		0–10	Eyl.20	19–30	29–40	39–50	11.3
IV Environmental Biodiversity Index (EBI)		Very bad	Bad	Moderate	Good	Very good	Bad
		0–20	19–40	39 60	59 80	79–100	29.5
9. Territories under protection (TUP)		Very bad	Bad	Moderate	Good	Very good	Bad
		<8	Tem.15	14–22	21–30	>29	10.1

(continued)

Table 6 (continued)

Parameter	Definition	Terms and its values					Azerbaijan
		Very bad	Bad	Moderate	Good	Very good	
10. Percentage of the country territory in the threatened eco regions (TTER)	%	Very bad	Bad	Moderate	Good	Very good	40
		>40	0–40	20–30	Eki.20	0–10	
11. National Biodiversity Index (NBI)	0–1	Very bad	Bad	Moderate	Good	Very good	Good
		<0.20	0.19–0.40	0.30–0.50	0.45–0.65	0.6–1	0.534
V 12. CO_2 and particulate emissions damage	MT per capita	Very high	High	Moderate	Low	Very low	High
		>4.5	03.05.2005	2.3–3.6	1.1–2.4	0–1.2	4.4 (2009)
VI 13. Capital investments for environmental protection programs	% of GDP	Very low	Low	Moderate	High	Very high	Very low
		0–1.2	1.1–2.3	2.2–3.5	03.04.2005	>4.9	0.5 (2009)
QNE		0–20	19–40	39–60	59–80	70–100	

After that, by means of the below-shown formula, membership degree of every point of support part of fuzzy number corresponding to linguistic variables were define:

$$\mu(x) = \prod_{j=1}^{n} exp\left[-\frac{1}{2}\left(\frac{x^i - c_j^i}{\sigma_j^i}\right)\right] \tag{1}$$

where n—number of input variables; x^i terms; i an index of term; c_j^i—a peak point of corresponding terms i; σ_j^i average square deviation of an interval of a corresponding term.

After that, weighted antecedent of initial rules is defined:

$$w_i^a = \frac{\mu_i(x)}{\sum_{i=1}^{n} \mu_i(x)} \tag{2}$$

$$\sum_{i=1}^{n} w_i^a = 1$$

where w_i^a—is weighted antecedent of initialed rules, $\mu_i(x)$—degree of fuzzy variables entered in the antecedent part of rules.

By substituting value of the formula (1) in the formula (2), we receive:

$$w_i^a = \frac{\prod_{j=1}^{n} exp\left[-\frac{1}{2}\left(\frac{x_j - c_j^i}{\sigma_j^i}\right)^2\right]}{\sum_{i=1}^{R} \prod_{j=1}^{n} exp\left[-\frac{1}{2}\left(\frac{x_j - c_j^i}{\sigma_j^i}\right)^2\right]} \tag{3}$$

In the following step by means of the mentioned below formula, w_i^c defines weighted values of the consequence part of rules:

$$w_i^c = \frac{\sum_{i=1}^{R} b_i \prod_{j=1}^{n} exp\left[-\frac{1}{2}\left(\frac{x_j^i - c_j^i}{\sigma_j^i}\right)^2\right]}{\sum_{i=1}^{R} \prod_{j=1}^{n} exp\left[-\frac{1}{2}\left(\frac{x_j^i - c_j^i}{\sigma_j^i}\right)^2\right]} \tag{4}$$

where b_i—peak point of corresponding terms of the consequence part of the rules.

Further, using the maximum values w_i^c, w_i^a we define new system of rules.

In new system of rules R1- are rule, which we find and R2 rule, which correspond to fixed meaning of input variables. By using composition operation are defined corresponding fuzzy numbers. At last, defuzzification of fuzzy numbers is carried out by means of Centroid method.

By using this algoritm meaning the ecological quality index has been obtained, equal to Bad. Very bad quality of air, bad quality of water, bad index of biodiversity and very low level of investment to environmental protection have contributed to the Bad value of the ecological quality.

4.2   Model of Green Economy

Green Economy is one of the most important concepts of the sustainable development of the country. UNEP defines green economy as one that results in improved human well-being and social equity, while significantly reducing environmental risks and ecological scarcities. It is—low carbon, resource, efficient, and socially inclusive [26].

The concept of a green economy has to replace brown economy as world economic development progress. Decades of creating new wealth through a brown economy model based on fossil fuels have not substantially addressed social marginalization and environmental degradation as well as resource depletion. In addition the world is still far from delivering on the Millennium Development Goals by 2015 [27].

United Nation Department of Economic and Social Affairs [28] having analyzed over 80 publications on the green economy and green growth concepts offer economic, social and ecological indicators to measure level of green economy development. Also, it is suggested to use Global Green Economy Index [28] GGEI and NASDAQ OMX Green Economy Benchmark Index (QGREEN), in order to estimate level of Green Economy. GGEI has been estimated by using following indicators such as Clean energy technology, Sustainable forms of tourism and Improved domestic environmental quality. QGREEN includes followings Energy efficiency, Clean fuels, Renewable energy generation, Natural resources, Water, Pollution mitigation and Advanced materials.

The green economy will emerge in different forms in different regions, depending on local economic strengths and weaknesses. This paper proposes National Green Economy Index (NGEI) to define level of development and methods estimation investments distribution to sectors of the green economy in Azerbaijan. To meet this objective we use following eleven indicators: Ecological quality ECQ, Renewable energy REE, Protection land PRL, Green tourism TOR, Quality of life QOL, Green GDP- EPP, Energy intensity—ENI, Organic agriculture ORA, Worldwide governance index WGI, International Innovation Index—III, Transport greenhouse gas emissions per capita—GHG.

In order to achieve this we have primarily applied data available from Azerbaijan and international organizations (UNEP, OECD). In order to solve problem of the National Green Economy Index (NGEI) we have used fuzzy set and fuzzy logic theory.

Table 7 Table Model of Green Economy

No	Categories	Source indicators	Development level					
			World indicators					
			Very low	Low	Medium	High	Very high	Azerbaijan
1	Ecological quality—ECQ	2010	0–0.2	0.18–0.4	0.38–0.6	0.58–0.8	0.78–1	L 0.25
2	Renewable energy—REE		0–0.2	0.18–0.4	0.38–0.6	0.58–0.8	0.78–1	VL 0.013
3	Protection land—PRL	2012	0–0.2	0.18–0.4	0.38–0.6	0.58–0.8	0.78–1	VL 0.102
4	Green tourism—TOR		0–0.2	0.18–0.4	0.38–0,6	0.58–0,8	0.78–1	VL 0.012
5	Quality of life—QOL	2011	0–0.2	0.18–0.4	0.38–0.6	0.58–0.8	0.78–1	M 0.548
6	Green GDP		0–0.2	0.18–0.4	0.38–0.6	0.58–0.8	0.78–1	VL 0.008

In order to model quality of the Green Economy the following terms have been used: Very Low (VL), Low (L), Medium (M), High (H) and Very High (VH), which were scaled in the interval [0, 1]. In the process of modeling we have also used terms very bad (VB), Bad (B), Moderate (M), Good (G), and Very Good (VG) (Table 7).

In order to estimate indices of the level of development of Green Economy used L. Zadehs composite rules of inference, which describe above.

At the task level, membership of 11 indicators of the terms is as follows:

Very Low (VL)	Low (L)	Medium (M)	High (H)	Very High (VH)
$\mu_{REE} = 0.03$	$\mu_{ECQ} = 0.55$	$\mu_{QOL} = 0.29$	$\mu_{ENI} = 0.05$	
$\mu_{PRL} = 1$	$\mu_{III} = 0.38$		$\mu_{GHG} = 0.008$	
$\mu_{TOR} = 0.03$		$\mu_{ORA} = 0.96$		
$\mu_{EEP} = 0.02$		$\mu_{WGI} = 0.06$		
min:0.02	min:0.38	min:0.06	min:0.005	min:0

The maximum value, which is equal to 0.38, is determined among the minimum values. This value corresponds to the term-"low", thus index of development level of Green Economy is defined.

Table 8 Indicators sub-indices of National Sustainable Development Aggregate index

Economical sustainability (ESI)	μ_{ES}	v_{ES}	Linguistic variables
Economic diversification level index (EDI)	0.4	0.6	Low
Financial stability index (FSI)	0.53	0.47	Low
Social sustainability (SSI)			
Social quality index (SQI)	0.9	0.1	Moderate
Human capital index (HCI)	0.9	0.1	Low
Environmental sustainability (ENI)			
Ecological quality index (EQI)	0.2	0.8	Low
Green economy index (GEI)	0.38	0.62	Low

5   Fuzzy Assessment of Aggregate Index National Sustainable Development

By using value of indices of economic diversification level, financial stability, social quality, human capital, ecological quality and green economy, demonstrated in following Table 8, is computed National Sustainable Development Aggregate Index (NSDAI).

For this purpose we use instruments of intuitionistic fuzzy set and generalized entropy measure.

In a fuzzy approach to construct the aggregate index of national sustainable development, the obtained values of sub-indices are classified into the following terms and corresponding fuzzy numbers:

- very low VL = (0, 0.1, 0.2);
- low LO = (0.19, 0.29, 0.4);
- moderate MD = (0.39, 0.49, 0.6);
- high H = (0.59; 0.69; 0.8);
- very high VH = (0.79, 0.89, 1).

In the first stage were defined entropy E(.) and weights W(.) of sub-indices, which results have been demonstrated below:

$$E(EDI) = 0.7, E(FSI) = 0.9, W(EDI) = 0.75, W(FSI) = 0.25$$
$$E(SQI) = 0.1, E(HCI) = 0.1, W(SQI) = 0.5, W(HCI) = 0.5$$

$$E(EQI) = 0.25, E(GEI) = 0.61, W(EQI) = 0.66, W(GEI) = 0.34$$

On the base of entropy value and sub-indices' weights of National Sustainable Development Aggregate Index have been defined:

ESI = 0.75 * (0.19, 0.29, 0.4) + 0.25 * (0.19, 0.29, 0.4) = (0.14, 0.22, 0.3) + (0.05, 0.07, 0.1) = (0.19, 0.29, 0.4) = low
SSI = 0.5 * (0.39, 0.49, 0.6) + 0.5 * (0.19, 0.29, 0.4) = (0.2, 0.25, 0.3) + (0.09, 0.15, 0.2) = (0.29, 0.45, 0.5) = moderate
ENI = 0.66 * (0.19, 0.29, 0.4) + 0.34 * (0.19, 0.29, 0.4) = (0.13, 0.19, 0.26) + (0.06, 0.09, 0.14) = (0.19, 0.28, 0.24) = low

Meaning of economical sustainability (ESI), social sustainability (SSI), environmental sustainability (ENI) gives us possibility to define National Sustainable Development Aggregate Index (NSDAI):

$$AINSD = \frac{LO + MD + LO}{3} = \frac{(0.19, 0.29, 0.4) + (0.29, 0.45, 0.5) + (0.19, 0.28, 0.24)}{3}$$
$$= \frac{0.67, 1.02, 1.3}{3} = (0.22, 0.34, 0.43) = L$$

6 Estimation Degree of Influence of Factors to Indicators of Socioeconomic System

In order to define the reasons of low level National Sustainable Development, relations between indicators of sub-indices need to be analyzed.

Results of the fuzzy analysis of the socioeconomic system quality show, that following indicators have a strong impact on development of this system:

A1—High level corruption; A2—Low level taxes revenue; A3—Low level natural environment; A4—Low productivity in economy; A5—Low index democracy.

Transparency International defines corruption as the misuse of entrusted power for private gain. Corruption takes many forms, from bribes to extortion to patronage. Corruption flourishes where there are few institutional checks on power, where decision making is obscure, where civil society is weak, where poverty is wide spread. There is a correlation between the attributes of democracy and the attributes of effective socioeconomic development. A large number of empirical studies have been undertaken in the past 10 years to investigate this question [29].

Indicators above have a lowering effect on the following indicators:

B1—Quality of Education; B2—Quality of Science; B3—Quality of Health care; B4—Quality of Housing; B5—Average Wages; B6—Level of Poverty; B7—Level of Unemployment; B8—Social Mobility; B9—Level of Employment of the Youth.

For estimating degree of influence of Ai factors to indicators of socioeconomic system—Bj, were used the method of the forgotten effects offered by professors A. Kaufmann and J. Gil Aluja [30]. The idea of this method consists in the following:

- The scale of influence is defined in the range of [0,1];
- The matrix of [M] direct impact of the elements Ai on Bj is defined;
- The matrix [And], describing extent of influence between the elements Ai is defined;
- The matrix [B] describing extent of influence between the elements Bj is defined;
- By max-min composition, is defined $[M^*] = [A] \circ [M] \circ [B]$;
- At the end using composition of three matrixes $[M^*]$, results of the forgotten effects are defined by $[O] = [M^*](-)[M]$.

We will use following fuzzy linguistic indicators, in order to define degree of effect.

Very low—VL (0 0.2); low—L(0.18 0.4); middle—M (0.38 0.6); high H (0.58 0.8); very high VH(0.78 1).

In order to define [O]—matrix of forgotten effects, we have used vertex method that calculates distance between two triangular fuzzy numbers, and consists of the following:

$$d(\widetilde{a}, \widetilde{b}) = \sqrt{\frac{1}{3} \left\{ (a_1 - b_1)^2 + (a_2 - b_2)^2 + (a_3 - b_3)^2 \right\}}$$

where $\widetilde{a} = (a_1, a_2, a_3)$ and $\widetilde{b} = (b_1, b_2, b_3)$ are two triangular fuzzy numbers.

Elements [A], [M], [B] matrixes, were defined by expert opinion and results of some investigation, which are given in Tables 9, 10 and 11 (Table 12).

Table 9 is constructed by composition of matrices A, M, B. Table 13 demonstrates forgotten effects among investigated indicators.

Table 9 Matrix of Effects

B	1	2	3	4	5	6	7	8	9
1	VH	VH	VH	VH	M	M	H	H	H
2	VH	VH	VH	L	M	VL	L	H	H
3	H	H	H	VL	M	VL	H	L	VL
4	M	M	M	VH	VL	VL	VL	L	VL
5	VH	VH	VH	VH	VH	VH	H	VH	H
6	VH	VH	VH	VH	VH	VH	VH	VH	VH
7	VL	VL	VL	VL	VL	VH	VH	H	VH
8	M	M	M	H	M	H	L	VH	M
9	L	L	L	VL	L	H	VH	H	VH

Table 10 Matrix of causes

A	1	2	3	4	5
1	VH	VH	VH	VH	M
2	L	VH	VH	H	VL
3	L	M	VH	H	VL
4	L	VH	VH	VH	VL
5	M	M	M	M	VH

Table 11 Matrix of direct incidents

M	1	2	3	4	5	6	7	8	9
1	VH	VH	VH	VH	H	H	H	VH	H
2	VH	VH	VH	H	VH	VH	VH	VH	H
3	VL	VL	VL	VL	VL	M	VL	VL	VL
4	VH	VH	VH	VH	VH	VH	VH	VH	VH
5	M	M	M	M	M	M	M	M	M

Table 12 Matrix of cumulated effect

M*	1	2	3	4	5	6	7	8	9
1	VH	VH	VH	VH	VH	VH	VH	VH	VH
2	VH	VH	VH	VH	VH	VH	VH	VH	VH
3	H	H	H	H	H	H	H	H	H
4	VH	VH	VH	VH	VH	VH	VH	VH	VH
5	M	M	M	M	M	M	M	M	M

Table 13 Matrix of forgotten effects

O	1	2	3	4	5	6	7	8	9
1	0	0	0	0	0.2	0.2	0.2	0	0.2
2	0	0	0	0.2	0	0	0	0	0.2
3	0.69	0.69	0.69	0.69	0.69	0.2	0.69	0.69	0.69
4	0	0	0	0	0	0	0	0	0
5	0	0	0	0	0	0	0	0	0

7 Conclusion

This investigation has methodological character. Proposed sub-indices, developed on the base of problems, solutions, do not fully cover the national indicators of sustainable development. In the future, it is necessary to include in the block of Economic Stability index of macroeconomic stability, in the block of Social Stability—the indices of social mobility, the level of populations satisfaction with the objects of life, in the block of Environmental Sustainability Index—an index of environmental security.

References

1. Mitlin, D.: Sustainable development: a guide to the literature. Environ. Urbanization **4**(1), 111–124 (1992)
2. ICSU, ISSC: Review of the Sustainable Development Goals: The Science Perspective, 88 pp. International Council for Science (ICSU), Paris (2015)
3. Good Practices in the National Sustainable Development Strategies of OECD Countries: OECD 2006, 35 pp
4. Azerbaijan 2020: Look into the Future Concept of Development, 41 pp. (2011)
5. Baumgrtner, S., Quaas, M.: What is sustainability economics? Ecol. Econ. **69**, 445–450 (2010)
6. Tress, R.C.: Unemployment and the diversification of industry. Manchester Sch. **9**, 140–152 (1938)
7. Jacquemin, A., Berry, C.: Entropy measure of diversification and corporate growth. J. Ind. Econ. **27**, 359–369 (1979)
8. Hirschman, A.O.: The Paternity of an index. Am. Econ. Rev. 761–762 (1964)
9. Kosko, B.: Fuzziness vs probability. Int. J. Gen. Syst. **17**(2–3), 211–240 (1990). United Kingdom
10. Borio, C., Drehmann, M.: Towards an operational framework for financial stability: fuzzy measurement and its consequences, BIS Working Papers No 284, 50 pp., June 2009
11. Atanassov, K.: Intuitionistic fuzzy sets. Fuzzy Sets Syst. **20**(1), 87–96 (1986)
12. Szmidt, E., Kacprzyk, J.: Entropy for intuitionistic fuzzy sets. Fuzzy Sets Syst. **118**, 467–477 (2001)
13. Beck, W., van der Maesen, L., Walker, A.: Social quality: from issue to concept. In: Beck, W., van der Maesen, L., Walker, A. (eds.) The Social Quality of Europe. Kluwar Law International, The Hague, Netherlands, pp. 263–296 (1997)
14. Beck, W., van der Maesen, L., Thomese, F., Walker, A. (eds.): Social Quality: A Vision for Europe. Kluwer Law, The Hague, Netherlands, pp. 307–360 (2001)
15. Van der Maeson, I., Walker, A., Keizer, M.: European Network Indicators of Social Quality ENIQ, Social Quality the final report. European Foundation on Social Quality, 105 pp., April 2005
16. Zadeh, L.A.: Outline of a new approach to the analysis of complex systems and decision processes. IEEE Trans. Syst. Man Cybern. SMC **3**, 28–44 (1973)
17. Poulsen, J.R.: Fuzzy time series Forecasting, AAUE, 67 pp., Nov 2009
18. Schults, T.W.: Investment in human capital. Am. Econ. Rev. **LI**(1), 1–17 (1961)
19. Becker, G.: Investments in human capital: a theoretical analysis. J. Polit. Econ. **70**, 9–44 (1962)
20. The Well-being of Nations: The role of human and social capital, OECD, Paris, 118 pp. (2001)
21. Kwon, D.-B.: Human capital and its measurement. In: The 3rd OECD World Forum on Statistics, Knowledge and Policy Charting Progress, Building Visions, Improving Life, Busan, Korea, 15 pp., 27–30 Oct 2009

22. World development indicators: World Bank, 430 pp. (2012)
23. Knowledge for development: World Bank. http://info.worldbank.org (2012)
24. The Conference board Total Economy Database TM: http://www.conference-board.org, Jan 2013
25. Morelli, J.: Environmental sustainability: a definition for environmental professionals. J. Environ. Sustain. 1(1), (Article 2), 9 (2011)
26. Ross, T.: Fuzzy Logic with Engineering Application, 2nd edn. Wiley, England (2004)
27. United Nations Environment Programme (UNEP): Towards a green economy. In: Pathways to Sustainable Development and Poverty Eradication (2011)
28. United Nations Department of Economic and Social Affairs (UN-DESA): A guidebook to the green economy, Issue 1 (2012)
29. Transparency International the global coalition against corruption: Annual Report (2009), 68 pp
30. Kaufmann, A., Gil-Aluja, J.: Modelospara la investigacion de effectosplvidados Ed. Milladoiro, Espana (1989)

A Dynamic Game for a Sustainable Supply Chain Management

Massimiliano Ferrara and Bruno Antonio Pansera

Abstract In this paper, we establish a dynamic game to allocate CSR (Corporate Social Responsibility) to the members of a supply chain. We propose model of three-tier supply chain in decentralized state that is including supplier, manufacturer and retailer. For analyzing supply chain performance in decentralized state and the relationships between the members of supply chain, we use Stackelberg game and we consider in this paper a hierarchical equilibrium solution for a two-level game. Specially, we formulate a model that crosses through multiperiods by a dynamic discreet Stackelberg game. We obtain an equilibrium point at where both the profits of members and the level of CSR taken by supply chains are maximized.

Keywords Dynamic game · Supply chain · CSR · Stackelberg game

1 Introduction

In recent years, companies and firms show an ongoing interest for accepting CSR. This is chiefly because, increasing consumer awareness of several CSR issues, e.g. the environment, human rights and safety. In addition, they are forced to accept CSR by government policies and regulations. According to previous studies, the

M. Ferrara (✉) · B. A. Pansera
Department of Law and Economics, University "Mediterranea" of Reggio Calabria,
Reggio Calabria, Italy
e-mail: massimiliano.ferrara@unirc.it

B. A. Pansera
e-mail: bruno.pansera@unirc.it

M. Ferrara · B. A. Pansera
Decisions Lab, University Mediterranea of Reggio Calabria, Reggio Calabria, Italy

M. Ferrara
ICRIOS - The Invernizzi Centre for Research in Innovation,
Organization, Strategy and Entrepreneurship Department of Management and Technology,
Bocconi University, Milan, Italy

© Springer International Publishing AG 2018   331
C. Berger-Vachon et al. (eds.), *Complex Systems: Solutions and Challenges
in Economics, Management and Engineering*, Studies in Systems,
Decision and Control 125, https://doi.org/10.1007/978-3-319-69989-9_19

long term investment on CSR is beneficial to supply chain. Sustainable supply chain needs a consideration of the social aspects of business [13]. Moreover, the CSR is an effective tool for supply chain management for coordination, purchasing, manufacturing, distribution, and marketing functions [8]. The members of supply chain make their decisions based on maximizing of their individual net benefits. In addition, when they need to take a level of CSR; this situation leads to an equilibrium status. Game theory is one of the most effective tools to deal with such a kind of management problems.

A growing number of research papers, use game theoretical applications in supply chain management. Cachon et al. [3] discuss Nash equilibrium in a noncooperative cases in a supply chain where there are one supplier and multiple retailers. Hennet et al. [7] presented a paper to evaluate the efficiency of different types of contracts between the industrial partners of a supply chain. They applied game theory method for decisional purposes. Tian et al. [14] presented a system dynamics model based on evolutionary game theory for green supply chain management as well.

In this paper, we formulate a model for decentralized supply chain network in CSR conditions in a long term with one leader and two followers. The Stackelberg game model is recommended and applied here to find an equilibrium point in which we maximize the profit of members of supply chain and the level of CSR taken by the supply chain. In this research, the supplier as a leader, can know the optimal reaction of his followers, and regards such processes to maximize his own profit. The manufacturer and the retailer as followers, try to maximize their profits by considering all conditions. We propose a Hamiltonian matrix to solve the optimal control problem to obtain the equilibrium in this game. The paper is organized as follows: Sect. 2 is devoted to problem description and assumptions. Objective functions, constraints and solution of the game are illustrated in Sect. 3. A numerical example is showed in Sect. 4. A conclusion is provided in Sect. 5.

2 Problem Description and Assumptions

We consider a three stage Stackelberg differential game involving three players playing the game over a fixed finite horizon model. This model is a three-tier, decentralized vertical control supply chain network (Fig. 1).

All retailers and suppliers at the same level make same decision therefor, we assume that only one supplier, one manufacturer and one retailer. The simplified model is shown in the following (Fig. 2).

We consider a dynamic game that goes through multi-periods as a repeated game with complete information. A long term Stackelberg game is played between the members of the decentralized supply chain through two levels in which all members take CSR into consideration. We use a discrete time dynamic model involves a sequence of decisions that each called period, and the game goes through finite periods. There are two different types of information structures in a differential

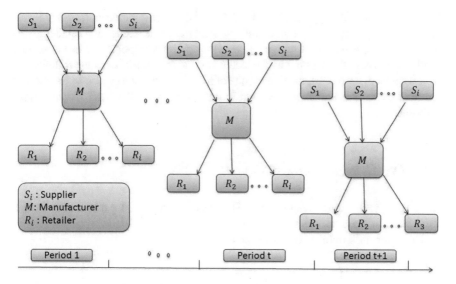

Fig. 1 Three-tier supply chain network

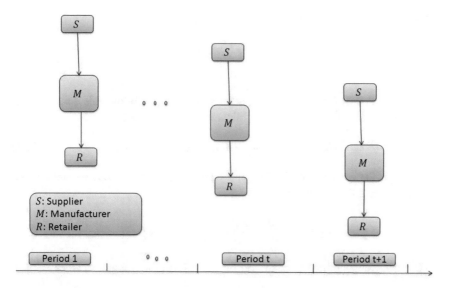

Fig. 2 The simplified model of three-tier supply chain network

game, open-loop and feedback information structures. In the open-loop strategy, the players choose their decisions at time t, with information of the state at time zero [5]. In this paper using an open-loop Stackelberg game. We formulate the model by selecting the supplier as the leader and both of the manufacturer and retailer as the follower in Stackelberg game. This model can be solved by considering two levels of Stackelberg game. At first level the manufacturer as the leader and retailer as the follower are considered. In this level, we find equilibrium point and in the second level, we consider the supplier as the leader and the manufacturer as the follower. In fact, we substitute the response functions of followers into the objective function of the leader and we find the final equilibrium point and all of the players make decisions.

2.1   The General Formulation

He et al. [6], illustrate an open-loop Stackelberg differential game model over a fixed finite horizon, as detailed below:

The follower's optimal control problem is:

$$Max_{r(.)}\left\{ J_R(X_0, r(.); w(.)) = \int_0^t e^{-pt}\pi_R(X(t), w(t), r(t))dt \right\},\qquad(2.1)$$

subject to the state equation

$$\dot{X}(t) = F(X(t), w(t), r(t)),$$
$$X(0) = X_0.\qquad(2.2)$$

where the function F represents the rate of sales, ρ is the followers's discount rate, and X_0, is the initial condition. The follower's Hamiltonian is

$$H_R(X, r, \lambda_R, w) = \pi_R(X, w, r) + \lambda_R F(X, w, r),\qquad(2.3)$$

where λ_R, is the vector of the shadow prices associated with the state variable X; and it satisfies the adjoin equation

$$\lambda_R = \rho\lambda_R - \frac{\partial H_R(X, r, \lambda_R, w)}{\partial X}, \quad \lambda_R(T) = 0.\qquad(2.4)$$

The necessary optimality condition for the follower's problem satisfies

$$\frac{\partial H_r}{\partial r} = 0 \Rightarrow \frac{\partial \pi_R(X, w, r)}{\partial r} + \lambda_R \frac{\partial F(X, w, r)}{\partial r} = 0.\qquad(2.5)$$

We derive the follower's best response are $r^*(X, w, \lambda_R)$.

The leaders problem is

$$Max_{w(.)}\left\{J_M(X_0, w(.)) = \int_0^t e^{-\mu t}\pi_M(x, w, r(x, w, \lambda_R))dt\right\},$$
$$\dot{X} = F(X, w, r(x, w, \lambda_R)),$$
$$X_0(0) = X_0.$$
$$\dot{\lambda}_R = \rho\lambda_R - \frac{\partial H_R(x, r(x, w, \lambda_R), \lambda_R, w)}{\partial x}, \quad \lambda_R(T) = 0,$$

(2.6)

where μ is the leader's discount rate and the above differential equations are obtained by substituting the follower's best response $r^*(X, w, \lambda_R)$. in the state equation and the adjoin equation of the follower respectively. We formulate leader's Hamiltonian, as follow:

$$H_M = \pi_M(x, \lambda_R, w, r(X, w, \lambda_R), \lambda_M, \varphi) + \lambda F(X, w, r(X, w, \lambda_R))$$
$$- \mu\frac{\partial H_R(X, r(X, w, \lambda_R), \lambda_R, w)}{\partial X},$$

(2.7)

where λ_M and μ are the shadow associated with X and λ_R respectively, and they satisfy the adjoint equations

$$\dot{\lambda}_M = \mu\lambda_M - \frac{\partial H_M(X, \lambda_R, w, r(X, w, \lambda_R), \lambda_M, w)}{\partial X}$$
$$= \mu\lambda_M - \frac{\partial \pi_M(X, w, r(X, w, \lambda_R))}{\partial X}$$
$$- \lambda_M\frac{\partial F(X, w, r(X, w, \lambda_R))}{\partial X} - \mu\frac{\partial^2 H_R(x, r(X, w, \lambda_R), \lambda_R, w)}{\partial X^2},$$
$$\dot{\varphi} = \mu\varphi - \frac{\partial H_M(X, \lambda_R, w, r(X, w, \lambda_R), \lambda_M, \varphi)}{\partial \lambda_R}$$
$$= \mu\varphi - \lambda_M\frac{\partial F(x, w, r(X, w, \lambda_R))}{\partial \lambda_R} - \mu\frac{\partial^2 H_R(X, r(X, w, \lambda_R), \lambda_R, w)}{\partial \lambda \partial X},$$

(2.8)

where $\lambda_M(T) = 0$ and $\varphi(0) = 0$ are the boundary conditions.

We apply the algorithm of the above general model on our model.

2.2 Notations and Definitions

To facilitate the model, some parameters and decision variables are used.

Table 1 shows notations and definitions that we use in our model. Variables

Table 1 Notations and definitions

t	Period t
T	Planning horizon
q_t	Demand quantity at period t
a	Market potential
b	Price sensitivity
x_t	State variable, degree of taking SR
H^S	Hamiltonian function of the supplier
H^M	Hamiltonian function of the manufacturer
H^R	Hamiltonian function of the retailer
J_t^S	Objective function of the supplier
J_t^M	Objective function of the manufacturer
J_t^R	Objective function of the retailer
$B^M(x_t)$	Social benefit of the supplier
$B^S(x_t)$	Social benefit of the manufacturer
$B^R(x_t)$	Social benefit of the retailer
$T^S(x_t)$	Tax return of the supplier
$T^M(x_t)$	Tax return of the manufacturer
$T^R(x_t)$	Tax return of the retailer
I_t^M	The amount of investment of the manufacturer
I_t^S	The amount of investment of the supplier
I_t^R	The amount of investment of the retailer
d	The percentage of investment of the supplier payoff
$\hat{d}$	The percentage of investment of the manufacturer payoff
w	Price of the supplier's raw material
z	Price of sold product by the retailer
δ	Parameter of the supplier's social benefit
δ	Parameter of the manufacturer's social benefit
$\hat{\delta}$	Parameter of the retailer's social benefit
λ	Quantity discount parameter of the price of raw material
α	Deteriorating rate of the level of current social responsibility
τ	The rate of individual post tax return on investment (ROI)
θ	The rate of supply chain's post tax return on investment (ROI)
β_1	The rate of converting the supplier's capital investment in CSR to the amount of CSR taken by the supply chain
β_2	The rate of converting the manufacturer's capital investment in CSR to the amount of CSR taken by the supply chain
β_3	The rate of converting the retailer's capital investment in CSR to the amount of CSR taken by the supply chain

2.3 Assumptions

This model has a state variable and control variables as any dynamic game. We define the state variable as the level of social responsibility taken by companies, and the control variables are the capital amounts invested in taking social responsibility. Specifically, all of the social responsibilities taken by firm j at period t can be expressed as investment I_t^j. We suppose that x_t evolves according to $x_{t+1} = f(I_t, x_t)$.

We have more specifically assumption:

Function $B_t(x_t) = \delta x_t$, represents social benefit which is proportionate to social responsibility taken by the supply chain system [1].

The following functions $T_t = \tau I_t[1 + \theta(I_t)]$, measures the value of the tax return to the members of supply chain [4]. Both τ and θ are tax return policy parameters. Specifically, τ is the rate of individual post tax return on investment (ROI), and θ is the rate of supply chain's post tax return on investment (ROI).

The market inverse demand is $P^M(q_t) = a - bq_t$ [9].

The accumulation of level of social responsibility taken by the firms is given by $x_{t+1} = ax_t + \beta_1 I_t^S + \beta_2 I_t^M + \beta_3 I_t^R$.

Here, β_1 is the rate of converting the supplier's capital investment in CSR to the amount of CSR taken by the supply chain; β_2 is the rate of converting the manufacturer's capital investment in CSR to the amount of CSR taken by the supply chain and β_3 is the rate of converting the retailer's capital investment in CSR to the amount of CSR taken by the supply chain as well [12].

3 Objective Function and Constraints

The objective functions are made to depend on the control vectors and the static variable. The member of supply chain attempt to optimize their net profits, that include minimize the cost of raw materials and investment in taking social responsibility, and maximize sale revenues and benefits from taking social responsibility as well as tax returns. Thus, the objective function of the supplier is

$$J^S = \sum_{t=1}^{T} P_t^S q_t - cq_t + B_t^S(x_t) + T_t^S(I_t^S, I_t) - I_t^S + dI_t^M$$

$$= \sum_{t=1}^{T} wq_t - cq_t + \delta x_t^2 + \tau I_t^S[1 + \theta(I_t^S + I_t^M + I_t^R)] - I_t^S + dI_t^M,$$

where, P_t^S is the price of the supplier's raw material. Let $P_t^S = w$. $B_t^S(x_t)$ is the social benefit of the supplier, δ is the parameter of the supplier's social benefit and $T_t^S(I_t^S, I_t)$ is the tax return of the supplier.

Similarly, the objective function of the manufacturer is

$$J^M = \sum_{t=1}^{T} P_t^M(q_t)q_t - P_t^S q_t + B^M(x_t) + T^M(I_t^M, I_t) - I_t^M + \hat{d} I_t^R$$

$$= \sum_{t=1}^{T} (a - bq_t)q_t - wq_t + \delta x_t^2 + \tau I_t^M\left(1 + \theta\left(I_t^S + I_t^M + I_t^R\right)\right) - I_t^M + \hat{d} I_t^R,$$

where $P_t^M(q_t)$ is the retail price of the product of the manufacturer. $B_t^M(x_t)$ is the social benefit of the manufacturer, δ is the parameter of the manufacturer's social benefit and $T_t^M(I_t^M, I_t)$ is the tax return of the manufacturer.

The objective function of the retailer is

$$J^R = \sum_{t=1}^{T} P_t^R q_t - P_t^M(q_t)q_t + B^R(x_t) + T^R(I_t^R, I_t) - I_t^R$$

$$= \sum_{t=1}^{T} zq_t - (a - bq_t)q_t + \hat{\delta} x_t^2 + \tau I_t^R\left(1 + \theta\left(I_t^S + I_t^M + I_t^R\right)\right) - I_t^R,$$

where P_t^R is the price of the product the retailer sells to the consumer. Let $P_t^R = Z$. $B_t^R(x_t)$ is the social benefit of the retailer, $\hat{\delta}$ is the parameter of the retailer's social benefit and $T_t^R(I_t^R, I_t)$ is the tax return of the retailer.

3.1 Mathematical Model: Level One

In this level, we establish a Stackelberg game between manufacturer as the leader and retailer as the follower. To calculate the equilibrium in this level, first we calculate the best reaction function of retailer, then we determine the manufacturer's optimal decisions based on the retailer' best reactions.

Since our dynamic differential game is an optimal control theory, we can apply the Hamiltonian function to find the equilibrium of the game [11].

Suppose the time interval is [1, T]. For any fixed I_t^S and I_t^M the retailer solves

$$\arg\max_{I_t^R} \sum_{t=1}^{T} P_t^R q_t - P_t^M(q_t)q_t + B^R(x_t) + T^R(I_t^R, I_t) - I_t^R,$$

subject to $x_{t+1} = \alpha x_t + \beta_1 I_t^S + \beta_2 I_t^M + \beta_3 I_t^R$.

We define the retailer's Hamiltonian for fixed I_t^S and I_t^M as

$$H_t^R = J_t^R + P_{t+1}^R(x_{t+1}).$$

By using the conditions for a maximization of this Hamiltonian, we get after some algebras:

$$I_t^R = \frac{1 - P_{t+1}^R \beta_3 - \tau\theta(I_t^M + I_t^S) - \tau}{2\theta\tau}. \tag{3.1}$$

The equation of I_t^R according to I_t^S and I_t^M, says any announced strategy of I_t^S and I_t^M, there is a unique optimal response of the I_t^R.

$$x_{t+1} = \frac{\partial H_t^R}{\partial P_{t+1}^R} = \alpha x_t + \beta_1 I_t^S + \beta_2 I_t^M + \beta_3 I_t^R, \tag{3.2}$$

by substituting (3.1) in (3.2), we obtain

$$x_{t+1} = (\beta_1 - \beta_3/2)I_t^S + (\beta_2 - \beta_3/2)I_t^M + \alpha x_t + \beta_3 \frac{1 - P_{t+1}^R \beta_3 - \tau}{2\tau\theta}. \tag{3.3}$$

As well as we have

$$P_t^R = \frac{\partial H_t^R}{\partial x_t} = 2\delta x_t + \alpha P_{t+1}^R. \tag{3.4}$$

The above sets of equations define the reaction function of the retailer. For any fixed I_t^S the manufacturer solves

$$\arg\max_{I_t^M} \sum_{t=1}^{T} P_t^M(q_t)q_t - P_t^S q_t + B^M(x_t) + T^M(I_t^M, I_t) - I_t^M + \hat{a}I_t^R,$$

subject to $x_{t+1} = \alpha x_{t+1} + \beta_1 I_t^S + \beta_2 I_t^M + \beta_3 I_t^R$.
Now, we integrate the value of I_t^R in (3.1) into J_t^M, and we obtain

$$J_t^M = (a - bq_t)q_t - wq_t + \delta x_t^2 + \frac{\tau\theta - \hat{d}}{2}I_t^S I_t^M + \frac{\tau - 1 - \beta_3 P_{t+1}^R - \hat{d}}{2}I_t^M - + \frac{\hat{d}}{2\theta}. \tag{3.5}$$

The Hamiltonian function of the manufacturer for fixed I_t^S is

$$H_t^M = J_t^M + P_{t+1}^M(x_{t+1}) + u_t(P_t^R), \tag{3.6}$$

consequently, we can obtain the unique optimal response of the follower from the equations in the following.

$$\frac{\partial H_t^M}{\partial I_t^M} = \tau\left(1 + \theta\left((1-\tau\theta)I_t^S + I_t^M\right)\right) - 1 + \frac{1 - \beta_3 P_{t+1}^R - \tau - \tau\hat{d}}{2} + (\beta_2 - \beta_3/2)P_{t+1}^M,$$

so, we have

$$I_t^M = \frac{-\tau\theta I_t^S + \beta_3 P_{t+1}^R + (1+\hat{d}-\tau)}{2\tau\theta} - \frac{(\beta_2 - \beta_3/2)P_{t+1}^M}{\tau\theta}. \tag{3.7}$$

Other constraints are,

$$P_t^M = \frac{\partial H_t^M}{\partial x_t} = 2\delta x_t + \alpha P_{t+1}^M + 2\widehat{\delta}u_t. \tag{3.8}$$

$$u_{t+1} = \frac{\partial H_t^M}{\partial P_{t+1}^R} = -\beta_3/2I_t^M - \frac{\hat{d}\beta_3}{2\tau\theta} - \frac{P_{t+1}^M\beta_3^2}{2\tau\theta} + \alpha u_t. \tag{3.9}$$

The equation of I_t^M is according to I_t^S, and we obtain the final equilibrium in the next section, level two.

3.2 Mathematical Model: Level Two

In the previous level, the manufacturer's optimal function was calculated by reaction function of retailer. In this level, the reaction functions of two followers (retailer and manufacturer) are placed in objective function of leader (supplier), and we can find the final equilibrium point.

The problem facing the supplier is simply given by

$$\arg\max_{I_t^S} \sum_{t=1}^{T} P_t^S q_t - cq_t + B_t^S(x_t) + T_t^S(I_t^S, I_t) - I_t^S + dI_t^M,$$

subject to $x_{t+1} = \alpha x_t + \beta_1 I_t^S + \beta_2 I_t^M + \beta_3 I_t^R$.

The Hamiltonian function of the supplier is defined by

$$H_t^S = J_t^S + P_{t+1}^S(x_{t+1}) + \mu_t(P_t^M) + u_t(p_t^R). \tag{3.10}$$

Substitute (3.1) and (3.7) into (3.10), we get the value of I_t^S, x_{t+1} and μ_{t+1}

$$\frac{\partial H_t^S}{\partial I_t^S} = 0, \tag{3.11}$$

therefore

$$I_t^S = \frac{(\beta_3/2 + \beta_2 - 2\beta_1)}{\tau\theta} P_{t+1}^S + \frac{(\beta_2 - \beta_3/2)}{\tau\theta} P_{t+1}^M + \frac{\beta_3}{2\tau\theta} P_{t+1}^R + \frac{3 - 3\tau - \hat{d} + 2d}{2\tau\theta}. \tag{3.12}$$

We have

$$x_{t+1} = \frac{\partial H_t^S}{\partial P_{t+1}^S}, \tag{3.13}$$

therefor we obtain

$$x_{t+1} = \alpha x_t + (\beta_1 - \beta_2/2 - \beta_3/4)I_t^S + \frac{(-2\beta_2 + \beta_3)(\beta_2 - \beta_3/2)}{2\tau\theta} P_{t+1}^M$$
$$+ \frac{(2\beta_2\beta_3) - (3\beta_3^2)}{4\tau\theta} P_{t+1}^R + \frac{(2\beta_2 - \beta_3)(1 - \tau + \hat{d}) + 2\beta_3(1 - \tau)}{4\tau\theta}. \tag{3.14}$$

As well as we have

$$\mu_{t+1} = \frac{\partial H_t^S}{\partial P_{t+1}^M} = \alpha\mu_t - \frac{(\beta_2 - \beta_3/2)}{2} I_t^S + \frac{(\beta_2 - \beta_3/2)(\beta_3 - 2\beta_2)}{2\tau\theta} P_{t+1}^S$$
$$- \frac{(\beta_2 - \beta_3/2)d}{\tau\theta}. \tag{3.15}$$

And we obtain

$$P_t^S = \frac{\partial H_t^S}{\partial x_t} = 2\delta x_t + \alpha P_{t+1}^S + 2\delta u_t. \tag{3.16}$$

Since we use the open-loop information structure variables depend on the time variable and the initial state variables. The initial x_1 is given parameter, $u_1 = 0$ and $\mu_1 = 0$. Furthermore, the boundary condition are $P_{t+1}^R = 0, P_{t+1}^M = 0$ and $P_{t+1}^S = 0$. It is well-know that the open-loop Stackelberg solution is time inconsistent, since a re-optimization latter in time, at period k for example, will give again to set $\mu_k = 0$ and $u_k = 0$ although initially calculated, at period 1, we have $\mu_k \neq 0$ and $u_k \neq 0$ [15].

3.3 Augmented Discrete Hamiltonian Matrix

In this section for solving the optimal control problem formulated in (3.1) and (3.2), we chose an algorithms given by Medanic and Radojevic which is an augmented discrete Hamiltonian matrix [10]. First, we assume

$$\begin{bmatrix} \tilde{x}_{t+1} \\ \tilde{P}_t \end{bmatrix} = \begin{bmatrix} A & B \\ C & D \end{bmatrix} \begin{bmatrix} \tilde{x}_t \\ \tilde{P}_{t+1} \end{bmatrix} + \begin{bmatrix} D \\ E \end{bmatrix} = \begin{bmatrix} A\tilde{x}_t + B\tilde{P}_{t+1} + D \\ C\tilde{x}_t + A\tilde{P}_{t+1} + E \end{bmatrix},$$

where $\tilde{x}_{t+1} = \begin{bmatrix} x_{t+1} \\ u_{t+1} \\ \mu_{t+1} \end{bmatrix}$ and $\tilde{P}_{t+1} = \begin{bmatrix} p^S_{t+1} \\ p^M_{t+1} \\ p^R_{t+1} \end{bmatrix}$, A, B, and C are 3×3 matrices, D and E are 3×1 matrices.

Such that

$$\tilde{x}_{t+1} = \begin{bmatrix} x_{t+1} \\ \mu_{t+1} \\ u_{t+1} \end{bmatrix} = \begin{bmatrix} A\tilde{x}_t + B\tilde{P}_{t+1} + D \end{bmatrix}$$

$$= \begin{bmatrix} a_{11} & a_{12} & a_{13} \\ a_{21} & a_{22} & a_{23} \\ a_{31} & a_{32} & a_{33} \end{bmatrix} \begin{bmatrix} x_t \\ \mu_t \\ u_t \end{bmatrix} + \begin{bmatrix} b_{11} & b_{12} & b_{13} \\ b_{21} & b_{22} & b_{23} \\ b_{31} & b_{32} & b_{33} \end{bmatrix} \begin{bmatrix} P^S_{t+1} \\ P^M_{t+1} \\ P^R_{t+1} \end{bmatrix} + \begin{bmatrix} d_1 \\ d_2 \\ d_3 \end{bmatrix} \quad (3.17)$$

$$= \begin{bmatrix} a_{11}x_t + a_{12}\mu_t + a_{13}u_t + b_{11}P^S_{t+1} + b_{12}P^M_{t+1} + b_{13}P^R_{t+1} + d_1 \\ a_{21}x_t + a_{22}\mu_t + a_{23}u_t + b_{21}P^S_{t+1} + b_{22}P^M_{t+1} + b_{23}P^R_{t+1} + d_2 \\ a_{31}x_t + a_{32}\mu_t + a_{33}u_t + b_{31}P^S_{t+1} + b_{32}P^M_{t+1} + b_{33}P^R_{t+1} + d_3 \end{bmatrix}.$$

The boundary conditions are $\tilde{x}_1 = \begin{bmatrix} 1 \\ 0 \\ 0 \end{bmatrix}$ and $\tilde{P}_{t+1} = \begin{bmatrix} 0 \\ 0 \\ 0 \end{bmatrix}$.

$$A = \begin{bmatrix} a_{11} & a_{12} & a_{13} \\ a_{21} & a_{22} & a_{23} \\ a_{31} & a_{32} & a_{33} \end{bmatrix},$$

where

$$a_{11} = \alpha, a_{12} = 0, a_{13} = 0, a_{21} = 0, a_{22} = \alpha, a_{23} = 0, a_{31} = 0, a_{32} = 0, a_{33} = \alpha.$$

$$B = \begin{bmatrix} b_{11} & b_{12} & b_{13} \\ b_{21} & b_{22} & b_{23} \\ b_{31} & b_{32} & b_{33} \end{bmatrix},$$

where

$$b_{11} = \frac{(\beta_1 - \beta_2/2 - \beta_3/4)(-2\beta_1 + \beta_2 + \beta_3/2)}{\tau\theta}.$$

$$b_{12} = \frac{(\beta_1 - \beta_2/2 - \beta_3/4)(\beta_2 + \beta_3/2)}{\tau\theta} + \frac{(\beta_2 - \beta_3/2)(-2\beta_2 + \beta_3)}{2\tau\theta}.$$

$$b_{13} = \frac{(\beta_1 - \beta_2/2 - \beta_3/4)(\beta_3)}{2\tau\theta} + \frac{(2\beta_2\beta_3)(-3\beta_3^2)}{4\tau\theta}.$$

$$b_{21} = \frac{(\beta_2 - \beta_3/2)(2\beta_1 - 3\beta_2 + \beta_3/2)}{2\tau\theta}.$$

$$b_{22} = -\frac{(\beta_2 - \beta_3/2)^2}{2\tau\theta}.$$

$$b_{23} = -\frac{\beta_3(\beta_2 - \beta_3/2)}{4\tau\theta}.$$

$$b_{31} = -\frac{\beta_3(-2\beta_1 + \beta_2 + \beta_3/2)}{4\tau\theta}.$$

$$b_{32} = \frac{-\beta_3^2 + 3/2\beta_3(\beta_2 - \beta_3/2)}{2\tau\theta}.$$

$$b_{33} = \frac{-\beta_3^2}{8\tau\theta}.$$

$$D = \begin{bmatrix} d_1 \\ d_2 \\ d_3 \end{bmatrix}, \text{ where}$$

$$d_1 = \frac{(\beta_1 - \beta_2/2 - \beta_3/4)(3 - 3\tau - \hat{d} + 2d)}{2\tau\theta} + \frac{(2\beta_2 - \beta_3)(1 - \tau + \hat{d}) + 2\beta_3(1 - \tau)}{4\tau\theta}.$$

$$d_2 = \frac{(-\beta_2 - \beta_3/2)(6d - \hat{d} - 3\tau + 3)}{4\tau\theta}.$$

$$d_3 = \frac{-\beta_3(-7\widehat{d} + 2d - \tau + 1)}{8\tau\theta}.$$

Similarly, we can get the value of matrices C and E

$$\tilde{P}_t = \begin{bmatrix} P_t^S \\ P_t^M \\ P_t^R \end{bmatrix} = \begin{bmatrix} c_{11} & c_{12} & c_{13} \\ c_{21} & c_{22} & c_{23} \\ c_{31} & c_{32} & c_{33} \end{bmatrix} \begin{bmatrix} x_t \\ \mu_t \\ u_t \end{bmatrix} + \begin{bmatrix} a_{11} & a_{12} & a_{13} \\ a_{21} & a_{22} & a_{23} \\ a_{31} & a_{32} & a_{33} \end{bmatrix} \begin{bmatrix} P_{t+1}^S \\ P_{t+1}^M \\ P_{t+1}^R \end{bmatrix}$$

$$+ \begin{bmatrix} e_1 \\ e_2 \\ e_3 \end{bmatrix} \begin{bmatrix} c_{11}x_t + c_{12}\mu_t + c_{13}u_t + a_{11}P_{t+1}^S + a_{12}P_{t+1}^M + a_{13}P_{t+1}^R + e_1 \\ c_{21}x_t + c_{22}\mu_t + c_{23}u_t + a_{21}P_{t+1}^S + a_{22}P_{t+1}^M + a_{23}P_{t+1}^R + e_2 \\ c_{31}x_t + c_{32}\mu_t + c_{33}u_t + a_{31}P_{t+1}^S + a_{32}P_{t+1}^M + a_{33}P_{t+1}^R + e_3 \end{bmatrix} \quad (3.18)$$

Therefore $C = \begin{bmatrix} 2\delta & 2\delta & 2\widehat{\delta} \\ 2\delta & 0 & 2\widehat{\delta} \\ 2\widehat{\delta} & 0 & 0 \end{bmatrix}$ and $E = \begin{bmatrix} 0 \\ 0 \\ 0 \end{bmatrix}$

3.4 Resolution

We solve the above problem by sweep method [2], by assuming a linear relation between $\tilde{p}_t$ and $\tilde{x}_t$;

$$\tilde{p}_k = S_k \tilde{x}_k - g_k. \tag{3.19}$$

Thus, we can compute

$$\tilde{x}_{k+1} = (I_{2*2} - BS_{k+1})^{-1}(A\tilde{x}_t - Bg_{k+1} + D). \tag{3.20}$$

Then by substituting (3.19) and (3.20) into the definition of p_{k+1} as given by the augmented Hamiltonian matrix, and equating both sides we finally get the difference equations:

$$S_k = C + AS_{k+1}(I_{2*2} - BS_{k+1})^{-1}. \tag{3.21}$$

$$g_k = AS_{k+1}(I_{2*2} - BS_{k+1})^{-1} + Bg_{k+1} - D + Ag_{k+1} - E. \tag{3.22}$$

The boundary conditions are $\tilde{x}_1 = \begin{bmatrix} x_1 \\ 0 \\ 0 \end{bmatrix}$ and $\widetilde{P}_{t+1} = \begin{bmatrix} 0 \\ 0 \\ 0 \end{bmatrix}$. And then $S_{t+1} = 0_{3*3}$

and $g_{t+1} = 0_{3*1}$.

From the boundary conditions we get $S_t = C$ and $g_t = E$. Once we get the different values of S_k and g_k by the backward loop then, the values of $\tilde{x}_t$ and $\tilde{p}_t$ get by forward loop. From them we get values of $x_t, I_t^S, I_t^M, I_t^R, p_t^S, p_t^M, p_t^R, =$ for all points in time.

4 Numerical Example

In this section provides a numerical example. We run the following numerical simulations with mathematica 8. The results presented here are obtained for the following values of the parameters: $a = 6, w = 3.8, c = 2.4, q_t = 100000,,$ $d = 0.6, c = 0.00001, d = 0.4, \quad \hat{d} = 0.4, z = 6, \theta = 0.01, \quad$ and $\quad \tau = 0.2$. We set $\beta_1 = 0.3, \beta_2 = 0.5$ and $\beta_3 = 0.8$. The $B_t(x_t) = \delta x_t$, the potential benefits firms obtain from taking social responsibility, such as increased demand, better reputation and

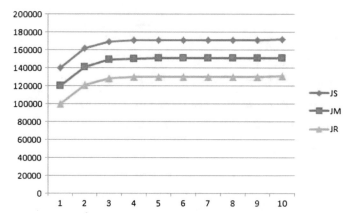

Fig. 3 Profits of supplier, manufacturer and retailer

so on. We set $\delta = 0.2$, $\delta = 0.2$ and $\widehat{\delta} = 0.2$. We assume that the time horizon is $T = 10$. The initial level of social responsibility is supposed to be $x_1 = 1$. We draw the results of the equilibrium from our model, three-stage Stackelberg dynamic game.

The Fig. 3 shows the trend of profits from periods one to ten in a Stackelberg game. JS is supplier's profit, JM is manufacturer's profit and JR is retailer's profit. We compare the profits of the supplier, manufacturer and retailer over a time horizon, first playing the game and then, without playing the game. Figures 4 shows the difference in supplier's profits when playing the game and without playing. JSO is supplier's profit without playing the game; JS is supplier's profit when playing the game. As in the first graph, the second and third one (Figs. 5 and 6) shows the difference in manufacturer's profit and retailer's profits when playing

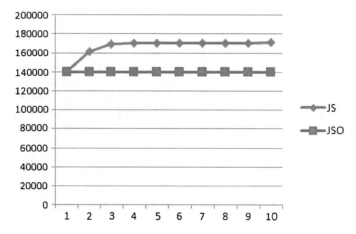

Fig. 4 Comparison of the supplier's profit, playing game one and without playing game

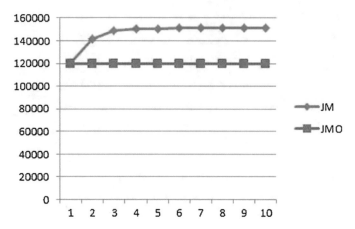

Fig. 5 Comparison of the manufacturer's profit, playing game one and without playing game

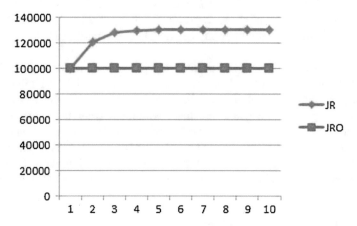

Fig. 6 Comparison of the retailer's profit, playing game one and without playing game

the game and without playing. *JMO* is manufacturer's profit without playing the game; *JM* is manufacturer's profit when playing the game. *JRO* is retailer's profit without playing the game; *JR* is retailer's profit when playing the game. Obviously, all of players gain extra profit from playing the games. Figure 7 shows compare the cumulated profits of the member's of supply chain, playing game one and without playing game.

In sum, the supplier, manufacturer and retailer are motivated to play the game because their benefits are increased and the supplier as the leader in the game earn more benefits than the followers. Of course, this results is obtained with a very specific dynamic game model. Another one may give other results.

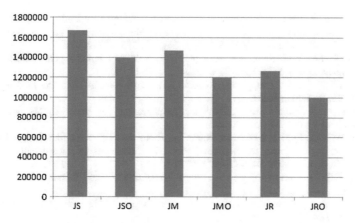

Fig. 7 Comparison of the cumulated profits of the member's of supply chain, playing game one and without playing game

5 Conclusion

In this paper we investigated a decentralized three-tire supply chain consisting of supplier, manufacturer and retailer for the allocating CSR to members of supply chain system in over time.

We considered two-level Stackelberg game consisting of two followers and one leader. The members of a supply chain play games with each other to maximize their own profits; thus, the model used was a long-term co-investment game model. The equilibrium point at which members make their decisions to maximize profits by implementing CSR among members of supply chain in a time horizon was determined. We applied control theory and used an algorithm (augmented discrete Hamiltonian matrix) to obtain an optimal solution for the dynamic game model. We presented a numerical example and we found that, the benefits of the player increased when they played game.

Acknowledgements This paper is devoted to Professor Jaime Gil-Aluja, who has inspired generations of theorists to relate their theory to interesting issues of economics and has convinced generations of experimentals to believe that his theories might provide useful insights.

The authors are grateful to the referees for their constructive input.

References

1. Batabyal, A.: Consisteny and optimality in a dynamic game of pollution control II: Monopoly. Environ. Resource Econ. **8**, 315–330 (1996)
2. Bryson, A.E., Ho, Y.C.: Applied Optimal Control. Wiley, New York (1975)
3. Cachon, G.P., Zipkin, P.H.: Competitive and cooperative inventory policies in a two-stage supply chain. Manage. Sci. **45**, 936–953 (1999)

4. Feibel, B.J.: Investment Performance Measurement. Wiley, New York (2003)
5. He, X., Gutierrez, G., Sethi, S.P.: A survey of Stackelberg differential game models in supply and marketing channels. J. Syst. Sci. Syst. Eng. **16**, 385–413 (2007)
6. He, X., Gutierrez, G., Sethi, S.P.: A review of Stackelberg differential game models in supply chain management. Serv. Syst. Serv. Manage. **1**, 9–11 (2007)
7. Hennet, J.C., Arda, Y.: Supply chain coordination: a game-theory approach. Eng. Appl. Artif. Intell. **21**, 399–405 (2008)
8. Hervani, A., Helms, M.: Performance measurement for green supply management. Benchmarking **12**, 330–353 (2005)
9. Mankiw, N.G.: Principles of Microeconomics. Thomson/South-Western, Mason, Ohio (2004)
10. Medanic, J., Radojevic, D.: Multilevel Stackelberg strategies in linear-quadratic systems. J. Optim. Theory Appl. **24**, 485–497 (1978)
11. Sethi, S.P.: Dimensions of corporate social responsibility. Calif. Manage. Rev. **17**, 58–64 (1975)
12. Shi, H.: A game theoretic approach in green supply chain management. M.S. thesis, University of Windsor (2011)
13. Svensson, G.: Aspects of sustainable supply chain management: conceptual framework and empirical example. Supply Chain Manage. **12**, 262–266 (2007)
14. Tian, Y., Govindan, K., Zhu, Q.: A system dynamics model based on evolutionary game theory for green supply chain management diffusion among Chinese manufacturers. J. Clean. Prod. **80**, 96–105 (2014)
15. Vallee, T.: Comparison of different Stackelberg solutions in a deterministic dynamic pollution control problem. Discussion Paper, LEN-C3E, University of Nantes, France (1998)

Pichat's Algorithm for the Sustainable Regional Analysis Management: Case Study of Mexico

Ingrid Nineth Pinto López, Anna María Gil Lafuente
and Guillermo Sánchez Flores

Abstract The group of regions with common similarities allow the improvement of its competences management, support its integration processes, as well as strengthen the social and economic development of the regions that form a group. The analysis of regions with multidimensional characteristics, make necessary the application of concepts, models and algorithms that allow the analysis of ambiguous variables. The presented model seeks to contribute to an informed decision making, closer to reality while identifying those regions that because of its characteristics, share common features to relatively higher levels. The goal of this work is to design a process of grouping regions from the analysis of sustainable indicators, this will allow us maximize the resources in amplitude and quantity with the purpose to generate synergies in order to take the maximum advantage of the capacities of development and well-being. To reach such a goal, a methodological approximation based on Pichat's algorithm is done using sustainable development indicators of each region as a reference and suggesting the degree of similarity among the regions so the determination of similar groups is possible.

Keywords Grouping · Diffused logic · Social well-being · Sustainable growth
Pichat's algorithm

I. N. Pinto López (✉)
Instituto de Estudios Universitarios, Puebla, Mexico
e-mail: ingrid.pinto@ieu.edu.mx

A. M. Gil Lafuente
Facultad de Economía y Empresa, Universidad de Barcelona, Barcelona, Spain
e-mail: amgil@ub.edu

G. Sánchez Flores
Tecnologías de Información y Comunicación, Universidad Tecnológica de Puebla, Puebla,
Mexico
e-mail: guillermo.sanchez@utpuebla.edu.mx

© Springer International Publishing AG 2018
C. Berger-Vachon et al. (eds.), *Complex Systems: Solutions and Challenges
in Economics, Management and Engineering*, Studies in Systems,
Decision and Control 125, https://doi.org/10.1007/978-3-319-69989-9_20

1 Introduction

The moment we are currently living, challenges us to keep the model of life we have gotten, and at the same time, provide security to future generations. More than a challenge, we have to see it as an opportunity to create a new sustainable model of development that sets the person in the center of the public policies [14].

There exist theories that show up the usefulness of taking into account the subjectivity of people and the similar groups, in decision making, the incorporation of them to the development founds on a need of social sustainability.

The analysis of regions through indicators with multidimensional characteristics, make necessary the application of concepts, models and algorithms that allow the analysis of undetermined variables. It is very usual for the decision makers to be trapped in short term goals, with new management techniques, it is possible to make short and long term goals compatible and move towards a healthy economy with steady levels of social well-being and sustainable development.

The presented model seeks to design a grouping process of regions from the analysis of sustainability indicators considering at first only the Economic dimension of sustainable development proposed in the inform of Bruntland. Because of the multidimensional nature of information, a methodological approximation based on Pichat's algorithm is suggested, this will allow to maximize the resources in amplitude and quantity with the purpose to generate synergies in order to take the maximum advantage of the capacities of development and well-being.

2 Grouping of Regions

Since always in human history, alliances or groups with a similar common goal, have been present. Among men and women, among groups or warriors, among kingdoms or imperial potencies, marriage, politic, military, religious or economic alliances have accompanied the man. They have been present in many advances and in many conflicts, and now they are more prevailing than ever before. The powerful alliances generated around warlike conflicts or the numerous strategic alliances among enterprises of all latitudes are made each day [20].

Next to those, the most renowned and with attractive effects, others more silent, but perhaps with deep impact in local and regional communities, are produced. The groups have a particularity that make them political and strategic interesting. They express a change in the approach of development problems, it is possible to state that the groups 'objective has to do with four kinds of purposes: to improve the quality of the citizens, to strengthen the social organization, to favor the life conditions, and to support the construction of the democracy (Ídem).

3 Sustainable Development

For (Pinto and Gil 2013), the sustainable development is in sight as one of the most important challenges of this century, in this regard, and according to the United nations (UN). In September, 2015, the countries had the opportunity to adopt a new set of global objectives to eradicate poverty, protect the planet, and ensure the prosperity for all as part of a new agenda of sustainable development.

Each objective has specific goals that must be achieved in the next 15 years. In order to achieve these goals, the whole world has to do its part: the government, the private sector, and the civil society, the strategies must be implemented by each country to an international, national, regional, and local level (UN 2016).

The sustainable development objectives that the United Nations present in the inform "The path towards dignity in 2030" are (ídem):

1. To eradicate the poverty in all possible ways in the whole world.
2. To end hunger, get alimentary security and a better nutrition, and promote the sustainable agriculture.
3. To guarantee a healthy life and to promote the well-being for all people of all ages.
4. To guarantee an inclusive and equitable education quality, and to promote permanent learning opportunities for everybody.
5. To reach equality among genders and to empower all women and girls.
6. To guarantee the availability and the sustainable management of water and sanitation for everybody.
7. To ensure the access to affordable, reliable, sustained and modern energies for all people.
8. To encourage the sustained economic growth, inclusive and sustainable, full and productive employment, and respectable job for everybody.
9. To develop resilient infrastructures, to promote the inclusive and sustained industrialization and to encourage innovation.
10. To lower inequalities among countries and inside them.
11. To have human settlements in the cities, inclusive, safe, resilient, and sustained.
12. To guarantee the sustainability of the consumption and production policies.
13. To take urgent actions in order to fight the climatic change and its effects.
14. To keep and give a sustained use to the oceans, seas, and marine resources in order to achieved a sustainable development.
15. To protect, restore and promote the sustainable use of the terrestrial ecosystems, to manage forests in a sustainable way, to fight decertification off, stop and revert the earth's degradation, and to brake the loss of the biological diversity.
16. To promote inclusive and pacific societies for the sustainable development, to facilitate the access to justice for everyone and found effective, responsible, and inclusive institutions at all levels.
17. To strengthen the execution means and to reignite the world alliance for the sustainable development.

4 Pichat's Algorithm

Along time, various diffused models have been developed, these have had application in very wide and diverse areas. These techniques are divided into two groups: numerical diffused models and non-numerical diffused models.

Among the numerical models we can find: The distance of Hamming, developed by Richard Wesley Hamming in 1950; The theory of expertons, introduced by Arnold Kaufmann in 1987; that same year, the model of Subjective preferences was proposed by Kaufmann and Gil Aluja, the OWA operators, model presented by R. Yager in 1988, the Pichat's algorithm proposed in 1970 by Pichat.

The Pichat's algorithm has been widely used, among the authors that recently refer to it we can find [1, 2, 4, 6–8].

The purpose of the path drew by Pichat is the obtaining of sub-matrixes or transitive graphs that end up in the obtaining the maximum similar sub relationships.

Similarities are defined as those homogeneous groups to determined levels, orderly structured, that link elements of two sets of distinct nature together, related by the own essence of the phenomena that they represent [13, 18].

There can be observed the existence of three shaper aspects of the affinity concept. The first makes reference to the fact that the homogeneity of each group is bounded to the chosen level. According to the experience of each characteristic (elements of one of the sets) there will be assigned a higher or lower level that defines the threshold from which there exist homogeneity. The second expresses the need that the elements of each set are bounded together by certain nature rules in some cases and by the human will in some others. The third demands the construction of a constitutive structure of a certain order which is likely to allow the subsequent decision (Ídem).

The purpose of the grouping is on the one hand, the type and the strength of the relationship among the elements of one and other set, on the other hand, from certain properties, they will unequivocally determine all possible groupings. Once the goal to achieve is specified, it is imperative to set a process that is susceptible of achieving the required results. For this, certain elements derived from the combinatorial analysis and reticular studies are used (Ídem).

This models typology has been used in diverse areas by authors such as: [5, 17]

4.1 Pichat's Algorithm

The steps to follow in order to obtain the maximum similar sub relationships are:

(1) The departure point is given by the existence or obtaining of a Boolean relation of similarity (symmetric and reflexive)

(2) Given the existence of symmetry, it is only considered the matrix part set above and from the main diagonal (this included)

(3) Zeros (empty) of each row operating in them are successively considered, one after the other as follows:

  (a) The elements of those columns in which there are zeros (empty) are multiplied.

  (b) The Boolean addition of the element of the corresponding row with the subsequent product is done

(4) The additions found for each row are collected through the Boolean product, in minimum terms, according the next rules:

  (a) The rows without zeros (there is no addition for them) are excluded from the process.

  (b) When the product of additions results in the repetition of an addend, only one of them is considered.
  So: $a + a = a$.

  (c) If in one of the resulting addends appear the same elements than in other, or the same plus one or some more, the one that has more number of elements is removed.
  So: $a + a * b * c = a$.

(5) In this way, there is found an addition of products of element. It is obtained a complement related to the referential for each of the addends. Each one of these complementary terms gives a maximum similar sub relationship.

5 Study Case

The study case presented refers to the Mexican Republic. The indicators taken for the analysis are obtained from the Computer and Indicators framework of the commission on sustainable development disclosed in the document "Indicators of Sustainable Development Guidelines and Methodolofies" published by the United Nations and shown in Table 1.

5.1 Study Data

The data used in this work were gotten from documents released by the Instituto Nacional de Estadística y Geografía (INEGI), which is an autonomous body of the

Table 1 Indicators of sustainable development CSD (2007)

Topic	Sub-topic	Indicator
Poverty	Income	Percentage of population living under the poverty line
	Inequality	Relationship of the national income, from the highest to the lowest quintile
	Sanitation	Proportion of the population with Access to improved sanitation services
	Drinking water	Population with Access to drinking water [14]
	Access to energy	Percentage of homes without electricity or energy services
	Life conditions	Proportion of the population that lives in slums
Government	Corruption	Percentage of the population that has paid a bribe
	Crime	Number of intentional homicides for every 1000,000 habitants
Health	Mortality	Mortality rate under 5 years
		Life expectancy at birth
	Health provision	Percentage of the population with access to primary health
		Immunization against child infectious diseases
	Nutritional status	Children nutritional status
	Health and risk status	Morbidity rate of the main diseases such as HIV AIDS, malaria, tuberculosis
Education	Educational level	Gross income rate to the last grade of primary education
		Net enrollment rate in primary education
		Scholar achievement in secondary level for adults
	Literacy	Literacy rate in adults
Demography	Population	Population growth rate
		Dependence relationship
Natural disasters	Natural disasters vulnerability	Percentage of population living in disaster-prone areas
Atmosphere	Climatic change	Carbon dioxide emissions
	Thinning of the ozone layer	Consumption of substances that deplete the ozone layer
	Air quality	Atmospherical pollution concentration in urban areas
Earth	Agriculture	Permanent growing are and arable
	Forests	Proportion of earth covered by forests
Oceans, seas and coasts	Coastal area	Percentage of the whole population living in coastal areas
	Fishing	Proportion of fish stuck that are inside the safe biological limits

(continued)

Table 1 (continued)

Topic	Sub-topic	Indicator
	Marine environment	Proportion of the protected maritime area
Fresh water	Water quantity	Total proportion of the water resources that are used
		Intensity of the water use for the economic activity
	Water quality	Presence of fecal waste in the water
Biodiversity	Ecosystem	Proportion of the terrestrial area protected, total of ecologic regions
	Species	Change in the threat status of species
Economic development	Macroeconomic results	Gross Domestic Product (GDP) per capita
		Investment share in GDP
	Sustainable public finance	Debt ratio—GDP
	Employment	Employment ratio—population
		Unit labor costs and labor productivity
		Women participation in the wage employment, without taking into account the agriculture sector
	Information and communication technologies	Internet users for every 100 habitants
	Research and development	Gross R&D indicator as a proportion of GDP
	Tourism	Tourism contribution to the GDP
Global economic development	Commerce	Current deficit as a percentage of GDP
	External financing	Official development assistance (ODA) given or gotten as a percentage of GNI
Consumption and production patterns	Materials consumption	Economy materials consumption
	Energy use	Total annual energy consumption and economic activity
		Total intensity of energy use and for economic activity
	Waste generation and management	Generation of dangerous waste
		Waste treatment and removal
	Transport	Modal distribution of passenger transport

Source Self-elaboration according to indicators of Sustainable Development: Guidelines and Methodologies

Mexican government, devoted to the management of the Statistic and geography National system of the country.

As a first approximation, the study is performed taking into account a dimension of the sustainable development, the economic dimension. They are considered as

Mexican Republic regions	
P1	Aguascalientes
P2	Baja California
P3	Baja California Sur
P4	Campeche
P5	Coahuila de Zaragoza
P6	Colima
P7	Chiapas
P8	Chihuahua
P9	Distrito Federal
P10	Durango
P11	Guanajuato
P12	Guerrero
P13	Hidalgo
P14	Jalisco
P15	México
P16	Michoacán de Ocampo
P17	Morelos
P18	Nayarit
P19	Nuevo León
P20	Oaxaca
P21	Puebla
P22	Querétaro
P23	Quintana Roo
P24	San Luis Potosí
P25	Sinaloa
P26	Sonora
P27	Tabasco
P28	Tamaulipas
P29	Tlaxcala
P30	Veracruz de Ignacio de la llave
P31	Yucatán
P32	Zacatecas

Indicadores del PIB en valores básicos de cada una de las Unidades Económicas	
C1	Agriculture, cattle raising, forestry, fishing and hunting
C2	Mining
C3	Electricity, water, gas supply by products to the final consumer
C4	Building
C5	Alimentary industry of drinks and tobacco
C6	Textiles, clothing and leather products
C7	Wood industry
C8	Paper industry, printing and related industries
C9	Petroleum and carbon products, plastic and rubber chemical industry
C10	Manufacturing of non-metallic mineral-based products
C11	Metallic industries
C12	Machinery and equipment
C13	Furniture manufacturing and related products
C14	Other manufacturing industries
C15	Commerce
C16	Transport, mailing and stocking
C17	Media information
C18	Insurance and financial services
C19	Real estate services and rental of movable and intangible property
C20	Professional services, scientific and technical
C21	Supporting services to the business, waste management and remediation services
C22	Educational services
C23	Social care and health services
C24	Services of cultural and sport recreations and other recreational services.
C25	Services of temporary accommodation, food and beverages preparation
C26	Other services except government activities
C27	Government activities

Fig. 1 Regions and indicators. *Source* Self-elaboration according to information taken from INEGI [INEGI 2011]

the set of indicators and sub-indicators that show up information related to the contribution of each economic unit of Gross Domestic Product corresponding to the year 2009 (INEGI 2011).

The Fig. 1 shows the regions that are going to be grouped and the set of indicators according to which the grouping is going to be performed.

Figure 2 Study data. The Fig. 2 gathers the information of the Mexican Republic States for each of the GDP indicators in basic values for every of the Economic Units (To see the complete data check [17]).

	P1	P2	P3	P4	P5	P6	P7	P8	P9	P10	P11	P12	P13
C1	5799428	10648689	2835334	3380938	11122985	3748123	18729543	22377383	1202972	15856081	18222202	9536580	8493014
C2	1345169	646227	2156210	4.84E+08	13007630	322269	14997468	5724654	83411	8377395	1468929	2607564	1725752
C3	553773	6347859	1328562	593027	4778950	1985122	7126691	5232325	9197319	2003940	3884719	5509416	3162940
C4	11613035	32755458	13676846	28713828	25931358	5412393	13667697	25960224	82701404	9840489	29352530	10007968	14022261
C5	10012608	13526856	1774730	1820514	12843957	2750103	10523182	12121175	54340433	18669632	37286751	6536730	15519907
C6	2851031	1414775	35244	665306	6951131	52883	264786	1816404	15348906	1000159	17044283	274010	4029698
C7	72515	510190	3159	112672	95313	85305	191017	7550719	617603	2420628	97805	455647	117243
C8	241038	3463555	21002	19872	2327185	27496	222104	1724181	11436020	1014432	2343631	57239	191554
C9	1579432	3006297	32703	164055	7094779	115752	5373545	2625128	76724665	536196	18824040	112215	15489359
C10	2124482	2694487	212527	118323	7928499	2206358	207344	5220449	2911070	629997	5061085	846383	11064044
C11	850216	6053364	67656	85677	28839345	157474	116305	3071641	28990837	1599194	6483583	2798955	991974
C12	18547046	26700258	5745	21	40358649	138048	388094	44013297	18593648	2603940	37253332	52324	3405251
C13	231636	1573704	18166	20907	553982	61161	93248	1034756	3725535	574750	342532	71186	136917
C14	119426	6369025	40065	14274	287804	31518	133139	4729307	614133	1039069	785125	1053453	862286
C15	17663450	52300310	12093494	10698903	43150267	9200748	32262590	52695926	344403920	19273994	65597665	26333235	19257640
C16	6025636	21398109	5898784	8152787	27758636	8343050	11941817	15644898	173637132	9689194	32599618	18248008	14069291
C17	3002873	11237023	1836193	2423942	5704261	1394793	5687685	10362704	140376909	2850636	8958651	4721121	3414672
C18	3529771	5816599	2173927	1327923	6106425	933734	1909368	6073376	231986119	3228867	8959224	1799285	2387826
C19	12206808	38406409	6510619	9566449	27991806	6333086	35141145	42667880	181514287	15822244	47539649	29755290	24186050
C20	2556131	9329216	1173988	6472910	10364302	787079	2849666	18972097	163983790	1579347	18970269	1203189	1386557
C21	2128389	4612310	1685411	6349231	7366247	936854	738478	4738370	105617627	3053653	6147119	1955554	1675391
C22	6814038	16636632	3155632	4244052	14354221	3499801	18637584	18435886	173637192	7965160	19442954	12821203	12258434
C23	4651662	10740823	2133962	2263712	11291912	2689160	6525853	11117671	81380678	4127250	11941104	6057583	4760283
C24	899988	1723753	480233	61629	712638	141678	358503	633484	10138457	139581	6672827	516352	331304
C25	1826753	9073597	8393543	3068224	6065826	2345358	4480081	9516371	36610177	2252882	8058920	10697430	1445506
C26	2600801	8320842	1680733	3143180	5808246	1487928	4053042	7311217	66029896	2356896	9167363	3310274	3289398
C27	6677402	15434880	4458795	4680747	10744694	4462870	1422409	12913975	137111397	7445733	14915225	12380121	7847774

Fig. 2 Information gathering *Source* INEGI 2011. GDP indicators of each one of the Economic Units

	P1	P2	P3	P4	P5	P6	P7	P8	P9	P10	P11	P12	P13
C1	0.144186	0.264749	0.070492	0.084057	0.276541	0.093186	0.465656	0.556350	0.029908	0.394216	0.453043	0.237100	0.211154
C2	0.002777	0.001334	0.004452	1.000000	0.026855	0.001904	0.030963	0.011819	0.000172	0.017297	0.003033	0.005383	0.003563
C3	0.039402	0.451660	0.094529	0.042195	0.340029	0.141244	0.507075	0.372288	0.654403	0.142583	0.276404	0.392003	0.225048
C4	0.140421	0.396069	0.165376	0.347199	0.313554	0.065445	0.165266	0.313903	1.000000	0.118968	0.354922	0.121013	0.169553
C5	0.098944	0.133671	0.017538	0.017990	0.126923	0.027176	0.103989	0.119780	0.536988	0.184492	0.368465	0.064595	0.153366
C6	0.167272	0.083006	0.002068	0.039034	0.407828	0.003103	0.015535	0.106570	0.900531	0.058680	1.000000	0.016076	0.236425
C7	0.009604	0.067568	0.000418	0.014922	0.012623	0.011298	0.025298	1.000000	0.081794	0.320582	0.012953	0.060345	0.015527
C8	0.017510	0.251602	0.001526	0.001444	0.169053	0.001997	0.016134	0.125249	0.830743	0.073691	0.170248	0.004158	0.013915
C9	0.020586	0.039183	0.000426	0.002138	0.092471	0.001509	0.070037	0.034215	1.000000	0.006989	0.245345	0.001463	0.201882
C10	0.094155	0.119417	0.009419	0.005244	0.351384	0.097784	0.009189	0.231365	0.129016	0.027921	0.224303	0.037511	0.490348
C11	0.019814	0.141073	0.001577	0.001997	0.672096	0.003670	0.002710	0.071584	0.675626	0.037269	0.151099	0.065229	0.023118
C12	0.297992	0.428988	0.000092	3.37E-07	0.648435	0.002218	0.006235	0.707153	0.298741	0.041837	0.598542	0.000841	0.054712
C13	0.060596	0.411678	0.004752	0.005469	0.144921	0.016000	0.024394	0.270691	0.974594	0.150354	0.089606	0.018622	0.034452
C14	0.012816	0.747846	0.004299	0.001532	0.030884	0.003382	0.014287	0.507502	0.659005	0.117941	0.084252	0.113047	0.092530
C15	0.051287	0.151857	0.035132	0.031065	0.125290	0.026715	0.093677	0.153006	1.000000	0.055963	0.190467	0.076478	0.055916
C16	0.034702	0.123235	0.033972	0.046953	0.159866	0.048049	0.068775	0.090101	1.000000	0.055801	0.187746	0.105093	0.081027
C17	0.021392	0.080049	0.013080	0.017267	0.040635	0.009936	0.040517	0.073821	1.000000	0.020307	0.063819	0.033632	0.024325
C18	0.015215	0.025073	0.009371	0.005724	0.026322	0.004025	0.008231	0.026180	1.000000	0.013918	0.038620	0.007756	0.010293
C19	0.067250	0.211589	0.035868	0.052704	0.154213	0.034890	0.193600	0.235066	1.000000	0.087168	0.261906	0.163928	0.133246
C20	0.015588	0.056891	0.007159	0.039473	0.063203	0.004800	0.017378	0.115695	1.000000	0.009631	0.115684	0.007337	0.008455
C21	0.020152	0.043670	0.015958	0.060115	0.069744	0.008870	0.006992	0.044863	1.000000	0.028913	0.058202	0.018515	0.015868
C22	0.071488	0.174539	0.033107	0.044525	0.150594	0.036717	0.195532	0.193416	1.000000	0.083565	0.203981	0.134511	0.128606
C23	0.057159	0.131982	0.026222	0.027816	0.138754	0.033044	0.080189	0.136613	1.000000	0.050715	0.146731	0.074435	0.058494
C24	0.088770	0.170021	0.047367	0.006079	0.070291	0.013994	0.035361	0.062483	1.000000	0.013767	0.658170	0.050930	0.032678
C25	0.049897	0.247844	0.229268	0.083808	0.165687	0.064063	0.122394	0.259938	1.000000	0.061537	0.220128	0.292198	0.039484
C26	0.039388	0.126016	0.025454	0.047602	0.087964	0.022534	0.061382	0.110726	1.000000	0.035694	0.138837	0.050133	0.049817
C27	0.048701	0.113009	0.032520	0.034138	0.078365	0.032549	0.103722	0.094230	1.000000	0.054304	0.108782	0.090292	0.057236

Fig. 3 Results of Step 1 of the algorithm

From the information gathering presented in Fig. 2, in order to get the regions grouping the next steps were followed:

Step 1. Description of each object from the sustainability indicators through a blurred subset. The result of the step 1 is showed in the Fig. 3. To see the data of the 32 regions check [17].

The relative distance of Hamming is a tool of the diffused logic that measures the relation variable to variable of a fact study, and how these fit a profile and calculate the difference between the ends of the intervals. These distances have been used in repeated occasions in the objects selection in which the intangibility characteristics prevail [19] (Trillini 2012; Pérez 2007). For the study presented, the relative distance of Hamming is appropriate.

Step 2. To identify the relative distance of Hamming in order to get the dissimilarities among each pair of regions of the Mexican Republic, in relation to the sustainability indicators. The matrix of distances or dissimilarity matrix is obtained from the Eq. 1.

Equation 1: The relative distance of Hamming between two objects given that Pj and Pk is:

$$\delta(P_j, P_k) = \frac{1}{n}\left(\sum_{r=1}^{n}|\mu_r^{(j)} - \mu_r^{(k)}|\right), \quad j, k = 1, 2, \ldots, m \tag{1}$$

For further information about this process it is recommendable to check [19].

	P1	P2	P3	P4	P5	P6	P7	P8	P9	P10	P11	P12	P13
P1	0	0.13548	0.04844	0.08546	0.12127	0.043204	0.077452	0.166348	0.714741	0.062089	0.174737	0.070822	0.0602
P2	0.13548	0	0.15846	0.19124	0.11847	0.162404	0.125363	0.100304	0.610659	0.141778	0.148293	0.112896	0.1478
P3	0.04844	0.15846	0	0.06099	0.15371	0.020496	0.06747	0.188996	0.7385	0.068078	0.204622	0.052322	0.078159
P4	0.08546	0.19124	0.06099	0	0.18156	0.064159	0.108274	0.223615	0.771074	0.104304	0.235789	0.104378	0.116578
P5	0.12127	0.11847	0.15371	0.18156	0	0.153137	0.126867	0.132652	0.6488	0.141926	0.117967	0.124519	0.111193
P6	0.0432	0.1624	0.0205	0.06416	0.15314	0	0.068804	0.193137	0.74413	0.059248	0.20797	0.057626	0.0689
P7	0.07745	0.12536	0.06747	0.10827	0.12687	0.068804	0	0.14603	0.711852	0.069104	0.166959	0.042911	0.074059
P8	0.16635	0.1003	0.189	0.22361	0.13265	0.193137	0.14603	0	0.691889	0.144474	0.160381	0.145044	0.169763
P9	0.71474	0.61066	0.7385	0.77107	0.6488	0.74413	0.711852	0.691889	0	0.731393	0.59953	0.701993	0.712681
P10	0.06209	0.14178	0.06808	0.1043	0.14193	0.059248	0.069104	0.144474	0.731393	0	0.184944	0.063141	0.071585
P11	0.17474	0.14829	0.20462	0.23579	0.11797	0.20797	0.166959	0.160381	0.59953	0.184944	0	0.174626	0.161433
P12	0.07082	0.1129	0.05232	0.10438	0.12452	0.057626	0.042911	0.145044	0.701993	0.063141	0.174626	0	0.067022
P13	0.0602	0.1478	0.07816	0.11658	0.11119	0.0689	0.074059	0.169763	0.712681	0.071585	0.161433	0.067022	0
P14	0.29739	0.20544	0.32648	0.35804	0.21539	0.330622	0.280967	0.229611	0.531596	0.292515	0.205556	0.279515	0.27333
P15	0.4667	0.33757	0.49579	0.527	0.36064	0.49993	0.443348	0.382548	0.419504	0.461356	0.338841	0.446874	0.434837
P16	0.11045	0.12897	0.11125	0.15141	0.12504	0.113044	0.07767	0.1261	0.702656	0.076085	0.160081	0.086989	0.111811
P17	0.03511	0.14099	0.04073	0.08275	0.13013	0.036981	0.067185	0.169548	0.7116	0.056763	0.175581	0.054815	0.044111
P18	0.04059	0.1597	0.01553	0.0599	0.15056	0.015344	0.059548	0.190474	0.745993	0.0558	0.205367	0.051119	0.068963
P19	0.35523	0.26159	0.38414	0.41532	0.26037	0.388589	0.350304	0.303681	0.523667	0.374044	0.287804	0.343933	0.328415
P20	0.06425	0.1365	0.05951	0.10543	0.12936	0.055515	0.046681	0.155163	0.714437	0.052541	0.168937	0.044904	0.059356
P21	0.12412	0.11295	0.1597	0.19353	0.07282	0.157356	0.119707	0.115833	0.654381	0.1267	0.094563	0.131893	0.116144
P22	0.05566	0.11186	0.08152	0.11379	0.16429	0.077656	0.084852	0.155733	0.671304	0.083696	0.162959	0.077741	0.08323
P23	0.08139	0.15678	0.05105	0.09776	0.16429	0.064556	0.092241	0.195567	0.687448	0.094107	0.204867	0.0711	0.104619
P24	0.07716	0.09527	0.09802	0.13521	0.0884	0.094433	0.070052	0.133237	0.666037	0.082422	0.154707	0.064007	0.073933
P25	0.08737	0.12267	0.07787	0.12256	0.12638	0.087559	0.054748	0.136837	0.713526	0.077207	0.15413	0.059407	0.089756
P26	0.08989	0.0922	0.11256	0.14351	0.08999	0.114211	0.0832	0.106511	0.669696	0.085081	0.135678	0.084193	0.100393
P27	0.06436	0.15554	0.05507	0.04761	0.14256	0.055315	0.067778	0.185911	0.738267	0.083059	0.199167	0.0688	0.082437
P28	0.11106	0.09154	0.12245	0.15135	0.11586	0.130963	0.080737	0.130448	0.638278	0.115144	0.136585	0.078322	0.112596
P29	0.04017	0.15918	0.0378	0.07408	0.14833	0.023514	0.083718	0.188325	0.737599	0.063473	0.203159	0.07251	0.070844
P30	0.20379	0.1719	0.20325	0.22982	0.17906	0.207389	0.146007	0.202615	0.619474	0.18957	0.172944	0.167837	0.18697
P31	0.03661	0.12994	0.04813	0.08752	0.11874	0.043904	0.053922	0.160441	0.707507	0.053137	0.169437	0.050196	0.04573
P32	0.03669	0.15539	0.03027	0.06829	0.14441	0.026796	0.055274	0.185393	0.749815	0.046807	0.200989	0.05103	0.065385

Fig. 4 Distance of hamming

The matrix of distances or dissimilarities gotten in shown in the Fig. 4.

Step 3. Conversión of the blurred dissimilarity relation of the Fig. 4 in a relation of similarity complementing the unit, each one of its elements. Therefore a symmetric and reflective blurred relation is obtained. The result of this process is shown in the Fig. 5.

Step 4. To find the next groupings by affinity a threshold is adopted which is considered to meet the homogeneity needed. For this reason, the threshold $\alpha \geq 0.70$ is suggested.

The threshold $\alpha \geq 0.70$. was chosen in search of balance, this is to say, taking into account that for a minor threshold there will be obtained less groups, but each of these with a bigger elements number, this means, there are more elements that share common characteristics with a lower similarity level, moreover the elements can be present in more groups, they can have common characteristics to lower levels with different group elements and for a major threshold there would be obtain more groups with less elements, this makes harder to find elements that share a high level of common characteristics and there would be more isolated elements inwardly. The resulted matrix of this process is presented in the Fig. 6.

Step 4.1. Following the next rule, the additions are identified.

	P1	P2	P3	P4	P5	P6	P7	P8	P9	P10	P11	P12	P13	P14	P15	P16	P17	P18	P19	P20	P21	P22	P23	P24	P25	P26	P27	P28	P29	P30	P31	P32
P1	1	0.9	1	1	1	1	0.9	1	0	0.9	0.8	0.9	0.9	0.7	0.5	0.9	1	1	1	0.9	1	0.9	1	1	0.9	0.9	1	1	1	0.8	1	1
P2	0.9	1	0.8	1	1	1	0.9	1	0	0.9	0.9	0.9	0.9	0.8	0.7	0.9	1	1	1	0.9	1	0.9	1	1	0.9	0.9	1	1	1	0.8	0.9	1
P3	1	0.8	1	1	1	1	0.9	1	0	0.9	0.8	0.9	0.9	0.7	0.5	0.9	1	1	1	0.9	1	0.9	1	1	0.9	0.9	1	1	1	0.8	1	1
P4	0.9	0.8	0.9	1	1	1	0.9	1	0	0.9	0.8	0.9	0.9	0.6	0.5	0.8	1	1	1	0.9	1	0.9	1	1	0.9	0.9	1	1	1	0.8	0.9	1
P5	0.9	0.9	0.8	1	1	1	0.9	1	0	0.9	0.9	0.9	0.9	0.8	0.6	0.9	1	1	1	0.9	1	0.9	1	1	0.9	0.9	1	1	1	0.8	0.9	1
P6	1	0.8	1	1	1	1	0.9	1	0	0.9	0.8	0.9	0.9	0.7	0.5	0.9	1	1	1	0.9	1	0.9	1	1	0.9	0.9	1	1	1	0.8	1	1
P7	0.9	0.9	0.9	1	1	1	1	1	0	0.9	0.8	1	0.9	0.7	0.6	0.9	1	1	1	1	1	0.9	1	1	0.9	0.9	1	1	1	0.8	0.9	1
P8	0.8	0.9	0.8	1	1	1	0.9	1	0	0.9	0.8	0.9	0.8	0.8	0.6	0.9	1	1	1	0.8	1	0.8	1	1	0.9	0.9	1	1	1	0.8	0.8	1
P9	0.3	0.4	0.3	0	0	0	0.3	0	1	0.3	0.4	0.3	0.3	0.5	0.6	0.3	0	0	0	0.3	0	0.3	0	0	0.3	0.3	0	0	0	0.4	0.3	0
P10	0.9	0.9	0.9	1	1	1	0.9	1	0	1	0.8	0.9	0.9	0.7	0.5	0.9	1	1	1	0.9	1	0.9	1	1	0.9	0.9	1	1	1	0.8	0.9	1
P11	0.8	0.9	0.8	1	1	1	0.8	1	0	0.8	1	0.8	0.8	0.8	0.7	0.8	1	1	1	0.8	1	0.8	1	1	0.8	0.8	1	1	1	0.8	0.8	1
P12	0.9	0.9	0.9	1	1	1	1	1	0	0.9	0.8	1	0.9	0.7	0.6	0.9	1	1	1	0.9	1	0.9	1	1	0.9	0.9	1	1	1	0.8	0.9	1
P13	0.9	0.9	0.9	1	1	1	0.9	1	0	0.9	0.8	0.9	1	0.7	0.6	0.9	1	1	1	0.9	1	0.9	1	1	0.9	0.9	1	1	1	0.8	1	1
P14	0.7	0.8	0.7	1	1	1	0.7	1	0	0.7	0.8	0.7	0.7	1	0.8	0.8	1	1	1	0.7	1	0.7	1	1	0.8	0.8	1	1	1	0.8	0.7	1
P15	0.5	0.7	0.5	0	1	1	0.6	1	1	0.5	0.7	0.6	0.6	0.8	1	0.6	1	1	1	0.5	1	0.6	1	1	0.6	0.6	1	1	1	0.7	0.5	1
P16	0.9	0.9	0.9	1	1	1	0.9	1	0	0.9	0.8	0.9	0.9	0.8	0.6	1	1	1	1	0.9	1	0.9	1	1	0.9	0.9	1	1	1	0.9	0.9	1
P17	1	0.9	1	1	1	1	0.9	1	0	0.9	0.8	0.9	1	0.7	0.5	0.9	1	1	1	0.9	1	0.9	1	1	0.9	0.9	1	1	1	0.8	1	1
P18	1	0.8	1	1	1	1	0.9	1	0	0.9	0.8	0.9	0.9	0.7	0.5	0.9	1	1	1	0.9	1	0.9	1	1	0.9	0.9	1	1	1	0.8	1	1
P19	0.6	0.7	0.6	1	1	1	0.6	1	0	0.6	0.7	0.7	0.7	0.8	0.7	0.7	1	1	1	0.6	1	0.7	1	1	0.6	0.7	1	1	1	0.8	0.7	1
P20	0.9	0.9	0.9	1	1	1	1	1	0	0.9	0.8	1	0.9	0.7	0.5	0.9	1	1	1	1	1	0.9	1	1	0.9	0.9	1	1	1	0.8	1	1
P21	0.9	0.9	0.8	1	1	1	0.9	1	0	0.9	0.9	0.9	0.9	0.8	0.7	0.9	1	1	1	0.9	1	0.9	1	1	0.9	0.9	1	1	1	0.8	0.9	1
P22	0.9	0.9	0.9	1	1	1	0.9	1	0	0.9	0.8	0.9	0.9	0.7	0.6	0.9	1	1	1	0.9	1	1	1	1	0.9	0.9	1	1	1	0.8	0.9	1
P23	0.9	0.9	0.9	1	1	1	0.9	1	0	0.9	0.8	0.9	0.9	0.7	0.5	0.9	1	1	1	0.9	1	0.9	1	1	0.9	0.9	1	1	1	0.8	0.9	1
P24	0.9	0.9	0.9	1	1	1	0.9	1	0	0.9	0.8	0.9	0.9	0.8	0.6	0.9	1	1	1	0.9	1	0.9	1	1	0.9	0.9	1	1	1	0.9	0.9	1
P25	0.9	0.9	0.9	1	1	1	0.9	1	0	0.9	0.8	0.9	0.9	0.8	0.6	0.9	1	1	1	0.9	1	0.9	1	1	1	0.9	1	1	1	0.9	0.9	1
P26	0.9	0.9	0.9	1	1	1	0.9	1	0	0.9	0.9	0.9	0.9	0.8	0.6	0.9	1	1	1	0.9	1	0.9	1	1	0.9	1	1	1	1	0.9	0.9	1
P27	0.9	0.8	0.9	1	1	1	0.9	1	0	0.9	0.8	0.9	0.9	0.7	0.5	0.9	1	1	1	0.9	1	0.9	1	1	0.9	0.9	1	1	1	0.8	0.9	1
P28	0.9	0.9	0.9	1	1	1	0.9	1	0	0.9	0.9	0.9	0.9	0.8	0.6	0.9	1	1	1	0.9	1	0.9	1	1	0.9	0.9	1	1	1	0.9	0.9	1
P29	1	0.8	1	1	1	1	0.9	1	0	0.9	0.8	0.9	0.9	0.7	0.5	0.9	1	1	1	0.9	1	0.9	1	1	0.9	0.9	1	1	1	0.8	0.9	1
P30	0.8	0.8	0.8	1	1	1	0.9	1	0	0.8	0.8	0.8	0.8	0.7	0.7	0.8	1	1	1	0.8	1	0.8	1	1	0.9	0.9	1	1	1	1	0.8	1
P31	1	0.9	1	1	1	1	0.9	1	0	0.9	0.8	0.9	1	0.7	0.5	0.9	1	1	1	0.9	1	0.9	1	1	0.9	0.9	1	1	1	0.8	1	1
P32	1	0.8	1	1	1	1	0.9	1	0	1	0.8	0.9	0.9	0.7	0.5	0.9	1	1	1	1	0.9	1	0.9	1	1	0.9	0.9	1	1	1	0.8	1

Fig. 5 Complementary matrix

	P1	P2	P3	P4	P5	P6	P7	P8	P9	P10	P11	P12	P13	P14	P15	P16	P17	P18	P19	P20	P21	P22	P23	P24	P25	P26	P27	P28	P29	P30	P31	P32
P1	1	1	1	1	1	1	1	1	0	1	1	1	1	1	0	1	1	1	0	1	1	1	1	1	1	1	1	1	1	1	1	1
P2		1	1	1	1	1	1	1	0	1	1	1	1	1	0	1	1	1	0	1	1	1	1	1	1	1	1	1	1	1	1	1
P3			1	1	1	1	1	1	0	1	1	1	1	1	0	1	1	1	0	1	1	1	1	1	1	1	1	1	1	1	1	1
P4				1	1	1	1	1	0	1	1	1	1	1	0	0	1	1	0	1	1	1	1	1	1	1	1	1	1	1	1	1
P5					1	1	1	1	0	1	1	1	1	1	0	1	1	1	0	1	1	1	1	1	1	1	1	1	1	1	1	1
P6						1	1	1	0	1	1	1	1	1	0	1	1	1	0	1	1	1	1	1	1	1	1	1	1	1	1	1
P7							1	1	0	1	1	1	1	1	0	1	1	1	0	1	1	1	1	1	1	1	1	1	1	1	1	1
P8								1	0	1	1	1	1	1	0	1	1	1	0	1	1	1	1	1	1	1	1	1	1	1	1	1
P9									1	0	0	0	0	0	0	0	0	0	0	0	0	0	0	0	0	0	0	0	0	0	0	0
P10										1	1	1	1	1	0	1	1	1	0	1	1	1	1	1	1	1	1	1	1	1	1	1
P11											1	1	1	1	0	1	1	1	1	1	1	1	1	1	1	1	1	1	1	1	1	1
P12												1	1	1	0	1	1	1	0	1	1	1	1	1	1	1	1	1	1	1	1	1
P13													1	1	0	1	1	1	0	1	1	1	1	1	1	1	1	1	1	1	1	1
P14														1	1	1	1	0	1	1	1	1	1	1	0	1	0	1	1	0	1	0
P15															1	0	0	0	1	0	0	0	0	0	0	0	0	0	0	0	0	0
P16																1	1	1	0	1	1	1	1	1	1	1	1	1	1	1	1	1
P17																	1	1	0	1	1	1	1	1	1	1	1	1	1	1	1	1
P18																		1	0	1	1	1	1	1	1	1	1	1	1	1	1	1
P19																			1	0	0	1	0	0	1	0	0	0	1	0	0	0
P20																				1	1	1	1	1	1	1	1	1	1	1	1	1
P21																					1	1	1	1	1	1	1	1	1	1	1	1
P22																						1	1	1	1	1	1	1	1	1	1	1
P23																							1	1	1	1	1	1	1	1	1	1
P24																								1	1	1	1	1	1	1	1	1
P25																									1	1	1	1	1	1	1	1
P26																										1	1	1	1	1	1	1
P27																											1	1	1	1	1	1
P28																												1	1	1	1	1
P29																													1	1	1	1
P30																														1	1	1
P31																															1	1
P32																																1

Fig. 6 Regions that meet for the threshold $\alpha \geq 0.70$

- The rows without zeros (there is no addition for them) are excluded from the process (Consult the Sect. 3.1).

$\alpha \geq 0.70$
P1 = P1 + P9P15P19
P2 = P2 + P9P15
P3 = P3 + P9P14P15P19
P4 = P4 + P9P14P15P19
P5 = P5 + P9P15
P6 = P6 + P9P14P15P19
P7 = P7 + P9P15P19
P8 = P8 + P9P15P19
P9 = P9 + P10P11P12P13P14P15P16P17P18P19P20P21P22P23P24P25P26P27P28P29P30P31P32
P10 = P10 + P15P19
P11 = P11 + P15
P12 = P12 + P15P19
P13 = P13 + P15P19
P14 = P14 + P18P27P29P32
P15 = P15 + P16P17P18P20P21P22P23P24P25P26P27P28P29P30P31P32
P16 = P16 + P19
P17 = P17 + P19
P18 = P18 + P19
P19 = P19 + P20P22P23P25P26P27P29P31P32

Step 4.2 The additions found for each row are gathered through the Boolean product, in minimum terms, according to the next rules (Consult Sect. 3):

- When the additions product is the repetition of an addend, only one of them is considered.
  So: a + a = a.
- If in one of the resulting addends appear the same elements than in other, or the same plus one or some more, the one that has more number of elements is removed.
  So: a + a * b * c = a.

The result of the union of each row through the Boolean product gave as a result 6 terms and they are:

P1P2P3P4P5P6P7P8P9P10P11P12P13P14P15P16P17P18

$\times$ P19P20P21P22P23P24P25P26P27P28P29P30P31P32

= P9P14P15P19 + P3P4P6P9P13P15P18P19P27P29P32

+ P1P2P3P4P5P6P7P8P9P10P11P12P13P15P16P17P18

$\times$ P20P22P23P25P26P27P29P32

+ P1P2P3P4P5P6P7P8P9P10P11P12P15P18P19P27P29P32

+ P1P2P3P4P5P6P7P8P9P10P11P12P13P16P17P18P19P20

$\times$ P21P22P23P24P25P26P27P28P29P30P31P32

+ P1P2P3P4P5P6P7P8P9P10P11P12P13P16P17P18P19P20

$\times$ P21P22P23P24P25P26P27P28P29P30P31P32

Step 5. Top similar sub relations, from obtaining the complement of each term with relation to the referential.

Sub relations	Regions grouped
Sub relation 1	*R1R2R3R4R5R6R7R8R10R11R12R13R16R17R18R20R21R22* *R23R24R25R26R27R28R29R30R31R32*
Sub relation 2	R1R2R5R7R8R10R11R12R14R16R17R20R21R22R23R24R25R26R28R30R31
Sub relation 3	R14R19R21R24R28R30R31
Sub relation 4	R13R14R16R17R20R21R22R23R24R25R26R28R30R31
Sub relation 5	R14R15
Sub relation 6	R9

According to the groupings derived from step 5, the groups are:

Sub relations	Regions grouped
Sub relation 1	*Aguascalientes, Baja California, Baja California Sur, Campeche, Coahuila de Zaragoza, Colima, Chiapas, Chihuahua, Durango, Guanajuato, Guerrero, Hidalgo, Michoacán de Ocampo, Morelos, Nayarit, Oaxaca, Puebla, Querétaro, Quintana Roo, San Luis Potosí, Sinaloa, Sonora, Tabasco, Tamaulipas, Tlaxcala, Veracruz de Ignacio de la Llave, Yucatán, Zacatecas*
Sub relation 2	Aguascalientes, Baja California, Coahuila de Zaragoza, Colima, Chihuahua, Durango, Guanajuato, Guerrero, Jalisco, Michoacán de Ocampo, Morelos, Oaxaca, Puebla, Querétaro, Quintana Roo, San Luis Potosí, Sinaloa, Sonora, Tamaulipas, Veracruz de Ignacio de la Llave, Yucatán
Sub relation 3	Jalisco, Nuevo León, Puebla, San Luis Potosí, Tamaulipas, Veracruz de Ignacio de la Llave, Yucatán
Sub relation 4	Hidalgo, Jalisco, Michoacán de Ocampo, Morelos, Oaxaca, Puebla, Querétaro, Quintana Roo, San Luis Potosí, Sinaloa, Sonora, Tamaulipas, Veracruz de Ignacio de la Llave, Yucatán
Sub relation 5	Jalisco, México
Sub relation 6	Distrito Federal

6   Conclusion

To achieve a sustained development has become the biggest challenge for the XXI century. In order to reach it, it is crucial to use new models that allow a more informed and closer to reality decision making. Through the methodology applied in this study, there have been obtained groupings of regions that share similar characteristics at a given level which share the singularities analyzed. It can be appreciated that the more exigent the process is raised, this is to say, the higher the level to which the accomplishment of characteristics in the groups is demanded, more groups of regions with less elements in each one appeared. Conversely, the less exigent the level to which the accomplishment of characteristics the regions share, less groups are formed but this turn out to be more numerous. The synergy of this information allows to guide the resources as well as the investments to empower the regions development and as a consequence, move towards an economy with stable levels of social welfare and sustained growth.

Bibliography

1. Alfaro, G., Alfaro, V.: La Afinidad de Socios en la Integración de Clúster: Caso Mipymes Morelianas. In: Global Conference on Business and Finance, vol. 10, no. 1 (2015)
2. Alfaro, G., Alfaro, V., Gómez, R.: Aplicación de la Teoría de Afinidades y Lógica Difusa en la Conformación de Agrupamientos Efectivos (2015)
3. Alfaro, G., Alfaro, V., González, F.: Aplicación de Recursos Fuzzy Logic para la Asociación de Hoteles de un Destino Turístico. In: Global Conference on Business and Finance, vol. 10, no. 2 (2015)
4. Alfaro, V., Gil Lafuente, A., Alfaro, G.: Methodological Structure for the Aggregation of Municipalities Under Uncertainty Conditions. The Case of Michoacán, México. Revista de Investigación en Ciencias de la Administración 7(13) (2015)
5. Barcellos, L.: Modelos de gestión aplicados a la sostenibilidad empresarial. Tesis para obtener el grado de Doctor en Empresa, Facultad de Economía y Empresa, Departamento de Economía y Organización de Empresas. Barcelona (2010)
6. Gil-Aluja, J., Gil-Lafuente, A., Merigo, J.: Using homogeneous groupings in portfolio management. Expert Syst. Appl. 38(9), 10950–10958 (2011)
7. Gil-Lafuente, A., Gil-Aluja, J., Barcellos De Paula, L.: Determining the composition of a portfolio management from a groupings model. Int. J. Uncertainty Fuzziness Knowl.-Based Syst. 21(4), 561–578 (2013)
8. Gil-Lafuente, J., Gil-Lafuente, A.: Affinity as basis of interchangeability between athletes. Decis. Making Knowl. Decis. Support Syst. 675, 95–106 (2015)
9. Gil-Aluja, J., Merigó, J., Gil-Lafuente, A.: Decision making with the induced generalized adequacy coefficient. Appl. Comput. Math. 10(2), 321–339 (2011)
10. Gil-Aluja, J., Merigó, J., Gil-Lafuente, A.: A new aggregation method for strategic decisión making and its application in assignment theory. Afr. J. Bus. Manage. 5(11), 4033–4043 (2011). ISSN 1993.8233
11. Gil-Aluja, J., Merigó, J., Gil-Lafuente, A.: Using homogeneous groupings in portfolio management. Expert Syst. Appl. Int. J. 38(9), 10950–10958 (2011). ISSN 0957-4174

12. Gil-Aluja, J., Merigó, J., Gil-Lafuente, A.: Soft computing techniques for decision making with induced aggregation operators. Inf. Int. Interdiscip. J. **14**(6), 2019–2039 (2011). ISSN 1343-4500
13. Kaufmann, A., Gil Aluja, J.: Técnicas especiales para la gestión de expertos. Ed. Milladoiro. Santiago de Compostela (1993)
14. Malfer, L.: Las alianzas locales para el desarrollo sostenible de la comunidad. Le alleanze local per lo sviluppo sostenibili di comunitá (2012)
15. ONU: Organización de las Naciones Unidas. El camino hacia la dignidad para el 2030: acabar con la pobreza y transformar vidas. Consultado el 2 de Mayo de 2016 en. http://onu.org.pe/destacados/informe-de-sintesis-del-secretario-general-el-camino-hacia-la-dignidad-para-2030/ (2016)
16. Organización de las Naciones Unidas: Indicators of Sustainable Development Guidelines and Methodologies, 3rd edn. Consultado en. http://www.un.org/esa/sustdev/natlinfo/indicators/guidelines.pdf (2007)
17. Pinto, I.: Modelo difuso de gestión para potencializar el desarrollo de regiones como contribución al Desarrollo Sostenible. Tesis para obtener el grado de Doctor en Planeación Estratégica y Dirección de Tecnología. Universidad Popular Autónoma del Estado de Puebla, México (2015)
18. Pichat, E.: Algorithm for finding the maximal elements of a finite universal algebra. Informe proveniente de la 68 Publ. North Holland (1969)
19. Soler, R., Martínez, L., Oñate, A.: Competence in uncertainty. Study case: Professors of Engineering of the Administrations School. Universidad y Sociedad, vol. 8, no. 1 (2016)
20. Villa, C.: Alianzas locales para el desarrollo. Lecciones de la experiencia colombiana. Territorios (1), 25–42 (1999)
21. Vizuete, E.: Algoritmos para el tratamiento y selección de productos financieros en la incertidumbre. Universidad de Barcelona, Tesis Doctoral (2014)

Companies' Selection Methods for Inclusion in Sustainable Indices: A Fuzzy Approach

Vicente Liern and Blanca Pérez-Gladish

Abstract Sustainability indices handle concepts which are both, of numerical and non-numerical nature. In this context, the use of Fuzzy Logic is highly useful as allows a more faithful representation of reality. Usually these indices follow a three-step process to define sustainable investment universes. First step consists of sustainability assessment. In the second step, assets are rated based on the previously assessed sustainability scores and finally, best assets are selected. This last step relies on the construction of a global score reflecting the performance of the assets in main sustainability dimensions. In this Chapter we are concerned with the third step of the selection process. We review the aggregation process used by sustainability indices to obtain overall sustainability scores and we propose the use of flexible aggregation operators for the obtaining of a global score describing the sustainability degree of a firm that takes into account the characteristics of the different dimensions to be aggregated. Assets are then ranked using this score from most to less sustainable. The proposed approach is be compared with the three-step selection process applied by Euronext in their selection process for inclusion of companies in the Euronext Vigeo family of sustainability indices.

Keywords Corporate social responsibility · Corporate sustainability
Sustainable responsible investment (SRI) · Aggregation operators
Induced ordered weighted geometric (IOWG) operator

V. Liern (✉)
Dpto. Matemáticas para la Economía y la Empresa, Universitat de València,
Avda. de Tarongers, s/n, 46022 Valencia, Spain
e-mail: vicente.liern@uv.es

B. Pérez-Gladish
Dpto. Economía Cuantitativa, Universidad de Oviedo, Avda. del Cristo s/n,
33006 Oviedo, Spain
e-mail: bperez@uniovi.es

© Springer International Publishing AG 2018
C. Berger-Vachon et al. (eds.), *Complex Systems: Solutions and Challenges
in Economics, Management and Engineering*, Studies in Systems,
Decision and Control 125, https://doi.org/10.1007/978-3-319-69989-9_21

365

1   Introduction

Sustainable Development was defined in the Brundtland Report of 1987 as the "development that meets the needs of the present without compromising the ability of future generations to meet their own needs" [1]. This is the most popular and used definition of sustainability in the world. Based on this definition, corporate sustainability is considered as the meeting of "the needs of a firm's direct and indirect stakeholders (such as shareholders, employees, clients, pressure groups, communities) without comprising its ability to meet the needs of future stakeholders as well" [2].

The Sixth Sustainable and Responsible Investment Study by the European Forum for Sustainable Investment [3], details the continued growth in assets under management in the European Sustainable and Responsible Investment (SRI) market. The study also highlights the growing diversity and sophistication of sustainable investment strategies in practice today [3]. In a context characterized by this growing interest on investing in sustainable assets, measurement of their sustainability degree is a key question.

However, the classical definition of sustainability is considered imprecise, vague and hardly measurable. In an attempt to make the concept more operational, Dyllick and Hockerts [2] affirm, "Firms have to maintain and growth their economic, social and environmental capital base while actively contributing to sustainability in the political domain" and this means "integrating the economic, ecological and social aspects in a *triple-bottom* line", as "economic sustainability alone is not sufficient condition for the overall sustainability of a corporation". One of the issues most discussed by practitioners and academics is whether or not it is adequate to obtain an overall sustainability indicator for financial assets and particularly, for firms. Sustainability is a concept with multiple dimensions and this implies aggregating a large amount of information and this opens the debate around how to aggregate and how to determine aggregation weights. As shown by J. Gil Aluja [4], fuzzy logic is helpful to combine non-numeric criteria and/or attributes with other which are.

Independent rating agencies supplying information about the Environmental, Social and Governance (ESG) performance of companies are interested on providing universal measures that are objective and valid for any decision maker. Some well-known examples are MSCI ESG STATS (known under the name of KLD Research & Analytics Inc.), Ethibel, Vigeo, Oekom Research, SAM (Sustainable Asset Management) or EIRIS, recently merged with Vigeo to become Vigeo EIRIS [5]. The aggregation process followed by these rating agencies usually relies in an arithmetic or geometric averages, weighted or not. In an attempt to avoid the use of subjective weights, the common practice is to assign equal weights to all the sustainability responsibility dimensions. The overall measures of sustainability are then used by sustainability indices such as the Down Jones Sustainability Index, FTSE4 Good, Stoxx Sustainability Index or Euronext Vigeo Family to select companies.

In this work, we are concerned with the proposal of a flexible aggregating method that overcomes some limitations of the usually used aggregation procedures. The method, based on the Induced Ordered Weighted Geometric operator, will allow us to rank firms using an overall measure of their sustainability taking into account the specific nature of the data and without the necessity of relying on the decision maker's preferences. Flexible aggregating operators as Ordered Weighted Aggregation, Induced Ordered Weighted Aggregation and Induced Weighted Geometric operators have been widely used by academics and practitioners due to their suitable characteristics for the resolution of real problems [6].

2 The Role of Fuzzy Sets Logic on Sustainability Assessment

Reality often shows facts that cannot be defined precisely, i.e. are not rigid or deterministic, so in order to handle them we need much more detailed data than those that a person can recognize, process and understand.

The majority of traditional tools for formal modeling, reasoning and data processing are rigid, deterministic and precise. It is usually assumed that the structures and parameters from the model are known and that they represent accurately our perception of the phenomenon addressed and the characteristics of the real system under study. Usually, precision implies that the problem has no ambiguity and, in fact, few times we explicitly question its valuations or its occurrence. However, managers know that we often have only partial information about reality and that normally, it cannot be associated to any concrete variable [7–9].

Aware of this, Lotfi A. Zadeh introduced, in the middle sixties, the Fuzzy Sets Theory to provide the basis of approximate reasoning using imprecise premises as instruments to formulate knowledge.

The main idea within a fuzzy set is that human thinking uses linguistic labels that allow objects to belong to one class or another in a soft and flexible manner. In practice, we talk about a "tall" or "short" person without stopping the interlocutor of having all the necessary information [7].

In a classic set values 0 or 1 are assigned to each element to indicate its belonging or not to the set. This function, called characteristic function, can be generalized such that the assigned values to the elements belonging to the set fall in a particular range (usually interval [0, 1]), and therefore they indicate the degree of membership to the set of the elements. A membership degree of 0 means no membership, of 1 membership in a Boolean sense and intermediate values mean uncertain membership.

A fuzzy number is a particular case of a fuzzy set. Let us consider number 0. Zero completely belongs to the set zero and there is no other number belonging to this set. However, what happens with those numbers which are very close to zero or almost zero? In the real world, it is reasonable to suppose that 0.0001 is "almost 0" whilst 35 is not. This leads us to the idea of fuzzy number which is a special case of a fuzzy sets for which certain conditions are required regarding its membership function.

Now we have fundamental elements for fuzzy logic, fuzzy sets and, in particular, fuzzy numbers, the following step is to be able to operate with them, to compare them, to order them etc. Following Gil Aluja [8, 10, 11], in what follows we will briefly describe the four actions in which all the previous operations and actions are included.

(a) *Relation*: fuzzy relations allow the establishment of different degrees or levels of relation. The analysis of intensity variations, levels of strengths of the relations over time, is fundamental in order to faithfully represent social, economic, environmental and managerial relations.
(b) *Assignment*: it is a type of relation in which three sets of physic or mental objects participate. The first one includes the elements to be assigned, the second one the elements which receive the assignation and the third one the assignment criteria.
    With the aim at obtaining the relations from which we can start the assignment process, we can use any index reflecting proximity as the notion of distance and similarity.
(c) *Group*: in social, economic and managerial activities situations in which it is necessary to collect and group "objects" of apparently different nature, are usual. Sometimes we need to select within the elements of the same group or to select a group among several groups. It is necessary to analyze the similarity relations between the elements of the referential and get to know a homogeneity degree that allows the obtaining of adequate groups.
(d) *Order*: in uncertain social and economic contexts, the concept of order occupies a privileged place. To order investments, financial sources or resources can be the prelude to success. When it is not possible to obtain all the valuations of the objects, appealing to a non-quantified order can be sufficient to make an adequate decision.

In the next sections we will address these four actions in order to obtain an overall score representing the sustainability degree of companies. This overall score is usually used by the sustainability indexes to select the best companies from a sustainable point of view.

3   Overall Valuation Based on the Geometric Mean with Equal Weights

In this section, we will describe the ranking procedure used by a social rating agency, Vigeo. In particular, we will analyze the procedure followed by the Advanced Sustainability Performance Eurozone Index ("ASPI Eurozone®") now

Euronext Vigeo Euro 120. The family of Euronext Vigeo indices are recognized as one of the leading sustainability indices families.

Vigeo is a leading European expert in the assessment of companies and organizations with regards to their practices and performance on ESG issues. Vigeo has developed Equitics®, a model based on internationally recognized standards to assess to which degree companies take into account social responsibility objectives in the definition and deployment of their strategy. Vigeo offers access to scores in 6 dimensions which are commonly used by the all the rating agencies: Human Rights; Human Resources; Environment; Business Behavior; Corporate Governance and Community Involvement. A description of these dimensions is presented in Table 1. Vigeo's database provides scores from 0 to 100, for each firm in each social dimension. Information about sectors' performance is also provided. Sectors are also scored from 0 to 100 in each dimension.

Euronext Vigeo indices are used by the growing community of responsible investors to define sustainable investment universes.

Euronext Vigeo Euro 120 follows a three-step process to select the constituents of the index. In a first step, sustainability assessment of the companies is done by Vigeo on the six sustainability dimensions following the scoring process previously described. Then, based on the obtained scores, companies' are rated based on sector

Table 1 List of Vigeo's evaluation criteria

HR	HUMAN RESOURCES: Continuous improvement of professional relations, labor relations and working conditions
ENV	ENVIRONMENT: Protection, safeguarding, prevention of damage to the environment, implementation of an adequate management strategy, eco-design, protection of biodiversity and coordinated management of environmental impacts on the entire lifecycle of products or services
C&S	BUSINESS BEHAVIOUR: Consideration of the rights and interests of clients, integration of social and environmental standards in the selection of suppliers and in the entire supply chain, effective prevention of corruption and respect for competitive practices
CG	CORPORATE GOVERNANCE: Effectiveness and integrity, guarantee of independence and efficiency of the Board of Directors, effectiveness and efficiency of auditing and control mechanisms, in particular the inclusion of social responsibility risks, respect for the rights of shareholders, particularly minority shareholders
CIN	COMMUNITY INVOLVEMENT: Effectiveness, managerial commitment to community involvement, contribution to the economic and social development of territories/societies within which the company operates, positive commitment to manage the social impacts linked to products or services and overt contribution and participation in causes of public or general interest
HRts	HUMAN RIGHTS AT THE WORKPLACE: Respect of freedom of association, the right to collective bargaining, non-discrimination and promotion of equally, elimination of illegal working practices such as child or forced labor, prevention of inhumane or degrading treatment such as sexual harassment, protection of privacy and personal data

Source http://www.vigeo.com

peers' comparison. In this second step, firms are classified as leaders, advanced, average, below average or not concerned. Finally, in the last step, a selection of the best 120 firms is conducted. In order to do this, for each company within the Euro Stoxx index, a Global Score is calculated using the geometric average of the six domain scores using equal weights:

$$\text{Euronext Vigeo Global Score} = \sqrt[6]{ENV \times HRts \times HR \times CIN \times C\&S \times CG} \quad (1)$$

The top 100-ranked companies are automatically selected for inclusion in the index. The rest are selected amongst the 101 to 140-ranked companies. A company is not eligible for inclusion in the ASPI Eurozone® if it is rated "0" in one domain and/or it is involved in the production or distribution of anti-personnel landmines or cluster munitions.

Let us consider Vigeo's 2012 scores. Table 2 displays an example of the scores obtained by some firms in each dimension and their overall score (OSF). We have ordered the firms following an alphabetical criterion and we have selected the first five companies. Last column shows the score obtained in average by the sectors of the firms (OSS). We can observe with this example, how companies performing well in some dimensions do not perform well in others. Allianz, for example, performs the best (together with ADIDAS) from a global point of view (it has an overall firm score of 54 points). However, although in Community Involvement (67 points) performs the best it performs the worst in Human Resources (29 points).

Comparison with sector average scores is a key question, as sectors tend to perform better in certain social responsibility dimensions depending on their type of activities. Inclusion in sustainability indices is, as mentioned in the introduction, based on comparison with the firm's sector performance. Only leaders or advanced firms (those performing better than the average of the sector) are included in the indices. Last column in Table 2 displays the overall score of the sectors of the firms (OSS). As we have seen before, ADIDAS and Allianz, for instance, obtain the same overall score (54 points) but the special retail sector has an overall score lower (37 points) than the insurance sector (40).

In Table 3, we have summarized the rankings in each dimension of the twenty-four sectors included in Vigeo's database.

Table 2 Vigeo's scores

Firm	Title	Sector	HR	ENV	C&S	CG	CIN	HRts	OSF	OSS
F1	Abertis	Trans and logistic	29	60	35	36	62	44	43	36
F2	Accor	Hotels and servic	42	46	67	57	40	58	51	39
F3	ADIDAS	Special. retail	49	75	43	63	54	47	54	37
F4	ADP	Trans and logistic	47	68	45	43	65	49	52	36
F5	Allianz	Insurance	29	66	57	58	67	58	54	40

Source Equitics®

Table 3 Sectors' ranking in each dimension

Sector	HR	ENV	C&S	CG	CIN	HRts
Aerospace	13	15	22	**11**	21	19
Automobiles	**1**	**2**	16	**24**	11	7
Banks	**6**	18	7	**20**	12	11
Beverage	16	12	14	16	8	18
...	...	...	...	...	...	...
Supermarkets	24	3	2	5	1	3
Telecommunications	12	4	10	6	5	6
Transport and logistics	8	14	21	23	14	20
Waste and water utilities	22	6	9	2	2	10

Source Own elaboration based on Equitics® database

We can observe how, for example, Automobiles is in average the best sector in Human Resources and is the worse in Corporate Governance. Supermarkets, on the other hand, ranks the last one in Human Resources and first one in Community Involvement.

If we now pay attention to the performance of the firms in each dimension, we can observe how the ranking changes depending on the firms and their sectors. Table 4 displays position of each firm above or below the average of the corresponding sector. A positive number means number of positions above the average of the sector (better performance than the sector) and a negative number represents the number of positions below the sector (worst performance than the sector), in each dimension. Within parenthesis the position of the firm in the ranking considering individually each dimension.

As we can observe, position changes depending on the considered dimension. For example, ADIDAS (F3) performs the best in Environment and the worst in Human Resources. We can also observe how there are firms performing worse than the average of their sectors in certain dimensions whereas being classified as leaders when taking into account the overall score aggregating all the dimensions (e.g. F4 and F52).

In sum, it seems convenient to take into account the performance of the firm compared to the performance of its sector and also the relevance of the different dimensions. In what follows we will present an aggregated measure of the sustainability of the firms that takes into account the variability of the scores in each

Table 4 Ranking based on the performance of firms with respect to their sector in each dimension

Firm	Sector	HR	ENV	C&S	CG	CIN	HRts
F3	Specialized retail	24 (6)	33 (1)	7 (8)	9 (2)	23 (8)	8 (10)
F4	Transport and logistics	20 (9)	27 (4)	12 (7)	−1 (8)	29 (4)	14 (9)
F52	Automobiles	36 (2)	21 (7)	16 (3)	−8 (10)	25 (7)	33 (2)

Source Own elaboration based on Equitics® database

dimension. The proposed aggregating operator will be the Induced Ordered Weighted Geometric (IOWG) operator. We will compare the ranking of the firms based on Vigeo's procedure (geometric mean with equal weights) and the ranking obtained by means of IOWG, which uses weights derived from the nature of the data in each dimension [12]. In order to compare the rankings we will calculate the difference in the position occupied by a firm in the two rankings proposing a flexible treatment of coincidences in the positions of the firms.

4   Overall Valuation Based on a Flexible Aggregation Operator

In previous section, we have described how a well-known social rating agency calculates an overall corporate social score and ranks firms accordingly. The proposed global indicator is obtained as the equally weighted geometric mean of the scores obtained by the firms in each dimension. In this section we will propose a more general global measure which will allow us to introduce objective weights for the dimensions taking into account the specific characteristics of the data in each of those dimensions. In particular, we propose the use of an Induced Ordered Weighted Geometric (IOWG) operator [13–15]. IOWG operators are aggregation operators that appear based on the initially proposed Ordered Weighted Averaging (OWA) operators. OWA operators provide a parameterized family of mean type aggregation operators that includes the minimum, the maximum, and the average [16]. For us, the most interesting characteristic of these operators is that the aggregation of the arguments is done based on a previous rearranging of them. Moreover, the aggregation weights are associated with a particular position in such reordering instead of being associated with a specific argument. This feature of OWA operators is especially desired in the context of the obtaining of corporate social overall scores. In what follows, we will give some basic definitions.

Definition 1 A vector $w = (w_1,\ldots,w_n)$ is called a weighting vector if the following two conditions are verified:

$$w_d \in [0, 1], d = 1, \ldots, n, \quad \text{and} \quad \sum_{d=1}^{n} w_d = 1. \tag{2}$$

In order to measure the similarity of other weighting vectors with the extreme weighting vectors we will introduce the concept of orness as follows:

Definition 2 The level of orness associated with the operator OWA_w is defined as

$$\alpha = \frac{1}{n-1} \sum_{d=1}^{n} (n-d)w_d, \quad 0 \le \alpha \le 1. \tag{3}$$

The level of orness measures the degree to which the aggregation behaves as the maximum operator or the minimum operator. Thus, degree 1 means that the operator is the maximum; degree 0 means that the operator is the minimum and in between we find all the other possibilities.

Ordered Weighted Geometric (OWG) operators [13–15] appeared when researchers realized that similar reasoning to the one done with the weighted sums were also valid for products weighted with powers [13].

Definition 3 Given a weighting vector w, the OWG operator OWG_w is defined to aggregate a list of values $a_1, a_2,..., a_n$, according to the following expression:

$$OWG_w(a_1, \ldots, a_n) = \prod_{d=1}^{n} a_{\sigma(d)}^{w_d},$$ (4)

where $a_{\sigma(d)}$ is the d-*th* largest element in the collection $a_1, a_2, \ldots, a_n$, i.e.,

$$a_{\sigma(1)} \geq \cdots \geq a_{\sigma(n)}.$$

In particular, for the weights $w_1 = (1, 0,..., 0)$, $w_2 = (0, 0,..., 1)$ and $w_3 = (1/n, 1/n,..., 1/n)$, we have $OWGw_1(a_1,..., a_n) = \max\{a_1,..., a_n\}$, $OWGw_2(a_1,..., a_n) = \min\{a_1,..., a_n\}$ and $OWGw_3(a_1,..., a_n) = \sqrt[n]{a_1 a_2 \ldots a_n}$, respectively.

The ordering of the arguments can be induced by another variable called the order-inducing variable [17]. When this happens, we have a different class of operators named IOWA operators. IOWA operators have been widely used in the literature due to their suitable properties [17–20].

The operator OGW can be generalized to an Induced Ordered Weighted Geometric (IOWG) operator, in which the arguments are not rearranged according to their magnitude but rather using a function of the arguments, i.e., by using an order-inducing variable, which is here denoted by z [13, 14].

Definition 4 Given a weighting vector $w = (w_1, w_2,..., w_n)$ and a vector of order-inducing variables $z = (z_1, z_2,..., z_n)$, the IOWG operator $IOWGw, z$ is defined to aggregate the second arguments of a list of 2-tuples $\{(z_1, a_1),..., (z_n, a_n)\}$ according to the following expression:

$$IOWG_{w,z}(<z_1, a_1>, \ldots, <z_n, a_n>) = \prod_{d=1}^{n} a_{\eta(d)}^{w_d},$$ (5)

where the arguments $<z_d, a_d>$ are rearranged in such a way that

$$z_{\eta(d)} \geq z_{\eta(d+1)}, \quad d = 1, \ldots, n-1.$$

In next section we will obtain overall social responsibility scores for a sample of firms using an IOWG operator. The order-inducing variable is chosen to quantify a certain property of the scores in each dimension. In our case, we are highly concerned about the variability of the scores within each social dimension. Therefore, our induced variable will be the variance of the scores obtained by the firms in each dimension. We will follow the procedure described in León et al. [19]. First, we will calculate an mxn matrix, M, with the scores of each firm, m, in each dimension, n. Then we will rearrange the columns in M according to the order-inducing variable. Third, we will determine the objective aggregation weights and finally, we will calculate the overall scores for each firm.

In order to calculate the weights we will use the method proposed by Wang and Parkan [21] in which they solve the so-called minimax disparity problem [19]:

$$
\begin{aligned}
&\text{Min} \quad \delta \\
&\text{s.t.} \quad \frac{1}{(n-1)} \sum_{k=1}^{n-1} (n-k)w_k = \alpha, \\
&\qquad w_1 + w_2 + \cdots + w_n = 1, \\
&\qquad w_k - w_{k+1} - \delta \le 0, \qquad k = 1, \ldots, n-1, \\
&\qquad w_k - w_{k+1} + \delta \ge 0, \qquad k = 1, \ldots, n-1, \\
&\qquad w_k \ge 0, \qquad\qquad\qquad k = 1, \ldots, n.
\end{aligned}
\tag{6}
$$

where $\alpha \in [0, 1]$ is the orness degree specified by the assets' manager. The overall score for each firm is the result of applying the IOWG operator to each element in a row with the aggregation weights.

5   Real Case Study

Let us illustrate the procedure described in Sect. 4 using a real case study. Our initial sample is composed of 73 firms. These firms are classified as leaders and advanced by Vigeo (therefore they perform better than their sector). First step consists of the rearranging of columns (social responsibility dimensions) according to the order-inducing variable, in our case, the variance of the scores in each dimension (see Table 5).

Once dimensions have been rearranged according to the order-inducing variable, we obtain the aggregation weights for different orness levels, α (see Table 6).

Finally, we calculate the overall scores for each firm applying the IOWG operator to each element in a row (obtained scores of the firm in each dimension) with the aggregation weights obtained in the previous step (see Table 7).

The resulting rankings for different orness levels are displayed in Table 8. Last row in Table 8 shows the number of coincidences in the ranking when comparing the rank obtained using the IOWG operator to the rank from Vigeo's overall scores. We have considered that there is a coincidence when a firm occupies the same position in both rankings.

Table 5 Rearranging of social dimensions based on the variance of their scores

Firm	HR	ENV	C&S	CG	CIN	HRts
F1	29	60	35	36	62	44
F2	42	46	67	57	40	58
...	...	...	...	...	...	...
F73	56	49	40	43	45	56
Variance	128.74	94.83	73.69	102.27	142.69	115.49

Source Own elaboration based on Equitics® database

Table 6 Aggregating weights for different orness levels

Weights	$\alpha = 0$	$\alpha = 0.25$	$\alpha = 0.5$	$\alpha = 0.75$	$\alpha = 1$
w_1	0	0.083333	0.166667	0.350	1
w_2	0	0.083333	0.166667	0.275	0
w_3	0	0.083333	0.166667	0.200	0
w_4	0	0.083333	0.166667	0.125	0
w_5	0	0.083333	0.166667	0.050	0
w_6	1	0.583333	0.166667	0	0

Source Own elaboration

Table 7 IOWG overall scores

Firm	$\alpha = 0$	$\alpha = 0.25$	$\alpha = 0.5$	$\alpha = 0.75$	$\alpha = 1$
F1	35	38.62	42.61	43.82	62
F2	67	58.32	50.77	45.96	40
...	...	...	...	...	...
F73	40	43.72	47.78	49.85	45

Source Own elaboration

Table 8 Ranks of the firms based on IOWG

Position	$\alpha = 0$	$\alpha = 0.25$	$\alpha = 0.5$	$\alpha = 0.75$	$\alpha = 1$
1	F2	F47	F25	F31	F62
2	F66	F66	F47	F52	F67
...	...	...	...	...	...
73	F38	F38	F50	F50	F50
# Coincidences	3	5	73	3	1

Source Own elaboration

As expected, the orness level for which the number of coincidences in the position of the firm is higher is 0.5. However, there are firms that ranking in different positions obtain similar overall scores. In order to take this into account,

Table 9 Difference between Vigeo's ranking and the IOWG based ranking

Firm	$\alpha = 0$	$\alpha = 0.25$	$\alpha = 0.5$	$\alpha = 0.75$	$\alpha = 1$
F1	23	−22	0	−6	−37
F2	−19	−19	0	−5	−20
F3	−30	−37	0	−8	−53
...	...	...	...	...	...
F73	12	12	0	0	0
# Coincidences	**17**	**21**	**73**	**13**	**12**

Source Own elaboration

we have calculated the difference between both rankings (see Table 10). We have taken into account the positions of the firms in the ranking based on the overall score of the firm (OSF) provided by Vigeo (see Table 2) and the positions of the firms in the ranking based on the IOWG operator (see Table 8). For each firm, j, we have calculated the difference between the position of the firm in the two rankings for the different orness levels:

$$D(\alpha, j) = \text{Position of } F_j \text{ in Vigeo} - \text{Position of } F_j \text{ in IOWG for } \alpha. \quad (7)$$

Table 9 displays these differences. For example, $D(0.75, 1) = -6$ means that for $\alpha = 0.75$, F1 is ranked 6 positions below in the IOWG based rank compared with its position in the Vigeo' rank.

Therefore, a positive $D(\alpha, j)$ means that the firm is worst ranked by Vigeo when compared to the IOWG-based ranking for that α. On the contrary, a negative $D(\alpha, j)$ means that Vigeo ranks the firm in a better position. The higher the $D(\alpha, j)$ in absolute terms the higher the difference between both rankings. In the last row of Table 10, we have indicated the number of coincidences in the positions of the firms taking into account the differences $D(\alpha, j)$ between the rankings. When determining if there is a coincidence or not we have distinguished between the first and last positions of the rank and the rest. For the first ten ranked firms we consider there is a coincidence in the position $D(\alpha, j) < 2$. For the last ten ranked positions we consider there is a coincidence in the position if $D(\alpha, j) < 5$. Finally, for the rest of positions we are more flexible and we permit $D(\alpha, j) < 10$.

In what follows we display the obtained results for all the companies in our sample. Column one shows the name of the company and column two its sector. The fourth column shows the ranking obtained by Vigeo and the fifth column the ranking obtained in this work based on IOWG operators for an orness level $\alpha = 0.75$. The difference in positions in both rankings is displayed, also for $\alpha = 0.75$, in the last column.

Table 10 Comparison of the rankings

Title	Country	Firm	VIGEO	IOWG	D (α, j)
Abertis infraestructuras	Spain	F1	25	31	−6
Accor	France	F2	47	52	−5
ADIDAS	Germany	F3	17	25	−8
ADP	France	F4	31	67	−36
Allianz SE	Germany	F5	52	17	35
Atlantia	Italy	F6	67	62	5
Banco Bilbao Vizcaya Argentaria (BBVA)	Spain	F7	45	32	13
Banco de Sabadell S.A.	Spain	F8	36	47	−11
Banco Espirito Santo E-R-	Portugal	F9	69	69	0
Beiersdorf	Germany	F10	54	36	18
BIC	France	F11	56	37	19
BMW	Germany	F12	62	45	17
BNP Paribas	France	F13	37	58	−21
Carrefour	France	F14	32	70	−38
CGG Veritas	France	F15	66	19	47
Coca-Cola Hellenic Bottling Co. S.A.	Greece	F16	71	30	41
Corio N.V.	The Netherlands	F17	3	66	−63
Credit Agricole SA	France	F18	58	20	38
CRH plc	Ireland	F19	20	8	12
Daimler	Germany	F20	5	77	−72
Danone	France	F21	28	56	−28
Deutsche Bank	Germany	F22	24	49	−25
Deutsche Post	Germany	F23	18	28	−10
Deutsche Telekom AG	Germany	F24	64	59	5
E.ON AG	Germany	F25	8	60	−52
EADS	The Netherlands	F26	65	4	61
EDF	France	F27	49	76	−27
ENI	Italy	F28	4	3	1
Erste Bank	Austria	F29	76	54	22
Essilor International	France	F30	60	18	42
France Telecom	France	F31	40	42	−2
GDF SUEZ	France	F32	70	29	41
Gecina NOM.	France	F33	23	10	13
Iberdrola	Spain	F34	41	41	0
Imerys	France	F35	53	68	−15
Intesa Sanpaolo S.p.A.	Italy	F36	30	74	−44
KBC	Belgium	F37	42	40	2
Klépierre	France	F38	77	6	71
Lafarge	France	F39	2	5	−3

(continued)

Table 10 (continued)

Title	Country	Firm	VIGEO	IOWG	D (α, j)
LEGRAND	France	F40	26	71	−45
L'Oreal	France	F41	43	64	−21
LVMH Moet Hennessy	France	F42	19	24	−5
Michelin	France	F43	6	55	−49
POSTNL	The Netherlands	F44	27	79	−52
PPR	France	F45	15	65	−50
PSA Peugeot Citroën	France	F46	59	43	16
Red Electrica De Espana	Spain	F47	10	23	−13
Reed Elsevier NV	The Netherlands	F48	63	53	10
Renault	France	F49	74	26	48
RWE AG	Germany	F50	68	11	57
Safran	France	F51	79	61	18
Saint-Gobain	France	F52	29	78	−49
Saipem	Italy	F53	55	72	−17
Sanofi-Aventis	France	F54	51	27	24
SBM OFFSHORE	The Netherlands	F55	22	15	7
Schneider Electric	France	F56	16	2	14
Siemens AG	Germany	F57	78	22	56
Snam Rete Gas	Italy	F58	75	44	31
Societe Generale	France	F59	72	16	56
Sodexo S.A.	France	F60	14	63	−49
Suez Environnement	France	F61	11	48	−37
Technip	France	F62	73	51	22
Telecom Italia	Italy	F63	44	1	43
Telefonica	Spain	F64	61	57	4
Terna	Italy	F65	48	14	34
Thales	France	F66	46	33	13
Unibail—Rodamco	France	F67	1	75	−74
UniCredit	Italy	F68	57	46	11
Unilever nv	The Netherlands	F69	38	34	4
Vallourec	France	F70	21	73	−52
Veolia Environnement	France	F71	34	38	−4
VINCI	France	F72	33	21	12
Volkswagen PREF	Germany	F73	50	50	0

Source Own elaboration

6 Conclusions

The construction and handle of sustainability indices require managing of concepts which by their own nature have been formulated imprecisely. In these circumstances, Fuzzy Logic and, in particular, their aggregation methods, are capable of handling this uncertainty inherent to some phenomena and they result very useful for the construction of operational sustainability indices.

Well-known rating agencies calculate overall sustainability scores based on the geometric mean with equal weights for the different social dimensions. In this work, we introduce a flexible aggregating method that extends the one used by the social rating agency Vigeo. Based on the geometric mean, the use of IOWG allows aggregation with objective weights and the incorporation of the differences in the nature of data corresponding to different dimensions. Different economic sectors have traditionally perform better in certain social dimensions due to their type of economic activities. Taking this fact into account when evaluating a firm in terms of its sustainable performance seems to be a key question.

References

1. Brundtland, G.H.: Our Common Future. World Commission on Environment and Development, Brussels (1987)
2. Dyllick, T., Hockerts, K.: Beyond the business case for corporate sustainability. Bus. Strategy Environ. **11**, 130–141 (2002)
3. Eurosif: European SRI Study 2014. Eurosif http://www.eurosif.org (2014)
4. Gil Aluja, J.: Elements for a Theory of Decision in Uncertainty. Kluwer Academic Publishers, Boston (1999)
5. Ballestero, E., Pérez-Gladish, B., García-Bernabeu, A. (eds.): Socially Responsible Investment. In: A Multi-criteria Decision Making Approach. International Series in Operations Research & Management Science, **219**, Springer (2015)
6. Gil Lafuente, A.M., Merigó, J.M.: Decision making techniques with similarity measures and OWA operators. Sort-Stat. Oper. Res. Trans. 81–102 (2012)
7. Kaufmann, A., Gil Aluja, J.: Técnicas Operativas de Gestión para el Tratamiento de la Incertidumbre. Hispano Europea, Barcelona (1987)
8. Gil Aluja, J.: Investment in Uncertainty. Kluwer Academic Publishers, Dordrecht (1998)
9. Gil Aluja, J.: The Interactive Management of Human Resources in Uncertainty. Kluwer Academic Publishers, Dordrecht (1998)
10. Gil Aluja, J.: La gestión interactiva de los recursos humanos en la incertidumbre. CEURA, Madrid (1996)
11. Gil Aluja, J.: Aproximación metodológica a la optimización en la incertidumbre. In: Escudero, L.F., Cerdá, E., Alonso-Ayuso, A., Sala Garrido, R. (eds.) En: Optimización bajo incertidumbre Rect@, **2**, 23–50 (2002)
12. Liern, V., Pérez-Gladish, B., Méndez-Rodríguez, P.: Measuring social responsibility: a multicriteria approach. In: Zopounidis, C., Doumpos, M. (eds.) Multiple Criteria Decision Making: Applications in Management and Engineering. Multicriteria Decision Making Book Series. Springer (in press) (2015)
13. Xu, Z.S., Da, Q.L.: The ordered weighted geometric averaging operators. Int. J. Intell. Syst. **17**(7), 709–716 (2002)

14. Herrera, F., Herrera-Viedma, E., Chiclana, F.: A study of the origin and uses of the ordered weighted geometric operator in multicriteria decision making. Int. J. Intell. Syst. **18**(6), 689–707 (2003)
15. Xu, Z.S.: On generalized induced linguistic aggregation operators. Int. J. Gen Syst **35**(1), 17–28 (2006)
16. Yager, R.R.: On ordered weighted averaging aggregation operators in multi-criteria decision making. IEEE Trans. Syst. Man Cybern. **18**(1), 183–190 (1988)
17. Yager, R.R., Filev, D.P.: Induced ordered weighted averaging operators. IEEE Trans. Syst. Man Cybern. Part B: Cybern. **29**, 141–150 (1999)
18. Yager, R.R.: Induced aggregation operators. Fuzzy Sets Syst. **137**(1), 59–69 (2003)
19. León, T., Ramón, N., Ruiz, J.L., Sirvent, I.: Using induced ordered weighted averaging (IOWA) operators for aggregation in cross-efficiency evaluations. Int. J. Intell. Syst. **29**, 1100–1126 (2014)
20. Merigó, J.M., Gil Lafuente, A.M., Zhou, L.G., Chen, H.Y.: Induced and linguistic generalized aggregation operators and their application in linguistic group decision making. Group Decis. Negot. Springer, Holland (2012)
21. Wang, Y.M., Parkan, C.: A minimax disparity approach for obtaining OWA operator weights. Inf. Sci. **175**(1–2), 20–29 (2005)

Part V
Financial Analyses

About Formal Construction of Financial Analysis

Alfonso M. Rodríguez

Abstract The conventional Financial Mathematics is, principally, financial calculi. Nevertheless, the financial phenomenon, the *preference* for *liquidity*, it would must be always present into the temporal economic studies. The *financial value* considers, joint the monetary amount, its temporal deferral, its liquidity. Both of them must be formalized in a binary vector (C, T), the *financial capital,* reclaiming different mathematic that financial conventional simply arithmetic (*scalar*). The financial analysis reclaims the *vector* Financial Mathematic that we propos. *Financing financial operations* (FFO) and *investment financial operations* (IFO) are very different *financial operations* (FO) by their purpose. FFO only pretends to get an *interest,* the financing price by its financial service. IFO intents to get an *investment yield*, its economic result. Interest and investment yield are different economic magnitudes. The interest is appointed by a financial market and it defines the financial *equilibrium* of FFO (finance equivalence). The investment yield is an economic result appointed by the financial *disequilibrium* of IFO. It is the reason why to Investigate an investment yield as a financial equilibrium, as an implicit interest, it is the grave financial mistake that IRR (Internal Rate of Return) commits, confusing investment yield with interest and an IFO with a FFO, with erroneous consequences to its definition and possible to investment decisions. The new methodological Financial Mathematics permits to revise conventional concepts as *interest, investment yield,* financial *productivity,* financial *profitability*, etc. also permitting to incorporate another unknown concepts as financial *degeneration,* financial *immunity,* etc, and to know other financial instruments, alternative to IRR, to investment decision.

Keywords Financial mathematics · Mathematic of financing · Mathematic of investment · Financial laws · Financial regimes · Financial operations (FO) Financing operations (FFO) · Investment operations (IFO) · Internal rate of return (IRR) · Financial immunity · Financial degeneration · Financial decision

A. M. Rodríguez (✉)
University of Barcelona, Barcelona, Spain
e-mail: alfonsomrr@telefonica.net

© Springer International Publishing AG 2018
C. Berger-Vachon et al. (eds.), *Complex Systems: Solutions and Challenges in Economics, Management and Engineering*, Studies in Systems, Decision and Control 125, https://doi.org/10.1007/978-3-319-69989-9_22

1 Introduction

This work pretends to show the necessary fundaments for a correct financial analysis. The financial analysis, even being not speculative discipline, it needs the sufficient fundaments for its formal construction, in order to get financial concepts and results more reliable that conventional analysis obtains.

It supposes a great satisfaction for me to dedicate this collaboration to Excm. Dr. Prof. Gil Aluja in your happy 80th anniversary, admired university-mate, respected investigator on common disciplines, especially old and excellent friend, the actual President of Spanish Royal Academic of Economics and Financial Sciences (RACEF), witch I belongs honored.

Consciously I pretends to differentiate methodological fundaments from their analytic development, foundations from their building, doctrinal analysis from theirs empirical applications. It is the reason why I put my principal attention on the basis permitting a formal construction of financial analysis, to revise the conventional and to consider its possible projection to news investigations and financial results. It is too frequent that any urgencies to get applications, before the deep knowledge of theirs fundaments, it is motivate by a precipitate inquiry, ¿what it is for?. Lamentably, that question incites to obviate—worst to devaluate—a previous concepts and principles without them soon they are closed the frontiers to scientific analysis. Nevertheless, we will here include any applications showing the *utility* of such principles.

Two lines explains the development of this work. The first, conceptual, it shows as formal object of financial analysis the economic phenomenon of *preference* for *liquidity* (for the full availabilities or for defer requirements). The second, methodological, intents to construct a financial-mathematical model sufficiently apt to the full description of *financial operations* (they were of *financing* or *investment*), according to their purpose. The first line is inserted into the Temporal Economical Theory that it considers the time as economic element to formation of economic value, the *financial value,* introducing the *preference* for *liquidity* for the economic valuation. The second line it constructs a Financial Mathematic, best to a simple financial calculi, founded on Definitions and Postulates configuring the necessary mathematic model permitting the logic-deductive application.

In the applied sciences, theory and reality are not opposed. Both them grow joint, nourished themselves in a permanent *popperiana falsation*. It corresponds to theory the conceptual abstraction of the phenomenon, with better analytical profundity than simple observation. That happens in financial analysis defining *financial laws* as an interpretation of the *preference* for *liquidity,* while *financial regimes* (simple or composed interest) only are their practice application. That is the reason why the financial laws sow different temporal character than financial regimes, *continuous* the first (as the time flow), practice *discreet* the last. Nevertheless, conventional financial analysis confuses financial regimes with financial laws, thus reverting financial mathematics in practice financial calculi.

The quantitative monetary valuation of a patrimonial mass, active or passive, the *account valuation,* only shows an amount in monetary units of immediate liquidity, thus forgetting the financial phenomenon.[1] Such *monetary valuation* is very different to a *financial valuation,* that it considers different economic values to the future monetary units. That is the reason why *financial capital* (expression of financial valuation) is formally represented by binary vector with two component magnitudes, *monetary amount* and *temporal deferral* (liquidity). The monetary amount coincides with accounting valuation an the temporal deferral exposes the wait to full availability (or defer requirement) trough a *temporal interval* that finishes in a punctual expiration.[2]

It implies that mathematical description of financial phenomenon must be, not a *scalar* mathematic, but a *vector* mathematic, doted with different algebra that scalar arithmetical. That is the reason why Financial Mathematic we defend differs to conventional. Nevertheless, Financial Mathematic can not be speculative discipline because the latest objective is the full economic analysis of financial operations, financing or investment, and financial decisions.

The *financial operations* (FO) can have different nature according to theirs distinct objectives. Any FO only can pretend to finance alien projects (consumption or production), being *financing financial operations* (FFO). FFO only pretend to get a *price* for their financing service, a *price certain,* appointed by financial market, the *interest.* Differently, other FO pretend to invert in projects for get and *result,* the uncertain *investment yield,* then being *investment financial operations* (IFO). The Investment yield can or do not accumulate an interest (auto-financing), by that being *brut* yield or *net* yield respectively. A financial *equilibrium* (financial equivalence) only exists in FFO appointed by the *interest.* In IFO only exists a financial *disequilibrium* origin of *investment yield.*

Finally, when we refer us to *financial operations* (FO), we do it in a wide sense that it includes, not only the assignations of monetary capitals, but also other patrimonial assignations whose values are financially represented by *financial capitals.*

2 Financial Mathematic Model

Mathematic model integrated by Definitions, Operations and Postulates.

Financial capital (C, T).—*Binary vector* (C, T), being C *monetary amount* and T *temporal deferral* (liquidity).

Financial equivalence (~).—*Equivalence relation* in financial capitals set, defined by a *financial factor* $f(T, T')$ such that

---

[1]*"Ensayo sobre la Contabilidad de la liquidez"*. International Premium "Antonio Rodríguez Sánchez, 1978". Rodríguez, A. M. Instituto de Censores Jurados de Cuentas de España. It shows the financial disequilibrium existing in an equilibrate accounting balance.

[2]Deferral T and expiration t must not be confused. Deferral T is a temporal term. Expiration t is a temporal punctual reference to deferral end (whose financial origin is not always the actual time). That is the reason why C_t (amount C *in* t) is a *scalar* number, while (C, T) (amount C *with* T deferral) is a complex binary *vector* representing *financial capital,* unlike the conventional analysis.

$$(C, T) \sim \left(C', T'\right) / C' = C \cdot f\left(T, T'\right).$$

Financial sum (Σ).—*Internal operation* in financial capital set, being the sum $(\mathbf{C}, \mathbf{T}) = \Sigma\{(C_r, T_r)\}$, $r = 1, 2, \ldots$ n, such that

$$\mathbf{C} = \sum C_r(aggregate\ amount)$$

$$\mathbf{T} / \sum C_r f(T_r, \mathbf{T}) = \mathbf{C}(average\ deferral).$$

Postulate I (of equivalence).—*"All financial capital set is financially equivalent to its sum,*

$$\{(C_r, T_r)\} \sim (\mathbf{C}, \mathbf{T})^{''}.$$

Financial operation (FO).—*Formal scheme* integrated by two financial capitals sets,

$$input; \quad \{(C_r, T_r)\}, \ r = 1, 2, \ \ldots \ n$$

$$output; \quad \left\{\left(C'_s, T'_s\right)\right\}, \ s = 1, 2, \ \ldots \ m.$$

Reduced financial (FO).—*Simple* FO whose *input* and *output* are the sums of *input* and *output* of the *complex* FO, being

$$input; \quad (\mathbf{C}, \mathbf{T})$$

$$output; \quad \left(\mathbf{C}', \mathbf{T}'\right)$$

It is *financial average term* of FO (FAT) the difference **t** between the *output* and *input deferral* of the reduced FO,

$$t = \mathbf{T}' - \mathbf{T}$$

Postulate II (of equivalence).—*"All FO is financially equivalent to its reduced one"*.
Financing financial operation (FFO).—FO whose *input* and *output* are financially equivalents,

$$\{(C_r, T_r)\} \sim \left\{\left(C'_s, T'_s\right)\right\}$$

Corollary I: *By the transitivity of equivalence relation,* input and output of reduced *FFO* are financially equivalents,

$$(\mathbf{C}, \mathbf{T}) \sim \left(\mathbf{C}, {}'\mathbf{T}'\right)$$

It is *interest* of FFO (**I**) the difference between *output* and *input* amount of the reduced FFO,

$$I = C' - C$$

Investment financial operation (IFO).—FO whose *input* and *output* are not financially equivalents,

$$\{(C_r, T_r)\} \neq \{(C'_s, T'_s)\}$$

Corollary II: *They aren't financially equivalents the* input and output of reduced *IFO*,

$$(\mathbf{C}, \mathbf{T}) \neq (\mathbf{C}', \mathbf{T}')$$

3 Financial Immobilization

Magnitude describing the *temporal permanence* of monetary amount in a FO.

In *simplex* FO, where *input* and *output* are unitary sets, *financial immobilization* is permanently constant, being defined by the binary vector [C, t],

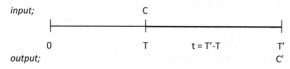

In *complex* FO, where the permanence of monetary amount is variable according to

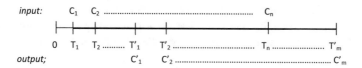

the financial immobilization is defined by its equivalent *reduced* FO one, [**C**, **t**].

4 Financing Cost (Interest)

Magnitude dependent of a *financial law* that it rules financial market, formally defined by a *financial factor* $f(T, T')$, that it signifies the necessary *cost* to finance financial immobilization.

In a FFO, the *interest* define the financial law of *equilibrium* between *input* and *output*,

$$\left(\mathbf{C'},\mathbf{T'}\right)\sim\left(\mathbf{C},\mathbf{T}\right)$$

being

$$\mathbf{I}=\mathbf{C'}-\mathbf{C}=\mathbf{C}\cdot f\left(\mathbf{T},\mathbf{T'}\right)-\mathbf{C}=\mathbf{C}\cdot\left[f\left(\mathbf{T},\mathbf{T'}\right)-1\right].$$

In IFO, other different amount, $\mathbf{C''}=\mathbf{C}\cdot f(\mathbf{T},\mathbf{T'})$, defines a market equilibrium with $(\mathbf{C},\mathbf{T})$,

$$\left(\mathbf{C''},\mathbf{T'}\right)\sim\left(\mathbf{C},\mathbf{T}\right)$$

then being

$$\mathbf{I}=\mathbf{C''}-\mathbf{C}=\mathbf{C}\cdot f\left(\mathbf{T},\mathbf{T'}\right)-\mathbf{C}=\mathbf{C}\cdot\left[f\left(\mathbf{T},\mathbf{T'}\right)-1\right].$$

Both $\mathbf{I}$ are monetary amount with deferral $\mathbf{T'}$, $(\mathbf{I},\mathbf{T'})$.

5 Financial Law (Financial Regimes)

A *financial law* is formally defined by a *financial factor* $f(\mathrm{T},\mathrm{T'})$. Discounting risk, the financial law interprets the *preference* for *liquidity* existing in a financial market, *equilibrium* (financial equivalence) appointed by a *strict interest rate* ρ (*strict financing price*).

A financial law can temporally defined variable (dynamic law) or static (stationary law). In dynamic law the financial factor determinates ρ as continuous function of time, $\rho(\tau)$. In stationary law it is ρ a constant, interpreting of the constant interest formation in the time, being then the financial factor,

$$f\left(\mathrm{T},\ \mathrm{T'}\right)\equiv f(\mathrm{t})=e^{\rho\cdot\mathrm{t}}$$

being

$$t=\mathrm{T'}-\mathrm{T}\text{ and }\rho=\frac{\ln f(\mathrm{t})}{\mathrm{t}}$$

Financial regimes (interest simple or compound) are practice financial rules to calculate interest with rate i. They are an empirical approximation to financial laws. The financial laws interpret *continuous* the interest generation on time, opposite the financial regimes that they consider *discrete* generation being simple approximation to continuous interest formation. That justifies a frequency normally inferior to year, also their i reference to a periodical frequency that ρ does not.

In compound-interest regime, with annual capitalization, the *empiric financial factor* is

$$f(\mathrm{T},\ \mathrm{T}') = (1+i)^{\mathrm{t}} = e^{\rho \cdot \mathrm{t}}$$

being the equivalent ρ,

$$\rho = \ln(1+i)$$

6 Implicit Financial Law

In equivalence of FFO,

$$\{(C_r, T_r)\} \sim \{(C'_s, T'_s)\}$$

it is not explicit the financial law existing. To a stationary financial law, with factor $f(\mathrm{T},\ \mathrm{T}') = e^{\rho \cdot \mathrm{t}}$, interest rate ρ must satisfy the equation between the financial actual values

$$\sum C_{r}, \cdot e^{-\rho \cdot \mathrm{Tr}} = \sum C'_{s} \cdot e^{-\rho \cdot \mathrm{T's}}$$

polynomial-exponential equation of indeterminate solutions to ρ^*.

To an annual composed regime, with empirical financial factor $(1 + i)^{\mathrm{t}}$, interest rate i must satisfy similar equation

$$\sum C_{r} \cdot (1+i)^{-\mathrm{Tr}} = \sum C'_{s} \cdot (1+i)^{-\mathrm{Ts'}}$$

polynomial equation of indeterminate solutions i^* also.

Postulate II permits to investigate solutions ρ^* in equivalent *reduced* FFO

$$(\mathbf{C}, \mathbf{T}) \sim (\mathbf{C}', \mathbf{T}')$$

being equation between actual values,

$$\mathbf{C} \cdot e^{-\rho \cdot \mathrm{T}} = \mathbf{C}' \cdot e^{-\rho \cdot \mathrm{Ts'}}$$

and also

$$\rho \cdot \mathrm{t} = \ln \frac{\mathbf{C}'}{\mathbf{C}} = k$$

equation of equilateral hyperbole parameter k, being **t** (FAT) function ρ,

$$\mathbf{t}(\rho) = \mathbf{T}'(\rho) - \mathbf{T}(\rho)$$

Solutions ρ^* are solutions of system,

$$\begin{cases} y = \mathbf{t}(\rho) \\ y = k/\rho \end{cases}$$

whose graphic representations they are intersections of theirs respective graphics.

Solutions ρ^* are, by particular analytic proprieties, none solution, only one, two, with different sign, finally three solutions, one with opposite sign.[3]

Being coherent solutions ρ^*, in sign and amount, with a market financial price, they define markets where it is possible the FFO. It is not possible if there is none solution. Such case, and possible multiplicity markets, it is due to a particular temporal distribution of amounts that integrating a complex FFO.[4]

Mistakenly the conventional analysis IRR attributes to implicit interest ρ^* the investment yield meaning in IFO. Thus confusing *implicit interest* with *investment yield*, the IFO with a FFO. IRR substitutes the financial *disequilibrium* of IFO for financial *equilibrium,* alone existing in FFO. Such mistake is more evident considering possibility none solution ρ^*, or multiple ρ^* solutions existing, absurd considering the only unique possible measure to the investment yield, on the other hand always existent.

We will show now two numerical FFO, one without possible solution ρ^*, the other with three solutions ρ^*_1, ρ^*_2 and ρ^*_3. We will add the equivalent rates I in annual composed interest.

1ª FFO.

input: $\{(100;1),(50;2),(50;3)\}$

output: $\{(100;0),(85;4),(12;5),(8;6)\}$

without solutions ρ^* (IRR).

Its graphic representation, without intersections, is.

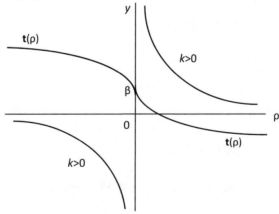

---

[3]Vid. *"Matemática de la Financiación"*. Cap. 9. Págs. 205–213. Rodríguez, A. M. Barcelona University (1994) [2].

[4]Vid. previous reference.

2ª FFO.

$$\text{input:} \quad \{(100;1),(90;2),(10;5)\}$$

$$\text{output:} \{(80;0),(40;3),(90;4)\}$$

with three solution ρ^* (IRR),

$$\rho^*_1 = -2.231567 \sim i = -89.26\%$$
$$\rho^*_2 = 0.095698 \sim i = 10.04\%$$
$$\rho^*_3 = 0.367966 \sim i = 44.00\%$$

being the graphic representation, with three intersections,

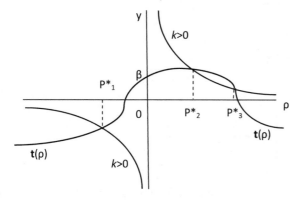

7 Improper IFO (Degeneration)

The FAT of *complex* FIO is negative when average deferral *input* is major than deferral *output*,

$$\mathbf{T}(\rho) > \mathbf{T}'(\rho)$$

It is possible due a particular temporal distribution of amounts existing in IFO (e.g. when IFO enjoys a previous subvention). Then, ρ^* (IRR) inverts its result sense, benefit or loss. In effect, being

$$\rho^* = \frac{k}{\mathbf{t}(\rho^*)}$$

and

$$t(\rho) = \mathbf{T}'(\rho) - \mathbf{T}(\rho) < 0$$

sign ρ^* is opposite to sign k.
  Being

$$k = \ln \frac{\mathbf{C}'}{\mathbf{C}}$$

it is positive k when $\mathbf{C}' > \mathbf{C}$, then signifying benefit being ρ^* (IRR) negative. Opposite, when $\mathbf{C}' < \mathbf{C}$ k is negative, then signifying loss being ρ^* (IRR) positive. Such grave contradiction to IRR even adds moreover confusion to meaning of IRR.
  Being $t(\rho)$ a continuous function (except $\rho = 0$ with evitable discontinuity, t $(0) = \beta^5$), an interval with extremes 0 and $\rho^{d)}$ ($t(\rho^{d)} = 0$, $\mathbf{T}'(\rho^{d)} = \mathbf{T}(\rho^{d)}$) defines the *degeneration interval* to $\rho > 0$,

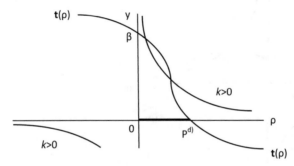

  To conventional analysis FAT of *complex* IFO is unknown magnitude. It substitutes FAT by the *duration,* magnitude without proprieties for a *financial reduction.* This is so conventional financial analysis ignores the possible *degeneration,* and the *improper* IFO, when FAT is negative.

8   Investment Yield (Profitability)

The *investment yield* of an IFO is an *absolute* magnitude that it defines the IFO result. It can be *brut* or *net* according to it integrates interest (self-financing) or not do. Only the *net investment yield* is a properly economic result, because it only discounts all necessary costs to an IFO (including opportunity cost).
  It is *brut investment yield,*

$$\bar{\mathbf{R}} = \mathbf{C}' - \mathbf{C} = \mathbf{C} \cdot [f(\mathbf{T}, \ \mathbf{T}') - 1]$$

---

[5]Vid. *"Matemática de la Inversión"*. Cap. 2. Págs. 27–34. Rodríguez, A. M. Barcelona University (1995) [3].

to a stationary financial law,

$$\bar{R} = C' - C = C \cdot [e^{\rho \cdot t(\rho)} - 1]$$

It is *net investment yield*,

$$\widehat{R} = \bar{R} - I = C' - C'' = C' - C \cdot f(T, T')]$$

to a stationary financial law,

$$\widehat{R} = C' - C \cdot e^{\rho \cdot t(\rho)}$$

The *profitability* of IFO is a *relative* magnitude that it relates investment yield **R** and financial immobilization $[C, t(\rho)]$. Also It can be *brut* $\bar{\rho}$ or *net* $\hat{\rho}$, according to investment yield $\bar{R}$ or $\widehat{R}$. To interest rate ρ they are,

$$\hat{\rho} = \bar{\rho} - \rho$$

Being,

$$C' = C \cdot e^{\rho \cdot t(\rho)}$$

it is

$$\bar{\rho} \cdot t(\rho) = \ln \frac{C'}{C} = k$$

finally,

$$\bar{\rho} = \frac{k}{t(\rho)} \tag{1}$$

It is grave conceptual mistake to confuse interest with investment yield. Interest is the financing cost of IFO, with positive sign as any cost (salving the economic politic). The nature of interest is *ex ante* to investment yield. Investment yield is the result of IFO, having nature *ex post* to interest. Its sign can be positive or negative, according to the result sense, benefit or loss. It is not admissible the inexistency (possible the nullity), neither the multiplicity. Nevertheless, both them are possible to the confuse $\rho*$ (IRR) meaning.[6]

---

[6]Vid. 4.

9 Financial Immunity and Duration

Formulation (1) permits to define $\rho = \rho^{i)}$ to min $\bar{\rho}$ (to max $\mathbf{t}(\rho)$), being then $\rho^{i)}$ the strict interest rate that it *immunizes* $\bar{\rho}$ facing a posterior volatility of interest ρ. In effect, being

$$\bar{\rho} = \frac{k}{\mathbf{t}(\rho^{i)})}$$

$\bar{\rho}$ has the minor value when $\rho = \rho^{i)}$, immunizing $\bar{\rho}$ to any posterior change in ρ. For to determine $\rho^{i)}$ it is necessary the nullity in the first derivate of $\mathbf{t}(\rho)$,

$$\frac{d\,\mathbf{t}(\rho^{i)})}{d\rho} = 0$$

Being,

$$\mathbf{t}(\rho) = \mathrm{T}'(\rho) - \mathrm{T}(\rho) = \frac{1}{\rho}\left(\ln\frac{\mathbf{C}'}{\mathrm{V'}_0} - \ln\frac{\mathbf{C}}{\mathrm{V}_0}\right) = \frac{1}{\rho}\left(\ln\frac{\mathbf{C}'}{\mathbf{C}} - \ln\frac{\mathrm{V'}_0}{\mathrm{V}_0}\right) = \frac{k - \Gamma(\rho)}{\rho}$$

where,

$$\mathrm{V'}_0 = \sum \mathrm{C}'_s \cdot e^{-\rho.\mathrm{T's}}$$
$$\mathrm{V}_0 = \sum \mathrm{C}_r \cdot e^{-\rho.\mathrm{Tr}}$$
$$\Gamma(\rho) = \ln\frac{\mathrm{V'}_0}{\mathrm{V}_0}$$

finally,

$$\rho \cdot \mathbf{t}(\rho) = k - \Gamma(\rho) \tag{2}$$

equation that it determines the *immunization rate* $\rho^{i)}$.

On the other hand, we have before referred that amount variable permanence in *complex* IFO it is studied by conventional analysis recurring to *duration* (substitutive of its unknown FAT). The *duration* weights the permanence of amounts in IFO with their proper imports. Particularly, Macauley[7] calculates the *duration* financial actualizing monetary amounts with market interest, thus defining *durations* of *input* and *output*,[8]

---

[7] 1.

[8] $\rho = \ln(1 + i)$.

$$D(\rho) = \frac{\Sigma T_r \cdot C_r \cdot e^{-\rho.T_r}}{\Sigma C_r \cdot e^{-\rho.T_r}} = -\frac{d \ln V_0(\rho)}{d\rho}$$

$$D'(\rho) = \frac{\Sigma T_s' \cdot C_s' \cdot e^{-\rho.T_s'}}{\Sigma C_s' \cdot e^{-\rho.T_s'}} = -\frac{d \ln V_0'(\rho)}{d\rho}$$

and *duration* of IFO,

$$\mathbf{d}(\rho) = D'(\rho) - D(\rho) = -\left(\frac{d \ln V_0'(\rho)}{d\rho} - \frac{d \ln V_0(\rho)}{d\rho}\right) = -\frac{d}{d\rho}\left(\ln \frac{V_0'(\rho)}{V_0(\rho)}\right) = -\frac{d\Gamma(\rho)}{d\rho}$$

Now, considering (2),

$$\rho \cdot \mathbf{t}(\rho) = k - \Gamma(\rho)$$

and deriving to ρ,

$$\mathbf{t}(\rho) + \rho \cdot \frac{d\mathbf{t}(\rho)}{d\rho} = -\frac{d\Gamma(\rho)}{d\rho} = \mathbf{d}(\rho)$$

finally it is,

$$\mathbf{d}(\rho) - \mathbf{t}(\rho) = \rho \cdot \frac{d\mathbf{t}(\rho)}{d\rho} \tag{3}$$

that it exposes the existing relation between Macauley¡s *duration* and FAT, both them coinciding to $\rho = 0$ and to $\rho = \rho^{i)}$.

Graphically,

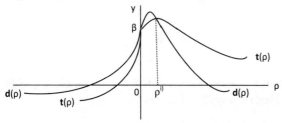

Being interest rate ρ external parameter to IFO, only dependent of financial market, in order to get the immunization of $\bar{\rho}$, facing the volatility of interest ρ, in origin the IFO would must be designed being $\rho^o = \rho^{i).9}$ Then, $\bar{\rho}$ and $\hat{\rho} = \bar{\rho} - \rho^o$,

---

[9] ρ^o represents the strict financing price of IFO in origin.

would be beneficed by any posterior variation of ρ. This is so the immunity would be as much effective as better were the approximation in origin between ρ^o and *immunity rate* $\rho^{i).10}$

10   The Financial Investment Decisions

At before paragraphs we have disqualified IRR (Internal Rate of Return) as acorrect financial instrument in order to get the better selection facing an investment alternative. We already have referred their conceptual mistakes, possible absurdities and possible confuse results. It demands to investigate other more correct financial instruments substitutive of IRR. We have defined already as possible $\bar{p}$ and $\hat{p}$, *financial yield rates* (FYR), as measures of *profitability* in IFO.

Now we want to go deeper into economic concepts of *productivity* and *profitability*. The *productivity* relates, *in physics units,* the obtained product with the employed factor. The *profitability* relates, *in monetary units,* the obtained product with the cost of employed factor. To translate these concepts to a IFO creates any confusion. In effect, investment yield **R** is the obtained product in IFO being always computed *in monetary units.* The financial immobilization [**c, t**] is the employed factor in FIO, being always computed *in monetary units for year.* Then, the *productivity* in a FIO is always expressed *in monetary units for year* by rates brut $\bar{p}$ and net $\hat{p}$.

On the other hand, strictly *profitability* in a FIO relates the obtained result **R** with the cost of the employed financial factor [**c, t**], that it is the interest **I**. Then, being strictly it is

$$brut\ profitability:\ \bar{\delta} = \frac{\bar{R}}{I}$$

$$net\ profitability:\ \hat{\delta} = \frac{\hat{R}}{I} = \frac{\bar{R} - I}{I} = \bar{\delta} - 1$$

Then they are three relative financial instruments that they permit to decide the better economic option facing an investment alternative. They are,

$$\rho * IRR\ (implicit\ interest)$$

$$\bar{p}\ or\ \hat{p}\ (strict\ productivity)$$

$$\bar{\delta}\ or\ \hat{\delta}\ (strict\ profitability)$$

---

[10]Vid. *"Matemática de la Inversión"*. Cap. 4. Págs. 48–78. Rodríguez, A. M. Barcelona University (1995) [3].

[11]The decision of the better option differs according to economic pronouncement.[12] Conventional IRR excluding, productivity or profitability must be selected to get the better option. In any case, it responds to previous inquiry, ¿*that it is for*?

Relational Bibliography

1. Macaulay, F.R.: Columbia University Press (1938)
2. Rodríguez, A.M.: Financial Mathematics. Barcelona University (1994)
3. Rodríguez, A.M.: Investment Mathematics. Barcelona University (1995)
4. Rodríguez, A.M.: Mistakes and dysfunctions of IRR. An alternative instrument, FYR. Tribuna Plural (READ) N° 9, 1/2016

---

[11]$\bar{\rho}$ and $\hat{\rho}$ differ In ρ°. $\bar{\delta}$ and $\hat{\delta}$ differ in 1. Both them differs in a constant being indifferent thus to better selection.

[12]Vid. "Matemática de la Inversión". Cap. 5. Págs. 79–87. Rodríguez, A. M. Barcelona University (1995) [3].

Stock Returns Forecast: An Examination By Means of Artificial Neural Networks

Martín Iglesias Caride, Aurelio F. Bariviera and Laura Lanzarini

Abstract The validity of the Efficient Market Hypothesis has been under severe scrutiny since several decades. However, the evidence against it is not conclusive. Artificial Neural Networks provide a model-free means to analize the prediction power of past returns on current returns. This chapter analizes the predictability in the intraday Brazilian stock market using a backpropagation Artificial Neural Network. We selected 20 stocks from Bovespa index, according to different market capitalization, as a proxy for stock size. We find that predictability is related to capitalization. In particular, larger stocks are less predictable than smaller ones.

1 Introduction

The Efficient Market Hypothesis (HME) is a theoretical framework, which constitutes the baseline of financial economics. It was developed following the guidelines of neoclassical economics [1]. According to the standard definition by Fama [2], a market is said to be informational efficient if prices conveys all relevant information. The definition immediately leads to the determination of the relevant information set. For this reason, Fama [2] classifies informative efficiency into three categories: (i) Weak efficiency, when the current price contains information from the past series of prices; (ii) Semi-strong efficiency, when the past price contains all the public information associated with that asset; and (iii) strong efficiency, when

M. Iglesias Caride (✉)
Master Program in Data Mining, Universidad de Buenos Aires, Buenos Aires, Argentina
e-mail: jmartiniglesias@yahoo.com.ar

A.F. Bariviera
Department of Business, Universitat Rovira i Virgili, Tarragona, Spain
e-mail: aurelio.fernandez@urv.cat

L. Lanzarini
Instituto de Investigación en Informática LIDI, Facultad de Informática,
Universidad Nacional de La Plata, La Plata, Argentina
e-mail: laural@lidi.info.unlp.edu.ar

© Springer International Publishing AG 2018
C. Berger-Vachon et al. (eds.), *Complex Systems: Solutions and Challenges in Economics, Management and Engineering*, Studies in Systems, Decision and Control 125, https://doi.org/10.1007/978-3-319-69989-9_23

the price reflects all public and private information relating to that asset. HME has been the unquestioned paradigm in economics until the 1980s. With the advent of personal computers and increased availability of data, HME testing became easier. Thus, in those years began to appear articles questioning the weak version of informational efficiency. The first articles inquired about seasonal effects on returns. Later, researchers began to look at return forecasts. In a review of his 1970 article, Fama [3] renames the first category of weak efficiency as *tests for return predictability*, where explanatory variables, in addition to past performance may be other financial variables such as interest rates, price-earning ratio, etc. If EMH is an adequate description of market behavior, the forecast of returns (whether based on past returns or adding other variables) is ruled out. In particular, all chartist theories that advocate the existence of more or less fixed patterns in the financial markets are discarded.

The aim of our paper is to test the EMH in its weak version for the Brazilian stock market, using sign prediction (upward or downward movement) through an Artificial Neural Network (ANN).

The rest of the paper is organized as follows: Sect. 2 makes a brief review of the use of ANN in economics; Sect. 3 describes the methodology; Sect. 4 describes data used in the empirical application carried out in Sect. 5. Finally, Sect. 6 draws the main conclusions of our research.

2 Stock Market Forecasts: Artificial Neural Networks

The use of ANNs to economic and financial problems is very prolific. ANNs, depending on their architecture and type, can be used to classify, to optimize and to forecast. This versatility, including their robustness and ability to deal with nonlinear dynamics, is undoubtedly part of their success. Usually, many business problems involve several variables, with an unknown functional (nonlinear) relationship. This situation enhances the use of data driven models that can be model using ANNs.

Göçken et al. [4] propose an hybrid model consisting in a harmony search or a genetic algorithm in combination with an ANN, in order to enhance the capability of return forecasts in the Turkish stock market. An hybrid approach was also selected by Qiu et al. [5]. They combine the ANN with a genetic algorithm or a simulated annealing, in order to filter the variables to be set in the input layer.

In addition to being used in stock market forecasting, ANNs are used also in credit scoring. For example, Lanzarini et al. [6] use and hybrid approach, that combines a neural network with an optimization technique, in order to classify customers of two public databases. Although results are rather similar in terms of accuracy, this novel approach allows to simplify the number of rules to obtain a given level of accuracy. The same technique was extended in [7] to a real and large database of customers of a financial institution in Ecuador. The outcome was a more intuitive customer classification, which helps managers to know with greater detail which variables really matters in a client credit application.

A recent survey by Tkáč and Verner [8], finds 412 articles in leading academic journals that apply ANN to business and economics. The survey covers the last two decades. The applications ranges from credit scoring, to stock forecast, to marketing. This is the reflect of an active research area, which applies data science to add value to business.

3 Methodology

Lanzarini et al. [9] studies the prediction ability of a backpropagation ANN, using daily values of an stock index. In this paper, instead of working with daily index values, we will work with high frequency intraday stock prices. We would like to test the ability of an ANN to forecast movements in times of minutes. The rationale for this study is the following: many investing firms perform *algorithmic trading*, i.e. firms operate very frequently in markets and their buy and sell decisions are made through machine learning algorithms. Consequently, it is an important topic for the financial industry.

The prediction was solved by means of an ANN [10]. This kind of construct tend to simulate the way human brain learns, by experience. Basically we can think of a neural network as a directed graph whose nodes are called neurons and the lines that connect them have an associated weight. This value in some sense represents the "knowledge" acquired. Neurons are organized into layers according to their function: the input layer provides information to the network, the output layer provides the answer and the hidden layers are responsible for carrying out the mapping between input and output [11].

In this paper we use a multilayered neural network. It is a feedforwad network, totally connected, organized in three layers: 10 input neurons, six neurons in a single hidden layer and one output neuron. This architecture coindices with the one used in [9, 12, 13].

Differently the learning algorithm is a resilient propagation instead of the more common backpropagation. The training is supervised. The resilient propagation otherwise of gradient descent performed by the backpropation updates each weight independently, and is not subject to the size of the derivative influence but only dependant on the temporal behaviour of its sign [14]. The network was designed to forecast the market return.

Let $P = (p_1, p_2, ..., p_L)$ be the sequence of stock quotes of Fig. 1, then the instantaneous return at the moment $t + 1$ is computed as indicated in (1):

$$r_{t+1} = \ln \left(\frac{p_{t+1}}{p_t} \right) \tag{1}$$

where p_{t+1} and p_t are stock quotes at period $t + 1$ and t, and r_{t+1} is the continuous compound return obtained for buying in period t and selling in period $t + 1$.

In order to help to improve network performance the trend of last nine returns which are used as input for the networks was calculated using ordinary least squared and used as the 10th input for the model. The fact that the training is supervised, implies knowing the expected value for each of the examples that will be used in the training.

Thus, it is required a set of ordered pairs $\{(X_1, Y_1), (X_2, Y_2), \ldots, (X_j, Y_j), \ldots, (X_M, Y_M)\}$ being $X_j = (x_{j,1}, x_{j,2}, \ldots, x_{j,10})$ the input vector, from which $x_{j,10}$ is the trend of last nine returns and Y_j the answer value that it is expected that the network learns for that vector. In this case:

$$x_{j,k} = r_{j+k} = \ln\left(\frac{p_{j+k}}{p_{j+k-1}}\right) \qquad k = 1, 2, \ldots 9 \tag{2}$$

$$Y_j = r_{j+1+N} = \ln\left(\frac{p_{j+1+N}}{p_{j+N}}\right) \tag{3}$$

The maximum number of pairs (X_j, Y_j) that can be formed with L stock quotes is $M = L - 9$. The first half of them were used to train the network and the second half to verify its performance. Once the network is trained, its answer for vector X_j is computed as indicated in Eq. (4).

$$Y_j' = G\left(a_0 + \sum_{i=1}^{6} a_i F\left(b_{0,i} + \sum_{k=1}^{10} x_{j,k} b_{k,i}\right)\right) \tag{4}$$

where $x_{j,k}$ is the value corresponding to the kth input defined in (2), $b_{k,i}$ is the weight of the arch that links the kth neuron of the input with the ith hidden neuron and a_i is the weight of the arch that links the ith hidden neuron with the single output neuron of the network. We should note that each hidden neuron has an additional arch, whose value is indicated in $b_{0,i}$. This value is known as *bias* or *trend term*. A similar thing happens to the output neuron with the weight a_0.

Finally, the obtained expected return Y_j' will be used to do the corresponding forecast. Instead of forecasting any positive (negative)value as a indicative that the return will positive (negative) and the stock should be bought (sold), a range was used. In case the predicted rate is greater than X, the stock should be bought. If the predicted rate is lower than $-X$, the stock should be sold. In case the predicted rate lays within this range, the prior position should be maintained. In case last decision was to buy, the rule dictates to keep the stock. On the contrary if last prediction was that the rate was going to decrease the short position should be maintained. Instead of using a fix range to determine where to buy or sell, the range was optimized during training.

3.1 Profitability Measures

In order to measure the profitablity of the ANN and our benchmark model (*naive buy-and-hold*), we selected some common metrics. The profitability measures used on pairs (X_j, Y_j) of the training set are the following:

- **ANN Buy** and **Hold Train**: is the return obtained by a simple *buy* and *hold* strategy and is computed by adding all the expected returns of the training set as indicated in (5)

$$bhTrain = \sum_{j=1}^{M/2} Y_j \tag{5}$$

- **ANN Buy** and **Hold Test**: is equivalent to the previous measure but applied to the testing set.

$$bhTest = \sum_{j=1+M/2}^{M} Y_j \tag{6}$$

- **Sign prediction ratio**: For rates sign predicted correctly we assign the value 1 whereas if the sign was not predicted correctly the -1 value corresponds.

$$SPR = \frac{\sum_{j=1+M/2}^{M} matches(Y_j, Y_j')}{M/2} \tag{7}$$

$$matches(Y_j, Y_j') = \begin{cases} 1 \ if \ sign(Y_j) = sign(Y_j') \\ 0 \ if \ not \end{cases} \tag{8}$$

where *sign* is the sign function that maps $+1$ when the argument is positive and -1 when the argument is negative.
- The **maximum return** is obtained by adding all the expected values in absolute value

$$MaxReturn = \sum_{j=1+M/2}^{M} abs(Y_j) \tag{9}$$

and represents the maximum achievable return, assuming perfect forecast.
- The **total return** is computed in the following way

$$TotalReturn = \sum_{j=1+M/2}^{M} sign(Y_j') * Y_j \tag{10}$$

where *sign* is the sign function. Notice that the better the network prediction the larger the total return.

- **Ideal Profit Ratio** is the ratio between the total return (10) and the maximum return (9).

$$IPR = \frac{Total\,Return}{Max\,Return} \tag{11}$$

- **Sharpe Ratio** is the ratio between the total return and its standard deviation. The different values of the total return arises from the several independent runs in the training of the network. This will give different networks that will generate different total returns.

$$SR = \frac{\mu_{Total\,Return}}{\sigma_{Total\,Return}} \tag{12}$$

4 Data

We used real time quotes for a subset of 20 stocks which are included within the Bovespa Index. These stocks were selected based on their market capitalization and are detailed in Table 1. The stocks which compose Bovespa were divided in quartiles and for each quartile the five stock with highest capitalization were selected in order to have a representative sample for each group. For each stock one thousand data points were used from 09-16-2015 at 2:00 pm and as late as 09-18-2015 08:16 pm in order to restrict the running time of the algorithms used to build the models used. Figure 1 shows the quotes' evolution during the sample period.

5 Empirical Results

We performed 30 independent runs for each stock in the sample described in Sect. 4. The maximum number of iterations was capped at 3000. The initials weight were randomly distributed between -1 and 1. Initial update value was set to 0.01, while the updates values were limited within e^{-6} and 50. Increase factor and decrease factor for weight updates was set to 1.2 and 0.5. The functions F and G used by the neural network are those defined in (13) and (14). Both are sigmoidal functions. The first one bounded between 0 and 1, and the second one is bounded between -1 and 1. In this way, F allows hidden neurons to produce small values (between 0 and 1), and G permits the net to split between negative (expecting negative returns) and positive (expecting positive returns).

$$G(n) = \frac{1}{1 + e^{-n}} \tag{13}$$

$$F(n) = \frac{2}{1 + e^{-2n}} - 1 \tag{14}$$

Table 1 Details of stocks in the sample

Company name	Sector	Market capitalization (thousand USD)	Percentile	Ticker
Ambev SA	Consumer staples	93,084,295	75th–100th	ABEV3 BS Equity
Itau Unibanco Holding SA	Financials	61,371,355	75th–100th	ITUB4 BS Equity
Petroleo Brasileiro SA	Energy	56,071,359	75th–100th	PETR3 BS Equity
Banco Santander Brasil SA	Financials	25,954,473	75th–100th	SANB11 BS Equity
Vale SA	Materials	25,871,589	75th–100th	VALE3 BS Equity
BMFBovespa SA	Financials	10,031,988	50th–75th	BVMF3 BS Equity
CCR SA	Industrials	9,295,719	50th–75th	CCRO3 BS Equity
WEG SA	Industrials	8,235,428	50th–75th	WEGE3 BS Equity
Engie Brasil Energia SA	Utilities	7,856,355	50th–75th	TBLE3 BS Equity
Lojas Americanas SA	Consumer discretionary	7,524,872	50th–75th	LAME4 BS Equity
Cosan SA Industria e Comercio	Energy	4,727,773	25th–50th	CSAN3 BS Equity
Gerdau SA	Materials	4,353,258	25th–50th	GGBR4 BS Equity
Natura Cosmeticos SA	Consumer staples	4,146,966	25th–50th	NATU3 BS Equity
Cia Brasileira de Distribuicao	Consumer staples	4,027,431	25th–50th	PCAR4 BS Equity
Fibria Celulose SA	Materials	3,762,646	25th–50th	FIBR3 BS Equity
EDP - Energias do Brasil SA	Utilities	2,678,135	0–25th	ENBR3 BS Equity
Localiza Rent a Car SA	Industrials	2,634,754	0–25th	RENT3 BS Equity
BR Malls Participacoes SA	Real state	2,312,292	0–25th	BRML3 BS Equity
Cia Paranaense de Energia	Utilities	2,264,843	0–25th	CPLE6 BS Equity
Usinas Sider. de Minas Gerais SA	Materials	2,164,639	0–25th	USIM5 BS Equity

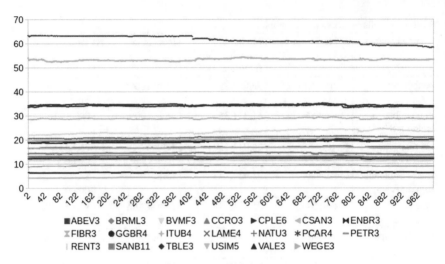

■ABEV3 ◆BRML3 ▽BVMF3 ▲CCRO3 ▶CPLE6 ◁CSAN3 ␾ENBR3
xFIBR3 ●GGBR4 +ITUB4 ×LAME4 +NATU3 ✱PCAR4 −PETR3
RENT3 ■SANB11 ◆TBLE3 ▽USIM5 ▲VALE3 ▶WEGE3

Fig. 1 Real time quotes for the twenty selected Bovespa's stocks

Table 2 Average and standard deviation of selected metrics after 30 independent runs. Stocks within 0–25th. capitalization percentile

Ticker	USIM5	CPLE6	BRML3	RENT3	ENBR3
Buy and Hold (train)	0.01830	0.00499	−0.01363	0.05550	−0.00081
Return rule (train)	0.04794	0.01800	0.02714	0.02624	0.01377
Std. dev	0.01274	0.02612	0.01884	0.03378	0.01027
Buy and Hold (test)	−0.00460	−0.01504	−0.00081	0.00299	−0.01888
Sign prediction (%)	73.58	57.13	59.75	54.34	65.29
Std. dev	0.00098	0.01171	0.05554	0.04627	0.11498
Total return	0.03354	0.01453	0.01142	−0.03183	0.01088
Std. dev	0.01096	0.02375	0.03158	0.03352	0.02149
Ideal profit ratio	0.04491	0.02990	0.02116	−0.05172	0.02575
Std. dev	0.01096	0.04888	0.05855	0.05445	0.05087
Sharpe ratio	3.05859	0.61174	0.36150	−0.94981	0.50614
Range	±0.0165	±0.005	±0.0083	±0.00516	±0.0116
Std. dev	0.00709	0.00000	0.00479	0.00091	0.00531

We show the results obtained for each stock in Tables 2, 3, 4 and 5 one for each stock quartile—by averaging the 30 runs and showing below the standard deviation when applicable. The sign prediction is always greater than 50%. We observe that

Table 3 Average and standard deviation of selected metrics after 30 independent runs. Stocks within 25th–50th capitalization percentile

Ticker	FIBR3	PCAR4	NATU3	GGBR4	CSAN3
Buy and Hold (train)	0.00485	−0.03741	0.01911	0.00155	0.02732
Return rule (train)	0.01939	0.04437	0.04626	0.04216	0.00613
Std. dev	0.02666	0.00719	0.07131	0.02489	0.01250
Buy and Hold (test)	−0.00728	−0.04381	0.00095	−0.01875	−0.02629
Sign prediction (%)	54.88	58.81	63.62	67.83	60.73
Std. dev	0.02602	0.02307	0.05573	0.05273	0.06667
Total return	0.02244	0.03985	0.02661	0.03626	0.04307
Std. dev	0.01912	0.00828	0.04353	0.02046	0.02895
Ideal profit ratio	0.05835	0.11278	0.08744	0.05288	0.07499
Std. dev	0.04972	0.02342	0.14303	0.02985	0.05041
Sharpe ratio	1.17350	4.81531	0.61135	1.77197	1.48775
Range	±0.01016	±0.0073	±0.005	±0.008	±0.00516
Std. dev	0.00713	0.00341	0.00000	0.00428	0.00091

Table 4 Average and standard deviation of selected metrics after 30 independent runs. Stocks within 50th–75th. capitalization percentile

Ticker	LAME4	TBLE3	WEGE3	CCRO3	BVMF3
Buy and Hold (train)	0.01427	0.01129	0.00538	−0.02447	0.00174
Return rule (train)	0.03965	0.00924	0.02033	0.05391	0.03464
Std. dev	0.03617	0.00990	0.02430	0.03173	0.03526
Buy and Hold (test)	−0.00177	−0.01947	−0.01804	−0.01482	−0.01575
Sign prediction (%)	59.28	52.67	64.69	65.65	58.72
Std. dev	0.05310	0.04114	0.09738	0.07825	0.06749
Total return	0.00640	0.02244	0.00766	0.01709	0.03197
Std. dev	0.01785	0.03648	0.01459	0.00815	0.02498
Ideal profit ratio	0.01475	0.03817	0.02548	0.04915	0.06104
Std. dev	0.04113	0.06206	0.04850	0.02344	0.04770
Sharpe ratio	0.35855	0.61495	0.52532	2.09646	1.27952
Range	±0.0056	±0.00916	±0.00516	±0.00916	±0.0055
Std. dev	0.00217	0.00789	0.00091	0.00475	0.00201

Table 5 Average and standard deviation of selected metrics after 30 independent runs. Stocks within 75th.–100th capitalization percentile

Ticker	VALE3	SANB11	PETR3	ITUB4	AMVEB
Buy and Hold (train)	0.03415	0.00757	0.04042	0.00419	−0.06319
Return rule (train)	0.00501	0.06283	0.00242	0.00252	0.65757
Std. dev	0.01122	0.04443	0.00671	0.00365	0.51873
Buy and Hold (test)	0.04247	−0.05064	−0.03265	0.00383	0.06991
Sign prediction (%)	54.59	56.42	52.89	54.77	63.63
Std. dev	0.03665	0.06154	0.08615	0.04399	0.05156
Total return	−0.02745	0.04937	0.01683	−0.00837	0.51680
Std. dev	0.02620	0.03814	0.02491	0.01501	0.63330
Ideal profit ratio	−0.04481	0.07002	0.02347	−0.01688	0.02664
Std. dev	0.04278	0.05410	0.03472	0.03028	0.03265
Sharpe ratio	−1.04739	1.29426	0.67585	−0.55764	0.81605
Range	±0.005	±0.0065	±0.005	±0.00516	±0.00616
Std. dev	0.00000	0.00418	0.00000	0.00091	0.00215

the average sign prediction diminishes when the group capitalization increases. This result is evident in Fig. 2.

When looking at the average total return we detect that for a few stocks, rule returns are negative. However when averaging results for each group, all returns are positive but we cannot affirm that returns are lower when capitalization increases. The same behaviour is appreciated when looking at the Ideal Profit Ratio or Sharpe Ratio.

Fig. 2 Relationship between log market capitalization and sign prediction. Colors represent stocks of the same quartile. Bigger color dots reflect the sample mean of each quartile

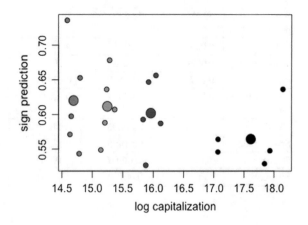

6 Conclusions

Stock markets have an important role not only as asset allocation, but also as a means through which systemic crisis spreads throughout the economy. So analysing whether markets are informationally efficient is relevant. If a rule based on an ANN (as in this paper), can outperform the market, means that information is not immediately conveyed into prices, as the EMH postulates. This situation arises some concerns regarding market regulations and policies recommendations.The aim of our research has been to carry out an intraday forecast on stock returns in the Brazilian stock market. Our study of the time series data has detected an important ability of the ANN to predict sign changes. In fact, the ANN outperforms in several cases a simple buy-and-hold strategy. Still,the performance is far from a perfect prediction. The ANN obtains only between −5 and 11.2% of the return that would be obtained if the investor had perfect forecast. The forecast ability is greater on stocks with lower capitalizations. However, this characteristic may be of particular importance for investor and policy makers.

According to our results, we can hightlight two conclusions: (i) the EMH is, albeit imperfect, a useful framework to analyse market efficiency (ii) there are some arbitrage opportunities that can be exploited by using some advanced tools as the ANN. Portfolios managed by big financial institutions are of huge magnitude. Consequently, small abnormal returns eventually represent millions of dollars. As a caveat, we should say that the true profitability of the technical rule suggested by the ANN can be substantially reduced if we consider transaction costs. In future research, it would be interesting to test other variables beside past prices and trends. For example, traded volume and volatility can be also taken into account in order to increase the performance of the forecast. Additionally, it would be valuable to analyse if the same size effect we appreciate in stocks is also present in other markets.

References

1. Ross, S.A.: Neoclassical finance. Princeton University Press, Princeton, N.J. (2005)
2. Fama, E.F.: J. Financ. **25**(2), 383 (1970)
3. Fama, E.F.: J. Financ. **46**(5), 1575 (1991)
4. Göçken, M., Özçalıcı, M., Boru, A., Dosdoğru, A.T.: Expert Syst. Appl. **44**, 320 (2016). https://doi.org/10.1016/j.eswa.2015.09.029
5. Qiu, M., Song, Y., Akagi, F.: Chaos. Solitons Fractals **85**(1) (2016). https://doi.org/10.1016/j.chaos.2016.01.004
6. Lanzarini, L., Villa-Monte, A., Fernández-Bariviera, A., Jimbo-Santana, P.: Obtaining classification rules using LVQ+PSO: an application to credit risk, vol. 377. Springer, 2015. https://doi.org/10.1007/978-3-319-19704-3_31
7. Lanzarini, L., Villa Monte, A., Bariviera, A., Jimbo Santana, P.: Kybernetes **46**(1), (2017) (in press)
8. Tkáč, M., Verner, R.: Applied Soft Computing **38**(788), (2016). https://doi.org/10.1016/j.asoc.2015.09.040

9. Lanzarini, L., Iglesias Caride, J.M., Bariviera, A.F.: World Congress of International Fuzzy Systems Association 2011 and Asia Fuzzy Systems Society International Conference 2011, pp. 21–25 (2011)
10. Isasi Viñuela, P., Galván León, I.M.: Redes de neuronas artificiales. Un enfoque práctico. Pearson, Pentice Hall (2004)
11. Freeman, J.A., Skapura, D.M.: Neural networks, algorithms, applications, and programming techniques. Addison-Wesley Publishing Company (1991)
12. Gencay, R.: J. Int. Econ. **47**(1), 91 (1999)
13. Fernández-Rodríguez, F., González-Martel, C., Sosvilla-Rivero, S.: Econ. Let. **69**(1), 89 (2000)
14. Riedmiller, M., Braun, H.: IEEE International Conference on Neural Networks, pp. 586–591 (1993)

Persistent Correlations in Major Indices of the World Stock Markets

Maciej Janowicz, Leszek J. Chmielewski, Joanna Kaleta, Luiza Ochnio, Arkadiusz Orłowski and Andrzej Zembrzuski

Abstract Time-dependent cross-correlation functions have been calculated between returns of the major indices of the world stock markets. One-, two-, and three-day shifts have been considered. Surprisingly high and persistent-in-time correlations have been found among some of the indices. Part of those correlations can attributed to the geographical factors, for instance, strong correlations between two major Japanese indices have been observed. The reason for other, somewhat exotic correlations, appear to be as much accidental as it is apparent. It seems that the observed correlations may be of practical value in the stock market speculations.

Keywords Stock market indices · Correlation functions · Pearson correlation · Technical analysis

1 Introduction

As is well known, the time-evolution of the stock markets, and, in particular, the stock market indices, exhibit considerable short-time correlations. These are sufficient to disprove claims of the reliability of the random-walk approximations to

M. Janowicz (✉) · L. J. Chmielewski · J. Kaleta · L. Ochnio · A. Orłowski · A. Zembrzuski
Faculty of Applied Informatics and Mathematics – WZIM, Warsaw University
of Life Sciences – SGGW, Nowoursynowska 159, bldg 34, 02-776 Warsaw, Poland
e-mail: maciej_janowicz@sggw.pl
URL: http://www.wzim.sggw.pl

L. J. Chmielewski
e-mail: leszek_chmielewski@sggw.pl

J. Kaleta
e-mail: joanna_kaleta@sggw.pl

L. Ochnio
e-mail: luiza_ochnio@sggw.pl

A. Orłowski
e-mail: arkadiusz_orlowski@sggw.pl

A. Zembrzuski
e-mail: andrzej_zembrzuski@sggw.pl

© Springer International Publishing AG 2018   411
C. Berger-Vachon et al. (eds.), *Complex Systems: Solutions and Challenges
in Economics, Management and Engineering*, Studies in Systems,
Decision and Control 125, https://doi.org/10.1007/978-3-319-69989-9_24

that evolution or strong versions of the efficient-market hypothesis [1, 2]. On the other hand, there exist a set of usually quite simple computational and visualization techniques, called technical analysis [3, 4], which aims to obtain the approximate predictions of trends and their corrections in market the data. It is so even though the data appear as realizations of a *random* process. It is to be noticed that some recent publications, e.g. [5–8] have lead to considerable revision of the previously ultra-critical stand of the many experts regarding technical analysis.

One of the many possible strategies of "beating the market" which are close to the spirit of technical analysis consists of identification of correlated pairs of stocks. In fact, if we find that, during a specific, long, time interval, the increase (decrease) of the value of one stocks has been followed by a corresponding change in another stock, we may attempt to guess that this relation may persist, at least for short additional time.

To quantify the above intuitive remarks, we have computed the time-dependent cross-correlation functions of returns of the major world stock market indices. Among the latter, the following indices have been included: ALL ORDINARIES (ALL-ORD), AMEX MAJOR (AMEX-MAJ), BOVESPA, B-SHARES, BUENOS, BUX, CAC40, DAX, DJIA, DJTA, DJUA, EOE, FTSE100, HANGSENG, MEX-ICIPS, NASDAQ, NIKKEI, RUSSELL, SASESLCT, SMI, SP500, TOPIX, and TSE300. Let us notice that in one of our previous work a similar research but involving *normalized values* has been reported. Here, however, we deal with correlations of daily *returns*.

The main body of this work is organized as follows. In Sect. 2 we describe our procedure; rhe description as it is, in fact, very simple, and involves well-known quantities. Section 3 is devoted to the presentation of results. Finally, Sect. 4 comprises some concluding remarks.

2   Correlation Functions of World Stock Market Indices

Let $K_n^{(a)}$ denote the closing value at the trading day n of a stock market index (a), and let $Z_n^{(a)}$ denote the corresponding relative return, i.e. $Z_n^{(a)} = (K_n^{(a)} - K_{n-1}^{(a)})/K_{n-1}^{(a)}$. We define the cross-correlation function of returns of two indices (a) and (b) as:

$$C^{(a,b)}(n, M, t) = \frac{Cov\left(Z_{n-M}^{(a)} Z_n^{(b)}\right)}{\sqrt{Var\left(Z_{n-M}^{(a)}\right) Var\left(Z_n^{(b)}\right)}}, \tag{1}$$

where t denotes the time interval over which the averaging in the calculations of the variances and covariance has been performed (unfortunately, we have no ensemble-averaging in our disposal here).

Thus, we have defined the correlation function in terms of the Pearson R coefficient [9] made of two sequences of the length t. A similar definition can be given in terms of, e.g., Spearman coefficient.

It is to be noted that $C^{(a,b)}(n, M, t)$ is a function of three time variables. In the following, we have considered only three values of M, $M = 1, 2, 3$. The averaging time t has been varied from $t = 60$ to $t = N_0$, with N_0 being the largest value of the trading sessions common to all indices. N_0 has been equal to 3452 enclosing the time interval from 10th of October, 2001 to 31st of October, 2016). The data for stock market indices have been downloaded from [11]. The correlation functions have been computed using `pearsonr` function from the Python (2.7) module *scipy.stats*, version 0.14.0 [10].

3 Results and Discussion

We have started our study with n equal to t. Firstly, we have considered Pearson's R coefficient using all available data.

In Tables 1, 2 and 3[1] we have provided pairs of indices with largest values of the correlation functions for $n = t = u$, $u = N_0 - 4$, for $M = 1, 2, 3$.

Even though the number of cross-correlation functions with the values greater than 0.4, say, has been rather small, in particular cases those correlations have been very significant for $M = 1$ as specified in Table 1. One might say, for instance, that during several years one can predict the behavior of the TOPIX index from the knowledge of the NIKKEI 225 index change one day before. Remarkably, it has worked only one way. What is more, for t significantly smaller than u a quite interesting structure in the n-dependence of the correlation function occurs, please see below.

The above relatively high values of the correlation functions between the returns of DJIA and the returns of other American (and Mexican) indices certainly cannot be called counterintuitive. However, it is quite remarkable that the pairs listed in Table 2

Table 1 List of indices with largest correlation functions for $n = t = N_0 - 4$, $M = 1$. In all cases p-value has been smaller than 0.05

Index (a)	Index (b)	$C^{(a,b)}(u, 1, u)$
NIKKEI	TOPIX	0.785
NASDAQ	SP500	0.694
NASDAQ	RUSSELL	0.690
NASDAQ	DJTA	0.651
DJIA	NASDAQ	0.474
DJIA	RUSSELL	0.411

---

[1]Tables will be properly formatted in the camera-ready version of the paper.

Table 2 List of indices with largest correlation functions for $n = t = N_0 - 4$, $M = 2$. In all cases p-value has been smaller than 0.05

Index (a)	Index (b)	$C^{(a,b)}(u, 2, u)$
DJIA	SP500	0.388
DJIA	DJUA	0.366
DJIA	DJTA	0.335
DJIA	RUSSELL	0.325
DJIA	MEXICIPC	0.256

Table 3 List of indices with largest correlation functions for $n = t = N_0 - 4$, $M = 3$. In all cases p-value has been smaller than 0.05

Index (a)	Index (b)	$C^{(a,b)}(u, 3, u)$
SMI	TSE-300	0.159
FTSE100	ALL-ORD	0.135
DJIA	MEXICIPC	0.105
EOE	SMI	0.092
SP500	TSE-300	0.091
AMEX-MAJ	TSE-300	0.090

do *not* appear in the list of highly correlated pairs for the shift $M \neq 2$. Please see also further comments following Fig. 2.

For $M = 3$ we have not observed any large values of the R coefficient, for $M > 3$ they become even smaller. Let us also report that we have not obtained any negative R coefficient with significant absolute value for any M. The indices definitely tend to be correlated rather than anti-correlated.

In the Tables 4, 5 and 6 we have listed the same pairs of indices as in Tables 1, 2 and 3 but accompanied with the numbers of correct and incorrect predictions. Our

Table 4 List of indices with largest correlation functions for $n = t = N_0 - 4$, $M = 1$ with the numbers of correct and incorrect predictions of the behavior one the second-column indices based on the change of the corresponding first-column ones

Index (a)	Index (b)	No. of correct preditions	No. of incorrect predictions
NIKKEI	TOPIX	2852	596
NASDAQ	SP500	2296	1089
NASDAQ	RUSSELL	2368	1080
NASDAQ	DJTA	2301	1147
DJIA	NASDAQ	2359	1089
DJIA	RUSSELL	2281	1167

Table 5 List of indices with largest correlation functions for $n = t = N_0 - 4, M = 1$ with the numbers of correct and incorrect predictions of the behavior one the second-column indices based on the change of the corresponding first-column ones

Index (a)	Index (b)	No. of correct preditions	No. of incorrect predictions
DJIA	SP500	2086	1362
DJIA	DJUA	1953	1495
DJIA	DJTA	2042	1406
DJIA	RUSSELL	1953	1495
DJIA	MEXICIPC	1887	1561

Table 6 List of indices with largest correlation functions for $n = t = N_0 - 4, M = 3$ with the numbers of correct and incorrect predictions of the behavior one the second-column indices based on the change of the corresponding first-column ones

Index (a)	Index (b)	No. of correct preditions	No. of incorrect predictions
SMI	TSE-300	1833	1516
FTSE100	ALL-ORD	1790	1658
DJIA	MEXICIPC	1828	1620
EOE	SMI	1777	1671
SP500	TSE-300	1885	1563
AMEX-MAJ	TSE-300	1813	1635

definition of a correct prediction is trivial: if the sign of the product of returns of index (a) at time t and index (b) at time $t + M$ is positive, we say that the prediction is correct; otherwise the prediction is incorrect.

The content of the Tables 4, 5 and 6 largely agrees with the conclusions one can draw from three previous tables. For the first five rows of Table 4 we can realize that the number of correct predictions is at least two times larger that the incorrect ones. Quite obviously, even such a ratio could not guarantee any successful trading on any index-based financial instruments. On the other hand, even the smallest advantage like that provided by Table 6 may sometimes be sufficient to get the edge on trading competitors.

Let us notice that the difference between correct and incorrect predictions in the case of EOE-SMI pair is so small that p-value in the G-test is ≈ 0.07. Thus, one cannot exclude the zeroth hypothesis that the difference between the numbers of predictions if purely accidental. In all other cases we have had p-value in the G-test smaller than 0.05.

We have also performed calculations for the values of n and t different from u. For instance, we have obtained the R coefficient from the last 480 trading sessions (approximately two years) preceding 31st of October, 2016. For the pair CAC40—EOE a record value of R equal to 0.94 has been obtained with the percent of correct predictions equal to 0.875. For $M = 2$ the most considerable correlation has been obtained between the DAX and SMI indices; R has been equal to 0.526, and the ratio of correct to incorrect predictions has almost exactly been equal to 2:1.

We have not, however, been satisfied with the above results for it has not been clear precisely how the correlation function depends on the time n if the averaging window t is smaller than n.

In the following figures we have displayed selected cross-correlation function for given M and t as functions of n. These have been compared with the time evolution of the indices themselves. Note that it is the closing values of the index (not the returns) which have been shown in the central and lower parts of each figure.

Naturally, the graphs in Fig. 1 provide additional insight into the results tabulated in Tables 1 and 4. In fact, we can see that correlation function involving NIKKEI

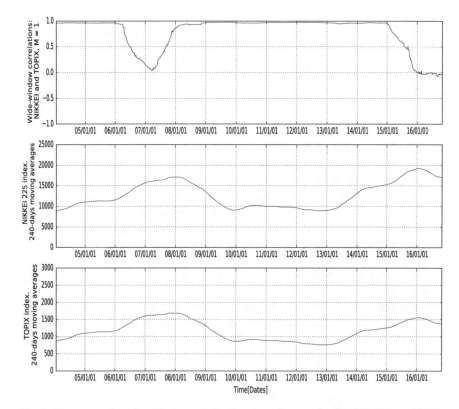

Fig. 1 Upper part: time-dependent correlation function of returns for NIKKEI 225 and TOPIX indices as a function of discrete time n for $M = 1$, $t = 240$; central part: simple moving average (with window 240) of closing values of the NIKKEI 225 index in the same time interval; lower part: corresponding values of the TOPIX index

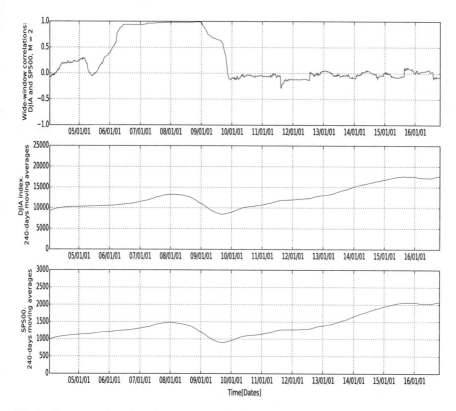

Fig. 2 Upper part: time-dependent correlation function of returns for DJIA and SP500 indices as a function of discrete time n for $M = 2$, $t = 240$; central part: simple moving average (with window 240) of closing values of the DJIA index in the same time interval; lower part: corresponding values of the SP500 index

and TOPIX has not been constant, and there have been two time intervals in which the correlation dropped quite significantly. The second one includes recent trading sessions. As can be seen from the central and lower parts of Fig. 1, there is nothing in the values of indices themselves which could correspond to the drops of correlations. The first period of low correlations may, perhaps, be attributed to the change of the weighting system with the help of which the TOPIX index has been calculated; that change was completed in January 2006. We have failed, however, to identify a possible reason of breaking down the correlation in the beginning of 2015.

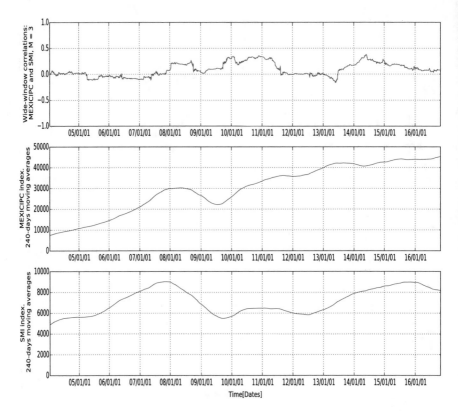

Fig. 3 Upper part: time-dependent correlation function of returns for MEXICIPC and SMI indices as a function of discrete time n for $M = 3$, $t = 240$; central part: simple moving average (with window 240) of closing values of the MEXICIPC index in the same time interval; lower part: corresponding values of the SMI index

The history of correlations of the indices DJIA and SP500 ($M = 2$), which, per-haps, might be considered as the most important and most obvious of all, is even more striking. It appears that the quite considerable value of R coefficient calculated during the last 14 years was built during the relatively short time interval 2006–2010. Recently, DJIA-SP500 have become almost completely "unreliable", so to say.

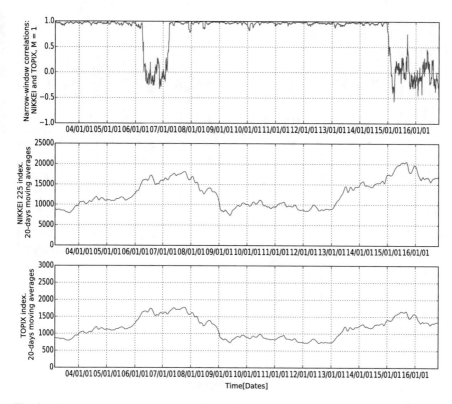

Fig. 4 The same as in Fig. 1 but for $t = 60$ and moving averages with the window length equal to 20

We have provided Figs. 4, 5 and 6 largely to confirm the results displayed in the three previous figures. One can convince oneself that the shorter intervals and smaller windows of moving averages do not not change the overall picture and provide, on this occasion, a little more than noise.

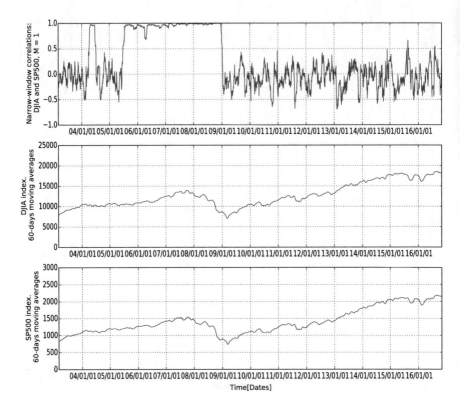

Fig. 5 The same as in Fig. 2 but for $t = 60$ and moving averages with the window length equal to 20

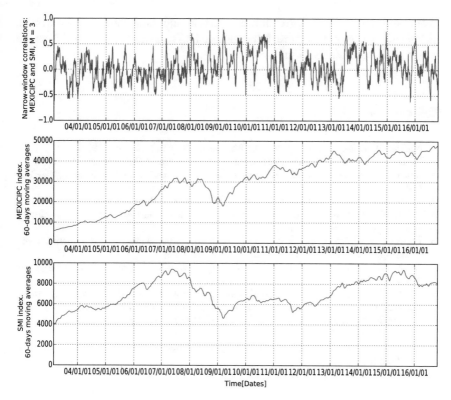

Fig. 6 The same as in Fig. 3 but for $t = 60$ and moving averages with the window length equal to 20

4 Concluding Remarks

In this work we have computed cross-correlation function of time series generated by the returns of important world stock market indices. We have identified pairs of indices such that correlations are both strong and persistent in time. Preliminary assessments of predictive power of the correlations have been performed. We believe that, with sufficient care, the knowledge of correlations between indices and individual stocks may be used in practice, possibly even to enhance the predictive power of technical indicators for trading purposes. Needless to say, extreme caution is required in any such attempts. In fact, they can break down at any time. For instance, when we used historical data up to the middle summer of 2016, we have observed the highest correlation function for $M = 3$, $n = t = u - 3$ in the pair MEXICIPC—SMI. During the subsequent mounts that correlation deteriorated quite spectacularly. One might think that it would be a valuable enterprise to find any indicators which could suggest that the correlations have just started to build up or are just about to get ruined, provided that such indicators exist. Finally, one may reasonably argue that the above very simple correlation analysis can and should be combined with the analysis of cointegration to have deeper insight into connection between shifted (i.e. with $M \neq 0$) and unshifted time series.[2]

References

1. Malkiel, B.: A Random Walk Down the Wall Street. Norton, New York (1981)
2. Fama, E.F., Blume, M.: Filter rules and stock-market trading. J. Bus. **39**, 226–241 (1966)
3. Murphy, J.: Technical Analysis of Financial Markets. New York Institute of Finance, New York (1999)
4. Kaufman, P.: Trading Systems and Methods. Wiley, New York (2013)
5. Brock, W., Lakonishok, J., LeBaron, B.: Simple technical trading rules and the stochastic properties of stock returns. J. Finan. **47**(5), 1731–1764 (1992)
6. Lo, A., MacKinley, A.: Stock market prices do not follow random walks: evidence from a simple specification test. Rev. Finan. Stud. **1**, 41–66 (1988)
7. Lo, A., MacKinley, A.: A Non-Random Walk down Wall Street. Princeton University Press, Princeton (1999)
8. Lo, A., Mamaysky, H., Wang, J.: Foundations of technical analysis: computational algorithms, statistical inference, and empirical implementation. J. Finan. **55**(4), 1705–1765 (2000)
9. Pearson, K.: Notes on regression and inheritance in the case of two parents. Proc. R. Soc. Lond. **58**, 240–242 (1895)
10. Scipy Community: Statistical functions in Python. http://docs.scipy.org/doc/scipy/reference/stats.html (2016)
11. Bossa.pl. http://bossa.pl/notowania/metastock. Accessed 15 Nov 2016

---

[2]References will be properly formatted in the camera-ready version of the paper.

Forgotten Effects in Exchange Rate Forecasting Models

Ezequiel Avilés Ochoa, Ernesto León Castro, Ana María Gil Lafuente
and José María Merigo Lindahl

Abstract The aim of the chapter is to identify with the methodology of forgotten effects, hidden variables that influence the behavior of the forward exchange rate US dollar/Mexican peso (USD/MXN) to then incorporate them into a structured model from the postulates of the theory of Purchasing Power Parity (PPP) and, thus, reduce forecast error. The research question that causes the problem is: Is it possible to decrease the prediction error of the PPP model, using the methodology of the forgotten effects to detect and include hidden or forgotten variables? Among the results it was found that the inclusion of hidden or forgotten variables in the PPP model, decreases forecast error for exchange rate USD/MXN in 2015.

Keywords Forgotten effects · Forecast exchange rate · Economy Mexican Peso

1 Introduction

Liberalization policies promoted by the Washington Consensus, from the eighties, have stimulated the growing internationalization of economics. Trade and financial flows, and therefore, operations in foreign exchange market have increased

E. A. Ochoa (✉) · E. L. Castro
University of Occidente, Blvd. Lola Beltrán s/n esq. Circuito Vial,
80020 Culiacán, Mexico
e-mail: ezequiel.aviles@udo.mx

E. L. Castro
e-mail: ernesto134@hotmail.com

A. M. G. Lafuente
University of Barcelona, Av. Diagonal 690, 08034 Barcelona, Spain
e-mail: amgil@ub.edu

J. M. M. Lindahl
University of Chile, Av. Diagonal Paraguay 257, 8330015 Santiago, Chile
e-mail: jmerigo@fen.uchile.cl

© Springer International Publishing AG 2018
C. Berger-Vachon et al. (eds.), *Complex Systems: Solutions and Challenges
in Economics, Management and Engineering*, Studies in Systems,
Decision and Control 125, https://doi.org/10.1007/978-3-319-69989-9_25

significantly. Nowadays, financial environment in which businesses operate, warn Gil-Aluja [14, 17], is volatile and uncertain. Exchange rate in Mexico, say Bazdresch and Werner [4], is characterized by a systematic positive trend, long periods of relative stability with sudden episodes of high volatility.

In scenarios of uncertainty, said Lustig et al. [24], global volatility increases risk in emerging economies. Interdependence of financial markets emphasizes exchange risk exposure in organizations located in developing countries, including Mexico. Thus, oriented to reduce the degree of exposure, corporate governance is an everyday practice [1, 5].

Exchange rate volatility, warn [19, 25], requires decision makers to structure models to predict in advance exchange rate, with an acceptable forecast error. Basic models based on variables such price index, interest rates and balance of payments, have not produced satisfactory results, having similar behavior to a random walk [8, 34].

Inefficiency of these models, said Pastor [29] and Krugman [22], is intrinsic since they are a simplification of reality. This, even if the phenomenon under study is based on a group of variables theoretically accepted, some are unknown or forgotten, so their impact and influence it is ignored in the model [16].

A methodology to reduce forecast error is the forgotten effects, which link cause and effect through incidence matrix to obtain or retrieve hidden or indirect items not directly considered by experts [5, 36]. From the above discussion, the main question of this research is clear: Is it possible to reduce forecast error in PPP model, using forgotten effects methodology to detect and include hidden or forgotten variables?

The objective of this article is to identify, using the methodology of forgotten effects, developed by Kaufmann and Gil-Aluja [15], the hidden variables that influence the behavior of the forward exchange rate USD/MXN to then incorporate them into a model structured from the postulates of the PPP theory and, thus, reduce forecast error.

In the research expertons are used to recognize the variables that remain hidden or generate indirect impact in determining the exchange rate, which should be included in an econometric model and, consequently, generate greater efficiency in forecasting forward exchange rate.

The remainder of the paper is organized as follows. In Sect. 2 we review the PPP model and some fuzzy techniques. Section 3 presents the process and the results of using expertons and forgotten effects methodology to identify hidden variables. Section 4 shows the results of the incorporation of the forgotten variables in the PPP model. Section 5 summarizes the main conclusions of the paper.

2 Methodological and Theoretical Approach

In this section, we review some of the basic concepts that will be used in the paper. We analyze the Purchase Power Parity (PPP) model for exchange rate forecasting, the experton technique and the forgotten effects methodology.

2.1 Exchange Rate Forecasting Techniques

Among the theories used to predict movements in exchange rate highlights the Purchasing Power Parity (PPP), which is defined as:

Definition 1 PPP model, in its relative version, postulates that variations in exchange rate in each period must be equal to inflation differential; that is, the weaker currency should depreciate [10, 31, 32].

$$\hat{e} = \pi_{MEX} - \pi_{EUA} \tag{1}$$

where $\hat{e}$ is the change in exchange rate and π denotes inflation.

An important aspect is that two variables were added in the model: the first one is the exchange rate with a lag and the second is the volatility that has been the behavior of the exchange rate [3, 33]. To generate the values of the variables that make up the model, the multiplicative decomposition time series method was used [11]. It is defined as follows.

Definition 2 To predict future value of the independent variables in forecast model, multiplicative decomposition time series method was used, is formulated as follows:

$$Y_t = T_t * S_t * C_t * I_t \tag{4}$$

where Y_t is observed value, T_t trend, S_t seasonality, C_t cycle and I_t irregularity.

2.2 Fuzzy Techniques for the Treatment of Uncertainty

All events that surround us are part of a system or subsystems, so any activity is subject to cause-effect incidence [18]. Despite the control systems, there is always the possibility to forget some causal relationships that are not explicit, obvious or visible [2]. In situations of uncertainty and volatility, Kaufmann and Gil-Aluja [21] indicates that there are variables that are not detectable immediately, because they are hidden variables that are the result of an accumulation of causes. To identify these forgotten elements, the application of the forgotten effects methodology and experton method allows to identify them. These techniques can be defined as follows.

Definition 3 The theory of expertons developed by Kaufmann [20] is a generalized notion of a random fuzzy event when the probability of a random event on each $\alpha - cut$ is replaced by the confidence interval. It is defined as follows.

Let D a finite set of certain alternatives. A group of q expert is requested to express their subjective opinion regarding each element from D in the form of a confidence interval

$$\forall d \in D: \left[\alpha^{j*}(a) \right], \left[\alpha^{j^*}(a) \right] \subset [0, 1]$$

where $\subset$ is a set of inclusion and j the expert. The property is monotonous, not horizontally and vertically downward, with a distribution network configuration $\wedge$ and $\vee$. So, that:

$$\forall \alpha \in [0, 1] a_1(\alpha) \ in \ [a_1(\alpha), a_2(\alpha)]$$
$$\forall \alpha, \alpha' \in [0, 1]: (\alpha' > \alpha) \Rightarrow \left(a_1(\alpha) \le a_1(\alpha'), a_2(\alpha) \le a_2(\alpha') \right)$$
$$(\alpha) = 0 \Rightarrow (a_1(\alpha) = 1, a_2(\alpha) = 1)$$

Definition 4 Forgotten effects methodology, developed by Kaufmann and Gil Aluja [21], is supported on the assumption of the existence of two sets:

$$A = \left\{ \frac{a_i}{i} = 1, 2, \ldots, n \right\}$$

$$B = \left\{ \frac{b_j}{j} = 1, 2, \ldots, m \right\}$$

It is conjectured that prevails incidence of a_i on b_j if the value of the membership function characteristic of (a_i, b_j) is valued in the range $[0, 1]$, i.e.:

$$\forall (a_i, b_j) \Rightarrow \mu(a_i, b_j) \in [0, 1]$$

The set of pair of evaluated elements is known as direct impact matrix, which shows the cause-effect relationship in different degrees, caused by the corresponding assembly A (causes) and the set B (effects)

$$M = \begin{array}{c|ccccc} & b_1 & b_2 & b_3 & \ldots & b_m \\ \hline a_1 & \mu_{a_1 b_1} & \mu_{a_1 b_2} & \mu_{a_1 b_3} & \ldots & \mu_{a_1 b_m} \\ a_2 & \mu_{a_2 b_1} & \mu_{a_2 b_2} & \mu_{a_2 b_3} & \ldots & \mu_{a_2 b_m} \\ a_3 & \mu_{a_3 b_1} & \mu_{a_3 b_2} & \mu_{a_3 b_3} & \ldots & \mu_{a_3 b_m} \\ \vdots & \ldots & \ldots & \ldots & \ldots & \ldots \\ a_n & \mu_{a_n b_1} & \mu_{a_n b_2} & \mu_{a_n b_3} & \ldots & \mu_{a_n b_m} \end{array}$$

The above matrix can be represented by a graph of incidence associated as follows:

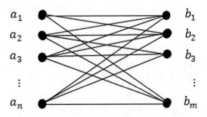

Once detected the direct impact between set A and B, it proceeds to the detection of the forgotten effects. Subsequently, it is assumed that there is a third set of elements, called C, expressed as follows:

$$C = \left\{ \frac{C_k}{k} = 1, 2, \ldots, k \right\}$$

This set consists of elements that are effects of set B, having an incidence matrix expressed as follows:

$$N = \begin{array}{c|ccccc} & c_1 & c_2 & c_3 & \cdots & c_m \\ \hline b_1 & \mu_{b_1 c_1} & \mu_{b_1 c_2} & \mu_{b_1 c_3} & \cdots & \mu_{b_1 c_m} \\ b_2 & \mu_{b_2 c_1} & \mu_{b_2 c_2} & \mu_{b_2 c_3} & \cdots & \mu_{b_2 c_m} \\ b_3 & \mu_{b_3 c_1} & \mu_{b_3 c_2} & \mu_{b_3 c_3} & \cdots & \mu_{b_3 c_m} \\ \vdots & \cdots & \cdots & \cdots & \cdots & \cdots \\ b_n & \mu_{b_n c_1} & \mu_{b_n c_2} & \mu_{b_n c_3} & \cdots & \mu_{b_n c_m} \end{array}$$

The incidence associated graph of this matrix takes the form:

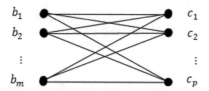

Now we have two incidence matrixes, which includes, as common element, set B, this relationship is expressed as follows:

$$M \subset A \times B \text{ and } N \subset B \times C$$

To detect relationship between sets A and C, using B, the max-min operator is used (expressed by symbol ∘), generating a new incidence matrix as follows:

$$M \circ N = P$$
$$P \subset A \times C$$

This new relationship is formulated

$$\forall (a_i, c_z \in A \times C)$$
$$\mu(a_i, c_z) M \circ N = \forall_{bj} \left(\mu M \left(a_i, b_j \right) \wedge \mu N \left(b_j, c_z \right) \right)$$

The resulting incidence matrix of performing the operation max-min is

	c_1	c_2	c_3	$\dots$	c_m
a_1	$\mu_{a_1 c_1}$	$\mu_{a_1 c_2}$	$\mu_{a_1 c_3}$	$\cdots$	$\mu_{a_1 c_m}$
a_2	$\mu_{a_2 c_1}$	$\mu_{a_2 c_2}$	$\mu_{a_2 c_3}$	$\cdots$	$\mu_{a_2 c_m}$
a_3	$\mu_{a_3 c_1}$	$\mu_{a_3 c_2}$	$\mu_{a_3 c_3}$	$\cdots$	$\mu_{a_3 c_m}$
$\vdots$	$\cdots$	$\cdots$	$\cdots$	$\cdots$	$\cdots$
a_n	$\mu_{a_n c_1}$	$\mu_{a_n c_2}$	$\mu_{a_n c_3}$	$\cdots$	$\mu_{a_n c_m}$

$P =$

The incidence graph associated is expressed as follows:

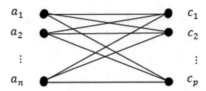

The P matrix defines the casual relationships between elements of sets A and C, in the intensity or degree that leads to consider those belonging to B.

To assign values located in the interval [0,1], Kaufmann and Gil-Aluja [21] propose an endecadarian scale constituted by 11 values $[0, 0.1, 0.2, 0.3, \dots, 1]$. This, point Luis and Gil-Lafuente [23], facilitates their adaptation and treatment since people are used to think and work in decimal form.

3   Detection of Forgotten Variables in Exchange Rate USD/MXN

In this paper, we consider as variable that explain exchange rate those include in Purchase Power Parity (PPP) model, Interest Rate Parity (IRP) model and Balance of Payment (BoP) model. Also, other macroeconomic variables that have been important in exchange rate such as growth expectations, monetary politics, country risk, global market and oil price are included [6, 7, 13, 30, 35, 38].

Based on that information, the variables considered causes are:

(a) Growth expectations
(b) Monetary politics
(c) Country risk
(d) Oil price
(e) Global markets

Also, the variables considered effects are

(a) Foreign inflation
(b) Domestic inflation
(c) Foreign interest rate

(d) Domestic interest rate
(e) Trade balance account
(f) Foreign domestic investment
(g) Foreign portfolio investment
(h) International reserves.

The impact of one variable over another is wholly subjective, and therefore hardly measurable and to evaluate this relationship an endecadarian scale was used as follows:

0: no impact
0,1: virtually no effect
0,2: almost no effect
0,3: very low effect
0,4: low effect
0,5: medium effect
0,6: significant effect
0,7: very significant effect
0,8: strong effect
0,9: very strong effect
1: the highest effect

To detect the hidden variables in exchange rate, we integrate the opinions of five experts who have knowledge and information of this topic. In this group three people are financial advisors in forex market and another two are academics that has previous work in this field. With the information provide, the experton technique and the results are presented in Tables 1, 2, and 3.

Table 1 Effect-effect

	PCI_{EUA}	PCI_{MEX}	r_{EUA}	r_{MEX}	BT	FDI	FPI	R
π_{USA}	1	0.4	0.6	0.3	0.3	0.3	0.3	0.3
π_{MEX}	0.4	1	0.6	0.6	0.4	0.4	0.4	0.4
r_{USA}	0.8	0.3	1	0.6	0.2	0.4	0.5	0.3
r_{MEX}	0.4	0.7	0.7	1	0.4	0.5	0.5	0.4
BT	0.3	0.5	0.3	0.4	1	0.6	0.6	0.5
FDI	0.2	0.4	0.5	0.4	0.5	1	0.6	0.5
FPI	0.3	0.3	0.4	0.5	0.5	0.6	1	0.4
R	0.3	0.4	0.4	0.5	0.4	0.6	0.6	1

Table 2 Cause-cause

	Growth expectations	Monetary politics	Country risk	Oil price	Global markets
Growth expectations	1	0.8	0.5	0.7	0.7
Monetary politics	0.8	1	0.7	0.6	0.6
Country risk	0.7	0.7	1	0.5	0.7
Oil price	0.7	0.5	0.4	1	0.8
Global markets	0.5	0.4	0.3	0.7	1

Table 3 Cause-effect

	PCI_{EUA}	PCI_{MEX}	r_{EUA}	r_{MEX}	BT	FDI	FPI	R
Growth expectations	0.4	0.6	0.6	0.7	0.8	0.7	0.7	0.5
Monetary politics	0.7	0.8	0.7	0.8	0.7	0.6	0.6	0.5
Country risk	0.3	0.6	0.5	0.8	0.5	0.7	0.7	0.5
Oil price	0.5	0.3	0.3	0.3	0.6	0.4	0.5	0.5
Global markets	0.6	0.4	0.6	0.4	0.4	0.6	0.6	0.4

PCI_{USA} price consumer index in USA; PCI_{MEX} price consumer index in México; r_{USA} interest rate in USA; r_{MEX} interest rate in Mexico; BT Balance of Trade in México; FDI foreign direct investment in México; FPI foreign portfolio investment in Mexico; R international reserves in Mexico

Once generated the matrices and unified the information obtained from experts, we proceed to use the forgotten effects methodology. The results are shown in Table 4.

From the information obtained in Table 4, the order of the hidden variables, according to their indirect effects, is as follows: oil price > country risk > global market > growth expectations > monetary risk.

Among the forgotten effects, there are two relationships with 0.4 of indirect effect, these are: (a) oil price and interest rate in Mexico, and (b) country risk and price consumer index in USA. The former has a second order effect via growth expectations, while the second through monetary policy (see Figs. 1 and 2).

Note: To develop these relationships, we use FuzzyLog Software. Free available in http://www.fuzzyeconomics.com/jaimegil.html.

Table 4 Forgotten effects in determining the forward exchange rate

Cause	Effect	M	$M \circ N$	$M^* = M \circ N \circ B$	$Q = M^* - M$
Growth expectations	PCI_{USA}	0.4	0.7	0.7	0.3
Growth expectations	PCI_{MEX}	0.6	0.8	0.8	0.2
Growth expectations	r_{USA}	0.6	0.7	0.7	0.1
Growth expectations	r_{MEX}	0.7	0.8	0.8	0.1
Growth expectations	BT	0.8	0.8	0.8	0
Growth expectations	FDI	0.7	0.7	0.7	0
Growth expectations	FPI	0.7	0.7	0.7	0
Growth expectations	R	0.5	0.5	0.6	0.1
Monetary politics	PCI_{USA}	0.7	0.7	0.7	0
Monetary politics	PCI_{MEX}	0.8	0.8	0.8	0
Monetary politics	r_{USA}	0.7	0.7	0.7	0
Monetary politics	r_{MEX}	0.8	0.8	0.8	0
Monetary politics	BT	0.7	0.8	0.8	0.1
Monetary politics	FDI	0.6	0.7	0.7	0.1
Monetary politics	FPI	0.6	0.7	0.7	0.1
Monetary politics	R	0.5	0.5	0.6	0.1
Country risk	PCI_{USA}	0.3	0.7	0.7	0.4
Country risk	PCI_{MEX}	0.6	0.7	0.7	0.1
Country risk	r_{USA}	0.5	0.7	0.7	0.2
Country risk	r_{MEX}	0.8	0.8	0.8	0
Country risk	BT	0.5	0.7	0.7	0.2
Country risk	FDI	0.7	0.7	0.7	0
Country risk	FPI	0.7	0.7	0.7	0
Country risk	R	0.5	0.5	0.6	0.1
Oil price	PCI_{USA}	0.5	0.6	0.6	0.1
Oil price	PCI_{MEX}	0.3	0.6	0.6	0.3
Oil price	r_{USA}	0.3	0.6	0.6	0.3
Oil price	r_{MEX}	0.3	0.7	0.7	0.4
Oil price	BT	0.6	0.7	0.7	0.1
Oil price	FDI	0.4	0.7	0.7	0.3
Oil price	FPI	0.5	07	0.7	0.2
Oil price	R	0.5	0.5	0.6	0.1
Global markets	PCI_{USA}	0.6	0.6	0.6	0
Global markets	PCI_{MEX}	0.4	0.5	0.6	0.2
Global markets	r_{USA}	0.6	0.6	0.6	0
Global markets	r_{MEX}	0.4	0.5	0.6	0.2
Global markets	BT	0.4	0.6	0.6	0.2
Global markets	FDI	0.6	0.6	0.6	0
Global markets	FPI	0.5	0.5	0.6	0.1
Global markets	R	0.4	0.5	0.6	0.2

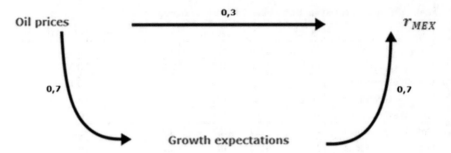

Fig. 1 Element interposed between oil prices and interest rate in Mexico

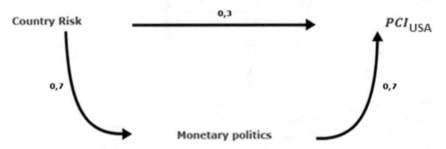

Fig. 2 Element interposed between country risk and price consumer index in USA

4 Incorporation of the Forgotten Variables in PPP Model

To determine the efficiency of the incorporation of forgotten effects in forecast exchange rate USD/MXN, the results of the PPP model is contrasted against the same model including the forgotten variable oil price. The models are as follow.

PPP model

Source	SS	df	MS			Number of obs =	251
						F(4, 246) =	6001.70
Model	4.28244058	4	1.07061014			Prob > F =	0.0000
Residual	.043882562	246	.000178384			R-squared =	0.9899
						Adj R-squared =	0.9897
Total	4.32632314	250	.017305293			Root MSE =	.01336

tcf	Coef.	Std. Err.	t	P>|t|	[95% Conf. Interval]	
v	.0234641	.002765	8.49	0.000	.018018	.0289101
tclag	.9141334	.0220617	41.44	0.000	.8706795	.9575873
pciUSA	-.0194534	.0398029	-0.49	0.625	-.0978513	.0589445
pciMEX	.0407992	.0221272	1.84	0.066	-.0027837	.0843821
_cons	.1066104	.070307	1.52	0.131	-.0318702	.245091

PPP model with forgotten effects

Source	SS	df	MS				
Model	4.28259392	5	.856518785				
Residual	.043729218	245	.000178487				
Total	4.32632314	250	.017305293				

		Number of obs	=	251
		F(5, 245)	=	4798.78
		Prob > F	=	0.0000
		R-squared	=	0.9899
		Adj R-squared	=	0.9897
		Root MSE	=	.01336

tcf	Coef.	Std. Err.	t	P>|t|	[95% Conf. Interval]	
v	.0228157	.0028529	8.00	0.000	.0171964	.0284349
tclag	.9127896	.0221156	41.27	0.000	.8692286	.9563505
pciUSA	.0456639	.0807508	0.57	0.572	-.1133904	.2047182
pciMEX	.0353876	.0228906	1.55	0.123	-.0096999	.0804751
mme	-.0098227	.0105975	-0.93	0.355	-.0306965	.011051
_cons	-.0161185	.1499267	-0.11	0.914	-.3114281	.2791911

where tc_F is forward exchange rate; tc_{lag} is exchange rate with one lag; v is volatility; pci_{USA} is price consumer index in USA; pci_{MEX} is price consumer index in Mexico and mme is mexican crude oil price. All variables are expressed in logarithms.

Econometric analysis

P value is used for significance test under the following assumptions:

1. H_0 is accepted if the P value is greater than 0.05
2. H_0 is rejected if the P value is less than 0.05

For PPP model, H_0 is rejected for tc_{lag} y pci_{MEX} and accepted for *constant*, v and pci_{USA} (see Annex 1). For PPP model with forgotten effects, H_0 is rejected for tc_{lag} and v, and accepted for *constant*, pci_{USA}, pci_{MEX} and mme (see Annex 2). Finally, for the analysis of the complete models,H_0 is rejected for all the variables (see Annex 3).

To detect autocorrelation, Durbin Watson statistic is used, appearing absence in both models (see Annex 4).

Finally, the analysis of Variance Inflation Factors (VIF) is performed to detect multicollinearity. In the PPP model, there is low for tc_{lag}, moderate to v and high for pci_{USA} and pci_{MEX} (see Annex 5). In the PPP model with forgotten effect is low for v and high for pci_{USA}, pci_{MEX}, tc_{lag} y mme (see Annex 6).

To analyze the results, we use the mean squared deviation (MSD) statistic, that is $MSD = \frac{\sum |y_t - \hat{y}_t|^2}{n}$, the results of both models are presented in Table 5.

Table 5 Exchange rate USD/MXN forecasting through econometric models and forgotten effects

Time	Spot exchange rate USD/MXN	Time series	Error	MSD	Time series and FE	Error	MSD
01–15	14.6808	15.6757	0.9949	0.9899	15.8839	1.2031	1.4473
02–15	14.9230	15.6272	0.7042	0.4958	15.8420	0.9190	0.8445
03–15	15.2136	15.2111	−0.0025	0.0000	15.4178	0.2042	0.0417
04–15	15.2208	15.2666	0.0458	0.0021	15.4674	0.2466	0.0608
05–15	15.2475	15.4326	0.1851	0.0343	15.6321	0.3846	0.1479
06–15	15.4692	15.7162	0.2470	0.0610	15.9168	0.4476	0.2003
07–15	15.9225	15.8889	−0.0336	0.0011	16.0907	0.1682	0.0283
08–15	16.5032	15.6731	−0.8301	0.6890	15.8782	−0.6250	0.3906
09–15	16.8519	15.6922	−1.1597	1.3450	15.8910	−0.9609	0.9233
10–15	16.5813	15.8567	−0.7246	0.5251	16.0594	−0.5219	0.2724
11–15	16.6325	15.7631	−0.8694	0.7558	15.9705	−0.6620	0.4382
12–15	17.0365	16.0779	−0.9586	0.9188	16.3045	−0.7320	0.5359
Average	15.8569	15.6568	−0.2001	0.4848	15.8629	0.0060	0.4443

5 Conclusions

The complexity of economic phenomena resulting in the non-inclusion in the forecast models, all statistically significant explanatory variables. To mitigate such failure expert knowledge is used to identify forgotten variables.

Hidden variables that influence the behavior of the forward exchange rate USD/MXN, with noticeable effects are the oil price and country risk. Among the finding it is detected that by including oil price as an additional variable to the PPP model, the forecast error decreases from −0.2001 to 0.0060.

In subsequent research, the aim is to incorporate the forgotten variables in parity of interest rate model and monetary approach to balance of payment model [9, 12, 26]. In this way, to analyze the information of experts will be used aggregator operators such as OWA, POWA or IOWA [27, 28, 37].

Appendix

See Annex 1, 2, 3, 4, 5, and 6.

Annex 1 Significance test for PPP model

Variable	P Value	Accept or reject H_0
Constant	0.131	Accept
tc_{lag}	0.000	Reject
V	0.625	Accept
pci_{USA}	0.066	Accept
pci_{MEX}	0.000	Reject

Annex 2 Significance test for PPP model with forgotten effects

Variable	P Value	Accept or reject H_0
Constant	0.914	Accept
tc_{lag}	0.000	Reject
V	0.000	Reject
pci_{EUA}	0.572	Accept
pci_{MEX}	0.123	Accept
mme	0.355	Accept

Annex 3 Significance test for complete models

Model	Valor P	Accept or reject H_0
PPP model	0.000	Reject
PPP model with forgotten effects	0.000	Reject

Annex 4 Durbin-Watson test for the residues of the models

Model	Durbin Watson	Result with 1% of significance	Result with 5% of significance
PPP model	1.9071	Without autocorrelation	Without autocorrelation
PPP model with forgotten effects	1.8988	Without autocorrelation	Without autocorrelation

Annex 5 Variance inflation factors for PPP model

Variable	VIF	Multicollinearity
tc_{lag}	1.077	Low
v	8.970	Moderate
pci_{EUA}	28.287	High
pci_{MEX}	12.424	High

Annex 6 Variance inflation factors for PPP model with forgotten effects

Variable	VIF	Multicollinearity
tc_{lag}	12.478	High
v	1.146	Low
pci_{EUA}	36.897	High
pci_{MEX}	30.255	High
mme	15.693	High

References

1. Allayannis, G., Ihrig, J., Weston, J.P.: Exchange-rate hedging: financial versus operational strategies. Am. Econ. Rev. **91**(2), 391–395 (2001)
2. Barcellos Paula, L., Gil Lafuente, A.M.: Algorithms applied in the sustainable management of human resources. Fuzzy Econ. Rev. **15**(1) (2010)
3. Bates, D.S.: Jumps and stochastic volatility: exchange rate processes implicit in deutsche mark options. Rev. Financial Stud. **9**(1), 69–107 (1996)
4. Bazdresch, S., Werner, A.: El comportamiento del tipo de cambio en México y el régimen de libre flotación: 1996–2001. In: Documento de investigación No. 2002-09 Direccion General de Investigacion Economica Banco de Mexico, pp. 1–18 (2002)
5. Beltrán, L.S., Gil-Lafuente, A.: Models for analyzing purchase decision in consumers of econologic products. Fuzzy Econ. Rev. **10**(1), 47 (2005)
6. Bergholt, D., Lujala, P.: Climate-related natural disasters, economic growth, and armed civil conflict. J. Peace Res. **49**(1), 147–162 (2012)
7. Blanchard, O., Adler, G.: Can Foreign Exchange Intervention Stem Exchange Rate Pressures from Global Capital Flow Shocks? National Bureau of Economic Research (2015)
8. Cheung, Y.W., Chinn, M.D., Pascual, A.G.: Empirical exchange rate models of the nineties: are any fit to survive? J. Int. Money Finance **24**(7), 1150–1175 (2005)
9. Dornbusch, R.: Monetary Policy Under Exchange Rate Flexibility. National Bureau of Economic Research (1979)
10. Dornbusch, R.: Purchasing power parity. New Palgrave Dict. Econ. **3**, 1075–1085 (1985)
11. Fischer, B., Planas, C.: Large scale fitting of regression models with ARIMA errors. J. Off. Statist. **16**(2), 173–184 (2000)
12. Flood, E., Lessard, D.: On the measurement of operating exposure to exchange rate. A conceptual approach. Financial Manag. Spring 25–36 (1986)
13. Ghosh, A.R., Ostry, J.D., Chamon, M.: Two targets, two instruments: monetary and exchange rate policies in emerging market economies. J. Int. Money Finance **60**, 172–196 (2016)
14. Gil-Aluja, J.: Towards a new concept of economic research. Fuzzy Econ. Rev. 5–23 (1995)
15. Gil-Aluja, J.: Towards a new paradigm of investment selection in uncertainty. Fuzzy Sets Syst. **84**(2), 187–197 (1996)
16. Gil-Aluja, J.: Elements for a Theory of Decision in Uncertainty. Springer, US (1999)
17. Gil-Aluja, J.: Handbook of Management Under Uncertainty. Springer, US (2001)
18. Gil-Aluja, J.: Fuzzy Sets in the Management of Uncertainty, vol. 145. Springer Science & Business Media (2004)
19. Jorion, P., Sweeney, R.: Mean reversion in real exchange rates: evidence and implications for forecasting. J. Int. Money Finance **15**(4), 535–550 (1996)
20. Kaufmann, A.: Theory of expertons and fuzzy logic. Fuzzy Sets Syst. **28**(3), 295–304 (1988)
21. Kaufmann, A., Gil Aluja, J.: Models for the Research of Forgotten Effects. (In Spanish). Milladoiro, Spain (1988)
22. Krugman, P.: Thinking about the liquidity trap. J. Jpn. Int. Econ. **14**, 221–237 (2000)

23. Luis, C., Gil-Lafuente, A.: Models for the analysis of attributes considered by the customers in a relational marketing strategy (In Spanish) PhD Thesis, University of Barcelona, Spain (2011)
24. Lustig, H., Roussanov, N., Verdelhan, A.: Common risk factors in currency markets. Rev. Financial Stud. **24**(11), 3731–3777 (2011)
25. Majhi, R., Panda, G., Sahoo, G.: Efficient prediction of exchange rates with low complexity artificial neural network models. Expert Syst. Appl. **36**(1), 181–189 (2009)
26. McCallum, B.: A reconsideration of the uncovered interest parity. J. Monet. Econ. **33**(1), 105–132 (1993)
27. Merigó, J.M., Gil-Lafuente, A.: The induced generalized OWA operator. Inf. Sci. **179**(6), 729–741 (2009)
28. Merigó, J.M.: Probabilities in the OWA operator. Expert Syst. Appl. **39**, 11456–11467 (2012)
29. Pastor, L.: Portfolio selection and asset pricing models. J. Finance **55**(1), 179–223 (2000)
30. Reboredo, J.C.: Modelling oil price and exchange rate co-movements. J. Policy Model. **34**(3), 419–440 (2012)
31. Rogoff, K.: The purchasing power parity puzzle. J. Econ. Lit. **34**(2), 647–668 (1996)
32. Taylor, A., Taylor, M.: The purchasing power parity debate. J. Econ. Perspect. **18**(4), 135–158 (2004)
33. Taylor, J.: The role of the exchange rate in monetary-policy rules. Am. Econ. Rev. **91**(2), 263–267 (2001)
34. Taylor, M.: The economics of exchange rates. J. Econ. Lit. **33**(1), 13–47 (1995)
35. Tsai, P.: Determinants of foreign direct investment and its impact on economic growth. J. Econ. Dev. **19**(1), 137–163 (1994)
36. Vizuete-Luciano, E., Gil-Lafuente, A.M., Garcia-Gonzalez, A., Boria-Reverter, S.: Forgotten effects of corporate social and environmental responsibility: a case study of Catalonian economy. Kybernetes **42**(5), 736–753 (2013)
37. Yager, R.R., Filev, D.: Induced ordered weighted averaging operators. IEEE Trans. Syst. Man Cybern. **29**(2), 141–150 (1999)
38. Zhang, J., Wang, L., Wang, S.: Financial development and economic growth: recent evidence from China. J. Comp. Econ. **40**(3), 393–412 (2012)

The Behaviour of Non-surviving Spanish Funds According to Their Investment Objectives

Antonio Terceño, M. Gloria Barberà-Mariné, Laura Fabregat-Aibar and Maraia Teresa Sorrosal-Forradellas

Abstract The main purpose of this paper is to determine if the characteristics which define the non-surviving funds are different according to their investment objectives. In Spanish market, the Spanish National Securities Market Commission (CNMV) classifies the mutual funds according to the type of assets in which each fund invests to form a portfolio. Thus, the investor could choose which fund is suitable based on the performance and the risk that he wants to assume. In this paper, Self-Organizing Maps (SOM) are used to cluster mutual funds that disappeared in 2013, 2014 and 2015, based on the variables that define its survival capacity and, as a result, to analyse if these variables, take similar values for all of them or, different depending on the funds' investment objectives.

1 Introduction

Currently, the savings of millions of families are channelled through mutual funds, which play a key role in the financing of companies and countries, as they are a financial investment with enormous economic and social impact. The assets of Spanish mutual funds have increased considerably because, by the end of 2015, mutual funds had reached 220.288 million of total net assets, representing an increase of 74,1% over the three previous years (the assets were 126.523 million in 2012).

A. Terceño (✉) · M. Gloria Barberà-Mariné · L. Fabregat-Aibar ·
M. T. Sorrosal-Forradellas
Department of Business Management, University Rovira i Virgili,
Campus Bellissens, Av. de la Universitat, 1, 43204 Reus, Spain
e-mail: Antonio.terceno@urv.cat

M. Gloria Barberà-Mariné
e-mail: gloria.barbera@urv.cat

L. Fabregat-Aibar
e-mail: laura.fabregat@urv.cat

M. T. Sorrosal-Forradellas
e-mail: mariateresa.sorrosal@urv.cat

© Springer International Publishing AG 2018   439
C. Berger-Vachon et al. (eds.), *Complex Systems: Solutions and Challenges
in Economics, Management and Engineering*, Studies in Systems,
Decision and Control 125, https://doi.org/10.1007/978-3-319-69989-9_26

The main purpose of this paper is to determine if the characteristics which define the non-surviving funds are different according to their investment objectives. In Spanish market, the Spanish National Securities Market Commission (CNMV) classifies the mutual funds according to the type of assets in which each fund invests to form a portfolio. Thus, the investor could choose which fund is suitable based on the performance and the risk that he wants to assume.

In this paper, Self-Organizing Maps (SOM) are used to cluster mutual funds that disappeared in 2013, 2014 and 2015, based on the variables that define its survival capacity and, as a result, to analyse if these variables, take similar values for all of them or, different depending on the funds' investment objectives.

We propose to analyse nine categories: bond funds, bond mixed funds, equity funds and equity mixed funds, distinguishing between those that invest their assets in national or international markets, and passive investing funds. Some studies consider that the investment objectives of the fund play an important role in its survival [2, 15, 17, 21] but they only focus on equity funds. Thus, this analysis is a novel aspect in the literature because we consider that it is important to know if the variables defined by the non-surviving funds are similar (or not) between the equity and bond funds, of which the type of assets and proportions held are very different.

In relation to the use of artificial neural networks techniques on mutual funds, Terceño et al. [25] use SOM to group mutual funds in surviving and non-surviving funds according to the similarities and differences between the variables analysed (size, age, return, risk and fees). They consider that the disappearance of mutual funds is explained by the fund size, age, 1-year return and expenses. Moreover, Moreno et al. [20] use SOM to evaluate the official classification of Spanish mutual funds by the CNMV and Inverco and they aimed to improve this classification with nonlinear techniques.

The paper shows the following structure. Section 2 defines which variables have an impact on survival capacity regarding the literature. It is followed by the description about the methodology used (Sect. 3). Section 4 describes the data and how it is processed. Section 5 sets out the results, and finally the last section presents the conclusions of the study.

2 Determinants of Mutual Funds Survivorship

Some studies consider the fund size is a key factor on survival capacity. They assert that smaller funds have a greater likelihood of disappearing [1, 3, 5–8, 14, 15, 17, 22, 25, 26].

Other studies maintain that the probability of disappearance is inversely related to the age of the fund and, therefore, those older portfolios are less likely to disappear [3, 5, 24, 26]. Nevertheless, Lunde et al. [19] and Cameron and Hall [6] differ in their conclusions regarding to the mentioned previously.

The performance is another explanatory variable examined in the literature. Numerous studies find that a poor performance increases the probability of disappearance [1, 3–8, 11, 14, 15, 17, 19, 22, 26]. Moreover, Brown and Goetzmann [5] observe that funds with negative returns for up to three consecutive years do not survive, while Carhart [7] considers a negative return for up to 5 years before their disappearance.

Closely related to the performance of the fund, some works analyse the risk based on the volatility of the fund. Elton and Gruber [14] and Carpenter and Lynch [9] assert that funds with higher risk are more likely to disappear. However, Asebedo and Grable [2] note that risk cannot be considered as a key variable in the survival capacity of mutual funds because higher risk does not imply better or worse performance and, consequently, it does not necessarily influence on a fund's disappearance.

Finally, other studies focus on the analysis of fees as a variable that influence the mortality of mutual funds, establishing that funds whit higher expenses ratios have a higher probability of disappearance [5–8, 14, 22].

3 Self-organizing Maps

The methodology that we use in this work is a particular kind of artificial neural network. Artificial neural networks are "massively parallel interconnected networks of simple (usually adaptive) elements and their hierarchical organizations which are intended to interact with the objects of the real world in the same way as biological nervous system do" [18]. Although its biological inspiration, they can be seen as non-linear regression techniques, that can be used, between others, for forecasting and clustering applications.

Artificial neural networks can be classified in two main groups: supervised and unsupervised ones. The first type try to extract relationships between inputs and outputs from a set of data, where the desired output is known. On the other hand, unsupervised neural networks do not need to know the output, they are used to classify or organize large data sets according to the similarity within its features.

Dr. Teuvo Kohonen was inspired by the ability of human brain to store internal representations of information in an organized spatially manner, and he developed the Kohonen's Self-Organizing Maps (SOM). A SOM is an unsupervised neural net that provides a two-dimension map by reducing the dimension of the original data. Because of its topology preserving property, two patterns that are near in the map are also near in the input space.

SOM have been applied to numerous problems in finance. Between them, in bankruptcy prediction [10, 12, 13], analysis of financial crises [16, 23], or in classification of mutual funds [20, 25]. In this study, SOM are used to analyze the classification of non-surviving Spanish mutual funds according to the main variables of survival.

Regarding SOM architecture, they are composed for two layers of processing units (or neurons): the input and the output layer. The input layer has as many neurons as variables defines the patterns to cluster in the map. Each pattern are represented in form of an n-component vector in the following way $X^p = \left(x_1^p, x_2^p, \ldots, x_i^p, \ldots, x_n^p\right)$, where the subscript i = 1, 2,..., i,..., n represents each one of the n variables used to characterize the pattern and the superscript p = 1, 2,... P represents each one of the P patterns analyzed in the study.

The output layer is a set of m neurons, distributed in rows and columns forming a bi-dimensional map or grid of processing units. The dimension of this map is decided by the user taking into account the amount of data, that is to say, the number of output neurons increases with the number of patterns. As we have implemented SOM using the SOM Toolbox for Matlab, the number of units by default of the output layer will be $m = 5 \cdot \sqrt{P}$.

Between the input and the output layer there are feed-forward connections. Each unit of the input layer i is connected with all the output neurons with a weight vector $W_{ki} = (w_{1i}, w_{2i}, \ldots, w_{mi})$. Initially, the components of W_{ki} are small random values.

The learning algorithm in SOM is competitive. It means that when an input is presented to the network, all the output neurons compete to match the input. The output neuron with the greatest similarity between the input pattern and its weight vector (usually, in terms of Euclidean distance) will be the Best-Matching Unit (BMU) or winner unit. The BMU indicates the location of the corresponding input pattern in the map. In mathematical terms, the BMU for a pattern p is the output neuron k* that satisfies:

$$\min\left(d_k^p\right) = \min\left(\sqrt{\sum_{i=1}^{n}\left(x_i^p - w_{ki}\right)^2}\right) \tag{1}$$

Once the BMU for a pattern is determined, the weight vector for k* and for its 'neighbor' neurons are modified. The neighborhood area is defined in the learning process (the process with which the weights change) and includes, from the k* unit, all adjacent neurons in a rectangular or hexagonal area. This area usually decreases with time.

The new weight vectors for these units are obtained from the following expression:

$$w_{ki}(t+1) = w_{ki}(t) + \alpha(t) \cdot \left[x_i^p - w_{ki}(t)\right] \tag{2}$$

where $\alpha(t)$ is the learning rate, a function that takes values between 0 and 1 and decreases with time.

In a schematic way, we can summarize the implementation of the SOM in the following steps:

- Pattern vectors are introduced into the system.
- The components of the vectors are normalized. In this specific case, we use the logistic transformation, that scales the values between 0 and 1.
- The output map is created. Its dimension is determined according to the rule described previously.
- By introducing the data, the BMU for each pattern is obtained from expression (1) and weight vectors of the BMU and its neighbor neurons are modified according to the expression (2).
- The preceding step is repeated a sufficient number of times to get the stability of the association between the different patterns with the same unit of the output layer.

Once the learning process ends, all the patterns are distributed in the bi-dimensional map. The more closely two patterns are in the map, the more similar in the input space they are. To interpret the specific location of a pattern in the map it is necessary to know the values that each component of the pattern vectors takes in each area of the map. This information is obtained from the SOM Toolbox through the component maps.

Finally, it is also possible to group the input patterns situated in the map in a particular number of groups. A pattern will belong to the closest group, according to the minimum distance between its characteristics and the mean of the characteristics in the group.

4   Data and Methodology

The databases are provided by the Spanish National Securities Market Commission (CNMV) and Morningstar Direct.

Our sample consists of 299 non-surviving funds from the Spanish market from 2013 until 2015. This sample is classified into nine different categories: bond funds, bond mixed funds, equity funds, equity mixed funds, distinguishing all of them between national and international, and the last one is passive investing funds. The description of each of these investment objectives is as follows:

- **Bond funds**: the percentage invested in Treasury bills and bonds is 100%. If the proportion invested in currencies different from euro is less than 10%, it is a **bond fund national**, while if the proportion invested in currencies different from euro is more than 10%, it is a **bond fund international**.
- **Bond mixed funds**: the percentage invested in Treasury bills and bonds is more than 70% and, therefore, the percentage invested in stocks is less than 30%. It is a **bond mixed fund national** if the proportion invested in currencies different

from euro is less than 30%. And it is a **bond mixed fund international**, if the proportion invested in currencies different from euro is more than 30%.

- **Equity funds:** the percentage invested in stocks is more than 75%. It is an **equity fund national** if the proportion invested in currencies different from euro is less than 30%, whereas if these funds invested in currencies different from euro is more than 30%, it is called an **equity fund international**.
- **Equity mixed funds**: the percentage invested in stocks is between 30–75%. If the proportion invested in currencies different from euro is less than 30%, it is an **equity mixed fund national**, while the proportion invested in currencies different from euro is more than 30%, it is an **equity mixed fund international**.
- **Passive investing funds:** This type of portfolio is an investment strategy that tracks a market index or benchmark.

Table 1 summarizes the variables included in our study and defines each variables and Table 2 shows a descriptive analysis of them.

Table 3 presents the correlation matrix. We find a strong correlation between variable 5 (1-year Standard Deviation) and 6 (3-year Standard Deviation). So, we

Table 1 List and definition of variables

Code	Variable	Definition
Var1	Age	Number of years since the creation of the fund until the year of the study
Var2	Size	Total net assets (TNA) to the date of disappearance, measured in euros
Var3	1-year return	Annual return obtained by the fund in the year of the study
Var4	3-year annualized return	Three years annualized return obtained by the fund in the year of the study
Var5	1-year standard deviation	Annual standard deviation calculated from quarterly returns
Var6	3-year standard deviation	Standard deviation in the three previous years calculated from quarterly returns
Var7	Total expense ratio (TER)	Total expense ratio charged by the fund in the year of the study

Table 2 Descriptive analysis

Code	Mean	Std deviation	Q1 (25%)	Q2 (50%)	Q3 (75%)
Var1	14,13	5,695	10	15	18
Var2	50.695.888	114.843.594	4.140.555	10.808.180	43.028.034
Var3	4,658%	6,814%	0,735%	2,463%	7,075%
Var4	1,891%	4,233%	0,000%	1,891%	3,662%
Var5	5,469%	6,527%	0,544%	2,469%	8,347%
Var6	7,346%	7,937%	0,938%	3,713%	12,320%
Var7	0,610%	0,885%	0,000%	0,610%	1,530%

Table 3 Correlation matrix

	Var1	Var1	Var1	Var1	Var1	Var1	Var1
Var1	1						
Var2	0,1168	1					
Var3	0,0288	−0,1364	1				
Var4	−0,0320	0,0315	0,3932	1			
Var5	0,0396	−0,2229	0,4020	−0,2713	1		
Var6	0,0030	−0,2460	0,4655	−0,1506	0,9340	1	
Var7	0,1439	−0,1997	0,2212	−0,3101	0,4542	0,4476	1

Table 4 Final selected variables

Code	Variable
Var1	Age
Var2	Size
Var3	1-year return
Var4	3-year annualized return
Var5	1-year standard deviation
Var6	Total expense ratio

decided to exclude variable 6 to avoid overweighting which could distort the results. After excluding the variable 6, we consider a total of 6 variables (Table 4).

5 Empirical Application

The network generates an output map of 11 × 8 (11 rows × 8 columns). The 88 cells have been grouped in 7 clusters.

Figure 1 shows the Kohonen map, where the corresponding patterns (non-surviving funds) have been numbered and their explanation is the following.

- "RFE": euro bond fund
- "RFME": euro bond mixed fund
- "RFI": international bond fund
- "RFMI": international bond mixed fund
- "RVE": euro equity fund
- "RVME": euro equity mixed fund
- "RVI": international equity fund
- "RVMI": international equity mixed fund
- "GP": passive investing fund.

We observe that groups situated at the bottom mostly contain the equity funds (included equity mixed funds); while those located at the top mainly represent

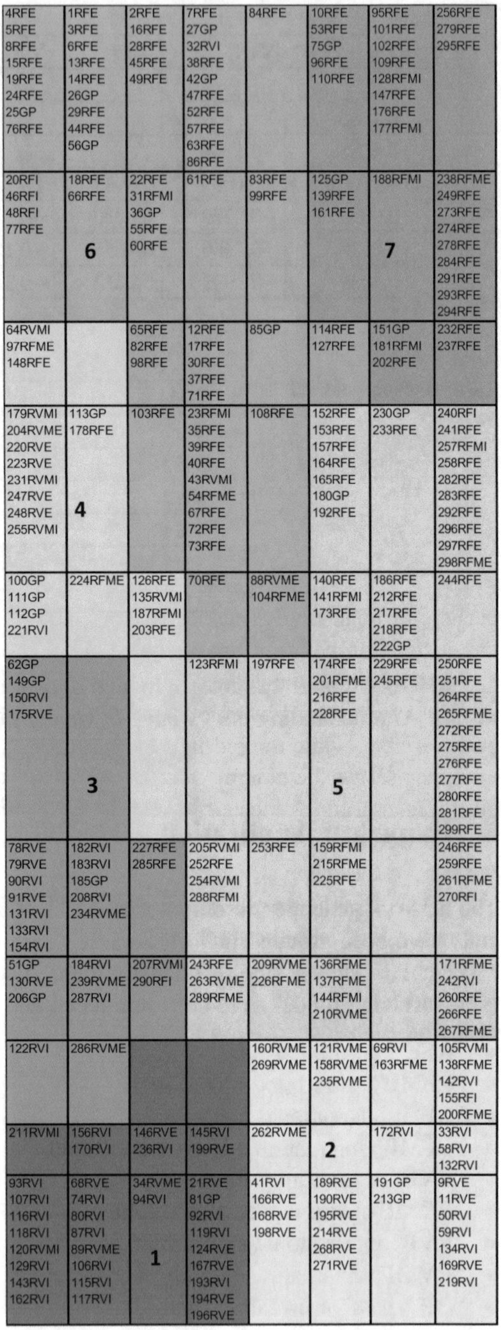

Fig. 1 Self-organizing map for Spanish funds

4RFE 5RFE 8RFE 15RFE 19RFE 24RFE 25GP 76RFE	1RFE 3RFE 6RFE 13RFE 14RFE 26GP 29RFE 44RFE 56GP	2RFE 16RFE 28RFE 45RFE 49RFE	7RFE 27GP 32RVI 38RFE 42GP 47RFE 52RFE 57RFE 63RFE 86RFE	84RFE	10RFE 53RFE 75GP 96RFE 110RFE	95RFE 101RFE 102RFE 109RFE 128RFMI 147RFE 176RFE 177RFMI	256RFE 279RFE 295RFE
20RFI 46RFI 48RFI 77RFE	18RFE 66RFE **6**	22RFE 31RFMI 36GP 55RFE 60RFE	61RFE	83RFE 99RFE	125GP 139RFE 161RFE	188RFMI **7**	238RFME 249RFE 273RFE 274RFE 278RFE 284RFE 291RFE 293RFE 294RFE
64RVMI 97RFME 148RFE		65RFE 82RFE 98RFE	12RFE 17RFE 30RFE 37RFE 71RFE	85GP	114RFE 127RFE	151GP 181RFMI 202RFE	232RFE 237RFE
179RVMI 204RVME 220RVE 223RVE 231RVMI 247RVE 248RVE 255RVMI	113GP 178RFE **4**	103RFE	23RFMI 35RFE 39RFE 40RFE 43RVMI 54RFME 67RFE 72RFE 73RFE	108RFE	152RFE 153RFE 157RFE 164RFE 165RFE 180GP 192RFE	230GP 233RFE	240RFI 241RFE 257RFMI 258RFE 282RFE 283RFE 292RFE 296RFE 297RFE 298RFME
100GP 111GP 112GP 221RVI	224RFME	126RFE 135RVMI 187RFMI 203RFE	70RFE	88RVME 104RFME	140RFE 141RFMI 173RFE	186RFE 212RFE 217RFE 218RFE 222GP	244RFE
62GP 149GP 150RVI 175RVE **3**			123RFMI	197RFE	174RFE 201RFME 216RFE 228RFE	229RFE 245RFE **5**	250RFE 251RFE 264RFE 265RFME 272RFE 275RFE 276RFE 277RFE 280RFE 281RFE 299RFE
78RVE 79RVE 90RVI 91RVE 131RVI 133RVI 154RVI	182RVI 183RVI 185GP 208RVI 234RVME	227RFE 285RFE	205RVMI 252RFE 254RVMI 288RFMI	253RFE	159RFMI 215RFME 225RFE		246RFE 259RFE 261RFME 270RFI
51GP 130RVE 206GP	184RVI 239RVME 287RVI	207RVMI 290RFI	243RFE 263RVME 289RFME	209RVME 226RFME	136RFME 137RFME 144RFMI 210RVME		171RFME 242RVI 260RFE 266RFE 267RFME
122RVI	286RVME			160RVME 269RVME	121RVME 158RVME 235RVME	69RVI 163RFME	105RVMI 138RFME 142RVI 155RFI 200RFME
211RVMI	156RVI 170RVI	146RVE 236RVI	145RVI 199RVE	262RVME	**2**	172RVI	33RVI 58RVI 132RVI
93RVI 107RVI 116RVI 118RVI 120RVMI 129RVI 143RVI 162RVI	68RVE 74RVI 80RVI 87RVI 89RVME 106RVI 115RVI 117RVI **1**	34RVME 94RVI	21RVE 81GP 92RVI 119RVI 124RVE 167RVE 193RVI 194RVE 196RVE	41RVI 166RVE 168RVE 198RVE	189RVE 190RVE 195RVE 214RVE 268RVE 271RVE	191GP 213GP	9RVE 11RVE 50RVI 59RVI 134RVI 169RVE 219RVI

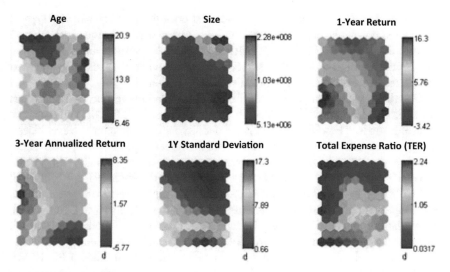

Fig. 2 Maps of features

clusters of bond funds (included bond mixed funds). There does not appear to be any distinction between funds invested in euro or international currency.

Figure 2 shows the value of all the variables in the map and their respective scale of values represented by colours. The colour blue indicates the minimum values of each variable, while the red indicates the highest ones. The scale of values next to each map shows the rank of values which are taken by the representative patterns of all the mutual funds located in one cell.

This information is useful to analyse the behaviour of each pattern through its position in the map.

We analyse the results of the SOM and the maps of features and we observe similarities between the funds included in each group.

Group 1 mainly comprises equity funds, specifically, 33 equity funds (RVE, RVI and RVME) and 1 passive investing fund. This group shows the following behavior: high 1-year return and 3-year annualized return, high expenses and high volatility. Moreover, these funds are small and mid-age.

Group 2 is formed by 36 funds, of which 31 are equity funds (RVE, RVI, RVME and RVMI) and only 5 belong to other categories (3 bond mixed funds and 2 passive investing funds). These funds show high volatility and fees but low 1-year return and low 3-year annualized return. As in group 1 they are also small and mid-age funds.

Group 3 contains 28 funds, specifically, 20 equity funds (RVE, RVI, RVME, and RVMI), 5 passive investing funds and 3 bond funds (RFE). This group is characterized by high 1-year return and 3-year annualized return but low expenses. In addition, they are mainly old and small.

Group 4 includes heterogeneous funds and does not predominate any type of investment objectives because we can observe 7 bond funds (RFE and RFME), 11 equity funds (RVE, RVME and RVMI) and 5 passive investing funds. They have low 1-year return, low volatility and low expenses unlike they show high 3-year annualized return. Moreover, they are small and mid-age.

Group 5 mainly comprises bond funds. In this group, we find 64 bond funds (RFE, RFME, RFI and RFMI), 6 equity funds (RVME, RVI, RVMI) and 3 passive investing funds. These funds have low volatility and 1-year return. Nevertheless, this group contains funds which show medium values regarding the 3-year annualized return and fees. Also, we can observe they are old and small.

Group 6 is formed by 72 funds, of which 62 are bond funds (RFE, RFME, RFI and RFMI) and only 10 are other category, specifically, 7 passive investing funds and 3 equity funds. These funds present the following behaviour: low 1 year-return, low volatility and low expenses. Nevertheless, this group contains funds which show high 3-year annualized return. Also, they are small and young.

Group 7 mainly contains bond funds, in particular, 31 bond funds (RFE, RFME and RFMI) and 3 passive investing funds. These funds show a similar values in the variables studied to the previous group. They have low 1 year-return, low volatility and low expenses. Moreover, this group has medium values concerning the 3-year annualized return. The main difference in this group is the size because these funds have high assets in their portfolios.

6   Conclusions

As could be appreciated, the groups that are obtained from SOM are not totally homogeneous, even so, there is an evident relation between the values of all the analysed variables and the investment objectives, with the exception of passive investing funds.

The SOM has clustered non-surviving funds regarding to investment objectives (equity and bond). If we analyse the values of each variable in the map, we find that they do not explain the disappearance according to the investment objectives, with the exception of risk and, to a lesser extent, the 3-year annualized return.

Some variables such as size, age, 1-year return and expenses, take similar values for all investment objectives and, as a result, we do not consider that it is different between the groups.

The non-surviving funds are small with the exception of some funds and, therefore, there is no differences between bond and equity funds.

Age is not a factor that characterizes any investment objectives.

Regarding to the performance, we find that 1-year return is not a reason for disappearance, since some groups such as 1 and 3 have funds which show a high return. Also, it is important to mention that groups 5, 6 and 7 basically contain bond funds which show a poor performance in short-term, even though they had obtained a good performance in long-term. This difference could be explained due to the high levels in sovereign bond spreads during the financial crisis. For example, the 3-year Spanish bond offered a return of 3,925% in 2011 and 3,159% in 2012. Expenses is not a factor that characterizes any investment objectives.

The only variable which distinguish between equity and bond is the risk. The fact that a bond fund (equity fund) has a low risk (high risk) could be consequence by the characteristics of each investment objectives and not the cause of fund's disappearance.

Given that there is not a common pattern for all categories of non-surviving funds, it is necessary to take into account its investment objectives in order to carefully analyse the survival capacity of mutual funds.

References

1. Andreu, L., Sarto, J.L.: Financial consequences of mutual fund mergers. Eur. J. Finance 22(7), 529–550 (2016)
2. Asebedo, G., Grable, J.: Predicting mutual fund over-performance over a nine-year period. J. Financial Couns. Plan. 15(1), 1–11 (2004)
3. Blake, D., Blake, D., Timmermann, A., Timmermann, A.: Mutual fund performance: evidence from the UK. Eur. Finance Rev. 2(1), 57–77 (1998)
4. Boubakri, N., Karoui, A., Kooli, M.: Performance and Survival of Mutual Fund Mergers: Evidence from Frequent and Infrequent Acquirers. Working Paper (2012)
5. Brown, S., Goetzmann, W.: Performance persistence. J. Finance 50(2), 679–698 (1995)
6. Cameron, C., Hall, A.D.: A survival analysis of Australian equity mutual funds. Aust. J. Manag. 28(2), 209–226 (2003)
7. Carhart, M.M.: On persistence in mutual fund performance. J. Finance 52, 57–82 (1997)
8. Carhart, M.M., Carpenter, J.N., Lynch, A.W., Musto, D.K.: Mutual fund survivorship. Rev. Financial Stud. 15(5), 1439–1463 (2002)
9. Carpenter, J.N., Lynch, A.W.: Survivorship bias and attrition effects in measures of performance persistence. J. Financial Econ. 54(3), 337–374 (1999)
10. Chen, N., Ribeiro, B., Vieira, A., Chen, A.: Clustering and visualization of bankruptcy trajectory using self-organizing map. Expert Syst. Appl. 40, 385–393 (2013)
11. Cogneau, P., Hübner, G.: The prediction of fund failure through performance diagnostics. J. Bank. Finance 50, 224–241 (2015)
12. Deboeck, G., Kohonen, T. (eds.): Visual explorations in finance: with self-organizing maps. Springer Science & Business Media (2013)
13. Du Jardin, P., Séverin, E.: Predicting corporate bankruptcy using a self-organizing map: an empirical study to improve the forecasting horizon of a financial failure model. Decis. Support Syst. 51, 701–711 (2011)
14. Elton, E.J., Gruber, Martin J.: Survivorship bias and mutual fund performance. Rev. Financial Stud. 9(4), 1097–1120 (1996)
15. Filip, D.: Survivorship bias and performance of mutual funds in Hungary. Period. Polytech. Soc. Manag. Sci. 22(1), 47–56 (2014)
16. Fioramanti, M.: Predicting debt crises using artificial neural networks: a comparative approach. J. Financial Stab. 4, 149–164 (2008)
17. Jayaraman, N., Khorana, A., Nelling, E.: An analysis of the determinants and shareholder wealth effects of mutual fund mergers. J. Finance 57(3), 1521–1551 (2002)
18. Kohonen, T.: Self-organization and Associative Memory. Springer, Berlin (1989)
19. Lunde, A., Timmermann, A., Blake, D.: The hazards of mutual fund underperformance: a Cox regression analysis. J. Empir. Finance 6(2), 121–152 (1999)

20. Moreno, D., Marco, P., Olmeda, I.: Self-organizing maps could improve the classification of Spanish mutual funds. Eur. J. Oper. Res. **174**(2), 1039–1054 (2006)
21. Namvar, E., Phillips, B.: Commonalities in investment strategy and the determinants of performance in mutual fund mergers. J. Bank. Finance **37**(2), 625–635 (2013)
22. Rohleder, M., Scholz, H., Wilkens, M.: Survivorship bias and mutual fund performance: relevance, significance, and methodical differences. Rev. Finance **15**(2), 441–474 (2011)
23. Sarlin, P., Marghescu, D.: Visual predictions of currency crisis using self-organizing maps. Intell. Syst. Account. Finance Manag. **18**, 15–38 (2011)
24. Ter Horst, J.R., Nijman, T.E., Verbeek, M.: Eliminating look-ahead bias in evaluating persistence in mutual fund performance. J. Empir. Finance **8**(4), 345–373 (2001)
25. Terceño, A., Fabregat-Aibar, L., Sorrosal-Forradellas, M.T., Barberà-Mariné, M.G.: Mutual funds survival in Spain using self-organizing maps. In: Modeling and Simulation in Engineering, Economics and Management, pp. 61–68. Springer (2016)
26. Zhao, X.: Exit decisions in the U.S. mutual fund industry. J. Bus. **78**(4), 1365–1402 (2005)

Part VI
Applications in Business and Technology

Management System for Agricultural Enterprise on the Basis of Its Economic State Forecasting

Igor Atamanyuk, Yuriy Kondratenko and Natalia Sirenko

Abstract Management system (MS) for agricultural enterprise on the basis of its economic state forecasting was developed. MS allows to estimate the results of enterprise's work in future under the realization of certain reorganization (change of land resources, labour resources, fixed assets). Calculating method for forecasting economic indices of agricultural enterprises on the basis of vector polynomial exponential algorithm for extrapolation of the realizations of random sequences is worked out. The model of prognosis allows to estimate the results of enterprise functioning (to estimate future gross profit, gross production) after its reorganization. Prognostic model does not impose any restrictions on the forecast random sequence (linearity, stationarity, Markov behavior, monotone, etc.) and thus allows fully to take into consideration stochastic peculiarities of functioning of agricultural enterprises. The simulation results confirm high efficiency of introduced calculating method. The scheme for reflecting the peculiarities of the forecast model functioning are also presented in the chapter. The method can be realized in the decision support systems for agricultural and non-agricultural enterprises with various sets of economic indices.

Keywords Management system · Calculation method · Random sequence
Canonical decomposition · Forecasting economic indices

I. Atamanyuk (✉) · N. Sirenko
Mykolaiv National Agrarian University, Georgy Gongadze Str. 9, Mykolaiv
54010, Ukraine
e-mail: atamanyuk_igor@mail.ru

N. Sirenko
e-mail: sirenko@mdau.mk.ua

Y. Kondratenko
Petro Mohyla Black Sea National University, 68th Desantnykiv Str. 10, Mykolaiv
54003, Ukraine
e-mail: y_kondrat2002@yahoo.com

© Springer International Publishing AG 2018
C. Berger-Vachon et al. (eds.), *Complex Systems: Solutions and Challenges in Economics, Management and Engineering*, Studies in Systems,
Decision and Control 125, https://doi.org/10.1007/978-3-319-69989-9_27

1 Introduction

Many decision-making processes in different area of economy (management of enterprises, transport logistics, finance forecasting, investment under uncertainty and so on) are based on the different mathematical models [1–4], experienced theoretical methods [5–8] and modern intelligent algorithms [9, 10]. For guaranteeing efficient performance of an enterprise on the market, it is necessary to form the strategy and tactics of enterprise development correctly, to ground the plans and management decisions. To do this is possible only based on effective diagnostics and prognostication of current and future economical situation at the enterprise. Western specialists have the priority in the investigation of the possibilities of the management on the basis of the forecasting of enterprise economic state. Bever (USA) started theoretical development and building of prognostic models, then it was continued in the works of Altman (USA) [11, 12], Alberichi (Italy), Misha (France) and others [13, 14]. More contemporary trend in the building of the algorithms of economic indices forecasting is the usage of stochastic methods of extrapolation. The relevance of such approach is explained with the influence of great number of accidental factors on the results of enterprise functioning (weather conditions, accidental variations of demand and supply, inflation etc.), under the influence of which the change of economic state indices obtains accidental character. It is especially important to take into account stochastic peculiarities of economic indices during the solving of the problems of prognostication of the state of agricultural enterprises.

But the existing models of prognosis impose considerable limitations on the accidental sequence describing the change of economic indices [15–20] (Markovian property, stationarity, monotony, scalarity etc.). Thereupon the problem of the building of the forecast model under the most general assumptions about the stochastic properties of the accidental process of the change of the indices of enterprise economic state arises.

2 Aim and the Raising of Problem

The aim of this work is the development of the system for agricultural enterprise management on the basis of the forecasting algorithm for its economic indices. The main requirement to the forecasting algorithm is the absence of any essential limitations on the stochastic properties of the accidental process of economic indices change.

3 Theoretical Conception of the Proposed Forecasting Method

The most universal method (from the point of view of the requirements to the investigated accidental sequence) is a method that based on the mechanism of canonical expansions [21–23]. The main primary indices of the economic state of agricultural enterprises are the gross profit, gross output, land resources, labour resources, fixed assets that is why the object of the investigation is the vector accidental sequence with five dependant constituents (if necessary the number of figures and their qualitative composition may be changed). Preliminary investigations (the check of dependence of accidental values on the basis of statistical data about the work of agricultural enterprises in Mykolaiv region) showed that the accidental sequences describing the change of the economic state of the enterprises which relate to the intensive [24] type of the development during the interval of eleven years that corresponds to the processing of twelve annual indices for the great number of the enterprises of the mentioned type have the most stable and significant stochastic relations. For such vector accidental sequence the canonical expansion has the following look [25, 26]:

$$X_h(i) = M[X_h(i)] + \sum_{\nu=1}^{i} \sum_{\lambda=1}^{5} V_\nu^{(\lambda)} \varphi_{h\nu}^{(\lambda)}(i), \ i = \overline{1,12}, \ h = \overline{1,5}, \tag{1}$$

where

$X_1(i), \ i = \overline{1,12}$ gross profit;

$X_2(i), \ i = \overline{1,12}$ gross output;

$X_3(i), \ i = \overline{1,12}$ land resources;

$X_4(i), \ i = \overline{1,12}$ labour resources;

$X_5(i), \ i = \overline{1,12}$ fixed assets

The elements of canonical expansion are the accidental coefficients $V_\nu^{(\lambda)}, \ \nu = \overline{1,12}, \ \lambda = \overline{1,5}$ and nonrandom coordinate functions $\varphi_{h\nu}^{(\lambda)}(i)$, $\nu = \overline{1,12}, \ \lambda = \overline{1,5}$:

$$V_\nu^{(\lambda)} = X_\lambda(\nu) - M[X_\lambda(\nu)] - \sum_{\mu=1}^{\nu-1} \sum_{j=1}^{H} V_\mu^{(j)} \varphi_{\lambda\mu}^{(j)}(\nu) - \sum_{j=1}^{\lambda-1} V_\nu^{(j)} \varphi_{\lambda\nu}^{(j)}(\nu), \nu = \overline{1,12}; \tag{2}$$

$$D_\lambda(\nu) = M\left[\left\{V_\nu^{(\lambda)}\right\}^2\right] = M\left[\{X_\lambda(\nu)\}^2\right] - M^2[X_\lambda(\nu)]$$
$$- \sum_{\mu=1}^{\nu-1} \sum_{j=1}^{H} D_j(\mu)\left\{\varphi_{\lambda\mu}^{(j)}(\nu)\right\}^2 - \sum_{j=1}^{\lambda-1} D_j(\nu)\left\{\varphi_{\lambda\nu}^{(j)}(\nu)\right\}^2, \ \nu = \overline{1,12}; \tag{3}$$

$$\varphi_{h\nu}^{(\lambda)}(i) = \frac{M\left[V_\nu^{(\lambda)}(X_h(i) - M[X_h(i)])\right]}{M\left[\left\{V_\nu^{(\lambda)}\right\}^2\right]} = \frac{1}{D_\lambda(\nu)}(M[X_\lambda(\nu)X_h(i)]$$

$$- M[X_\lambda(\nu)]M[X_h(i)] - \sum_{\mu=1}^{\nu-1}\sum_{j=1}^{H} D_j(\mu)\varphi_{\lambda\mu}^{(j)}(\nu)\varphi_{h\mu}^{(j)}(i) \tag{4}$$

$$- \sum_{j=1}^{\lambda-1} D_j(\nu)\varphi_{\lambda\nu}^{(j)}(\nu)\varphi_{h\nu}^{(j)}(i), \ \lambda = \overline{1,5}, \nu = \overline{1,i}.$$

Coordinate functions $\varphi_{h\nu}^{(\lambda)}(i), h, \lambda = \overline{1,5}, \ \nu, i = \overline{1,12}$ have the following properties:

$$\varphi_{h\nu}^{(\lambda)}(i) = \begin{cases} 1, & h = \lambda \ \& \ \nu = i; \\ 0, & i < \nu \ \text{or} \ \text{h} \ < \lambda \& \nu \ = i. \end{cases} \tag{5}$$

On Fig. 1 the algorithm block-diagram of the canonical model (1) parameters determination is represented.

The algorithm of extrapolation on the basis of canonical expansion has the look [26]:

$$m_h^{(\mu,l)}(i) = \begin{cases} M[X_h(i)], \mu = 0, \\ m_h^{(\mu,l-1)}(i) + \left[x_l(\mu) - m_l^{(\mu,l-1)}(\mu)\right]\varphi_{h\mu}^{(l)}(i), l \neq 1, \\ m_h^{(\mu,5)}(i) + \left[x_1(\mu) - m_1^{(\mu-1,5)}(\mu)\right]\varphi_{h\mu}^{(1)}(i), l = 1, \end{cases} \tag{6}$$

where $m_h^{(\mu,l)}(i) = M[X_h(i)/x_\lambda(\nu), \ \lambda = \overline{1,5}, \ \nu = \overline{1,\mu-1}; \ x_j(\mu), j = 1, l], \ h = \overline{1,5}, \ i = \overline{k,12}$—is the linear optimal quantity by the criterion of the minimum of the average square of the error of the prognosis I s the estimation of the future values of the investigated sequence under the condition that the values are known $x_\lambda(\nu), \ \lambda = \overline{1,5}, \ \nu = \overline{1,\mu-1}; \ x_j(\mu), j = \overline{1,l}.$

As it follows from (4) the values $\varphi_{h\nu}^{(\lambda)}(i), h, \lambda = \overline{1,5}, \nu, i = \overline{1,12}$ are determined through auto- and mutually correlated functions of the investigated vector accidental sequence. In the Table 1 the values of autocorrelated function are presented $(M[\overset{\circ}{X}_1(\nu)\overset{\circ}{X}_1(i)], \nu = \overline{1,12}, \ i = \overline{1,12})$ for the first constituent $x_1(i), i = \overline{1,12}.$

For the period of 2004–2014 the values of the autocorrelated functions $M[\overset{\circ}{X}_h(\nu)\overset{\circ}{X}_h(i)], \nu = \overline{1,11}, i = \overline{1,11}, \ \text{h} = \overline{1,5}$ determined by means of the processing of statistic data (indices of the activity of Nikolaev region agricultural enterprises during 2004–2014). For 2015 $M[\overset{\circ}{X}_h(\nu)\overset{\circ}{X}_h(12)], \nu = \overline{1,11}, \ \text{h} = \overline{1,5}$ are calculated on the basis of the determinate models:

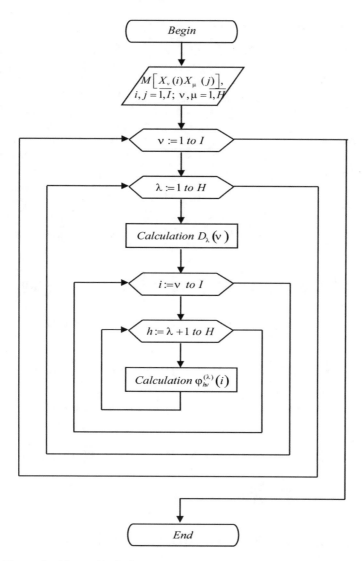

Fig. 1 The algorithm block-diagram of calculating the parameters $D_\lambda(\nu)$, $\varphi_{h\nu}^{(\lambda)}(i), h, \lambda = \overline{1,5}, \nu, i = \overline{1,12}$ of vector canonical decomposition (1)

$$M\left[\overset{o}{\underset{1}{X}}(\nu)\overset{o}{\underset{1}{X}}(12)\right] = 0,828\left[\overset{o}{\underset{1}{X}}(\nu)\overset{o}{\underset{1}{X}}(11)\right] - 0,143M\left[\overset{o}{\underset{1}{X}}(\nu)\overset{o}{\underset{1}{X}}(10)\right]$$
$$+0,345M\left[\overset{o}{\underset{1}{X}}(\nu)\overset{o}{\underset{1}{X}}(9)\right] - 0,233M\left[\overset{o}{\underset{1}{X}}(\nu)\overset{o}{\underset{1}{X}}(8)\right], \nu = \overline{1,11},$$

(7)

Table 1 Autocorrelated function of the component $X_1(i)$, $i = \overline{1, 12}$

	2004	2005	2006	2007	2008	2009	2010	2011	2012	2013	2014	2015
2004	1	0,99	0,70	0,42	0,79	0,74	0,44	0,76	0,65	0,42	0,53	0,47
2005	0,99	1	0,75	0,47	0,78	0,72	0,54	0,75	0,66	0,47	0,58	0,46
2006	0,70	0,75	1	0,57	0,67	0,58	0,70	0,69	0,70	0,66	0,78	0,60
2007	0,42	0,47	0,57	1	0,35	0,36	0,45	0,21	0,41	0,36	0,19	0,18
2008	0,79	0,78	0,67	0,34	1	0,81	0,55	0,91	0,80	0,72	0,53	0,41
2009	0,74	0,72	0,58	0,36	0,81	1	0,72	0,73	0,92	0,81	0,51	0,44
2010	0,44	0,54	0,70	0,45	0,55	0,72	1	0,51	0,74	0,73	0,49	0,41
2011	0,76	0,75	0,69	0,21	0,91	0,73	0,51	1	0,77	0,80	0,74	0,55
2012	0,65	0,66	0,70	0,41	0,80	0,92	0,74	0,77	1	0,91	0,60	0,59
2013	0,42	0,47	0,66	0,36	0,72	0,81	0,73	0,80	0,91	1	0,71	0,46
2014	0,53	0,58	0,78	0,19	0,53	0,51	0,49	0,74	0,60	0,71	1	0,71
2015	0,47	0,46	0,60	0,18	0,41	0,44	0,41	0,55	0,59	0,46	0,71	1

$$M\left[\overset{o}{\underset{2}{X}}(\nu)\,\overset{o}{\underset{2}{X}}(12)\right] = 2,363M\left[\overset{o}{\underset{2}{X}}(\nu)\,\overset{o}{\underset{2}{X}}(11)\right] - 1,344M\left[\overset{o}{\underset{2}{X}}(\nu)\,\overset{o}{\underset{2}{X}}(10)\right]$$
$$+ 0,123M\left[\overset{o}{\underset{2}{X}}(\nu)\,\overset{o}{\underset{2}{X}}(9)\right] - 0,215M\left[\overset{o}{\underset{2}{X}}(\nu)\,\overset{o}{\underset{2}{X}}(8)\right],\ \nu = \overline{1,11}, \tag{8}$$

$$M\left[\overset{o}{\underset{3}{X}}(\nu)\,\overset{o}{\underset{3}{X}}(12)\right] = 0,876M\left[\overset{o}{\underset{3}{X}}(\nu)\,\overset{o}{\underset{3}{X}}(11)\right] - 0,044M\left[\overset{o}{\underset{3}{X}}(\nu)\,\overset{o}{\underset{3}{X}}(10)\right]$$
$$+ 0,076M\left[\overset{o}{\underset{3}{X}}(\nu)\,\overset{o}{\underset{3}{X}}(9)\right] - 0,033M\left[\overset{o}{\underset{3}{X}}(\nu)\,\overset{o}{\underset{3}{X}}(8)\right],\ \nu = \overline{1,11}, \tag{9}$$

$$M\left[\overset{o}{\underset{4}{X}}(\nu)\,\overset{o}{\underset{4}{X}}(12)\right] = 0,911M\left[\overset{o}{\underset{4}{X}}(\nu)\,\overset{o}{\underset{4}{X}}(11)\right] + 0,022M\left[\overset{o}{\underset{4}{X}}(\nu)\,\overset{o}{\underset{4}{X}}(10)\right] +$$
$$- 0,004M\left[\overset{o}{\underset{4}{X}}(\nu)\,\overset{o}{\underset{4}{X}}(9)\right] - 0,013M\left[\overset{o}{\underset{4}{X}}(\nu)\,\overset{o}{\underset{4}{X}}(8)\right],\ \nu = \overline{1,11}, \tag{10}$$

$$M\left[\overset{o}{\underset{5}{X}}(\nu)\,\overset{o}{\underset{5}{X}}(12)\right] = 0,786M\left[\overset{o}{\underset{5}{X}}(\nu)\,\overset{o}{\underset{5}{X}}(11)\right] - 0,056M\left[\overset{o}{\underset{5}{X}}(\nu)\,\overset{o}{\underset{5}{X}}(10)\right] +$$
$$- 0,017M\left[\overset{o}{\underset{5}{X}}(\nu)\,\overset{o}{\underset{5}{X}}(9)\right] + 0,059M\left[\overset{o}{\underset{5}{X}}(\nu)\,\overset{o}{\underset{5}{X}}(8)\right],\ \nu = \overline{1,11}, \tag{11}$$

The parameters of the Eqs. (7)–(11) satisfy the minimum of the average error of approximation (relative error of the forecast is not more than 1%).

In the Table 2 coordinate function $\varphi_{1\nu}^{(1)}(i)$, $\nu, i = \overline{1, 12}$ corresponding to auto-correlated function $M\left[\overset{o}{\underset{1}{X}}(\nu)\,\overset{o}{\underset{1}{X}}(i)\right]$, $\nu = \overline{1, 12}$, $i = \overline{1, 12}$ and determining the degree of the influence of former values of gross profit for future values is presented.

Table 2 Coordinate function $\varphi_{1\nu}^{(1)}(i)$, $\nu, i = \overline{1,12}$

	2004	2005	2006	2007	2008	2009	2010	2011	2012	2013	2014	2015
2004	1	0,89	0,54	0,55	0,62	0,43	0,45	0,89	0,858	0,90	2,36	2,65
2005	0	1	2,25	-1,46	-2,40	0,27	5,47	-2,71	3,55	1,83	2,85	4,70
2006	0	0	1	5,09	1,17	-1,53	-0,03	-2,77	-0,23	-5,52	2,34	5,07
2007	0	0	0	1	0,17	0,26	0,94	0,18	0,77	1,05	-0,57	-1,17
2008	0	0	0	0	1	0,48	1,27	1,06	1,05	2,01	-2,37	0,69
2009	0	0	0	0	0	1	-1,81	0,74	3,53	0,37	9,31	2,86
2010	0	0	0	0	0	0	1	-0,68	1,44	3,18	-6,74	-3,39
2011	0	0	0	0	0	0	0	1	1,29	2,21	-3,30	0,93
2012	0	0	0	0	0	0	0	0	1	3,88	0,19	-8,44
2013	0	0	0	0	0	0	0	0	0	1	1,99	-4,96
2014	0	0	0	0	0	0	0	0	0	0	1	0,50
2015	0	0	0	0	0	0	0	0	0	0	0	1

Additionally to the Table 2 in the model (6) the values $\varphi_{h\nu}^{(\lambda)}(i), h, \lambda = \overline{1,5}$, $h \neq \lambda$, $\nu, i = \overline{1,12}$ which allow to consider mutual stochastic relations between the constituents $X_h(i)$, $h = \overline{1,5}$ (for example, the influence of land resources on gross profit, labour resources on gross output etc.) are used.

The future values of the mathematical expected value of the investigated vector accidental sequence $\{X\}$ are estimated with the usage of the determinate model

$$
\begin{aligned}
M[X_h(12)] &= 3,456M[X_h(11)] - 2,673M[X_h(10)] \\
&+ 1,345M[X_h(9)] - 0,145M[X_h(8)], \quad h = \overline{1,5}.
\end{aligned}
\tag{12}
$$

The parameters of the Eq. (12) as well as in (7)–(11) are determined from the condition of the minimum of the average error of approximation. For agricultural enterprises of Nikolaev region related to intensive type of development the mathematical expectations are $M[X_1(12)] = 5789,4$, $M[X_2(12)] = 16534,8$.

In all in the algorithm of the prognosis (6) 55 entrance values $x_h(i), h = \overline{1,5}$, $i = \overline{1,11}$ and 1775 that are not equal to zero balance coefficients $\varphi_{h\nu}^{(\lambda)}(i), h, \lambda = \overline{1,5}$, $\nu, i = \overline{1,12}$ are used.

For the increase of the effectiveness of the calculating processes during the prognosis by extrapolator (6) it is advisably to use the calculating procedure the substance of which is the fulfillment of the following stages:

Step 1.
For the fixed point t_ν (initially $\nu = 1$) the dispersions $D_\lambda(\nu)$ (initially $\lambda = 1$) of the accidental coefficients $V_\nu^{(\lambda)}$ with the help of the expression (3) are determined;

Step 2.
Using the obtained at the previous step value $D_\lambda(\nu)$ coordinate functions $\varphi_{h\nu}^{(\lambda)}(i)$ for $h = \overline{\lambda,5}$; $i = \overline{\nu,12}$ by the formula (4) are calculated;

Step 3.
The condition $\lambda < 5$ is checked. If the outcome is positive, λ is increased by one $\lambda = \lambda + 1$ and the transition to Step 1 is fulfilled. Otherwise the calculating process is continued by the transition to the next Step 4.

Step 4.
The check of $\nu < 12$ is fulfilled. If the condition is performed, the value ν is increased by one $\nu = \nu + 1$, the parameter λ is given the value one $\lambda = 1$ and the transition to Step 1 is fulfilled. If the condition is not carried out, it means that the parameters of the extrapolator are determined for all points of discretization in which accidental process is viewed and the transition to Step 5 is fulfilled;

Step 5.
The estimation of the future value of the investigated process is specified by the introduction into the calculating process the next value $x_l(\mu)$, $l = \overline{1,5}$ (initially $\mu = 1$). For $l = 1$ the third expression of the formula (6) is used, for $l = \overline{2,5}$ the second one is used;

Step 6.

It is checked whether all values are used for the forecast: $\mu = 11$. If the condition is fulfilled, the process of calculations is finished, otherwise the value μ. Is increased by one $\mu = \mu + 1$ and the transition to Step 5 is fulfilled.

The block diagram in Fig. 2 illustrates the work of the algorithm.

Model (6) gives the possibility to estimate gross profit $x_1(12)$ and gross output $x_2(12)$ for 2015 for a certain enterprise basing on the data $x_h(i), h = \overline{1,5}, i = \overline{1,11}$ of its work for eleven previous years.

Essential deficiency of the forecast model (6) is the assumption of existence of only linear stochastic relations in the sequence $X_h(i), h = \overline{1,5}, i = \overline{1,12}$, describing the process of change of economic indices of agricultural enterprises. The analysis of statistical data about the work of agricultural enterprises of Nikolaev region showed that the stochastic relations till the fourth order

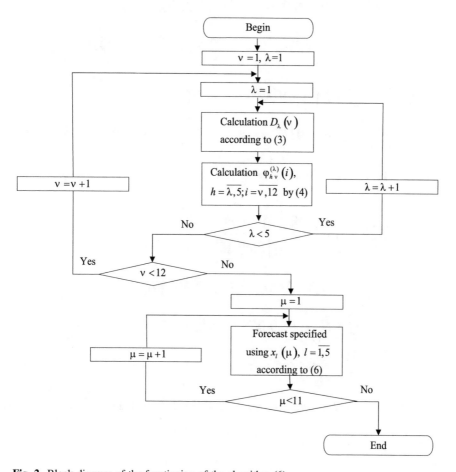

Fig. 2 Block diagram of the functioning of the algorithm (6)

$M\left[\overset{o}{X_h^l}(\nu)\,\overset{o}{X_m^s}(i)\right]\neq 0,\ \nu,i=\overline{1,12},\ l+s\le 4,\ h,m=\overline{1,5}$ are essential for such a sequence. Non-linear canonical model of the investigated sequence with taking account of non-linear relations takes the form [27]:

$$X_h(i)=M[X_h(i)]+\sum_{\nu=1}^{i-1}\sum_{l=1}^{5}\sum_{\lambda=1}^{3}W_{\nu l}^{(\lambda)}\beta_{l\lambda}^{(h,1)}(\nu,i)+\sum_{l=1}^{h-1}\sum_{\lambda=1}^{3}W_{il}^{(\lambda)}\beta_{l\lambda}^{(h,1)}(i,i)+W_{ih}^{(1)},\ i=\overline{1,12}.$$

$$(13)$$

Random coefficients $W_{\nu l}^{(\lambda)},\ \nu=\overline{1,12},\ 1=\overline{1,5},\ \lambda=\overline{1,3}$ and nonrandom coordinate functions $\beta_{l\lambda}^{(h,s)}(\nu,i),\ \nu,i=\overline{1,12},\ 1,h=\overline{1,5},\ \lambda,s=\overline{1,3}$ are determined with the help of expressions:

$$\begin{aligned}W_{\nu l}^{(\lambda)}=&X_l^{\lambda}(\nu)-M\left[X_l^{\lambda}(\nu)\right]-\sum_{\mu=1}^{\nu-1}\sum_{m=1}^{5}\sum_{j=1}^{3}W_{\mu m}^{(j)}\beta_{mj}^{(l,\lambda)}(\mu,\nu)\\&-\sum_{m=1}^{l-1}\sum_{j=1}^{3}W_{\nu m}^{(j)}\beta_{mj}^{(l,\lambda)}(\nu,\nu)-\sum_{j=1}^{\lambda-1}W_{\nu l}^{(j)}\beta_{lj}^{(l,\lambda)}(\nu,\nu),\ \nu=\overline{1,12};\end{aligned}$$

$$(14)$$

$$\begin{aligned}D_{l,\lambda}(\nu)=&M\left[\left\{W_{\nu l}^{(\lambda)}\right\}^2\right]=M\left[X_l^{2\lambda}(\nu)\right]-M^2\left[X_l^{\lambda}(\nu)\right]\\&-\sum_{\mu=1}^{\nu-1}\sum_{m=1}^{5}\sum_{j=1}^{3}D_{mj}(\mu)\left\{\beta_{mj}^{(l,\lambda)}(\mu,\nu)\right\}^2-\sum_{m=1}^{l-1}\sum_{j=1}^{3}D_{mj}(\nu)\left\{\beta_{mj}^{(l,\lambda)}(\nu,\nu)\right\}^2\\&-\sum_{j=1}^{\lambda-1}D_{lj}(\nu)\left\{\beta_{lj}^{(l,\lambda)}(\nu,\nu)\right\}^2,\ \nu=\overline{1,12};\end{aligned}$$

$$(15)$$

$$\begin{aligned}\beta_{l\lambda}^{(h,s)}(\nu,i)=&\frac{M\left[W_{\nu l}^{(\lambda)}\left(X_h^s(i)-M[X_h^s(i)]\right)\right]}{M\left[\left\{W_{\nu l}^{(\lambda)}\right\}^2\right]}\\=&\frac{1}{D_{l\lambda}(\nu)}\left(M\left[X_l^{\lambda}(\nu)X_h^s(i)\right]-M\left[X_l^{\lambda}(\nu)\right]M\left[X_h^s(i)\right]\right.\\&-\sum_{\mu=1}^{\nu-1}\sum_{m=1}^{5}\sum_{j=1}^{3}D_{mj}(\mu)\beta_{mj}^{(l,\lambda)}(\mu,\nu)\beta_{mj}^{(h,s)}(\mu,i)\\&-\sum_{m=1}^{l-1}\sum_{j=1}^{3}D_{mj}(\nu)\beta_{mj}^{(l,\lambda)}(\nu,\nu)\beta_{mj}^{(h,s)}(\nu,i)\\&-\sum_{j=1}^{\lambda-1}D_{lj}(\nu)\beta_{lj}^{(l,\lambda)}(\nu,\nu)\beta_{lj}^{(h,s)}(\nu,i),\ \lambda=\overline{1,h},\ i=\overline{1,12},\nu=\overline{1,i}.\end{aligned}$$

$$(16)$$

Vector algorithm of extrapolation [28–30] for the considered quantity of the components and order of stochastic relations on the basis of canonical expansion (13) takes the form:

$$
m_{j,h}^{(\mu,l)}(s,i) = \begin{cases} M[X_h(i)], \ \mu = 0; \\ m_{j,h}^{(\mu,l-1)}(s,i) + \left(x_j^l(\mu) - m_{j,j}^{(\mu,l-1)}(l,\mu)\right)\beta_{j,l}^{(h,s)}(\mu,i), \ l > 1, j < 5; \\ m_{j,h}^{(\mu,3)}(s,i) + \left(x_{j+1}(\mu) - m_{j,j+1}^{(\mu,1)}(3,\mu)\right)\beta_{j+1,1}^{(h,s)}(\mu,i), \ l = 1, j < 5; \\ m_{5,h}^{(\mu,3)}(s,i) + \left(x_1(\mu+1) - m_{5,1}^{(\mu,3)}(3,\mu+1)\right)\beta_{1,1}^{(h,s)}(\mu+1,i), l = 1, j = 5. \end{cases}
$$

(17)

$$
m_{j,h}^{(\mu,l)}(1,i) = M\left[X_h(i)/x_\lambda^n(\nu), \ \lambda = \overline{1,5}, n = \overline{1,3}, \nu = \overline{1,\mu-1}; x_\lambda^n(\mu), \lambda = \overline{1,j}, n = \overline{1,l}\right]
$$

is optimal by the criterion of minimum of mean-square error of prognosis estimation of future values of economic index with ordinal number h provided that for the prognosis values $x_\lambda^n(\nu)$, $\lambda = \overline{1,5}$, $n = \overline{1,3}$, $\nu = \overline{1,\mu-1}$; $x_\lambda^n(\mu)$, $\lambda = \overline{1,j}$, $n = \overline{1,l}$ are used.

Altogether 165 values $x_h^\lambda(i), h = \overline{1,5}, i = \overline{1,11}, \lambda = \overline{1,3}$ and 5220 not equal to zero weight coefficients $\beta_{l\lambda}^{(h,s)}(\nu,i), \nu, i = \overline{1,12}, \ l, h = \overline{1,5}, \lambda, s = \overline{1,3}$ are used in the algorithm of prognosis (17).

The expression for mean-square error of extrapolation with the help of algorithm (11) by known values $x_j^n(\mu)$, $\mu = \overline{1,k}; j = \overline{1,5}; n = \overline{1,3}$ is in the form:

$$
E_h^{(k,3)}(i) = M\left[X_h^2(i)\right] - M^2\left[X_h(i)\right] - \sum_{\mu=1}^{k} \sum_{j=1}^{5} \sum_{n=1}^{3} D_{jn}(\mu)\left\{\beta_{jn}^{(h,1)}(\mu,i)\right\}^2, i = \overline{k+1,12}.
$$

(18)

This expression is equal to dispersion of a posteriori casual sequence $\{X_h(i)/x_\lambda^n(\nu), \lambda = \overline{1,5}, n = \overline{1,3}, \nu = \overline{1,\mu-1}; x_\lambda^n(\mu), \lambda = \overline{1,j}, n = \overline{1,l}\}$.

In Fig. 3 the scheme reflecting the peculiarities of functioning of the forecast model (17) is represented.

A significant advantage of the predictive model (17) as compared to the algorithm (6) is the account of non-linear relationships studied vector random sequences.

Method of prognostication of future values of economic indices on the basis of the forecast model (17) presupposes realization of the following stages:

Stage 1.
Gathering of statistical data about the results of enterprises functioning;

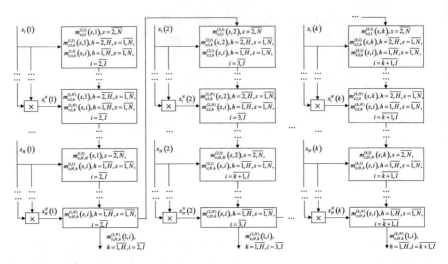

Fig. 3 Scheme of functioning of the forecast model (17) ($N = 3$, $H = 5$)

Stage 2.

Estimation of moment functions $M[X^s(i)]$, $M\left[X_j^l(\nu)X_h^s(i)\right]$ on the basis of cumulated realizations of random sequence describing the process of change of economic indices;

Stage 3.
Calculating of the parameters of the algorithm of extrapolation (17);

Stage 4.
Estimation of future values of economic indices on the basis of the forecast model (17);

Stage 5.
Estimation of the quality of the solving of the forecast problem for investigated sequence with the help of the expression (18).

Forecasting method is approbated on the basis of statistical data of functioning of agricultural enterprises in Nikolaev region during the period 2004–2015 (74 enterprises with gross profit 200–900 thousands grivnas). Moment functions $M[X_h^s(i)]$, $M\left[X_j^l(\nu)X_h^s(i)\right]$ were estimated by known formulae of mathematical statistics for sections 2004, 2005, …, 2014. Data about the work of the enterprises for 2015 were supposed to be unknown and the estimation of moment functions $M[X_h^s(12)]$, $M\left[X_j^l(\nu)X_h^s(12)\right]$ for the last section (corresponding to 2015) was carried out on the basis of determinate models with the use of four previous years (2011–2014) in tabular processor Microsoft Excel (instrument "Search for solutions"). For example, in Table 3 the values of autocorrelated function

Table 3 Autocorrelated function of the component $X_2(i)$, $i = \overline{1,12}$

	2004	2005	2006	2007	2008	2009	2010	2011	2012	2013	2014	2015
2004	1	0,99	0,70	0,42	0,79	0,74	0,49	0,72	0,63	0,46	0,55	0,43
2005	0,99	1	0,72	0,42	0,74	0,74	0,52	0,70	0,64	0,48	0,59	0,46
2006	0,70	0,72	1	0,57	0,67	0,58	0,701	0,69	0,70	0,66	0,78	0,60
2007	0,42	0,42	0,57	1	0,38	0,36	0,45	0,21	0,41	0,36	0,19	0,18
2008	0,79	0,74	0,67	0,38	1	0,81	0,55	0,91	0,80	0,72	0,53	0,41
2009	0,74	0,74	0,58	0,36	0,81	1	0,72	0,73	0,92	0,81	0,51	0,44
2010	0,49	0,52	0,70	0,45	0,55	0,72	1	0,51	0,74	0,73	0,49	0,41
2011	0,72	0,70	0,69	0,21	0,91	0,73	0,51	1	0,77	0,80	0,74	0,55
2012	0,63	0,64	0,70	0,41	0,80	0,92	0,74	0,77	1	0,91	0,60	0,59
2013	0,46	0,48	0,66	0,36	0,72	0,81	0,73	0,80	0,91	1	0,71	0,46
2014	0,55	0,59	0,78	0,19	0,53	0,51	0,49	0,74	0,60	0,71	1	0,71
2015	0,43	0,46	0,60	0,18	0,41	0,44	0,41	0,55	0,59	0,46	0,71	1

$M\left[\overset{o}{\underset{2}{X}}(\nu)\,\overset{o}{\underset{2}{X}}(i)\right]$, $\nu = \overline{1,12}$, $i = \overline{1,12}$ for the component $X_1(i)$, $i = \overline{1,12}$ (gross output) are represented.

For 2015 values $M\left[\overset{o}{\underset{2}{X}}(\nu)\,\overset{o}{\underset{2}{X}}(12)\right]$, $\nu = \overline{1,11}$ are obtained on the basis of determinate model:

$$
M\left[\overset{o}{\underset{2}{X}}(\nu)\,\overset{o}{\underset{2}{X}}(12)\right] = 0,522M\left[\overset{o}{\underset{2}{X}}(\nu)\,\overset{o}{\underset{2}{X}}(11)\right] - 0,253M\left[\overset{o}{\underset{2}{X}}(\nu)\,\overset{o}{\underset{2}{X}}(10)\right]
$$
$$
+ 0,3928M\left[\overset{o}{\underset{2}{X}}(\nu)\,\overset{o}{\underset{2}{X}}(9)\right] - 0,095M\left[\overset{o}{\underset{2}{X}}(\nu)\,\overset{o}{\underset{2}{X}}(8)\right], \quad \nu = \overline{1,11}, \tag{19}
$$

Coordinate function $\beta_{21}^{(2,1)}(\nu,i)\,\nu, i = \overline{1,12}$ (Table 4) corresponds to correlated function $M\left[\overset{o}{\underset{2}{X}}(\nu)\,\overset{o}{\underset{2}{X}}(i)\right]$, $\nu = \overline{1,12}$, $i = \overline{1,12}$.

In Table 5 weight coefficients $\beta_{23}^{(2,1)}(\nu,i)\,\nu, = \overline{1,11}$, $i = \overline{2,12}$ determining the influence of values $x_2^3(i), i = \overline{1,11}$ of gross profit in high-order third degree on future values of this parameter are represented.

As it can be seen in Table 5 values $\beta_{23}^{(2,1)}(\nu,i)\,\nu, = \overline{1,11}$, $i = \overline{2,12}$ are relatively small but this doesn't mean that given weight coefficients don't influence on the forming of the estimation of future value as $\beta_{23}^{(2,1)}(\nu,i)\,\nu, = \overline{1,11}$, $i = \overline{2,12}$ are multiplied in the process of calculations by values $x_2^3(i), i = \overline{1,11}$ (values of the sixth-seventh order).

For functioning of the forecast model (17) on the basis of statistical data 25 tables of weight coefficients analogous to Tables 4 and 5 were calculated.

Table 4 Coordinate function $\beta_{21}^{(2,1)}(\nu, i) \; \nu, i = \overline{1,12}$

	2004	2005	2006	2007	2008	2009	2010	2011	2012	2013	2014	2015
2004	1	0,89	0,54	0,55	0,62	0,43	0,45	0,89	0,858	0,90	2,36	2,65
2005	0	1	2,25	-1,46	-2,40	0,27	5,47	-2,71	3,55	1,83	2,85	4,70
2006	0	0	1	5,09	1,17	-1,53	-0,03	-2,77	-0,23	-5,52	2,34	5,07
2007	0	0	0	1	0,17	0,26	0,94	0,18	0,77	1,05	-0,57	-1,17
2008	0	0	0	0	1	0,48	1,27	1,06	1,05	2,01	-2,37	0,69
2009	0	0	0	0	0	1	-1,81	0,74	3,53	0,37	9,31	2,86
2010	0	0	0	0	0	0	1	-0,68	1,44	3,18	-6,74	-3,39
2011	0	0	0	0	0	0	0	1	1,29	2,21	-3,30	0,93
2012	0	0	0	0	0	0	0	0	1	3,88	0,19	-8,44
2013	0	0	0	0	0	0	0	0	0	1	1,99	-4,96
2014	0	0	0	0	0	0	0	0	0	0	1	0,50
2015	0	0	0	0	0	0	0	0	0	0	0	1

Table 5 Values of coordinate function $\beta_{23}^{(2,1)}(\nu,i)\,\nu, =\overline{1,11},\ i=\overline{2,12}$

	2005	2006	2007	2008	2009	2010	2011	2012	2013	2014	2015
2004	$9*10^{-6}$	$-1*10^{-7}$	$-3*10^{-7}$	$-5*10^{-7}$	$-3*10^{-7}$	$-2*10^{-7}$	$-8*10^{-6}$	$-3*10^{-6}$	$-1*10^{-6}$	$-3*10^{-7}$	$-4*10^{-7}$
2005	0	$-1*10^{-4}$	$-1*10^{-4}$	$1*10^{-5}$	$-3*10^{-4}$	$3,5*10^{-5}$	$7*10^{-5}$	10^{-4}	$6*10^{-5}$	$-2*10^{-4}$	$8*10^{-5}$
2006	0	0	$-1*10^{-5}$	$2*10^{-6}$	$-3*10^{-6}$	$-9*10^{-7}$	$2*10^{-8}$	$6*10^{-6}$	$-6*10^{-6}$	$-9*10^{-6}$	$-7*10^{-7}$
2007	0	0	0	$2*10^{-8}$	$-9*10^{-8}$	$2*10^{-9}$	$3*10^{-8}$	$7*10^{-8}$	$8*10^{-9}$	$4*10^{-9}$	$3*10^{-7}$
2008	0	0	0	0	$-8*10^{-8}$	$2*10^{-8}$	$2*10^{-8}$	$7*10^{-7}$	$3*10^{-7}$	$4*10^{-7}$	$3*10^{-8}$
2009	0	0	0	0	0	$-8*10^{-7}$	$-6*10^{-7}$	$-7*10^{-7}$	$-4*10^{-8}$	$-3*10^{-7}$	$-9*10^{-8}$
2010	0	0	0	0	0	0	$8*10^{-9}$	$8*10^{-6}$	$5*10^{-7}$	-10^{-8}	$-4*10^{-7}$
2011	0	0	0	0	0	0	0	$2*10^{-5}$	$8*10^{-7}$	$-2*10^{-6}$	$8*10^{-6}$
2012	0	0	0	0	0	0	0	0	$2*10^{-8}$	$4*10^{-7}$	$-3*10^{-6}$
2013	0	0	0	0	0	0	0	0	0	$3*10^{-8}$	$-7*10^{-6}$
2014	0	0	0	0	0	0	0	0	0	0	$9*10^{-6}$

During the application of the method of economic indices prognostication for 2016 optimal order of non-linear relations of the investigated random sequence is unknown. But taking into consideration that $N=4$ is invariable during 11 years there is quite high probability that given parameter will remain on the same level.

Values in Table 6 reflects the change of relative error of prognostication of gross profit of enterprise (component $X_1(i), i=\overline{1,12}$) during 2015 depending on the order of stochastic relations used in model (11).

Thus the results of the experiment showed (Table 6) that application of nonlinear relations in the forecast model allows to increase considerably the quality of economic indices prognostication.

The diagram of the computer system functioning on the basis of the developed technology of management of an agricultural enterprise is presented in Fig. 4.

Table 6 Relative error of prognostication of gross profit

Order of stochastic relations	2	3	4
Relative error	6,9%	3,3%	1,5%

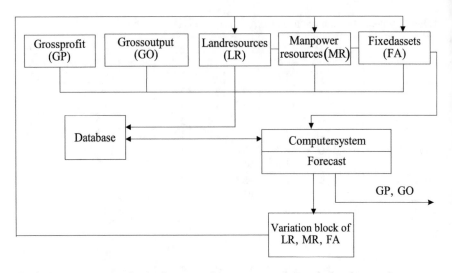

Fig. 4 Computer system for the forecast and management of an agricultural enterprise

Management system gives the possibility to estimate gross profit and gross output for a certain enterprise basing on the data of its work for previous years.

4   Conclusion

Management system for agricultural enterprises on the base of the estimation of future values of economic indices is obtained in this chapter. The algorithm for extrapolation of vector random sequence based on the nonlinear polynomial canonical expansion is assumed as a basis of the proposed method. The optimal algorithm for the extrapolation of the economic indices of agricultural enterprises (as well as canonical expansion put into its base) doesn't impose any essential limitations on the stochastic properties of economic indices. In addition to pre-aggregate indicators (gross output, land resources, manpower, plant and equipment), a range of parameters can be used as the components of investigated random sequence (weather conditions, prices of resources, etc.) influencing the effectiveness of the agricultural enterprises functioning. The results of the numerical experiment showed that the forecast model possesses high accuracy characteristics at the expense of maximal taking into consideration of stochastic qualities of random sequence of economic indices change. Schemes for calculation of the parameters of the forecast model and estimations of future values of economic indices on its basis are introduced in the chapter. Expression for the mean-square error of extrapolation allows to estimate the quality of the forecast problem solving.

References

1. Drozd, J., Drozd, A.: Models, methods and means as resources for solving challenges in co-design and testing of computer systems and their components. In: Proceedings of 9th International Conference on Digital Technologies, Zhilina, Slovak Republic, pp. 225–230 (2013)
2. Gil-Aluja, J.: Investment in Uncertainty. Kluwer Academic Publishers, Dordrecht, Boston, London (1999)
3. Gil-Lafuente, A.M.: Fuzzy Logic in Financial Analysis. Studies in Fuzziness and Soft Computing, vol. 175. Springer, Berlin (2005)
4. Trunov, A.: An adequacy criterion in evaluating the effectiveness of a model design process. Eastern-Eur. J. Enterprise Technol. 1, 4(73), 36–41 (2015)
5. Kauffman, A., Gil-Aluja, J.: Introduction of fuzzy sets theory to management of enterprises (Translated from Spanish). In: Krasnoproshin, V.V., Lepeshinskiy, N.A. (eds.). Minsk, Higher School (1992) (in Russian)
6. Kauffman, A., Gil-Aluja, J.: Models for the study of hidden influences (Translated from Spanish). In: Krasnoproshin, V.V., Lepeshinskiy, N.A. (eds.). Minsk, Higher School (1993) (in Russian)
7. Kauffman, A., Gil-Aluja, J.: Iterative methods of decision-making based on the models with uncertainty (Translated from Spanish). In: Krasnoproshin, V.V., Lepeshinskiy, N.A. (eds.). Minsk, Skarina (1995) (in Russian)
8. Trunov, A.: Recurrent approximation as the tool for expansion of functions and models of operation of neural networks. Eastern-Eur. J. Enterprise Technol. 5, 4(83), 41–48 (2016)
9. Kondratenko, Y.P., Encheva, S.B., Sidenko, E.V.: Synthesis of intelligent decision support systems for transport logistic. In: Proceeding of the 6th IEEE International Conference on Intelligent Data Acquisition and Advanced Computing Systems: Technology and Applications, IDAACS'2011, vol. 2, Prague, Czech Republic, pp. 642–646 (2011)
10. Kondratenko, Y.P., Sidenko, Ie.V.: Decision-making based on fuzzy estimation of quality level for cargo delivery. In: Zadeh, L.A., et al. (eds.) Recent Developments and New Directions in Soft Computing. Studies in Fuzziness and Soft Computing, vol. 317, pp. 331–344. Springer International Publishing, Switzerland (2014)
11. Altman, E.I., Marco, G., Varetto, F.: Corporate distress diagnosis: comparisons using linear discriminant analysis and neural networks. J. Bank. Finance 18, 505–529 (1994)
12. Altman, E.I., Narayanan, P.: An international survey of business failure classification models. J Financial Markets Inst. Instrum. 6(2), 81–130 (1997)
13. Granger, C.W.J., Newbold, P.: Forecasting Economic Time Series. Academic Press (1986)
14. Hall, S.G.: Applied economic forecasting techniques. Harvester Wheatsheaf (1994)
15. Ryabushkin, B.T.: Application of Statistical Methods in an Economic Analysis and Prognostication. Finances and Statistics, Moscow (1987)
16. Trifonov, YuV, Plehanova, A.F., Yurlov, F.F.: Choice of Effective Decisions in an Economy in the Conditions of Vagueness. Publishing House NNGU, Nizhniy Novgorod (1998)
17. Połoński, M.: Prognozowanie czasu zakończenia inwestycji na podstawie jej bieżącego zaawansowania. J. Metody ilościowe w badaniach ekonomicznych, Tom XIII/3, pp. 169–179 (2012)
18. Prędki, A.: Subsampling approach for statistical inference within stochastic DEA models. J. Metody ilościowe w badaniach ekonomicznych XIV(2), 158–168 (2013)
19. Teyl, G.: Economy Prognosis and Making Decision. Statistics, Moscow (1971)
20. Szmuksta-Zawadzka, M., Zawadzki, J.: Modele harmoniczne ze złożoną sezonowością w prognozowaniu szeregów czasowych z lukami systematycznymi. J. Metody ilościowe w badaniach ekonomicznych, Tom XIV/3, 81–90 (2013)
21. Pugachev, V.S.: The Theory of Random Functions and its Application. Fitmatgiz, Moscow (1962)

22. Kudritsky, V.D.: Filtering, Extrapolation and Recognition Realizations of Random Functions. FADA Ltd., Kyiv (2001)
23. Shebanin, V., Atamanyuk, I., Kondratenko, Y.: Simulation of vector random sequences based on polynomial degree canonical decomposition. Eastern-Eur. J. Enterprise Technol. 5, 4(83), 4–12 (2016)
24. Sirenko, N.N.: Management of strategy innovative development of agrarian sector of economy of Ukraine. MNAU, Mykolaiv (2010)
25. Atamanyuk, I.P., Kondratenko, Y.P., Sirenko, N.N.: Forecasting economic indices of agricultural enterprises based on vector polynomial canonical expansion of random sequences. In: Ermolayev, V., et al. (eds.) Proceedings of the 12th International Conference on Information and Communication Technologies in Education, Research, and Industrial Application. Integration, Harmonization and Knowledge Transfer, 21–24 June, 2016, Kyiv, Ukraine, ICTERI'2016, CEUR-WS, vol. 1614, pp. 458–468 (2016)
26. Atamanyuk, I., Kondratenko, Y., Sirenko, N.: Management of an agricultural enterprise on the basis of its economic state forecasting. J. Metody ilościowe w badaniach ekonomicznych: Quant. Methods Econ. XV(2), 7–16 (2014)
27. Atamanyuk, I., Kondratenko, Y.: Calculation method for a computer's diagnostics of cardiovascular diseases based on canonical decompositions of random sequences. ICT in education, research and industrial applications: integration, harmonization and knowledge transfer. In: Batsakis, S., et al. (eds.) Proceedings of the 11 International Conference ICTERI-2015, CEUR-WS, vol. 1356, pp. 108–120 (2015)
28. Atamanyuk, I.P.: Algorithm of extrapolation of a nonlinear random process on the basis of its canonical decomposition. J Cybern. Syst. Anal. 2, 131–138 (2005)
29. Atamanyuk, I.P., Kondratenko, V.Y., Kozlov, O.V., Kondratenko, Y.P.: The algorithm of optimal polynomial extrapolation of random processes. In: Engemann, K.J., Gil-Lafuente, A. M., Merigo, J.L. (eds.) Modeling and Simulation in Engineering, Economics and Management. International Conference MS 2012, New Rochelle, NY, USA, Proceedings. Lecture Notes in Business Information Processing, vol. 115, pp. 78–87. Springer (2012)
30. Atamanyuk, I.P.: Optimal polynomial extrapolation of realization of a random process with a filtration of measurement errors. J. Autom. Inf. Sci. (Begell House, USA) 41(8), 38–48 (2009)

Partner Selection in Green Innovation Projects

Marina Z. Solesvik

Abstract In this paper, we consider issues related to green innovation strategies. Notably, we focus on the issues related to R&D strategic alliances aimed to develop green technologies in maritime sector. Still managers often use their gut feelings to select partners from the prospective candidates. However, expensive R&D projects aimed at developing radical green innovations need thorough preliminary analysis of collaborators. We apply fuzzy logic approach to facilitate decision making of management teams who are responsible for selection partners for collaborative green innovation projects. Namely, in this study the approach of formal concept analysis is used to facilitate partner selection.

Keywords Formal concept analysis · Fuzzy logic · Green innovation
Partner selection

1 Introduction

Green innovation is a popular concept and hot topic in academic and political debate. "Does it pay to be green?" ask Ambec and Lanoie [1] in highly cited paper published in Academy of Management Perspective journal. Probably, the same question is asked by executives of firms that have to formulate an innovation strategy for the future of their companies. The significant pressure from the governmental and international authorities as well from the stakeholders towards the development and implementation of green technologies is observed in many industries [13]. For example, the significant reductions of CO_2 and NO_x emission are introduced in shipping from 2020. The vessels that do not comply the requirements simply will not be allowed to enter European ports. The cost of compliance to governmental environment regulations are significant. By 2020, the operational costs will be doubled for shipping firms due to stricter environment

M. Z. Solesvik (✉)
Nord University Business School, Nord University, 8049 Bodø, Norway
e-mail: mzs@hsh.no

© Springer International Publishing AG 2018
C. Berger-Vachon et al. (eds.), *Complex Systems: Solutions and Challenges in Economics, Management and Engineering*, Studies in Systems, Decision and Control 125, https://doi.org/10.1007/978-3-319-69989-9_28

471

regulations related to emission of sulphur dioxide [36]. In order to comply with regulations as well as stakeholders' ethical expectations, more and more shipping companies engage into R&D to develop environmentally friendly vessels that both help to save environment and reduce costs for owners of shipping companies [6]. Recent developments in shipping market are spectacular, i.e. electrical-driven ferries and fishing vessels, hybrid vessels combining diesel and electrical engines, and vessels driven by liquid natural gas allow both save costs and prevent pollution of air, ocean and port environments. However, the success of these R&Ds is a result of multidiscipline groups of scholars, engineers, entrepreneurs, in other words, it is a result of successful interfirm collaboration. It is hardly possible to produce such advanced R&D products by companies alone nowadays. Thus, firms should be good in selection of appropriate partners for their R&D alliances [34].

The task of partner selection is even more difficult since managers often do not have full information about possible partners and have to make decisions under the conditions of uncertainty in the modern business world [18]. Fuzzy logics is an excellent tool to facilitate decision-making under the conditions of legal, environmental, market, and political uncertainty that face modern firms [19, 24, 45]. Thus, the purpose of this paper is to apply a novel mathematical method of formal concept analysis (FCA) to partner selection for interfirm collaboration related to green innovation. The novel contribution of this research is an appliance of mathematical method of formal concept analysis that can be used by company managers when firms select partners for green innovation to warrant successful attainment of an alliance purposes and achieving competitive advantage.

The remainder of the paper is structured as follows. First we present theoretical background of the paper related to green innovation strategies and partner selection. Then we introduce the formal concept analysis. The illustrative example will demonstrate how the FCA can be applied to facilitate a selection of partner for green innovation R&D collaboration. The paper terminates with discussion and conclusions.

2   Theoretical Background

2.1   Green Innovation

Research related to green innovation is steadily more popular. Green innovation strategy is defined "as a class of manufacturing practices that include source reduction, pollution prevention, and the adoption of an environmental management system" [13]. Introduction of green innovation might improve business performance in several ways. First, firms adopting green innovation strategies use raw materials more efficiently, they also produce less waste and utilize it effectively [45]. Second, implementation of new green technologies allows to reduce emissions to the atmosphere and water, something that is diminishing company's conformity and liability costs [20]. Third, firms use previously unused resources or waste to create value engaging in bricolage activities [2]. New technologies are often more efficient, however, the capital cost of innovation development are significant [5].

Starting from 1970s, governments started to introduce strict environmental regulations, forcing firms to employ green technologies. However, not all measures imposed by the governmental environment regulations were positively and significantly related to firms' performance. In about fifty per cent of cases firms' performance improved, however, in the rest of cases performance has not changed or performance decreased [21]. The adoption of green innovation strategy might serve as a proactive measure for the firms to adjust faster and in more efficient way to new environmental regulations. Notably, not only governments influence firms' management decision to foster innovations. Stakeholders' perceived pressure and managerial environment concerns [13] are two other important factors influencing selection of green innovation path. Governments can use regulative measures (e.g. bans, prohibitions, emission standards) and economic measures (tax schemes, R&D grants) to promote green innovation. It has been argued that economic measures are more effective in promoting green innovations [29].

2.2 Partner Selection

Despite a growing volume of research focusing on partner selection for R&D alliances, there is still a gap in the knowledge relating to the mechanism of partner selection. Partner evaluation and selection is a part of an alliance life cycle that goes after the opportunity recognition phase. The next phases associated with the functioning of an alliance and its eventual termination [30]. The results of R&D alliance determined by the choice of a partner who provides specific skills, abilities and resources [3]. The decision-making procedure that lie beneath the selection of an alliance partner has been explored from several theoretical perspectives, i.e. the resource-based view of the firm, the knowledge-based view, game theory, the organizational learning perspective, the resource-dependence perspective, transaction costs economics, and the competence-based view of the firm. Studies exploring trust and loyalty of partners revealed that trust can also affect the creation of alliances. However, the existence of trust between parties is not enough for successful interfirm cooperation [4].

Extensive research relating to partner selection for R&D alliances has been conducted with reference to several industrial contexts. Issues relating to the maritime industry have generally been neglected [33, 35]. In the same time, the correct selection of partners for an alliance is one of the core success factors of the alliance [32].

Geringer [17] has presented a widely respected classification of partner selection criteria, and he made a key distinction between partner-related and task-related dimensions. The partner-related criteria include reputation, strategic fit, trust between the top management teams, financial stability of the partner, position within the industry, and enthusiasm for the project. The task-related criteria include knowledge of the local and international markets, competence in new product development, knowledge of partner's culture and internal standards, links with major buyers, suppliers and distribution channels, product-specific knowledge, capital and finance, local regulatory knowledge, political influence and other criteria relating to the industry and alliance goals.

In some situations, decision criteria are qualitative and subjective. Quantitative methods facilitating decision-making at different situations in business are steadily more prevalent [14, 23, 25]. Formal concept analysis applied in this study is aimed to facilitate the decision making process, as other mathematical techniques proposed by prior research, such as the mixed integer linear programming [22, 41], optimization modeling [8, 15], the analytic hierarchical process [27, 37], and the analytic network process approaches [10, 31, 43]. Notably, quantitative techniques are not substitutes for qualitative methods of decision-making [28, 42]. The aim of FCA is to compliment subjective perceptions of decision-makers when they evaluate an appropriate partner for green innovation R&D alliance.

2.3   Formal Concept Analysis

Information processing is a complicated cognitive process. Conversion of information into the concepts makes the process of information analysis and handling easier. The lattice theory is complicated mathematical theory that can be used to analyse concepts. However, it was difficult to apply for non-mathematician community. Wille [38] developed the formal concept analysis in order to make application of the lattice theory easier. Define FCA as "a method to visualize data and its inherent structures, implications and dependencies".

FCA is grounded on an idea of a *concept* which consists of two parts *extension* and *intension*. A concept is defined as "a set of specific objects, symbols, or events which are grouped together on the basis of shared characteristics and which can be referenced by a particular name or symbol" [26]. The *extension* embraces all objects appropriate to the concept, whereas the *intension* includes all attributes related to particular objects [40]. A *context* is a triple (G, M, I) where G and M are sets and $I < G \times M$. The elements of G and M are called *objects* and *attributes* respectively. I is a relation between G and M. Sometimes the relation I is called the incidence relation of the context [9]. If the sets are finite, the context can be specified with a help of a cross-table. An example of a simple context is shown in Table 1.

A *concept* of a context (G, M, I) is defined as a pair (G', M'), where set of objects which members of M' have in common is G' and the set of attributes that the members of G' have in common is M' [7]. From Table 1, $(\{g_1, g_3\}, \{m_1\})$ is a concept of the context, where g_1 and g_3 are the members of the extension and m_1 is a member of intension.

Table 1 A formal context

	m_1	m_2	m_3
G_1	×		×
G_2		×	
G_3	×		
G_4		×	

Fig. 1 A concept lattice corresponding to a context in Table 1

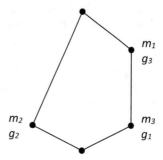

A lattice is a network-like classification structure that can be generated automatically from a term-document indexing relationship. Such a network structure outperforms hierarchical classification structure since the former enables many paths to a particular node while the latter restricts each node to possess only one parent [33]. Hence, the lattice navigation provides an alternate browsing-based approach which can overcome the weakness of hierarchical classification browsing [11]. The set of all concepts of the context (G, M, I) is a complete lattice and it is identified as the *concept lattice* of the context (G, M, I). The concept lattice which matches to the context in Table 1 is illustrated in Fig. 1.

The major drawback of the prior mathematical techniques used in partner selection is that their implementation needs knowledgeable computer programmers [12]. FCA outperforms other mathematical techniques applied in earlier studies on alliance partner selection by the relative simplicity for practitioners, possibility to be integrated into ICT applications and to visualise the analysis.

3 Illustrative Example

In this illustrative example, we would like to show how the FCA can be applied to facilitate decision making in the process of partner selection for a strategic alliance for green innovation. This strategic alliance is a part of green innovation strategy of a shipping company. The aim of this collaboration is to create a new type of the vessel that uses hydrogen fuel cells as fuel. A shipping company that initiates a project is owned and managed by businessmen with clear environment friendly mindsets. They are concerned to prevent environment from pollution caused by the vessels they operate. Managers are experienced in developing and implementing radical innovations in shipping. Earlier, they have designed and introduced to the market an offshore vessel driven by liquid natural gas. It is worth noting that the seaborne transport is quite polluting for the environment. Governments and stakeholders in Norway wish to reduce dangerous emissions from the vessels and seek to develop new technology that would be better for environment. Notably, it is not typical yet for shipping companies to be directly involved into radical innovations development related to new vessels creation. There are several reasons for this. First, shipping companies are often too small and cannot afford a full range competence in-house related to radical innovations development. Second, R&D competence is out of core competence scope for the shipping companies. For example, new engine technologies that are under development nowadays (e.g., hydrogen-, solar-, electrical-driven) need competence from disciplines that are not typical for shipping at all. Thus, maritime firms need to collaborate with actors both from the maritime industry and from other industries to create innovative vessels.

Hydrogen fuel cells technology seemed to be interesting and promising for the shipping company managers. This technology allows to use hydrogen as a source of fuel. The final emission product of fuel cell engines is pure water. Thus, having such an engine on board can, on one hand, satisfy continuously strict government regulation acts related to environment. On the other hand, the possible fuel savings compared to traditional diesel or LNG engines are enormous. Taking into account this considerations, Firm A's management decided to attract partners to a new type vessel creation with fuel cell engine. Analysis showed that they need to attract competence from the following firms: ship design, engine manufacturer, shipyard and fuel cell equipment producer. Some of them (ship design firms, ship engine manufactures, and shipyards) they knew through previous collaboration projects. Producers of fuel cell equipment were all new. Firm A's managers screened potential partners. There are four probable collaborators (i.e. firms 1, 2, 3, and 4). In order to build a lattice, it is necessary to have a context that is a binary connection between objects and attributes. The attributes that prospective firms possessed are listed below and registered in Table 2:

S   strategic fit
O   organizational fit
T   trust between the top management teams
F   stable financial position to finance R&D work for many years
R   experience with offshore shipping market
D   R&D competence
A   access to advanced technology
H   competence in hydrogen fuel cells technology

Table 2 is a cross-table where cells include crosses ($\times$) or empty cells. Value ($\times$) means that a firm has a certain attribute, and an empty cell means that it does not possess necessary attribute.

	S	O	T	F	R	D	A	H
F 1	×	×		×		×		×
F 2		×	×			×		×
F 3			×		×		×	×
F 4	×				×			×

Table 2 Context—attributes of prospective partners

F1, F2, F3, F4 (objects)—firms 1, 2, 3, and 4 respectively; (×)—a firm possess an attribute; an empty cell—a firm does not possess and attribute

For further analysis, we present data in visual form. We use a Hasse diagram to visualize a lattice. Nodes represent formal concepts and edges show the subconcepts. With the help of a concept lattice, we can explore relations between concepts, objects, and attributes. The concepts presented hierarchically (i.e. the nearer a concept is to the supremum (the peak node), the more attributes it has). Shifting from one vertex to a related vertex that is nearer to the supremum signifies moving from a more general to a more precise portrayal of the attributes, if an object appears in both concepts.

Concepts are displayed by the tags attached to the nodes of the lattice (Fig. 2). The connotation of the used tags is as follows:

- Node number 23 has a tag $I = \{H, S\}$, $E = \{F1, F4\}$. This means that firms $F1$ and $F4$ possess two characteristics in common, H and S (i.e. competencies in hydrogen fuel technology and have strategic fit with a shipping company that looks for partners).
- Node number 27 has a label $I = \{D, H, O\}$, $E = \{F1, F2\}$. This means that firms $F1$ and $F2$ have three traits in common, D, H, O (i.e. experience with fuel cells, R&D and organizational fit).

4 Conclusions

The partner selection method for interfirm cooperative arrangements aimed to develop green innovations can impact the alliance's performance [4]. The selection of the 'right' partner is important for the success of a strategic alliance aimed at developing green innovations and, thus, for ensuring a firm's competitive advantage. The purpose of this study was to introduce a quantitative approach of formal concept analysis to assist decision making during selection of partners to implement green innovation strategy. FCA is a method of data analysis, an important technique of the graphical representation of knowledge and information management. Recently, FCA has been growing as a research field with a wide range of applications. Numerous applications of FCA are reported in different research areas, such as economics, marketing, psychology, biology, sociology, computer sciences,

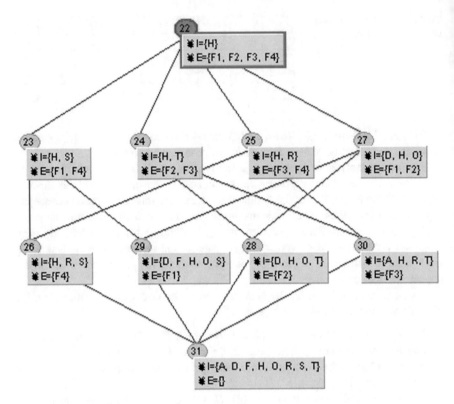

Fig. 2 Concept lattice corresponding to data in Table 2

and industrial engineering [16, 39]. This study demonstrated that the FCA can be also applied by managers when they select partners for collaborative green innovation projects.

References

1. Ambec, S., Lanoie, P.: Does it pay to be green? A systematic overview. Acad. Manage. Perspect. **22**(4), 45–62 (2008)
2. Baker, T., Nelson, R.E.: Creating something from nothing: resource construction through entrepreneurial bricolage. Adm. Sci. Q. **50**(3), 329–366 (2005)
3. Berg, S., Friedman, P.: Corporate courtship and successful joint ventures. Calif. Manag. Rev. **22**(2), 85–91 (1980)
4. Bierly III, P.E., Gallagher, S.: Explaining alliance partner selection: fit, trust and strategic expediency. Long Range Plan. **40**(2), 134–153 (2007)
5. Borch, O.J., Solesvik, M.Z.: Innovation on the open sea: examining competence transfer and open innovation in the design of offshore vessels. Technol. Innov. Manage. Rev. **5**(9), 17–22 (2015)

6. Borch, O.J., Solesvik, M.Z.: Partner selection versus partner attraction in R&D strategic alliances: the case of the Norwegian shipping industry. Int. J. Technol. Mark. **11**(4), 421–439 (2016)

7. du Boucher-Ryan, P., Bridge, D.: Collaborative recommending using formal concept analysis. Knowl.-Based Syst. **19**, 309–315 (2006)

8. Cao, Q., Wang, Q.: Optimizing vendor selection in a two-stage outsourcing process. Comput. Oper. Res. **34**(12), 3757–3768 (2007)

9. Carpineto, C., Romano, G.: Concept Data Analysis: Theory and Applications. Wiley, Hoboken, NJ (2004)

10. Chen, S.-H., Lee, H.-T., Wu, Y.-F.: Applying ANP approach to partner selection for strategic alliance. Manage. Decis. **46**(3), 449–465 (2008)

11. Cheung, K., Vogel, D.: Complexity reduction in lattice-based information retrieval. Inf. Retrieval **8**(2), 285–299 (2005)

12. Choy, K.L., Lee, W.B., Lo, V.: An enterprise collaborative management system—a case study of supplier relationship management. J. Enterp. Inf. Manage. **17**(3), 191–207 (2004)

13. Eiadat, Y., Kelly, A., Roche, F., Eyadat, H.: Green and competitive? An empirical test of the mediating role of environmental innovation strategy. J. World Bus. **43**(2), 131–145 (2008)

14. Encheva, S., Kondratenko, Y., Solesvik, M.Z., Tumin, S., Simos, T.E., Psihoyios, G.: Decision support systems in logistics. In: AIP Conference Proceedings, vol. 1060, no. 1, pp. 254–256 (2008)

15. Fuqing, Z., Yi, H., Dongmei, Y.: A multi-objective optimization model of the partner selection problem in a virtual enterprise and its solution with generic algorithms. Int. J. Adv. Manuf. Technol. **28**(11–12), 1246–1253 (2006)

16. Ganter, B., Stumme, G., Wille, R.: Formal Concept Analysis: Foundations and Applications. LNAI 3626. Springer, Heidelberg (2005)

17. Geringer, J.M.: Strategic determinants of partner selection criteria in international joint ventures. J. Int. Bus. Stud. **22**(1), 41–62 (1991)

18. Gil-Aluja, J.: Elements for a Theory of Decision in Uncertainty, vol. 32. Springer Science & Business Media (1999)

19. Gil-Aluja, J.: Fuzzy Sets in the Management of Uncertainty, vol. 145. Springer Science & Business Media (2004)

20. Hart, S.L.: A natural-resource-based view of the firm. Acad. Manage. Rev. **20**(4), 986–1014 (1995)

21. Jaffe, A.B., Peterson, S.R., Portney, P.R., Stavins, R.N.: Environmental regulation and the competitiveness of US manufacturing: what does the evidence tell us? J. Econ. Lit. **33**(1), 132–163 (1995)

22. Jarimo, T., Salo, A.: Multicriteria partner selection in virtual organisations with transportation costs and other network interdependencies, Technical report. Helsinki University of Technology, Helsinki (2008)

23. Kondratenko, Y.P., Klymenko, L.P., Sidenko, Ie.V.: Comparative analysis of evaluation algorithms for decision-making in transport logistics. In: Jamshidi, M., Kreinovich, V., Kazprzyk, J. (eds.) Advance Trends in Soft Computing. Studies in Fuzziness and Soft Computing, vol. 312, pp. 203–217 (2014)

24. Kondratenko, Y., Kondratenko, V.: Soft computing algorithm for arithmetic multiplication of fuzzy sets based on universal analytic models. In: Ermolayev, V. et al. (eds.) Information and Communication Technologies in Education, Research, and Industrial Application. Communications in Computer and Information Science 469. ICTERI'2014, pp. 49–77. Springer International Publishing, Switzerland (2014)

25. Kondratenko, Y.P., Sidenko, Ie.V.: Decision-making based on fuzzy estimation of quality level for cargo delivery. In: Zadeh, L.A. et al. (eds.) Recent Developments and New Directions in Soft Computing. Studies in Fuzziness and Soft Computing 317, pp. 331–344. Springer International Publishing Switzerland (2014)

26. Merrill, M.D., Tennyson, R.D.: Concept Teaching: An Instructional Design Guide. Educational Technology, Englewood Cliffs, NJ (1977)

27. Mikhailov, L.: Fuzzy analytical approach to partnership selection in formation of virtual enterprises. Omega **30**(5), 393–401 (2002)
28. Ordoobadi, S.M.: Fuzzy logic and evaluation of advanced technologies. Ind. Manage. Data Syst. **108**(7), 928–946 (2008)
29. Pearce, D.W., Markandya, A., Barbier, E.:. Blueprint for a Green Economy, vol. 1. Earthscan (1989)
30. Pidduck, A.B.: Issues in supplier partner selection. J. Enterp. Inf. Manage. **19**(3), 262–276 (2006)
31. Sarkis, J., Talluri, S., Gunasekaran, A.: A strategic model for agile virtual enterprise partner selection. Int. J. Oper. Prod. Manage. **27**(11), 1213–1234 (2007)
32. Reuer, J.J., Ariño, A., Olk, P.M.: Entrepreneurial Alliances. Prentice Hall (2011)
33. Solesvik, M.Z., Encheva, S.: Partner selection for interfirm collaboration in ship design. Ind. Manage. Data Syst. **110**(5), 701–717 (2010)
34. Solesvik, M., Gulbrandsen, M.: Partner selection for open innovation. Technol. Innov. Manage. Rev. **3**(4), 6–11 (2013)
35. Solesvik, M.Z., Westhead, P.: Partner selection for strategic alliances: case study insights from the maritime industry. Ind. Manage. Data Syst. **110**(6), 841–860 (2010)
36. Sysla: New requirements can make fuel costs twice as high. http://sysla.no/2016/11/07/snappet/nye-krav-dobler-drivstoffkostnadene_172956/ (2016)
37. Wang, T.-C., Chen, Y.-H.: Applying consistent fuzzy preference relations to partnership selection. Omega **35**(4), 384–388 (2007)
38. Wille, R.: Restructuring lattice theory: an approach based on hierarchies of concepts. In: Rival, I. (ed.) Ordered Sets, pp. 445–470. Reidel, Dordrecht (1982)
39. Wille, R.: Conceptual knowledge processing in the field of economics. In: Ganter, B., Stumme, G., Wille, R. (eds.) Formal Concept Analysis: Foundations and Applications, vol. 3626, pp. 226–249. Springer, Heidelberg, LNAI (2005)
40. Wolff, K.E.: A first course in formal concept analysis: how to understand line diagrams. In: Faulbaum, F. (ed.) Advances in Statistical Software, pp. 429–438. Gustav Fischer Verlag, Stuttgart (1993)
41. Wu, N., Su, P.: Selection partners in virtual enterprise paradigm. Robot. Comput.-Integr. Manuf. **21**(2), 119–131 (2005)
42. Wu, W.-H., Lu, H.-L., Huang, D.-Y.: Government expenditure decision-making under fuzzy relations. Int. J. Serv. Stand. **3**(3), 289–298 (2007)
43. Wu, W.Y., Shih, H.-A., Chan, H.-C.: The analytic network process for partner selection criteria in strategic alliances. Expert Syst. Appl. **36**(3), 4646–4653 (2009)
44. Young, J.: Reducing waste, saving materials. In: Brown, L., Flavin, C., Postel, S., Starke, L. (eds.) State of the World, pp. 39–55. Norton, New York (1991)
45. Zadeh, L.A.: Toward a theory of fuzzy information granulation and its centrality in human reasoning and fuzzy logic. Fuzzy Sets Syst. **90**(2), 111–127 (1997)

Evaluation of Structural Integrity of Metal Plates by Fuzzy Similarities of Eddy Currents Representation

Mario Versaci and Francesco Carlo Morabito

Abstract In this paper, we present a practical application of the methodologies introduced by Jaime Gil-Aluja and Lotfi Zadeh in Civil and Electrical Engineering. Metallic plates bi-axially loaded deform producing dangerous mechanical stresses that are not visually appreciable. Being the representation construction of such stress conditions by $2D$ images extremely complex, in this work, we propose to generate suitable Eddy Currents (*ECs*) images to translate the information content of mechanical stresses into representative electric signals easier to image. By grouping the produced images in different classes related to different bi-axial loads and in a single class all the images referring to plates in absence of loads, the evaluation of the integrity of a plate is transformed into a problem of classification/decision. This further step is carried out by means of the measure of Fuzzy Similarities (*Ss*) between the $2D$ EC signal at hand and the prototypical classes. The achieved performance are comparable to more established approaches that are commonly plagued by a higher computational load. The proposed methodology is also shown to be able to manage uncertainty in an application of relevant industrial interest.

1 Backgrounds

1.1 Uncertainty, Fuzzy Logic and Engineering

Fuzzy logic and fuzzy set theory are today considered important theoretical tools as almost all classes and concepts around us can be described as fuzzy rather than crisp. They form the core of a novel approach to manage data, linguistic variables, and the related algorithms, which is referred to as Soft Computing. Lotfi Zadeh, the founder

M. Versaci (✉) · F. C. Morabito
DICEAM Department, Universitá Mediterranea degli Studi di Reggio Calabria,
Cittadella Universitaria, Via Graziella Feo di Vito, Reggio, Calabria, Italy
e-mail: mario.versaci@unirc.it

F. C. Morabito
e-mail: morabito@unirc.it

© Springer International Publishing AG 2018
C. Berger-Vachon et al. (eds.), *Complex Systems: Solutions and Challenges in Economics, Management and Engineering*, Studies in Systems, Decision and Control 125, https://doi.org/10.1007/978-3-319-69989-9_29

481

of fuzzy logic, proposed its theoretical framework by considering the approximate nature of data, from words and language to signals, images and physical measurements. Jaime Gil-Aluja and Arnold Kaufmann had the strong idea to apply the theoretical instruments for representing uncertainty in the relevant field of Economics, thus achieving remarkable success [1–4]. The fundamental principle they introduced (*simultaneidad gradual*) basically affirms that an object/element can belong to more classes at the same time provided we can assign a degree to its belongingness (membership) to the different classes [5]. This strong principle changed the perspective of researches in every field, including engineering. Together with the Zadeh fuzzy extension principle, their work enabled us to work on fuzzy sets by using the standard tool of crisp set theory. In contrast, in the crisp world the partitions assign each element to a single subset (class), thus ignoring the possibility that each element may also belong to other subsets (classes). This is a clear limitation of crisp sets and partitions, which ultimately affects their ability to process different kinds of data in a way consistent with human intuition but also with their actual physical properties. Indeed, in the crisp realm we are unable to represent the uncertainty or the imperfect, imprecise or incomplete knowledge often encountered in many applications, as the one reported in this paper. Fuzzy reasoning became essentially a practical subject. Its main role is to assist the decision-maker in a real world situation, taking into account the practical meaning of the concepts involved.

1.2   Introduction to the Problem

In the field of the finite elasticity, bi-phase materials are derived from the solid-to-solid transformations. Starting from Ericksen pioneering work [6], where a two-dimensional model with no-convex energy representing a stable solution with discontinuous deformation was developed, the scientific community tried to prove the validity of the model over different elasticity problems such as the necking and the shear band [7]. In addition, this model has been shown to be an excellent starting point for the study of both shape-memory-alloy models and micro-structures in crystalline solids [8, 9]. However, multidimensional modeling in finite elasticity entails considerable analytical and computational difficulties both in obtaining closed solutions and, in case of $2D$ problems, to ensure the uniqueness of the solution [10]. Furthermore, in the case of symmetric bi-axial loads, stable deformations with different stretches induced by the loads have been observed [5]. Finally, multiple and asymmetric deformations have also been observed in the presence of symmetric loads. In these conditions, it has been further shown that symmetric loads can generate multiple asymmetrical deformations [11, 12] even in non-linear elasticity conditions [13, 14].

1.3   Theoretical Aspects in Continuum

According to Chen's theory [10], energy is a key concept to approach the study of stress-deformation problems is energy. Let Ω be a homogeneous, regular and bounded body whose boundary is $\partial\Omega$, with z be an isochoric deformation in the set defined as follows:

$$H = \{z \in C^1(\Omega, \mathbf{R}^3) : det\mathbf{F} = 1\} \tag{1}$$

where $\mathbf{F} = \nabla z$ is the deformation gradient. The stored energy can be expressed in the form $\mathbf{W} = \mathbf{W}(\lambda_1, \lambda_2)$ with $\mathbf{W} \in C^\infty$ and λ_1, λ_2 principal stretches (eigenvalues of the matrix $(\mathbf{FF}^T)^2$). If the body is subject to tractions on the physical boundaries, in equilibrium conditions, on $\partial\Omega$ $\mathbf{t} = \mathbf{Tn}$ occurs where $\mathbf{t}$ represents the vector tensor, $\mathbf{T}$ is stress tensor and $\mathbf{n}$ is the outward normal on $\partial\Omega$. Finally, functional $\mathbf{E}$ over $\mathbf{H}$ is built as follows:

$$\mathbf{E} = \int_\Omega \mathbf{W}(\lambda_1, \lambda_2) - \mathbf{T} \cdot \nabla \mathbf{z} \tag{2}$$

$\mathbf{E}$, in equilibrium conditions, admits minimum value. In the related literature it has been shown that the problem (with its restrictions) admits both equilibrium states and acceptable stability. However, the model does not describe in detail the physics of the problem because it neglects effects related to the internal microstructures. Under the hypothesis of finite elasticity, many materials support fields with strong gradients of discontinuity with gradual variation of the deformation except for specific surface-sections, where a jump of the gradient occurs. In this context of plastic deformation studies, the focus is on the twins formations, that is a phenomenon of the plastic-phase which anticipates shearband formations [7].

1.4   Brief Remarks on the Computational Aspects and Minimization Problems

Assuming that the bi-phase deformation gradient represents a local minimum for all the points of discontinuity, then the deformation corresponding to $\mathbf{F}^+$, $\mathbf{F}^-$ is a global minimum. Thus, it holds:

$$\mathbf{W}(\mathbf{F}^+) - \mathbf{W}(\mathbf{F}^-) = \mathbf{S}(\mathbf{F}^\pm) \cdot (\mathbf{F}^+ - \mathbf{F}^-) \tag{3}$$

where $\mathbf{S}(\mathbf{F})$ is the Kirchhoff-Piolas tensor. In two dimensions, and taking into account the geometrical compatibility conditions according to the Hadamard's theory, if $\mathbf{F}^+$, $\mathbf{F}^- \mathbf{M}^{n \times m}$, then

$$\exists \mathbf{z} : \mathbf{R}^3 \to \mathbf{R}^3 \Rightarrow \nabla \mathbf{z} = \mathbf{F}^+, \nabla \mathbf{z} = \mathbf{F}^- \tag{4}$$

if $\mathbf{F}^+$, $\mathbf{F}^-$ were rank1-connected, that is the following condition is valid

$$\mathbf{F}^+ - \mathbf{F}^- = \mathbf{a} \otimes \mathbf{n}, \quad \forall \mathbf{a} \neq 0 \in \mathbf{R}^3 \tag{5}$$

then in a subset Γ of Ω characterized by:

$$\Gamma = \{\mathbf{x} \in \Omega : \mathbf{x} \cdot \mathbf{n} = \gamma_1\} \quad \gamma_1 < \gamma_m \tag{6}$$

you can build the continuous deformation $\mathbf{z}(\mathbf{x})$, such that:

$$\mathbf{z}(\mathbf{x}) = \mathbf{F}^+ + \left[\int_0^{\tau \cdot \mathbf{n}} X(\Gamma) \right] \tag{7}$$

with X suitable function. Then, $\mathbf{z}(\mathbf{x})$ is a microstructure that satisfies the following propriety:

$$\mathbf{z}(\mathbf{x}) = \mathbf{F}^+ + X(\mathbf{x} \cdot \mathbf{n})\mathbf{a} \otimes \mathbf{n} = \mathbf{F}^+ \ or \ \mathbf{F}^- \tag{8}$$

A deformation $\bar{\mathbf{z}}$ admits global stability if:

$$E(\bar{\mathbf{z}}) \leq E(\mathbf{z}) \quad \forall \mathbf{z} \in H \tag{9}$$

where:

$$H = \{\mathbf{z} \in \mathbf{W}^{1,p}(\Omega) \in' \ det\nabla\mathbf{z} \geq 0\} \tag{10}$$

with $\mathbf{W}^{1,p}$ Sobolevs space. In other words, $\bar{\mathbf{z}}$ minimizes E over the set of admissible deformations. In [15] it was demonstrated that fine-phase mixtures can be interpreted as a limit of the minimizer sequences for the functional E. If it makes sense to consider the $lim_{n\to\infty}\mathbf{z}_n = \bar{\mathbf{z}}$ then $\mathbf{z}$ is interpreted as homogeneous deformations macro-resultant from more and more fine-mixtures. Although this procedure is very interesting because of its strict adherence to the real problem, it implies a high computational complexity, which makes it unattractive for real-time application.

1.5   The Problem of Metallic Plates

Following the logical steps described in the previous paragraph, metallic plates under bi-axial loads deform producing inner mechanical stresses; this lead to a complex model extremely difficult to implement that makes in practice impossible to derive related 2D maps. To assess the integrity of these structural elements, it is necessary to exploit techniques which, starting from the mapping of the mechanical stress, are able to translate the relevant information content into different representing 2D maps easily represented through an electric equivalent. Eddy Currents (ECs) have been shown to be useful to this purpose. Indeed, applied to bi-axially loaded plates, they produce electrical 2D maps easily obtainable equivalent to the mechanical one.

However, different mechanical loads produce different deformations and, consequently, different mechanical stresses; it seems obvious to quantify them in order to assess the integrity of the plate. We propose here a new approach to face the problem of quantifying the mechanical stresses from 2D EC images with the Fuzzy Similarities (Ss) that allows us to take into account uncertainties and/or inaccuracies present in any industrial process. Specifically, by collecting images produced by applying symmetric bi-axial loads (or no loads) to metallic plates, we generate a database of different classes of images; the assessment of the integrity of a fresh plate is achieved by comparing the Ss associated to its EC 2D signal with the different reference classes. Formally, if $F(X)$ is the set of all possible fuzzy sets, Ss is a function defined over $F(X)^2$ with range [0, 1]. If NS (No-Stress) is the fuzzy set representing the class of the signals derived from plates without loads and WS_j (With-Stress) the fuzzy set of the jth class of the signals derived from plates subjected to known loads, we compute the Ss of a signal derived from a plate subjected to an unknown load (US). This value is compared to the reference value of the classes NS or WS_j, $Ss(NS/WS_j, US)$; this way, we have an indication of how much the unknown signal is similar (that is, close in a fuzzy sense) to each class (with or without bi-axial load). In these terms, the evaluation of structural integrity of a plate becomes a fuzzy classification operation.

This paper is structured as follows. The next session will detail the proposed procedure of image classification by means of Ss. Then, we will present the experimental work: starting from electrical considerations regarding the ECs and describing the characteristics of the experimental database, we will discuss the main numerical results and compare them with the ones achieved by using the technique of Self-Organized Maps (SOM). Finally, some concluding remarks will be given.

2 Image Classification Using Fuzzy Similarities

In the context of digital image processing, the application of fuzzy logic concepts is well-established. The usefulness of fuzzy logic applied to digital image processing has been remarkably demonstrated, in particular in industrial processes that are invariably affected by uncertainty and/or imprecision [16, 17]. Fuzzy image processing, based on fuzzy theory, exploits both fuzzy logic and its relational aspects. In this paper, the fuzzy similarity relation is exploited to address and solve the above described classification problem.

2.1 Fuzzy Similarity Measure

Many measures of similarity among fuzzy sets have been proposed in the literature, and some have been incorporated into linguistic approximation procedures. The motivations behind these measures are both geometric and set-theoretic. This kind of metrics simply categorize pairs of fuzzy concepts as either *similar* or *dissimilar* [18]. For distinguishing between degrees of similarity or dissimilarity, different kind of measures show different performance as well as different properties. Typically, the best measures were ones that focus on only one slice of the membership function. Such measures are easiest to compute and may provide insight into the way human brain judges similarity among fuzzy concepts. The notion of similarity plays a fundamental role in both theories of knowledge and behavior. Geometric models have dominated the theoretical analysis at the basis of similarity relations. These models generally represent objects as points in some coordinate space such that the observed dissimilarity among objects corresponds to the metric distance between the respective points. However, similarity may be better modeled by a function that is conceptually different from a geometric distance (e.g., a set-theoretic function). Several authors have proposed similarity indexes for fuzzy sets that can be viewed as generalizations of the classical set-theoretic similarity functions [5]. These generalizations rely heavily on the definitions of cardinality and difference in fuzzy set theory. Definitions of the cardinality of fuzzy subsets have been proposed by several authors. A systematic investigation of this notion was performed in [18]. In this work, we propose a specific metric for fuzzy similarity that fits the requirements of our problem.

2.2 Some Basic Concepts

Usually, an $M_1 \times M_2$ image I with L gray levels can be represented by a matrix of pixels of the same size. Here, to each pixel (i, j), $i = 1, 2, ..., M_1$ and $j = 1, 2, ..., M_2$, is associated its gray level, y_{ij}. From a fuzzy point of view, a digital image and a fuzzy set can be similarly defined. In particular, on an image I, similar to a collection of elements having a membership degree, it is important establish the membership value of each pixel to I by means a membership function, $\mu_I(y_{ij})$, defined on I, ranging on $[0, 1]$ thus defining the membership of y_{ij} to I. In particular, if $\mu_I(y_{ij}) = 1$, y_{ij} totally belongs to I in a fuzzy sense; if $\mu_I(y_{ij}) = 0$, y_{ij} does not belong to I in fuzzy terms. Finally, if $\mu_I(y_{ij}) \in (0, 1)$ partially belongs to I. From the point of view of the information content of y_{ij} into I, $\mu_I(y_{ij})$ can define an image I by means of the following relation:

$$\sum_{i=1}^{M_1} \sum_{j=1}^{M_2} \{y_{ij}, \ \mu_I(y_{ij}), \ i = 1, 2, ..., M_1, \ j = 1, 2, ..., M_2\} \tag{11}$$

where $\mu_I(y_{ij})$ represents the grey level of y_{ij}.

2.3 Fuzzy Similarity: A Brief Overview

The formalization of the concept of fuzzy Similarity (Ss) of $2D$ signals is obtained by means of a binary relation on $F(I)$, the universe of discourse of the grid points of an image I. On $F(I)^2$, Ss is quantified by using the following formal expression:

$$S : F(I) \times F(I) \to [0, 1] \tag{12}$$

with the following properties:

1. Ss is *reflexive*. This is evident considering the fact that each image is quite similar to itself or, more generally, for two identical image, Ss must be equal to unity;
2. Ss is *symmetric*. In other words, Ss should not depend on the order in which the images are taken into account;
3. Ss should be robust to noise (*monotony*). If an image I is corrupted by noise, getting a new image $I_{corrupted}$, their Ss with respect to a reference image I_{ref} is quite similar. Formally:

$$S(I, I_{ref}) \simeq S(I_{corrupted}, I_{ref}). \tag{13}$$

Very good formulations of Ss satisfying the properties 1. 2. and 3. have been proposed in the literature, [19–21]. However, as the aim of this work is to consider a useful approach for real-time applications, the attention was paid to three formulations characterized by a reduced computational load. In particular, if A and B are two digital gray-scale images and L is the maximum value of gray level in the images, the three following Ss have been considered [22, 23]:

$$S_1(A, B) = \frac{1}{L} \sum_{n=1}^{L} \frac{min(\mu_A(n), \mu_B(n))}{max(\mu_A(n), \mu_B(n))} \tag{14}$$

$$S_2(A, B) = 1 - \frac{\sum_{n=1}^{L} |\mu_A(n), \mu_B(n)|}{\sum_{n=1}^{L} (\mu_A(n) + \mu_B(n))} \tag{15}$$

$$S_3(A, B) = \frac{\sum_{n=1}^{L} min(\mu_A(n), \mu_B(n))}{\sum_{n=1}^{L} (min(\mu_A(n), \mu_B(n)) + \alpha\Omega + \beta\Psi)} \tag{16}$$

where $\mu_A(n)$ and $\mu_B(n)$ are the membership values of the nth gray levels related to the histograms of the two images and $\Omega = min(\mu_A(n), 1 - \mu_B(n))$, $\Psi = min(1 - \mu_A(n), \mu_B(n))$. We explicitly point out that S_3 is an extension of the Generalized Tverskys Model [22].

2.4 The Fuzzy Similarity Approach to Solve the Classification Problem

The proposed procedure computes the Ss values between two images A e B by initially dividing the images into N sub-parts, A_i and B_i (10×10 *pixels*). The membership functions extraction is done by normalizing the histograms of the N sub-images through the use of a normalizing factor. Let h and z be the maximum values of the histogram of the N sections of the two images; the normalizing factor, *norm factor*, equals $max(h, z)$. Then, for each pair of the corresponding sub-parts, A_i and B_i, $Ss_i(A_i, B_i)$, $i = 1, 2, \ldots, N$ is computed, and finally the global Ss between A and B is obtained by averaging the individual values of Ss_i:

$$S(A.B) = \frac{1}{N} \sum_{i=1}^{N} S_i(A_i, B_i) \qquad (17)$$

The logical steps of the proposed procedure are shown in Fig. 1. In particular, the approach includes an initial pre-processing step of the images in which any filtering operations, contrast enhancement and thresholding may be required to improve the image quality. The next step requires the segmentation of the images into N sub-parts as above specified. The membership functions are obtained by means of the histogram normalization of the sub-parts carried out by the segmentation step. In the next step, the Ss are computed for the corresponding section and, finally the global Ss value is determined, that is used as basis for the image classification.

2.5 Structural Integrity Assessment Through Ss in Terms of Image Classification

The goal of this work is the design a procedure capable of assessing the integrity of a plate under bi-axial loads. To this aim, Step 5 of Fig. 1 deserves an additional level of detail (Fig. 2). By considering an EC image which is related to a plate with an

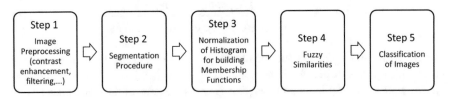

Fig. 1 The proposed procedure for classifying images via the Ss approach

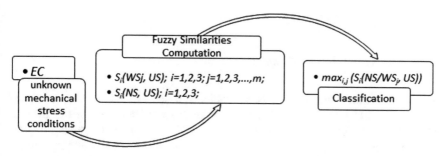

Fig. 2 A pictorial view of the Step 5 of the proposed classification procedure

unknown mechanical stress (US), and considering $m + 1$ different classes (m classes WS_j related to m different bi-axial loads and a class NS related to a set of unstressed plates) all similarities $Ss_i(NS/WS_j, US)$ are evaluated. The class WS_j or NS is determined by computing the $max_{ij}(Ss_i(NS/WS_j, US))$ that classifies US as a 2D signal belonging to that class.

3 Fuzzy Similarities and SOM

To assess the quality of the results obtained by the proposed procedure, the classification has been carried out also through SOM in which the parameters of similarity among neurons are also fuzzy [24]. SOM is an unsupervised neural network implemented by a grid of knots and artificial neurons completely connected to the input layer whose weights are adjusted according to the input data using special features and similarities. The typical learning algorithms for these structures is the so called *the winner takes all* where neurons having high similarity win the competition. When a knot wins the competition, the weights of the neighboring knots are also modified and the process is repeated, for a number of cycles, for each input pattern. Here, the more complex operation concerns the choice of the parameters of similarity between an input pattern and the vector of the weights of a single knot of the map. However, by using a Gaussian function, $G(h, k) = 2^{\frac{|z_h - z_k|^2}{\sigma^2}}$, the algorithm is able to determine the knot (or neuron) with a higher similarity to the input and to update the weights of the neighboring knots. In this paper, to emphasize the fuzzy nature of the problem, the parameters are set according to the fuzzy similarities values.

4 Experimental Work

4.1 The Experimental Setup

The experimental campaign has been developed at the Laboratory of Electrotechnics and Non-Destructive Testing, of the *Mediterranea* University of Reggio Calabria. A $200 \times 200 \times 5$ steel plate under symmetrical gradually increasing bi-axial loads has been investigated. For the training step, the specimen was subjected to the load starting from 150 kN to the final load of 180 kN, with steps of 10 kN. Because of the microscopic structure of the material, each load modifies locally the magnetic properties of the sample; accordingly, its state of degradation was investigated by analyzing the magnetic changes induced in the structure of the material when the deformations take place. The sensor exploited is a FLUXSET®type probe [25] whose pick-up voltage provides a measure which is proportional to the component of the magnetic field parallel to the longitudinal axis of the sensor. In addition, both an *AC* sinusoidal exciting field (1 kHz) and an electric current (120 mA RMS) have been applied. With regard to the driving signal, the choice fell on a triangular shape at the frequency of 100 kHz with 2 Vpp amplitude. The sensor was assembled to an automatic step-by-step scanning system moved on a square portion (70 mm side) at the middle of the specimen. After each symmetrical bi-axial load application (see Fig. 3), the specimen was investigated by means of FLUXSET®to obtain four bidimensional signals (*BSs*) representatives of the real part, the imaginary part, the module and the phase of the pick-up voltage ($[mV]$) at each point of the surface. Since similar loads produce in turn similar deformations (and therefore similar magnetic properties), it was considered to apply to the plate a series of gradually increasing loads, thus creating classes of loads with continuously variable mechanical and magnetic properties. Thus, four classes of loads have been created, namely, $Class_{150kN}$, $Class_{160kN}$, $Class_{170kN}$ and $Class_{180kN}$. To each class, they belong the image produced by the nominal load corresponding to the class and the ones generated by loads that are very close to the nominal one. Similarly, the class of images related to the plate in absence of loads (i.e., in absence of stress states) has been obtained. Regarding the testing database, an appropriate set of images has been produced by subjecting a specimen of the same characteristics with different symmetrical bi-axial loads that, for the purposes of the present work, are supposed unknown. Table 1 summarizes the characteristics of the database while Fig. 4 shows the real part of the pick-up voltage for both a plate not subjected to loads (Fig. 4a) and a plate subjected to a bi-axial load (Fig. 4b).

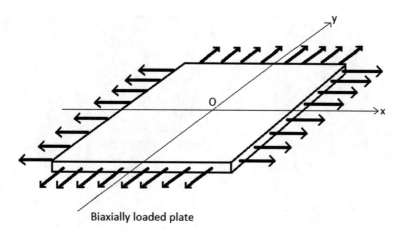

Biaxially loaded plate

Fig. 3 A simple schematization of a plate subjected to a bi-axial load

Table 1 Experimental dataset divided into classes

Training database			
Classes	$N°$ BSs	Classes	$N°$ BSs
$Class_{150kN} = WS_1$	50	$Class_{160kN} = WS_1$	50
$Class_{170kN} = WS_1$	50	$Class_{180kN} = WS_1$	50
Class of signals relative to the plates without applied loads $= NS$	40		
Testing database			
$N°$ 50 BSs			

5 Numerical Results

Before performing the computational analysis in terms of Ss, it is necessary to check in advance the fuzziness of each signal affected by uncertainty by means of a specific index. To this aim, we exploited both the Index of Linearity (LI):

$$LI = \frac{2}{N} \sum_{n=1}^{N} min(\mu_A(n), 1 - \mu_A(n)) \tag{18}$$

and the Fuzzy Entropy index (FE):

$$FE = \frac{1}{N} \sum_{n=1}^{N} [-\mu_A(n)log[\mu_A(n)] - [1 - \mu_A(n)]log[1 - \mu_A(n)]] \tag{19}$$

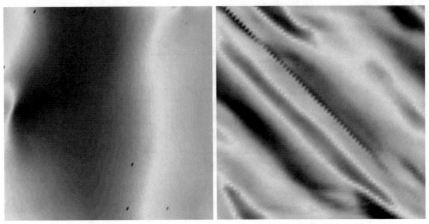

(a) Map of real part of the measured signal   (b) Map of real part of the measured signal
without loads.   with bi-axial load.

Fig. 4 Typical *EC 2D* images extracted from plates without loads and with bi-axial loads by means of FLUXSET® sensor (*AC* sinusoidal exciting field (1 kHz), electrical current 120 mA RMS, triangular driving signal (100 kHz) and 2 Vpp amplitude)

Table 2 Performance of classification compared to *SOM*

Technique	Classification (%)
S_1	99.3
S_2	99.2
S_3	99.7
SOM	99.6

High values of both indices implies high fuzziness of the data included in each class thus justifying the applicability of the procedure reported in Fig. 2. An *EC 2D* image with unknown mechanical stress is attributed to a class WS_j by similarity measured through its *Ss*: the highest value of *Ss* determines the class. A test case is reported as example in Fig. 5: the image under analysis belongs to the class WS_3 (Fig. 5c). As reported in Table 2, all the available images were successfully classified (percentage greater than 99% of cases) and the results are comparable with those obtained by applying the *SOM* type technique where the parameters of similarity among neurons have been considered of fuzzy nature but with considerably higher computational complexity.

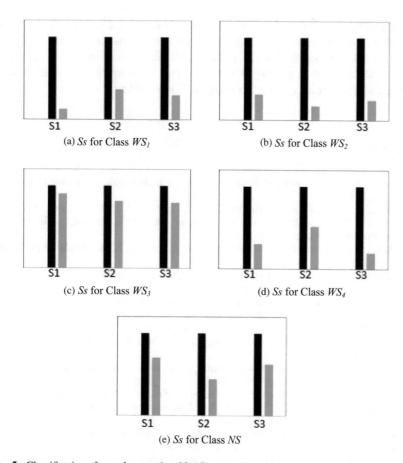

(a) *Ss* for Class *WS₁*

(b) *Ss* for Class *WS₂*

(c) *Ss* for Class *WS₃*

(d) *Ss* for Class *WS₄*

(e) *Ss* for Class *NS*

Fig. 5 Classification of an unknown signal by *Ss* comparisons

6 Concluding Remarks

In this paper, a new approach for assessing the integrity of metallic plates under bi-axial loads through the joint use of the concept of fuzzy similarities and of the inspection procedure based on eddy currents was proposed. An experimental laboratory campaign of measurements has been carried out on metallic plates aiming to obtain 2*D* images of electrical maps that are equivalent to the mechanical stress. Three classes of bi-axial loads and a class of no loads have been considered. After the fuzziness evaluation of each image by means of two specific index, three fuzzy similarities formulations were chosen and implemented to classify unknown stress states. The proposed classification procedure, characterized by a reduced computational complexity, is able to classify more than 99% of the cases under study and could be considered as an excellent candidate for both real-time application and hardware implementation. Regarding this two aspects, the comparison with established *SOM* techniques encourages further study in this direction.

References

1. Aluja, J.G.: Towards a new paradigm of investment selection in uncertainty. Fuzzy Sets Syst. **84**(2) (1996)
2. Aluja, J.G.: Investment in Uncertainty. Kluwer Academic Publishers, Dordrecht (1998)
3. Aluja, J.G.: The Interactive Management of Human Resources in Uncertainty. Kluwer Academic Publishers, Dordrecht (1986)
4. Aluja, J.G.: Elements for a Theory of Decision in Uncertainty. Kluwer Academic Publishers, Dordrecht (1999)
5. Dubois, D., Prade, H.: Fundamental of Fuzzy Sets. Springer (2000)
6. Ericksen, J.L.: Equilibrium of bar. J. Elast. **25**, 191–201 (1975)
7. Nadai, A.: Theory of Flow and Fracture of Solids. Mc Graw-Hill, New York (1950)
8. Hartl, D.J., Lagoudas, D.C.: Thermomechanical Characterization of Shape Memory Alloy Materials, Shape Memory Alloys, pp. 53–119. Springer, Berlin (2008)
9. Song, G., et al.: Applications of shape memory alloys in civil structures. Eng. Struct. **28**(9), 1266–1274 (2006)
10. Chen, Y.C.: Bifurcation and stability of homogeneous of an elastic body under load tractions with Z2 symmetry. J. Elast. **25**, 117–136 (1991)
11. Metzger, M., Seifert, T.: Computational assessment of the microstructure-dependent plasticity of lamellar gray cast iron. Part I: Methods and microstructure based models. Int. J. Solids Struct. **66**, 184–193 (2015)
12. Bechle, N.J., Kyriakides, S.: Evolution of phase transformation fronts and associated thermal effects in a *NiTi* tube under a biaxial stress state. Extreme Mech. Lett. **8**, 55–63 (2016)
13. Xue, X., Liao, J., Vincze, G., Pereira, A.B., Barlat, F.: Experimental assessment of nonlinear elastic behaviour of dual-phase steels and application to springback prediction. Int. J. Mech. Sci. **117**, 1–15 (2016)
14. Droniou, J., Lamichhane, B.P.: Gradient schemes for linear and non-linear elasticity equations. Numerische Mathematik **129**(2), 251–277 (2015)
15. Ball, J.M., James, R.D.: Proposed experimental test of a theory of fine microstructures and the two-wells problem. Philos. Trans. R. Soc. Lond. **38**, 389–450 (1992)
16. Pal, S.K., King, R.A., Hishim, A.: Index of area coverage of fuzzy image subset and object extraction. Pattern Recogn. Lett. **11**, 831–841 (1992)
17. Rosenfeld, A.: The fuzzy geometry of image subsets. Pattern Recogn. Lett. **2**, 311–317 (1984)
18. Dubois, D., Prade, H.: Ranking fuzzy numbers in the setting of possibility theory. Inf. Sci. **30**(3), 193–224 (1983)
19. Tversky, A.: Features of similarity. Psycol. Rev. **84**(4), 327–352 (1977)
20. Versaci, M., Calcagno, S., Cacciola, M., Morabito, F.C., Palamara, I., Pellican, D.: Innovative fuzzy techniques for characterizing defects in ultrasonic nondestructive evaluation. In: Ultrasonic Nondestructive Evaluation Systems. Industrial Application Issues, pp. 201–232. Springer, Heidelberg (2015)
21. Pellicanó, D., Palamara, I., Cacciola, M., Calcagno, S., Versaci, M., Morabito, F.C.: Fuzzy similarity measures for detection and classification of defects in CFRP. IEEE Trans. Ultrason. Ferroelectr. Freq. Control **60**, 1917–1927 (2013)

22. Chaira, T., Ray, A.K.: Fuzzy Image Processing and Applications with MatLab. CRC Press, Taylor & Francis Group, Boca Raton, New York (2010)
23. Versaci, M.: Fuzzy approach and Eddy currents NDT/NDE devices in industrial applications. Electron. Lett. **52**(11), 943–945 (2016)
24. Kohonen, T.: Self-Organizing Maps, Series 30. Springer, Heidelberg (2001)
25. Pavo, J., Gasparics, A., Sebestyen, I., Vertesy, G., Darczi, C.S., Miya, K.: Eddy Current Testing with Fluxset Probe. Applied Electromagnetics and Mechanincs. JSAEM, Tokyo (1996)

Cochlear Implants: Consequences of Microphone Aging on Speech Recognition

C. Berger-Vachon, P. A. Cucis, E. Truy, H. Thai Van and S. Gallego

Abstract Aging is a general phenomenon which affects everything and everybody in this world. Designed for the rehabilitation of profound deafness, cochlear implants (CI) do not escape to this general rule. One very insidious effect concerns the microphone as an ongoing drift occurs over the time. In this work we wish to assess the consequences of this evolution on speech recognition. In order to perform this task, a general population of CI users and NH subjects (using a CI simulator) participated to this study. They listened to French dissyllabic words and we recorded recognition percentages. Words were presented to the listeners in noise with a variable signal to noise ratio (SNR) and the percentages ranged from 0 to 100%. For the CI simulator, the drift was simulated from data coming from figures measured on regular hearing aids. This choice seems relevant as CIs and hearing aids use the same microphones. Also, the CI simulator we used, picked up the general principles of a vocoder to represent the classical coding strategies used in CIs (CIS-like and n-of-m). With CI users, the results were compared before and after cleaning the microphones; also, in a subgroup of CI users, we performed the replacement of the head filter protecting the microphone and the recognition percentages were compared with those coming from the standard "Brush and Blow" cleaning procedure. The results have been revisited and quantified after a curve

C. Berger-Vachon (✉) · P. A. Cucis · E. Truy · H. Thai Van · S. Gallego
University Claude-Bernard Lyon1, 43 bd du 11 Novembre,
69622 Villeurbanne-Cedex, France
e-mail: christian.berger-vachon@univ-lyon1.fr

C. Berger-Vachon · E. Truy · H. Thai Van
Lyon Neurosciences Research Centre, (CRNL), INSERM U1028, 95 bd Pinel, 69675
Bron-Cedex, France

C. Berger-Vachon
LBMC-IFSTTAR, 25 Avenue François Mitterrand, 69500 Bron, France

S. Gallego
Audition-Conseil Centre, 34 Avenue Lacassagne, 69003 Lyon, France

E. Truy · H. Thai Van
CRIC, ORL Building, Edouard-Herriot Hospital, Place d'Arsonval,
69437 Lyon-Cedex 03, France

© Springer International Publishing AG 2018
C. Berger-Vachon et al. (eds.), *Complex Systems: Solutions and Challenges in Economics, Management and Engineering*, Studies in Systems,
Decision and Control 125, https://doi.org/10.1007/978-3-319-69989-9_30

497

fitting. The outcomes indicated that the CIS-like coding schemes were less sensitive to aging than the n-of-m strategies. Also, cleaning ameliorated the recognition performances, but the increase was not dramatically high. Furthermore, the improvement mainly occurred in the middle of the SNR range where the noise was not too intense. We made these observations with CI users and with NH subjects so it indicates that the results should be linked to the properties of the signal. Finally, as we cannot stop the consequences of aging, we can set up an action plan to reduce its effect. And this is true in everyday life. In the case of CIs, a lot of solutions are available, among them the choice of the sound coding strategy and the periodicity of the clinical check and device setting.

Keywords Cochlear implants · Microphone aging · CI and NH listeners Coding strategies · Cleaning procedures · Syllable recognition in noise

1 Cochlear Implants and Aging

Cochlear implants (CI) have successfully restored partial hearing to over 400,000 deaf persons in the world and the rhythm of implantations is high in the world with a minimal rate of about 30,000 implantations [1].

Economically, CI is a good investment for the society. The machine is expensive (about $25,000), the surgery also and the follow up of the patient during the first year (speech therapy, audiology) pushes this figure to $50,000. Nevertheless, if we consider the cost of a handicapped persons during his life (education in dedicated centers, perturbed familial life, professional limitations and so on) fitting a deaf person with a CI is worthwhile to be done and many social systems cover the expense mostly when it is for children.

1.1 Reliability

Designed to last for more than 10 years, CI suffers from failures and some of them come from aging [2, 3] and one of the goal of engineering is to reduce their impact. In this paper we wish to discuss this issue.

Cochlear implants are biomedical prostheses which restore some hearing in profoundly deaf subjects. Basically, they are made with two main components, an external device which is for speech reception and processing, and an implant which decodes the information transmitted by the external device and distributes it to the ends of the auditory nerve in the inner ear. All the elements are subjected to aging and patients are deeply affected when failures occur in the system. There are two kinds of failure, the hard failure easy to detect as the machine breakdown is seen immediately and the soft failure when the failure is intermittent. Another kind of burden, more insidious, is the progressive loss of efficiency on one of the components of the machine. Among them the attack of the microphone is one of the hazards affecting the machine. Several alerts have been raised and some dramatic stories about alterations on the opening in hearing aids have issued warnings [4] which must not be ignored. A check of the microphone entrance often discovers debris accumulated in undesirable locations and the input sound pathway is more or less obliterated.

1.2 Microphone Aspect

The microphone repair issue is well ranked into the external failures of a cochlear implant [5, 6]. It comes in the fourth or fifth position in the classification after the headpiece (antenna), the battery compartment and the speech processor and it represents nearly 11% of the repairs.

But the progressive drift of the microphone properties is often not noticed and everybody should keep in mind that the microphone works in very adverse environments which can range from very cold and dry to very humid and hot, with a common exposure to sweat, dirt, dust, grease and so on. This is a very aggressive situation and moisture and grease are likely to affect the sensitivity of the microphone and its frequency response.

Consequently this behavior should be studied and this aspect of aging should not be ignored. In a previous approach Razza et al. [7, 8] indicated that speech tracking scores are correlated significantly with the loss of the microphone sensitivity. Responses curves were compared to those of fully working microphones and the consequences on speech recognition need to be studied.

This situation is not very simple as the use of cochlear implants occurs often in noise which is part of our usual environment.

1.3 Coding Strategies

CI allows several coding strategies which can be implanted on almost all the machines, and the choice of a relevant coding strategy is something which must be discussed.

Basically, in CI, the speech input spectrum is divided into frequency bands, and this "simplified" spectrum is distributed to the electrodes distributed along the cochlea which is the hearing organism.

Two main schemes are usually taken to program the speech processors by the CI manufacturers:

- n-of-m: the speech spectrum is divided into m bands, and the n with the highest energy are kept and transmitted to the cochlea. This scheme is popular, because the presentation of selected acoustic features corresponds to an approach based on the phonetic theory. In this circumstance, only a subset of the available electrodes is stimulated. Also, as the number of active electrodes is limited, it is expected that the overlap between two successive stimulations is small, but this may be not true if two consecutive electrodes are selected [9].

• CIS (Continuous Interleaved Sampling): it is a sound processing scheme designed to enhance the delivery of temporal fine structure to the implant, in order to beat the poor intelligibility of speech in a competing noise. All the band pass filters are kept and the number of electrodes is reduced, leading to a wide spacing, between two consecutive electrodes, which is likely to reduce the interaction. Also, the stimulation rate is high and it enhances the delivery of temporal fine structure to the implant, and this is an advantage in a competing noise [10].

1.4   Noise Influence

The choice of the noise is also a major issue. The effect of listening conditions determines the performances of the listeners. The signal to noise ratio (SNR) is an important parameter for the assessment of the recognition performances. Lowering the SNR spoils the incoming speech signal and this action is worthy to be measured. Also, with different values of the SNR it is easy to avoid the ceiling and the floor effects, enabling the determination of some characteristic parameters traditionally used in the evaluation of speech recognition [11].

In our work, the choice of the Fournier's lists (a French equivalent to the spondee lists) led to recognition percentages ranging from 0 to 100%, according to the SNR, and it was suitable to the aims of our experiment.

1.5   NH Versus CI

Two aspects are generally considered when coding strategies are examined in cochlear implants: the recognition seen with patients using the coding schemes, and the behavior of normal hearing (NH) subjects facing the signal (a simulation is done for NH subjects). Both performances are complementary as in one case (CI users) we have the direct results with the group of persons concerned by the prosthesis and in the second group we can test the signal action using a more comprehensive approach [12]. In other words, one matter is more oriented to the patient and the other tests the signal. Both groups receive the same amount of information about speech, but the outcomes may be different or similar [13, 14]. Furthermore, the CI users group is very heterogeneous and the standard deviation in the measures is large making the statistical interpretation of the results more difficult. On the contrary, with the NH listeners, several strategies and microphone drift are evaluated on the same people and intra-subject studies can be done making the evaluation more accurate.

1.6 Organization of This Study

The aim of this study is to assess the behavior of two coding strategies (n-of-m and CIS-like) in front of the aging of the CI microphone, and to give some indication to the audiologists in charge of CI settings in the specialized clinical centers.

After an introduction of the CI issues about aging and the main elements which will be manipulated to carry out the study, the second section introduces the tools for the simulation (CI and microphone drift), the subjects who participated to the experiment and the mathematical fitting of the recognition results. Then in the third section, the results (recognition percentages and audiological features) obtained with the CI users and the NH subjects are presented.

The fourth section is a discussion about the different elements which were used in this work, The comparison of the performances of CI users and NH subjects, the choice of the noise and evaluation of the strategy and how to reduce the influence of the microphone drift with aging. Finally, a conclusion sums up the main contributions of this work.

2 Materials and Methods

2.1 Stimulation

2.1.1 Cochlear Implant Principles

We will not expose the details of signal processing in cochlear implants here, but only the essential principles (Fig. 1). Cochlear implant can be divided into two main parts:

- The external block (signal processor) processes the signal and separates the sound in multiple frequency bands (channels).
- The internal block (the implant) receives the information from the processor and distributes the energy of the channels to the electrodes along the cochlea.

The information is transmitted from the processor to the internal part through the skin by a radiofrequency signal.

Sounds in the environment are captured by the microphones that are very impacted by soiling and aging. Then the processor fits the signal received from the microphones to the patient's physiology and separates the signal into multiple

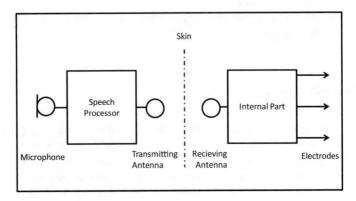

Fig. 1 Shematic view of a cochlear implant

frequency bands. The frequency bands correspond to the treatment channels scaled using a logarithmic scale. We chose to use the Bark scale for our CI simulator [15].

Most of the existing sound coding strategies are based on two main principles: there are CIS-like strategies and "n-of-m" strategies. It is interesting to notice that, for the n-of-m strategies, if n = m it is very similar to CIS. So the value of n is important to be considered.

2.1.2 Vocoder

The CI simulator used for testing NH subjects is a basic vocoder fitted for the purpose of the study (Fig. 2).

The different steps of the signal processing are:

- The input signal goes through a pre-emphasis filter which is a high-pass filter (cutoff frequency 1.2 kHz and slope 3 dB/octave).
- The signal is then sampled (16 kHz sampling frequency, 16-bit quantization), a short term fast Fourier transform (FFT) is applied to the samples, frame length is 128 points (8 ms), leading to 64 spectral beams (amplitude and phase), ranging from 0 to 8 kHz. The step between two beams is 125 Hz.
- Then, the spectral beams are grouped into frequency bands which are logarithmically distributed, according to the ear physiology (Bark scale). Considering the usual values taken in cochlear implants we used 20 bands (leading to 20 channels).
- In each band, the energy is calculated using the Parseval's formula (the squares of the amplitude of each beam are added). In the CIS-like coding all the channels are taken. For the n-of-m coding only the 8 more energetic channels

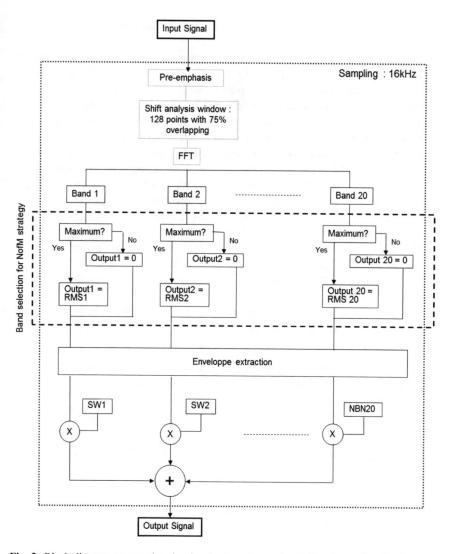

Fig. 2 Block diagram representing the signal processing performed by the n-of-m simulator

are kept. The value n = 8 is classical in cochlear implants. There is a frame overlap of 6 ms (75% overlap), and a set of pulses is calculated for each frame.
• Each channel is represented by a narrow-band spectrum coming from a white noise spectrum. The amplitude of the narrow band follows the energy detected in the corresponding channel. The first two-channels, which are very narrow, were represented by sine waves. The output signal is obtained by summing the selected channels (8 for n-of-m; 20 for CIS-like).

2.1.3   Evaluation of the Frequency Drift of a Hearing Aid Microphone

In this study we tested speech recognition in noise with CI users, before and after a microphone cleaning process. For NH subjects we used a CI simulator integrating a low-pass filter mimicking the microphones frequency drift due to soiling. The low-pass filter approximated a measured median attenuation. We based our work on a study about microphone frequency drift due to soiling on conventional hearing-aids. Knowles electronics (Itasca, Illinois, USA) is the manufacturer which leads the market of microphones for hearing-aids and cochlear implants. That is why we considered that this study matched for cochlear implants' microphones too.

The study of the microphones sensitivity loss had been conducted, during a routine maintenance appointment in a hearing care center, on regular hearing aids worn by n = 129 hearing impaired patients [16]. Two steps were necessary in order to check the microphone frequency drift. First the receiver and the ear mold were deeply cleaned leading to the state "clean receiver and dirty microphone". Secondly, the microphone was carefully cleaned, leading to the state "clean earphone and clean microphone". The subtraction of the two states gave the attenuation due to the microphone soiling.

Then the microphone transfer function was evaluated after a frequency warble sweep ranging from 200 to 8,000 Hz (at 60 dB SPL).

Four degrees of attenuation have been defined:

- Case 1: clean microphone (no loss); the response was not modified,
- Case 2: medium loss corresponding to the attenuation seen on 50% of the microphones (percentile 50%)
- Case 3: strong loss, corresponding to the 20% worse microphones (percentile 80%),
- Case 4: very strong loss, corresponding to the 10% worse microphones (percentile 90%).

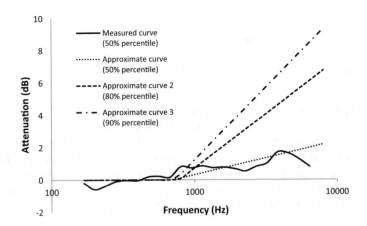

Fig. 3 Simulation of the microphone loss of sensitivity

The clean signal was modified by the different attenuation curves and then passed through the vocoder.

On Fig. 3, the three degrees of loss of sensitivity, corresponding to the 50, 80 and 90% percentiles, are indicated. Each degree was represented by two straight lines for simplification. The signal attenuation started at 800 Hz and then three slopes were introduced, respectively, 2 dB, 7 dB and 9 dB per decade.

2.2 Tests

2.2.1 Participants

The work presented in this paper follows a previous pilot study [17] and it was approved by the French Ethics Committee "Sud Est 2" (August, 27, 2014), under the supervision of the HCL (Civil Hospices of Lyon). All the participants filled an agreement form prior entering the study and all the recordings were made by an audiologist.

CI Users

Sixty-one CI patients were included in this study. Their ages ranged from 18 to 60 years (average 37 years old). Nineteen subjects were fitted with CIS like strategies and 42 had the n-of-m strategies.

The population of the CI users had been parted into four groups:

- Subpopulation 1 (SP1): Med-El® and Advanced-Bionics® subjects (19 patients): "brush and blow" cleaning procedure and CIS-like coding strategies,
- Subpopulation 2 (SP2): Neurelec® subjects (13 patients); "brush and blow" cleaning and n-of-m strategies,
- Subpopulation 3 (SP3): Cochlear® subjects (14 patients): "brush and blow" cleaning and n-of-m strategies,
- Subpopulation 4 (SP4): Cochlear® subjects (15 subjects): replacement of the protective filter and n-of-m strategies.

NH Subjects

Twenty NH subjects participated to this experiment. Their age ranged from 18 to 33 years (average 25 years). All the subjects had an ORL examination before entering the study, in order to eliminate previous pathologies or deafness which may corrupt the results. The auditory thresholds were measured; they were always below 20 dB HL for all the frequencies between 250 and 8,000 Hz. According to the BIAP (International Bureau for Audio-Phonology) all the subjects were considered as NH listeners.

2.2.2 Session

Material

In this work we used the French dissyllabic Fournier's lists. These lists were well adapted to the patients' recognition ability in this environment. They were uttered by a male voice and were the vocal part of the signal. They are similar to the spondees lists used in English. Each list contains 10 disyllabic words (for instance "le bouchon" = the cork). Forty lists are available, and the recognition unit was the syllable. So the recognition step was 5%.

A cocktail-party noise has been used in this study. This noise was a mix of voices from 8 speakers, 4 males and 4 females.

Test Sessions

The signal delivered to the subjects was made of the Fournier's lists mixed with the cocktail-party noise. The mixing was managed by a Madsen Orbiter 922 audiometer and the signal to noise ratio (SNR) was perfectly adjusted.

*CI users

For CI users, tests were made in free field in an audiometric booth. In all cases only one ear was stimulated. When a subject was fitted with two implants, only the best ear was kept.

For NH subjects, the Fournier's lists and the cocktail-party noise were first processed by the CI simulator and then passed through the audiometer and finally delivered on the right ear by a TDH 39 headset.

For the two groups of subjects the experiment consisted in a speech audiometry in noise. The maximum level delivered was below 65 dB SPL. According to the conditions requested by the ethics committee it did not exceed the 80 dB SPL limitation recommended for professional noises exposure.

The recognition scores of the CI users were collected before and after cleaning the microphone, leading to two different situations ("dirty and clean"). Our experiment took place at the beginning of the periodical clinical check and device setting occurring periodically at the CRIC (Cochlear Implant Setting Center) located in the ORL department of the Edouard-Herriot University hospital of Lyon. This check-up consists of an appointment with a speech therapist, a setting of the implant parameters and a clinical examination. Consequently, in our study the microphone cleaning occurred before the classical check-up. This device check follow-up is carried out at least once a year.

We used two different procedures to clean the microphones ports depending on the type of speech processor: A "Brush and Blow" technique (commonly used in French hearing care centers) and the replacement of the protective filters. The "brush and blow" procedure consists in brushing and blowing a flow of dry air on the microphones ports to remove dirt and dust.

The following tasks were realized in our work:

- Verification of the patient's medical file,
- Short training session to help the patient to understand the instructions,
- First test with the Fournier's lists, before cleaning the microphone. The lists were presented to the patient with an increasing level of difficulty (SNR decreased from 18 dB to −3 dB; step −3 dB),
- Microphone cleaning,
- Second test with different Fourier's lists after cleaning the microphone.

The full session lasted about 30 min. Sixteen Fourier's lists were used in this experiment (8 before cleaning and 8 after) and the lists were not repeated.

*NH subjects

For NH subjects, stimuli were delivered to only one ear similar to the experiment with CI users. Furthermore, we chose to test the subjects on the right ear considering that the left brain auditory cortex is specialized in processing speech.

Eight conditions have been considered:

- 2 coding schemes (CIS-like and n-of-m)
- 4 degrees of microphone sensitivity loss.

Five levels of SNR were tested for each condition leading to 40 (2 * 4 * 5) combinations. Each combination was assigned to a Fournier's list so that the lists were not repeated. Each session started with a short training period to help the listener to understand the instructions. Then the 40 lists were randomly presented to each subject. The session lasted about 45 min.

2.3   Mathematical Representation

The recognition percentages versus the SNR can be represented by a sigmoid regression curve (Fig. 4).

Three features were considered here:

- the SNR corresponding to 50% of the maximum recognition plateau denoted here $x_{50\%}$;
- the "slope" (SNR interval, given in dB, between 25 and 75% of the maximum recognition) and denoted here $\Delta_{75-25\%}$;
- the top asymptote y_{max} shows the maximum recognition score.

These analytical values are represented on the sigmoid curve. The minimum recognition was 0% (seen for SNR = −3 dB). Thus, the sigmoid equation is:

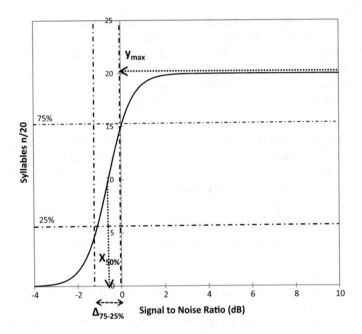

Fig. 4 Fitting of the recognition percentages by a sigmoid curve

$$y = \frac{a}{1 + e^{-b(x-c)}}$$

where:

- y is the recognition percentage
- x is the SNR
- a is y_{max}
- c is $x_{50\%}$
- b is linked to the slope: b = $2.2/\Delta_{75-25\%}$ => $\Delta_{75-25\%}$ = 2.2/b

3 Results

3.1 Percentage of Intelligibility

3.1.1 NH Subjects

Figure 5 shows the percentage of correctly repeated syllables, function of the SNR, for the simulated strategies (a) CIS-like (b) n-of-m. The intelligibility is represented for the 4° of simulated soiled microphones (no attenuation, 800 Hz low pass filtering at −2, −7 and −9 dB per decade)

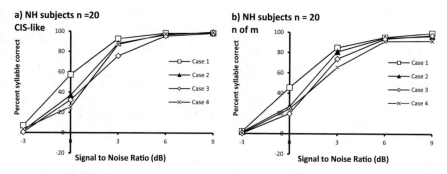

Fig. 5 Intelligibility with respect to the SNR in four cases of simulated soiled microphones, **a** for the CIS-like strategy, **b** for the n-of-m strategy

- *N-of-m coding scheme; ANOVA and post hoc analyses*

A two-factor repeated measure ANOVA was used to analyze the results. It indicated a significant effect of SNR [$p < 10^{-4}$] and of the degree of soiling [$p < 10^{-4}$]. Furthermore a significant interaction was seen between them [$p < 10^{-4}$].

Table 1 shows the results of Wilcoxon's tests taken for the paired post hoc comparisons, between soiling conditions, with respect to the SNR; significant differences are highlighted.

Largest differences were located at 9, 3 and 0 dB of SNR (Table 1). The results indicated that Case 1 was better than Cases 2 and 4 at the highest SNR (9 dB). For the SNR 3 dB the intelligibility for the Case 1 was significantly higher than for the Cases 3 and 4 and also Case 2 was better than Case 4. Finally, Case 1 was better than Cases 2, 3 and 4 for the SNR 0 dB.

- *CIS-like coding scheme; ANOVA and post hoc analyses*

The two-factors repeated measures ANOVA indicated a significant effect of SNR [$p < 10^{-4}$] and of the degree of soiling [$p < 10^{-4}$]. Furthermore a significant interaction was seen between them [$p < 10^{-4}$]. Table 2 shows the results of Wilcoxon's tests for the paired post hoc comparisons with respect to the SNR. The significant differences are highlighted.

Table 1 n-of-m, paired comparisons between the 4 cases with respect to the SNR

NH n-of-m p-value	C1–C2	C1–C3	C1–C4	C2–C3	C2–C4	C3–C4
SNR9	*0.0345*	0.0982	*0.0033*	0.5556	0.2805	0.1654
SNR6	0.2638	0.6714	0.0948	0.8834	0.1579	0.3092
SNR3	0.5549	*0.0067*	*0.0145*	0.0725	*0.0048*	0.1900
SNR0	*0.0058*	*0.0004*	*0.0019*	0.3367	0.6307	0.2575
SNR-3	0.6675	0.0650	0.1048	0.1736	0.1736	0.7728

Table 2 CIS-like, intelligibility paired comparisons between the 4 cases with respect to the SNR

NH CIS-like p-value	C1–C2	C1–C3	C1–C4	C2–C3	C2–C4	C3–C4
SNR9	0.2031	0.8514	0.5862	0.2837	0.3458	0.5160
SNR6	0.2423	0.0688	0.7970	0.4087	0.4991	0.3044
SNR3	0.1369	*0.0006*	*0.0187*	*0.0088*	0.5145	0.0238
SNR0	*0.0035*	*0.0028*	*0.0006*	0.4973	0.0925	0.2754
SNR-3	0.0649	*0.0080*	0.2770	0.2342	0.5067	0.1003

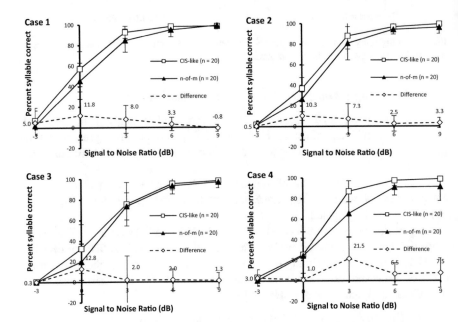

Fig. 6 Intelligibility with respect to the SNR in four cases of simulated soiled microphones, comparison of CIS-like and n-of-m

Largest differences are located between 3 and −3 dB of SNR (Table 2). The results indicate that Case 1 was better than Cases 3 and 4 and that Case 2 was better than Case 3 at the SNR 3 dB. For the SNR 0 dB the intelligibility for Case 1 was significantly higher than for the Cases 2, 3 and 4. Finally, Case 1 was better than Case 3 for the SNR −3 dB.

Strategies are compared on Fig. 6.

3.1.2 CI Users

The effect of cleaning the microphone ("Brush and Blow" procedure) was evaluated on the recognition percentages; it is represented on Fig. 7 according to the

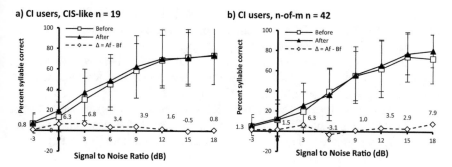

Fig. 7 Intelligibility with respect to the SNR before and after cleaning the microphones' port ("Brush and Blow" cleaning procedure), **a** CIS-like strategies, **b** n-of-m strategies

strategies. It indicates the efficiency of cleaning. The heterogeneity in the results did not allow definitive conclusions and only indicated clues.

- *N-of-m coding schemes; ANOVA analysis*

For the n-of-m strategies the improvements (Fig. 7b) were mostly noticed for the high values of the SNR (12, 15 and 18 dB) and there was an average improvement of 2.7% in syllable recognition after cleaning (corresponding to "½ syllable" per list).

The two-factors repeated measures ANOVA indicated a significant effect of SNR [$p < 10^{-4}$] but no difference between before and after cleaning [$p = 0.056$].

- *CIS-like coding schemes; ANOVA analysis*

For the CIS-like strategy (Fig. 7a), the best improvements were observed in the SNR range 0–9 dB (0, 3, 6, and 9 dB) and the average improvement after cleaning was 2.9%.

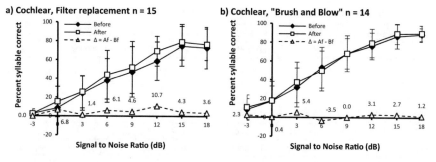

Fig. 8 Intelligibility with respect to the SNR before and after cleaning the microphones' port, **a** cochlear processors, protective filters replacement, **b** cochlear processors, "Brush and Blow" cleaning procedure

The two-factors repeated measures ANOVA indicated a significant effect of SNR [$p < 10^{-4}$] but no difference between before and after cleaning [$p = 0.12$].

Protective filter replacement; ANOVA analysis

In the case of the replacement of the protective filter, there was an average improvement of 4.7% in syllable recognition after replacement (corresponding to one syllable per list); the results are represented on Fig. 8.

The two-factors repeated measures ANOVA indicated a significant effect of SNR [$p < 10^{-4}$] but no difference was seen between before and after cleaning [$p = 0.43$] and no interaction occurred between the two factors [$p = 0.91$]

3.2 Audiological Values

3.2.1 NH Subjects

- *N-of-m coding schemes; ANOVA and post hoc analyses*

A one-factor repeated measures ANOVA indicated a significant effect of the soiling degree on $x_{50\%}$ [$p < 10^{-4}$] but did not reveal a difference for the features $\Delta_{75-25\%}$ [$p = 0.52$] and y_{max} [$p = 0.32$].

We can see on Table 3 that the feature $x_{50\%}$ was better for Case 1 compared to Cases 2, 3, 4. Also, $x_{50\%}$ for Case 2 was significantly better than for Case 4.

- *CIS-like coding scheme; ANOVA and post hoc analyses*

Again we used a one-factor repeated measure ANOVA. It indicated a significant effect of the soiling degree on the feature $x_{50\%}$ [$p < 10^{-4}$] but did not reveal a difference for the features $\Delta_{75-25\%}$ [$p = 0.48$] and y_{max} [$p = 0.43$].

We can see on Table 4 that the feature $x_{50\%}$ was better for Case 1 compared to Cases 2, 3, 4. Also, $x_{50\%}$ for Case 2 was significantly better than for Case 3.

Table 3 N-of-m, feature $x_{50\%}$ paired comparisons between the 4 cases

NH n-of-m p-value	C1–C2	C1–C3	C1–C4	C2–C3	C2–C4	C3–C4
$x_{50\%}$	0.0076	0.0003	0.0008	0.1403	0.0458	0.3411

Table 4 CIS-like, feature $x_{50\%}$ paired comparisons between the 4 cases

NH CIS-like p-value	C1–C2	C1–C3	C1–C4	C2–C3	C2–C4	C3–C4
$x_{50\%}$	0.0038	0.0002	0.0012	0.0349	0.1978	0.1851

3.2.2 CI Users

- *n-of-m coding schemes; Wilcoxon's tests*

The Wilcoxon's tests indicated no significant effect of the cleaning on the curve features $x_{50\%}$ [p = 0.66], $\Delta_{75-25\%}$ [p = 0.85] and y_{max} [p = 0.35].

- *CIS-like coding schemes; Wilcoxon's tests*

The Wilcoxon's tests indicated no significant effect of the cleaning on $x_{50\%}$ [p = 0.63], $\Delta_{75-25\%}$ [p = 0.29] and y_{max} [p = 0.41].

3.2.3 Filter Replacement Procedure

- *Wilcoxon's tests for the procedure "filter replacement"*

The Wilcoxon's tests indicated no significant effect of the cleaning on the curve features $x_{50\%}$ [p = 0.96], $\Delta_{75-25\%}$ [p = 0.62] and y_{max} [p = 0.19].

4 General Comments

4.1 About the Present Study

Several indications came out of this work:

- The higher the SNR, the better the recognition (this is obvious). What is less obvious is the SNR for the ceiling effect. For NH subjects it was 6 dB for CIS-like, and n-of-m whatever the microphone drift. Also, the floor effect (0% recognition) was always reached at −3 dB of SNR. For CI users, the ceiling was seen for 15 dB, before and after cleaning for CIS-like, and n-of-m.
- Cleaning improves the recognition (this is sensible). For NH subjects the improvement could not be seen for the extreme values of the SNR range (−3 dB and +9 dB). The importance of cleaning was maximum for the medium values of the SNR range (0 and 3 dB). For CI users, the results were similar to the NH subjects; only the improvement was reduced.
- Considering the sound coding strategies, the recognition percentages obtained with CIS-like, are better than those seen with n-of-m, except for the extreme values (floor and ceiling effects).

Some exceptions can be pointed out:

For NH subjects, ceilings were not the same for CIS-like, and n-of-m, when the microphone was strongly dirty. For CI users when the microphone was clean, the n-of-m scheme led to better results than CIS-like, at the high values of the SNR. May be due to the heterogeneity of CI users, this result was not significant.

Consequently, the robustness of this finding has to be established and further studies should be done to clarify this aspect. So far we may suggest that the coding scheme could be adapted to the noisy environment. Also, a cleaning protocol should be inserted in the regular check of the machines.

Another aspect which was seen was the cleaning procedure: "Brush and Blow" compared to the microphone protective filter replacement.

Basically it seems that the replacement led to a better improvement of the recognition percentage than the "Brush and Blow" procedure. In our experiment, the "Brush and Blow" procedure was performed by a qualified audiologist; it may be different when audiologists which are not used to this procedure have to accomplish this task, mostly if they are worried about damaging the device. If the filter replacement is cheap and fast, it may be a safer approach to systematically change the head filter protecting the microphone, but it has to be seen in our work that it is not worthy to be done.

Going further, a replacement of the microphone can be suggested but other questions are open by this procedure. Obviously, a new microphone would be the best solution, but this is more costly and the connection can lead to contact problems; the junction can be fragile and could be a source of hard failure. Also, for reasonable and good values of the SNR, the difference of performances seen between a new microphone and a microphone with a small drift (50% of the users) is very light and not significant (result seen with the NH subjects). Also, cleaning the microphone in the case of CIS-like, users, did not lead to a dramatic increase of the performances (possibly, the microphones were only slightly soiled).

4.2   About the Microphone

Aging occurs for all components but the action is not the same for all. A breakdown is an acute failure, and it will be followed by an immediate repair. When the fault is irregular, it is more difficult to stop it, because no one knows when the incident will occur and this is a very worrying situation. When the trouble comes insidiously, it is often not perceived and the subject loses progressively his efficiency without being aware of it. This is more or less what happens during the life; someone gets older, in good health, and he hardly notices the decrease of this aptitude to perform many tasks in the life, mentally or physically. Slopes get steeper, mental activities get harder, but the subject can still cope, even if it is a little bit more difficult. This is typically what happens with the microphone; it does not break, but it gets less efficient and the user is not aware of it. This is why this evolution must be studied; what are the consequences and how we can beat it with rather inexpensive actions, in time and in money? The community surrounding a patient, a cochlear-implantee in our case, must be aware of this condition and should take the necessary steps to restore the good conditions attached to the machine. Several protective measures are easy to imagine. To get a good protection to moisture, grease, sweat, dust, debris, a head filter is a good shield for the microphone and it should be systematically checked by the audiologist each time the patient come for a follow up with the medical team. The degree of sensitivity reduction has its importance; this reduction can be beaten; we should lower the signal perturbation, with an attenuation and a systematic shift of the frequencies seen with an increase of the stiffness and of the dirt of the membrane [7].

A study, in simulation, can bring situations which are not often seen in real life; it emphasizes effects which are attached to extreme conditions and sheds more light on the issue we want to study. The influence of the modifications of the properties of the microphone membrane was tested in classical hearing aids checked by a hearing aid analyzer [16]. Then all the situations (coding scheme, microphone drift, signal to noise ratio) were presented to the same NH subjects, a rather homogenous group, and powerful paired student's tests could be conducted. Obviously, the study in simulation cannot ignore the behavior and the performances obtained with CI users as this group is the target for this technology.

4.3 About the Listening Conditions

Three approaches have been done in this work, simulation with NH subjects, simulation of the microphone drift and addition of noise. Noise needs to be introduced in this kind of work for two reasons. First noisy situations are very common in the usual environment, for CI users and for NH subjects. Secondly, to get understandable results, one must avoid the floor and the ceiling effects; the range of the noise, associated with a good choice of the speech signal, allows to fulfill this requirement. Experimental conditions are a real issue, and their choice is far from being easy; pilot studies are more than necessary [17].

Then, the assessment of a listening condition cannot be done only by the comparison of recognition percentages. Other speech tests are worthy to be considered as they throw another glance on the performance. These parameters help to understand a scatter plot, and we have calculated the SNR for the mean recognition, the slope representing the action of the noise on the percentages, and the ceiling indicating the maximum performance reached in a given situation. In order to do so, a mathematical fitting was necessary, and finally it was worth to be done regarding the data synthesis it provided.

Practically, it is necessary for hearing aid management and rehabilitation of CI recipients that their devices are functioning well and are set by an audiologist during the check which is periodically done in all the clinics. The decrease of the microphone sensitivity is part of this clinical evaluation and it should be corrected, as much as possible. Nevertheless the correlation between the decline of the microphone sensitivity and its consequences on speech recognition scores was not found to be dramatic.

4.4   Strategy

As indicated above, CIS-like, strategy was less sensitive to the microphone drift than the n-of-m coding, mostly in the middle of the SNR range. This is an indication to the audiologists in charge of CI settings and to the manufacturers.

Is it only due to the stimulation speed, the spacing between the electrodes which lowers the overlapping? Is the spectrum coverage more useful to the brain than a selection of the phonetic features presented to the brain? For many researchers, CIs work, "because there is a brain behind" and we should not forget the efficiency of this premier organ of our body. We give information to the brain, then let it work.

Then other settings on the strategy have their influence: sampling rate, filters settings, length of the time frame (window), and they have been studied by many authors [1, 18–20]. Most of the studies concluded to an advantage of the CIS coding over n-of-m. Then, the Advanced Combination Encoder (ACE) scheme is very efficient; it uses a very fast updating of the data and therefore provides a fine temporal analysis of the signal, indicating that spectrum is not the only component of speech.

Then the results are subjected to the patient's condition and mostly to the degree of nerve survival and to the intellectual background; it has consequences on the appropriate strategy for individual implant patient [21]. A good adjustment of a strategy to a subject is a goal to be reached, and it should be adapted throughout the time.

4.5   Noise

The robustness of the strategies should be also seen in noise. In a background noise, the performance of CI users is significantly better when using a full filter bank approach (CIS-like) in comparison to using a peak picking method. Consequently, when the signal is distorted (noise or microphone drift) CIS-like, coding seems to be more robust, but this is not unanimously accepted [22].

With CI users the performance in noise is very difficult to interpret and it is related to the patient's characteristics and to the wide variability in this group of people. The degree of deterioration in noise was similar with each strategy.

Also the number of channels needed to understand speech in noise is still an open question [23]. More channels are needed to understand speech in noise than in quiet, both for CI users and NH listeners, even if results are better in simulation with NH subjects. Again, the number of electrodes (channels) must be kept low to reduce the overlap between the electrodes.

A recommendation which is classically delivered when hard of hearing people are concerned is to provide good listening conditions (high SNR) and to speak slowly and distinctively. This situation is not always easy to meet and we should reduce the noise level as much as we can. Consequently many studies have been done to lower the noise, in other words to increase the SNR [24–26]. This is attractive, but not always easy to implement as hypotheses should be made about the noise.

Aging is an important issue for human, animal, vegetal and machines. "Time flows and we follow". Wisdom is to adapt and to find the best strategies to reduce the consequences; this is also true for the cochlear implants microphones and that was the aim of this dissertation: how can we lower the impact of time?

According to us, to lower that impact on the microphones several suggestions can be made to keep the microphone function:

- "Stop the time flow" (unfortunately this is not possible),
- Use the Brush a Blow cleaning procedure,
- Make a head filter replacement,
- Perform a microphone replacement.

The cheapest is to apply a "Bruch and Blow" cleaning procedure at every routine check and the results on the performance should be satisfactory. So we can recommend this procedure to the specialized teams in charge of the CI patient's care.

5 Conclusion

In this work we have studied the effect of soiling and aging of the microphones of cochlear implants on the recognition of syllables in the French Fournier's disyllabic lists of words.

Several parameters were studied in this work:

- The influence of the coding strategies (CIS-like and n-of-m),
- The action of the signal to noise ratio,
- The effect of cleaning the microphones.

Experiments were conducted with CI users and NH listeners as the two approaches are complementary. Also, two procedures to clean the microphones were considered, the classical "Brush and Blow" and the replacement of the protective filter of the microphone. With CI users we were in the true situation, and with NH subjects we could see more analytically the issue and we could perform a more precise analyses using data taken on the same person.

Results indicated that:

- The CIS-like coding strategies were more robust than the n-of-m strategies regarding the microphone drift; the difference could reach two syllables in a list (10%) and it was seen mainly in the middle of the SNR range,
- The SNR effect was very important, the recognition percentages ranging from 0% (SNR = −3 dB) to 100% (SNR = 9 dB or higher). Results were generally lower with CI users than with NH subjects. In some circumstances, the difference could be very small,

- Cleaning the microphones improved the percentages, mostly in the case of NH subjects in simulation as more extreme situations could be considered and they were tested on the same subjects (through paired comparisons).
- With CI users, it was difficult to get significant results, probably due to the heterogeneity of the group, and this is very general when pathology is considered. Also, the protective filter replacement seemed to be more efficient than the classical "Brush and Blow" cleaning procedure, but this effect was not seen everywhere on the SNR range, and this result was not statistically proven.

Finally, we recommend the use of the CIS-like strategies over the n-of-m coding strategies whenever it is possible. Also, in the regular checkup which is programmed in the follow-up of CI users, we suggest that the audiologists perform a cleaning of the microphone (the "Brush and Blow" procedure is not aggressive) as the drift is very insidious and often not noticed by the subjects. In some cases (very dirty microphones) it would be helpful for the patient.

We cannot stop aging, but we can attenuate its effects and it was the aim of this work on the microphone drift for cochlear implants ("a pin hole in a wide issue").

Acknowledgements The authors are grateful to the persons and the institutions who participated to the study; M. Kevin Perreaut who initiated the work, Dr. Fabien Seldran and Dr. Fabien Millioz for the scientific contribution and Ms. Evelyne Veuillet for the links with the ethic committee. We also wanted to thank the members of the CRIC Lyon and the staff of the Edouard-Herriot hospital for their collaboration, the subjects who listened to the Fourier's lists, the Hospitals of Lyon and the Polytechnic School of the University of Lyon.

References

1. Aguiar, D.E., Taylor, N.E., Li, J., Gazanfari, D.K., Talavage, T.M., Laflen, J.B., et al.: Information theoretic evaluation of a noiseband-based cochlear implant simulator. Hear. Res. (2015)
2. Battmer, R.-D., O'Donoghue, G.M., Lenarz, T.: A multicenter study of device failure in European cochlear implant centers. Ear Hear. **28**, 95S–99S (2007)
3. Balkany, T.J., Hodges, A.V., Buchman, C.A., Luxford, W.M., Pillsbury, C.H., Roland, P.S., et al.: Cochlear implant soft failures consensus development conference statement. Cochlear Implants Int. **6**, 105–122 (2005)
4. Schweitzer, C.: Mind the ports! The effect of severe microphone inlet occlusion: port disasters, or how everyday activities can lead to severe microphone occlusion in a directional hearing aid. Hear. Rev. **15**, 14 (2008)
5. Pereira, A.M., de Melo, T.M., Pereira, A.M., de Melo, T.M.: Repair issues associated with cochlear implants external components: the influence of age and time of use. Rev. CEFAC **16**, 1419–1425 (2014)
6. Silverman, C.A., Linstrom, C.J., Gilston, N., Schoepflin, J.R.: Repair issues associated with cochlear implants. Cochlear Implants Int. **11**, 469–472 (2010)
7. Razza, S., Burdo, S., Bonaretti, S.: Acoustical signal check: microphone integrity evaluation through a common hearing aid analyzer. In: 5th Object Measurement Symposium Cochlear Implants ABI, pp. 19–22 (2007)

8. Razza, S., Burdo, S.: An underestimated issue: unsuspected decrease of sound processor microphone sensitivity, technical, and clinical evaluation. Cochlear Implants Int. **12**, 114–123 (2011)

9. Wouters, J., McDermott, H.J., Francart, T.: Sound Coding in Cochlear Implants: from electric pulses to hearing. IEEE Signal Process. Mag. **32**, 67–80 (2015)

10. Wilson, B.S., Lawson, D.T., Zerbi, M., Finley, C.C., Wolford, R.D.: New processing strategies in cochlear implantation. Am. J. Otol. **16**, 669–675 (1995)

11. Garnham, C., O'Driscoll, M., Ramsden And, R., Saeed, S.: Speech understanding in noise with a Med-El COMBI 40+ cochlear implant using reduced channel sets. Ear Hear. **23**, 540–552 (2002)

12. Dorman, M.F., Loizou, P.C., Fitzke, J., Tu, Z.: Recognition of monosyllabic words by cochlear implant patients and by normal-hearing subjects listening to words processed through cochlear implant signal processing strategies. Ann. Otol. Rhinol. Laryngol. Suppl. **185**, 64–66 (2000)

13. Shannon, R.V., Fu, Q.-J., Galvin, J.: The number of spectral channels required for speech recognition depends on the difficulty of the listening situation. Acta Oto-Laryngol. Suppl. 50–54 (2004)

14. Dorman, M.F., Loizou, P.C.: The identification of consonants and vowels by cochlear implant patients using a 6-channel continuous interleaved sampling processor and by normal-hearing subjects using simulations of processors with two to nine channels. Ear Hear. **19**, 162–166 (1998)

15. Traunmüller, H.: Analytical expressions for the tonotopic sensory scale. J. Acoust. Soc. Am. **88**, 97–100 (1990)

16. Serra, P.-O.: Effet de l'entretien des aides auditives sur leurs performances [Audiology diploma dissertation]. University of Montpellier (2015)

17. Perreaut, K., Gallego, S., Berger-Vachon, C., Millioz, F.: Influence of microphone encrusting on the efficiency of cochlear implants preliminary study with a simulation of CIS and "n-of-m" strategies. AMSE J. Model. C **75–2**, 199–208 (2014)

18. Kerber, S., Seeber, B.U.: Sound localization in noise by normal-hearing listeners and cochlear implant users. Ear Hear. **33**, 445–457 (2012)

19. Hazrati, O., Loizou, P.C.: Comparison of two channel selection criteria for noise suppression in cochlear implants. J. Acoust. Soc. Am. **133**, 1615–1624 (2013)

20. Shannon, R.V., Zeng, F.G., Kamath, V., Wygonski, J., Ekelid, M.: Speech recognition with primarily temporal cues. Science **270**, 303–304 (1995)

21. Wilson, B.S., Finley, C.C., Farmer, J.C., Lawson, D.T., Weber, B.A., Wolford, R.D., et al.: Comparative studies of speech processing strategies for cochlear implants. Laryngoscope **98**, 1069–1077 (1988)

22. Nelson, P.B., Jin, S.-H., Carney, A.E., Nelson, D.A.: Understanding speech in modulated interference: cochlear implant users and normal-hearing listeners. J. Acoust. Soc. Am. **113**, 961–968 (2003)

23. Loizou, P.C., Dorman, M.F., Tu, Z., Fitzke, J.: Recognition of sentences in noise by normal-hearing listeners using simulations of speak-type cochlear implant signal processors. Ann. Otol. Rhinol. Laryngol. Suppl. **185**, 67–68 (2000)

24. Jeanvoine, A., Gnansia, D., Truy, E., Berger-Vachon, C.: Contribution of noise reduction algorithms: perception versus localization simulation in the case of binaural cochlear implant (BCI) coding. Emerg. Trends Comput. Biol. Bioinf. Syst. Biol. 307–324 (2015)

25. Kallel, F., Laboissiere, R., Ben Hamida, A., Berger-Vachon, C.: Influence of a shift in frequency distribution and analysis rate on phoneme intelligibility in noisy environments for simulated bilateral cochlear implants. Appl. Acoust. **74**, 10–17 (2013)

26. Hu, Y., Loizou, P.C.: A new sound coding strategy for suppressing noise in cochlear implants. J. Acoust. Soc. Am. **124**, 498–509 (2008)

Alternative Fuzzy Approaches for Efficiently Solving the Capacitated Vehicle Routing Problem in Conditions of Uncertain Demands

Brigitte Werners and Yuriy Kondratenko

Abstract This paper deals with the analysis of fuzzy models and fuzzy approaches for efficiently solving transportation and vehicle routing problems (VRP) with constrains on vehicle's capacity. Authors focused their research on VRP for marine bunkering tankers and planning and optimisation of tanker's routes in conditions of uncertain fuel demands at nodes. Triangular fuzzy numbers are proposed for modelling uncertain demands and the optimization problem is considered as multi-criteria problem with (a) minimizing total length of planned routes, (b) satisfying all orders at nodes (ships, ports), (c) maximizing total sales volume of unloaded fuel, (d) minimizing fleet size. Two alternative fuzzy approaches for efficiently solving such marine VRP are discussed. The first alternative deals with the development of a multi-stage iterative heuristic procedure and the second alternative concerns the development of a fuzzy decision-making system for the current evaluation of satisfaction values for uncertain order realizations.

Keywords Vehicle routing problem (VRP) · Capacitated vehicle routing problem (CVRP) · Fuzzy demands · Iterative heuristic · Decision-making Satisfaction value

B. Werners (✉)
Institute of Management, Ruhr University Bochum, Universitätsstraße 150,
44801 Bochum, Germany
e-mail: or@rub.de

Y. Kondratenko
Intelligent Information Systems Department, Petro Mohyla Black Sea
National University, 68th Desantnykiv Street 10, Mykolaiv 54003, Ukraine
e-mail: y_kondrat2002@yahoo.com

© Springer International Publishing AG 2018
C. Berger-Vachon et al. (eds.), *Complex Systems: Solutions and Challenges
in Economics, Management and Engineering*, Studies in Systems,
Decision and Control 125, https://doi.org/10.1007/978-3-319-69989-9_31

1 Introduction

Transportation by ships plays a significant role in international trade and passenger traffic between countries within marine regions as Mediterranean Sea, Black Sea etc. as well as between continents. For some countries with long shorelines, navigable rivers or with multiple islands, transportation by ships also plays an important role in regional or local trade [1, 16, 26]. Nowadays, among such traditional marine cargo as coal, oil, cement etc., natural gas is an important cargo in marine shipping taking into account that liquefaction ports transform gas into liquefied natural gas [12]. The most important and complex problems for all kinds of marine transportation are the routing and scheduling of vessels in marine environment, which is fuzzy and uncertain because of strong disturbances and different influences of weather, distance and others. Some special characteristics of ship routing and scheduling problems are [1, 12, 36, 42, 44]:

- a fleet is mostly heterogeneous, ships may have various deadweights and loading capacities, cruise speed and special constructions for different kind of cargoes (tankers, bulkers, dry-ships etc.);
- sometimes the distance between two ports is an uncertain value (parameter) and in some cases it is necessary to change destination while at sea to cope with increasing travel times between departure and arrival ports;
- the weather may have strong impact on ships during long trips, speed and travelling time usually depend on wind, current and wave influences;
- loading and unloading processes at the ports usually depend on specific working time windows and cost penalties are to be calculated as a function of the relation between opening and closing of cargo processing at a port on one side and the arrival time of a ship on the other side.

Maritime transportation planning problems can be classified [1] with respect to the corresponding planning horizon into strategic, tactical and operational problems and according to the up-dated classification for modes of ship's operations they can be divided into the three different categories: liner, tramp and industrial operations, but there are no clear bounds between the abovementioned categories.

Taking into account uncertain sea environment it is necessary to use efficient methods for mathematical formalization of marine transportation problems. In particular, for mathematical formalization of various processes and systems in uncertainty it is beneficial to use the theory of fuzzy sets, developed by Zadeh [48]. Specialists have a great interest in such intelligent approaches in terms of practical applications of its mathematical methods in different fields: business process management [2, 4–6, 8, 23], engineering [14], economics [7, 9,], finances [10, 30], decision-making [3, 11, 20, 31–33, 46, 47], medical and technical diagnostics, transport logistics [18, 22, 24], etc. There are multiple fundamental theoretical contributions to the development of fuzzy sets and fuzzy logic theory and their applications for solving optimization problems made by scientists all over the world [9, 19, 25, 29, 34, 38–40, 49, 50]. For example, the solution of a classical transportation problem with uncertain information about transportation cost of cargo unit was proposed by Arnold Kaufmann (France) and Jaime Gil-Aluja (Spain) in [15]. This solution is based on the simplex method of linear optimization and implementation of triangular fuzzy numbers for decreasing uncertainty. This proposed fuzzy approach [15] can be successfully applied to solve marine transportation problems under uncertainty.

In this article we focus on a real-world problem which deals with planning and optimization of a bunkering tanker routing problem in conditions when a priori orders in nodes are uncertain.

2 Problem Statement

The routing problem for bunkering tankers is one of the most important and complicated VRP in marine transportation [7, 15, 16]. Bunkering tankers should provide bunkering operations (transportation and unloading) for various ships to be served. These serviced ships can be located in different geographically distributed marine ports and open sea points. Marine practice shows that very often the information about fuel demands of serviced ships is uncertain. Usually the ship-owner sends an approximate order to the bunkering company. For example, the ship-owner sends an order for fuel supply using such uncertain terms as "approximately A", "about B", "between C and D", "at least R", "not less than S", "not more than K" and so on. It is possible to represent such kind of orders as fuzzy demands, for example, as fuzzy numbers with triangular membership function [2, 20, 21, 43]. The classical VRP [28, 37], taking into account the restricted fuel capacity Cap of each tanker, can be transformed to [13, 17, 37, 41, 42] a capacitated vehicle routing problem (CVRP) under uncertainty. Solving such CVRPFD (CVRP in the conditions of fuzzy demands) deals with (a) minimization of both total length of planned tanker routes and (b) used fleet size of tankers, (c) satisfaction of orders in all destinations and (d) maximization of total value of unloaded fuel. So we will consider a CVRPFD as bi- or multi-criteria optimization problem under uncertainty. Two alternative fuzzy approaches will be considered for solving marine CVRPFD.

3 First Alternative Fuzzy Approach Based on Iterative Heuristic Algorithm

3.1 Mathematical Model for Solving CVRPFD

A first suggested fuzzy multi-criteria approach for solving CVRPFD, where fuzzy demands at nodes are presented by fuzzy numbers $\tilde{d}_j$ with triangular membership functions $\tilde{d}_j = (\underline{d}_j, \hat{d}_j, \bar{d}_j), j = 1 \ldots N$, is based on a mixed integer linear mathematical programming model. The first compromise solution can be interactively modified to meet the decision makers' requirements with respect to different criteria [39, 40, 42]. Indices $i, j = 0, 1, \ldots, N$ are for the depot and different ports, and $k = 1, \ldots, K$ are for routes. The objective function of the considered CVRP is to minimize the total travel distance:

$$\min \sum_{i=0}^{N} \sum_{j=0}^{N} \sum_{k=1}^{K} c_{ij} x_{ijk} \tag{1}$$

and the capacity Cap of each tanker should be sufficient to meet the fuzzy demands $\tilde{D}_k = \sum_{j=1}^{N} \tilde{d}_j y_{jk}$, $k = 1, \ldots, N$ of all ports on the k-th planned route:

$$\sum_{j=1}^{N} \tilde{d}_j y_{jk} \leq \text{Cap}, \quad k = 1, \ldots, N \tag{2}$$

where c_{ij} is the distance from i-th port to j-th port, $i, j = 0, \ldots, N$,

$$x_{ijk} = \begin{cases} 1 & \text{if j follows i on route k} \\ 0 & \text{else} \end{cases},$$

$$y_{jk} = \begin{cases} 1 & \text{if port j belongs to route k} \\ 0 & \text{else} \end{cases},$$

$u_{jk} \geq 0$ is the sequence no. of port j on route k

Each port j belongs to exactly one of the routes except port 0 with the depot

$$\sum_{k=1}^{K} y_{jk} = 1, j = 1, \ldots, N. \tag{3}$$

Each port j on the route must be entered exactly once on the trip

$$\sum_{i=0}^{N} x_{ijk} = y_{jk}, j = 0, \ldots, N, k = 1, \ldots, K. \tag{4}$$

Each port i on the route must be exited exactly once on the trip

$$\sum_{j=0}^{N} x_{ijk} = y_{ik}, i = 0, \ldots, N, k = 1, \ldots, K. \tag{5}$$

No port can follow itself on the route

$$x_{iik} = 0, \quad i = 0, \ldots, N, k = 1, \ldots, K. \tag{6}$$

Sub-routes are forbidden

$$u_{jk} \geq u_{ik} + 1 - (1 - x_{ijk})N, \quad i = 0, \ldots, N, j = 1, \ldots, N, j \neq i. \tag{7}$$

Each trip starts in the depot

$$u_{0k} = 1, \quad k = 1, \ldots, N. \tag{8}$$

To model fuzzy demands it is suggested to consider the possibility that the actual demand of all ships on one route is less or equal to the capacity Cap of the tanker. If the demand is not known exactly, it is possible to find a solution for which the possibility to serve the demand is required at least [10] to a certain degree $\alpha \in [0, 1]$.

The decision maker has to determine α in advance. Considering a fuzzy number as a method of representing uncertainty in a given quantity by defining a possibility distribution for the quantity is analyzed in [13, 41, 42]

$$\text{Pos}\left(\sum_{j=1}^{N} \tilde{d}_j y_{ik} \leq \text{Cap}\right) \geq \alpha, \quad k = 1, \ldots, N; \quad \alpha \in [0, 1]. \tag{9}$$

An even stronger condition is to determine a certain degree of necessity β that the demand on the route can be served

$$\text{Nec}\left(\sum_{j=1}^{N} \tilde{d}_j y_{ik} \leq \text{Cap}\right) \geq \beta, \quad k = 1, \ldots, N; \quad \beta \in [0, 1]. \tag{10}$$

The requirements (9) and (10) for capacity can be transformed as follows

$$\text{Pos}\left(\text{Cap} - \sum_{j=1}^{N} \tilde{d}_j y_{ik} \geq 0\right) \geq \alpha, \quad k = 1, \ldots, N; \quad \alpha \in [0, 1], \tag{11}$$

$$\text{Nec}\left(\text{Cap} - \sum_{j=1}^{N} \tilde{d}_j y_{ik} \geq 0\right) \geq \beta, \quad k = 1, \ldots, N; \quad \beta \in [0, 1]. \tag{12}$$

and for triangular fuzzy numbers $\tilde{d}_j = (\underline{d}_j, \hat{d}_j, \bar{d}_j)$ the fuzzy number

$\left(\text{Cap} - \sum_{j=1}^{N} \tilde{d}_j y_{ik}\right)$ can be presented also as triangular fuzzy number as follows

$$\text{Cap} - \sum_{j=1}^{N} \tilde{d}_j y_{ik} = \left(\text{Cap} - \sum_{j=1}^{N} \bar{d}_j y_{ik}, \text{Cap} - \sum_{j=1}^{N} \hat{d}_j y_{ik}, \text{Cap} - \sum_{j=1}^{N} \underline{d}_j y_{ik}\right). \tag{13}$$

The mathematical models for possibility $\text{Pos}(\text{Serve}\,\tilde{D}_k)$ and necessity $\text{Nec}(\text{Serve}\,\tilde{D}_k)$ that the capacity Cap of the tanker is sufficient to serve all demands $\tilde{D}_k$ on k-th route are

$$
\text{Pos}(\text{Serve}\,\tilde{D}_k) = \begin{cases} 1, & \text{Cap} \geq \sum_{j=1}^{N} \hat{d}_j y_{jk} \\[2mm] \dfrac{\text{Cap} - \sum_{j=1}^{N} d_j y_{jk}}{\sum_{j=1}^{N} (\hat{d}_j - \underline{d}_j)\, y_{jk}}, & \sum_{j=1}^{N} \underline{d}_j y_{jk} < \text{Cap} \leq \sum_{j=1}^{N} \hat{d}_j y_{jk}, \\[2mm] 0, & \text{Cap} < \sum_{j=1}^{N} \underline{d}_j y_{jk} \end{cases} \tag{14}
$$

$$
\text{Nec}(\text{Serve}\,\tilde{D}_k) = \begin{cases} 1, & \text{Cap} \geq \sum_{j=1}^{N} \overline{d}_j y_{jk} \\[2mm] \dfrac{\text{Cap} - \sum_{j=1}^{N} \hat{d}_j y_{jk}}{\sum_{j=1}^{N} (\overline{d}_j - \hat{d}_j)\, y_{jk}}, & \sum_{j=1}^{N} \hat{d}_j y_{jk} < \text{Cap} \leq \sum_{j=1}^{N} \overline{d}_j y_{jk}. \\[2mm] 0, & \text{Cap} < \sum_{j=1}^{N} \hat{d}_j y_{jk} \end{cases} \tag{15}
$$

According to [42] the following relation applies

$$
\text{Pos}(\text{Serve}\,\tilde{D}_k) < 1 \Rightarrow \text{Nec}(\text{Serve}\,\tilde{D}_k) = 0 \tag{16}
$$

and it is more demanding to request the necessity to be greater than 0 than to request the possibility to be less or equal to 1.

For $\alpha > 0$ we can model the following constraints as crisp equivalents for the fuzzy constraint (2):

$$
\text{Pos}(\text{Serve}\,\tilde{D}_k) \geq \alpha \Leftrightarrow \sum_{j=1}^{N} \left(\alpha\,\hat{d}_j + (1-\alpha)\underline{d}_j\right) y_{jk} \leq \text{Cap}, k = 1, \ldots, K, \alpha \in (0, 1] \tag{17}
$$

and for $\beta > 0$, respectively:

$$
\text{Nec}(\text{Serve}\,\tilde{D}_k) < \beta \Leftrightarrow \sum_{j=1}^{N} \left(\beta\overline{d}_j + (1-\beta)\hat{d}_j\right) y_{jk} \leq \text{Cap}, k = 1, \ldots, K, \beta \in (0, 1] \tag{18}
$$

To solve this fuzzy mathematical programming model, it is suggested to determine the optimal solutions with respect to a given degree of possibility α or even stronger a given degree of necessity β that the capacity Cap is sufficient to meet the total demand $\tilde{D}_k$ of the served ships on each of the K routes.

3.2 Multi-criteria Optimization for CVRPFD Based on Fuzzy Approach

A bunkering company usually wants to serve all the demands and sell as much fuel as possible. Sales are restricted by the demands and by the capacity of the tanker for a route. A solution is preferable if the amount of the demand served is high. To maximize sales in this considered fuzzy context means to determine and maximize a fuzzy set which depends on the fuzzy demand, the route and the capacity of the tanker.

The fuzzy set sales $\tilde{S}_k$ on route k can be calculated as the minimum of demand $\tilde{D}_k$ and the capacity of the tanker. The membership results as

$$\mu_{\tilde{S}_k}(x) = \begin{cases} \mu_{\tilde{D}_k}(x), & x < Cap \\ Pos\ (\tilde{S}_k \geq Cap), & x = Cap. \\ 0, & x > Cap \end{cases} \tag{19}$$

It is suggested to use the following defuzzyfication method [40–42, 44] to determine the crisp approximation $D\tilde{S}_k$ for the sales on the k-th route

$$D\tilde{S}_k = \frac{1}{3}\left(\min\left\{\sum_{j=1}^{N} \underline{d}_j y_{jk}, Cap\right\} + \min\left\{\sum_{j=1}^{N} \hat{d}_j y_{jk}, Cap\right\} + \min\left\{\sum_{j=1}^{N} \bar{d}_j y_{jk}, Cap\right\}\right) \tag{20}$$

Let $\sum_{k=1}^{K} \tilde{S}_k$ be a fuzzy set of the total sales (depended on all K routes) with corresponding membership function

$$\mu_{\sum \tilde{S}_k}(z) = \sup_{\sum_{k=1}^{K} x_k = z} \min_{k=1}^{K}\left\{\mu_{\tilde{S}_k}(x_k)\right\}. \tag{21}$$

and $\sum_{k=1}^{K} D\tilde{S}_k$ is a crisp approximation of total sales $\sum_{k=1}^{K} \tilde{S}_k$.

The fuzzy multi-criteria approach for mathematical formalization of such kind of corresponding optimization problems for tankers CVRP and trucks CVRP are presented in [41, 42]. The multi-criteria model in a fuzzy context has to be considered in more detail. It is based on the minimization of total length of planned tanker's routes

$$\min Z^1(x, y, u) = \sum_{i=0}^{N} \sum_{j=0}^{N} \sum_{k=1}^{K} c_{ij} x_{ijk} \tag{22}$$

and the maximization of total value of unloaded cargo (fuel) on all routes

$$\max Z^2(x, y, u) = \sum_{k=1}^{K} , \min\left\{ \sum_{j=1}^{N} \tilde{d}_j y_{jk}, Cap \right\}, \tag{23}$$

according to constraints (3)–(8), fuzzy relations (11), (12) and their crisp equivalents (17), (18), where x, y and u stand for the vectors of variables in the considered model and operator $(\min)$ means the extended minimum of the two fuzzy sets $\tilde{D}_k$ and $Cap \Leftrightarrow C\tilde{a}p = (Cap, Cap, Cap)$. For this model we assume that the number of vehicles is not restricted, the fleet of vehicles is homogeneous and each single demand is less than the capacity of the vehicle Cap.

For finding the optimal solution of CVRPFD (22), (23), (11), (12), (3)–(8) the modified iterative method developed in [12, 13] is used.

Step 1. The optimization criterion is (22) and simultaneously criterion (23) can be transformed to constraint

$$\sum_{j=1}^{N} \underline{d}_j y_{jk} \le Cap, \quad k = 1, \ldots, K, \tag{24}$$

that means that at least the lower bound of the fuzzy demand $\sum_{k=1}^{K} \sum_{j=1}^{N} \underline{d}_j y_{jk}$ is served. The individual optimum of Z^1 (the minimal total distance) $Z^1(x^1, y^1, u^1) = Z^{1*}$ can be determined by the solution algorithm and corresponding optimal parameters are x_{ijk}^1, y_{jk}^1 and u_{jk}^1, where i, j = 0,...,N; k = 1,...,K.

Step 2. The value $Z^2(x^1, y^1, u^1) = Z_*^2$ should be calculated using (23) taking into account that

$$Z^2(x^1, y^1, u^1) \ge \sum_{k=1}^{K} \sum_{j=1}^{N} \underline{d}_j y_{jk}. \tag{25}$$

Step 3. Transform the second criterion (23) using the crisp approximation $D\tilde{S}_k$ (20) to the following form

$$\max Z^2(x, y, u) = \sum_{k=1}^{K} D\tilde{S}_k \ge Z^2(x^1, y^1, u^1). \tag{26}$$

To determine the individual optimum of Z^2 and fuzzy efficient solution it is necessary to consider all alternative solutions (x^2, y^2, u^2) and choose best solutions with the fulfilment of condition

$$\min Z^1(x^2, y^2, u^2) \le Z^1(x^1, y^1, u^1). \tag{27}$$

In this case optimal solutions according to criterion (23) will satisfy constraint (18) for the strongest condition: $\beta = 1$ and, practically, a fuzzy efficient individual optimum Z^2 can be found solving model (1) with constraints (3)–(8) and with constraint (2) modified as follows

$$\sum_{j=1}^{N} \bar{d}_j y_{jk} \leq Cap, \quad k = 1, \ldots, K. \tag{28}$$

As a result, the optimal solution is (x^2, y^2, z^2) for which $Z^2(x^2, y^2, z^2) = Z^{2*}$ and $Z^1(x^2, y^2, u^2) = Z^1_*$.

Step 4. Consider goal functions (22) and (23) as fuzzy sets and define their membership functions $\mu_{Z^1}(x)$, $\mu_{Z^2}(x)$ using individual optimal values Z^{1*}, Z^{2*} and pessimistic solutions Z^1_*, Z^2_*:

$$\mu_{Z^1}(x) = \begin{cases} 1, & Z^1(x, y, u) \leq Z^{1*} \\ \dfrac{Z^1_* - Z^1(x, y, u)}{Z^1_* - Z^{1*}}, & Z^{1*} < Z^1(x, y, u) < Z^1_*, \\ 0, & Z^1(x, y, u) \geq Z^1_* \end{cases} \tag{29}$$

$$\mu_{Z^2}(x) = \begin{cases} 1, & Z^2(x, y, u) \geq Z^{2*} \\ \dfrac{Z^2(x, y, u) - Z^2_*}{Z^{2*} - Z^2_*}, & Z^2_* < Z^2(x, y, u) < Z^{2*}. \\ 0, & Z^2(x, y, u) \leq Z^2_* \end{cases} \tag{30}$$

Step 5. Set the first compromise model in the following way

$$\max \lambda \tag{31}$$

subject to

$$\left(Z^1_* - Z^{1*}\right)\lambda + \sum_{i=0}^{N}\sum_{j=0}^{N}\sum_{k=1}^{K} c_{ij} x_{ijk} \leq Z^1_*, \tag{32}$$

$$\left(Z^2_* - Z^{2*}\right)\lambda + \sum_{k=1}^{K} D \tilde{S}_k \geq Z^2_*, \tag{33}$$

$$\lambda \in [0, 1] \tag{34}$$

and constraints (3)–(8).

Step 6. An equivalent linear model is given by constraints (31), (32), (34), (3)–(8) and substituting the following linear constraints for the nonlinear constraint (33)

$$\left(Z^2_* - Z^{2*}\right)\lambda + \sum_{k=1}^{K} \frac{\left(\underline{g}_k + \hat{g}_k + \bar{g}_k\right)}{3} \geq Z^2_*, \tag{35}$$

$$\underline{g}_k \leq \sum_{j=1}^{N} \underline{d}_j y_{jk}, \quad k = 1, \ldots, K, \tag{36}$$

$$\hat{g}_k \leq \sum_{j=1}^{N} \hat{d}_j y_{jk}, \quad k = 1, \ldots, K, \tag{37}$$

$$\bar{g}_k \leq \sum_{j=1}^{N} \bar{d}_j y_{jk}, \quad k = 1, \ldots, K, \tag{38}$$

$$0 \leq \underline{g}_k, \hat{g}_k, \bar{g}_k \leq \text{Cap}, \quad k = 1, \ldots, K. \tag{39}$$

Step 7. To solve the linear optimization model of CVRPFD in *Step* 6 with (31), (32), (34), (3)–(8) and (35)–(39) it is necessary to determine:

(a) all nodes (ports, served ships or destination points) on every route k, k = 1,..., K and the set of completed routes;
(b) the length L_k of each route k (total distance), k = 1,...,K;
(c) the calculated values of sales $D\tilde{S}_k$ for the realization of each route k, k = 1,..., K;
(d) the total length (distance) $L_{\text{total}} = \sum_{k=1}^{K} L_k = Z_1$ for the realization of all planned routes;
(e) total sales $\sum_{k=1}^{K} D\tilde{S}_k = Z_2$ for the realization of all planned routes.

Step 8. Solve the local TSP (Travelling Salesperson Problem) separately for the nodes of each preliminary planned route k, k = 1,...,K. So the length for each route is minimized with fixed values of sales by using any well-known TSP exact algorithms, TSP heuristics [27, 44] or evolutionary optimization algorithms [35] (genetic algorithm, bio-geography optimization algorithm, etc.) depending on the number of nodes at a route. As a result, the optimal or nearly optimal length L_k^{opt} of each route k of the corresponding TSP is ascertained.

Step 9. The total optimal length for the realization of all optimized routes is $L_{\text{total}}^{opt} = \sum_{k=1}^{K} L_k^{opt} = Z_1^{opt}$.

Step 10. If $L_{\text{total}}^{opt} < L_{\text{total}}$ then the sequence and consequences of port's visits for tankers have to be changed according to the optimization process in *Step* 8, providing additional minimization of total length of all planned routes, else for $L_{\text{total}}^{opt} = L_{\text{total}}$, no change is required.

3.3 Discussion of Modelling Results with Additional Route Optimization

Using the above-considered optimization models and the presented iterative algorithm, all routes and sales for the same input data as in the example published in

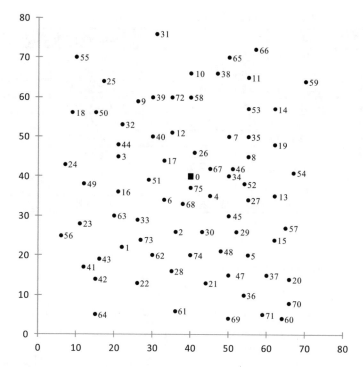

Fig. 1 Location of ports and depot

[41] are determined. The CVRPFD (Fig. 1) consists of 75 ports, which are situated at locations characterized by their corresponding coordinates in the plane $\left(x_j, y_j\right), j = 1, \ldots, 75$ and with Deport 0 with coordinates (40, 40). The tanker fleet is homogeneous with capacity Cap = 160 of each tanker and fuzzy demands are identical to those in [41].

The modelling results after the first seven steps of the suggested iterative algorithm are presented in Table 1. The number of planned routes is 10.

The total distance

$$L_{total} = \sum_{k=1}^{K} L_k = Z_1$$

and total sales

$$\sum_{k=1}^{K} D\tilde{S}_k = Z_2$$

Table 1 Modelling results for routes 1–10

Route no.	Port sequence	Remaining fuel $\left(\text{Cap} - \sum \tilde{d}_i\right)$
Route 1	0-16-3-44-32-9-39-12-0	$(-5,24,53)$
Route 2	0-75-45-29-48-30-2-6-0	$(-8,19,46)$
Route 3	0-52-27-13-54-19-53-38-58-0	$(-12,14,40)$
Route 4	0-26-7-8-46-34-4-0	$(18,35,52)$
Route 5	0-5-47-36-69-71-60-70-20-37-15-57-0	$(-15,15,45)$
Route 6	0-67-17-40-51-0	$(56,65,74)$
Route 7	0-68-74-21-28-62-1-73-33-0	$(-12,14,40)$
Route 8	0-63-23-56-43-41-42-64-22-61-0	$(-9,18,45)$
Route 9	0-35-14-59-66-65-11-0	$(-3,12,27)$
Route 10	0-72-10-31-25-55-18-50-24-49-0	$(-9,20,49)$

Table 2 Characteristics of the routes 1–10

Route no.	Length L_k	Fuzzy demands $\tilde{D}_k$			Sales $D\tilde{S}_k$
		$\underline{D}_k$	$\hat{D}_k$	$\bar{D}_k$	
Route 1	70.23	107	136	165	134.333
Route 2	57.91	114	141	168	138.333
Route 3	94.4	120	146	172	142.0
Route 4	44.38	108	125	142	125.0
Route 5	118.03	115	145	175	140.0
Route 6	46.37	86	95	104	95.0
Route 7	80.92	120	146	172	142.0
Route 8	126.82	115	142	169	139.0
Route 9	97.32	133	148	163	147.0
Route 10	139.98	111	140	169	137.0

are calculated based on the modelling data from each planned route, including length L_k, fuzzy demands $\tilde{D}_k = \left(\underline{D}_k, \hat{d}_k, \bar{D}_k\right)$ and sales $D\tilde{S}_k, k = 1, \ldots, N$ for each route (Table 2).

All modelling results are obtained for $\alpha = 1$ and different values of β. The resulting values after the first seven steps of the suggested iterative algorithm for CVRPFD are

$$L_{total} = 876.36 \approx 876,$$

$$\sum_{k=1}^{N} D\tilde{S}_k = 1339.67 \approx 1340.$$

Table 3 Final versions of improved routes after TSP-optimization on the steps 8–10 of the proposed iterative algorithm

Route No.	Port sequence	Length L_k^{opt}
Route 7	0-68-74-21-28-62-**73-1**-33-0	80.63
Route 8	0-63-23-56-**41-43**-42-64-22-61-0	126.02
Route 10	0-72-10-31-**55-25-50-18**-24-49-0	135.48

Then for continuing the decision making process with the goal to find the best solution it is necessary to realize the TSP-improvement by *Steps* 8–10.

Using a 2-opt heuristic [27, 44] the same node sequences for routes 1,2,3,4,5,6,9 result. But routes 7, 8 and 10 are improved in this case (Table 3).

The lengths of the improved routes are

$$L_7^{opt} = 80.63, L_8^{opt} = 126.02, L_{10}^{opt} = 135.48$$

and total optimized distance for all routes is

$$L_{total}^{opt} = 870.77$$

taking into account that

$$D\tilde{S}_7 = 142, D\tilde{S}_8 = 139, D\tilde{S}_{10} = 137,$$

$$\sum_{k=1}^{N} D\tilde{S}_k = 1339.67 = \text{const.}$$

Analyzing all modelling results it can be seen that after 2-opt TSP- improvement (*Steps* 8,9) (Fig. 2)

$$L_7^{opt} < L_7, \ L_8^{opt} < L_8, L_{10}^{opt} < L_{10}$$

and the above-mentioned condition of *Step* 10

$$L_{total}^{opt} < L_{total}$$

is satisfied. Finally, it is necessary to choose the optimized routes (Table 3) for the practical realization of CVRPFD taking into account that

$$\Delta L_{total}^{opt} = L_{total} - L_{total}^{opt} \approx 5.59 > 0.$$

This value illustrates the improved result compared to the modelling results in [14].

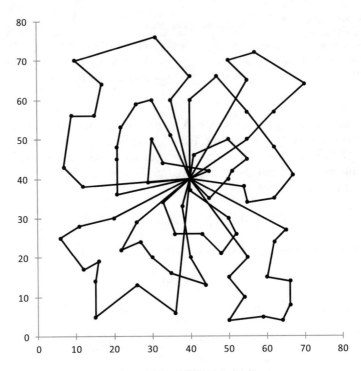

Fig. 2 Final CVRPFD solution after additional TSP-optimization

4 Second Alternative Fuzzy Approach Based on Evaluation of Satisfaction Value

4.1 Description of Conflict Situation in Route Planning

For implementation of the second alternative fuzzy approach it is suggested to form the intelligent model and planning algorithm for CVRPFD in the following way.

First of all, it is necessary to form mathematical models of fuzzy demands $q_j = \left(\underline{q}_j, \hat{q}_j, \bar{q}_j\right), j = 1 \ldots N$ at nodes by the expert evaluation method. Three examples of demands $q_j = \left(\underline{q}_j, \hat{q}_j, \bar{q}_j\right), j = 1 \ldots N$ are represented by fuzzy sets $\underset{\sim}{A}, \underset{\sim}{B}$ and $\underset{\sim}{C}$ and shown in Fig. 3, where fuzzy set $\underset{\sim}{A}$ is a model of fuzzy demand of type "not less than 150 ", fuzzy set $\underset{\sim}{B}$ is a model of fuzzy demand of type "not more than 500", fuzzy set $\underset{\sim}{C}$ is a model of fuzzy demand of type "about 350" or "between 250 and 450", $\mu_X(q)$ is a membership function, $X = \left\{\underset{\sim}{A}, \underset{\sim}{B}, \underset{\sim}{C}\right\}$.

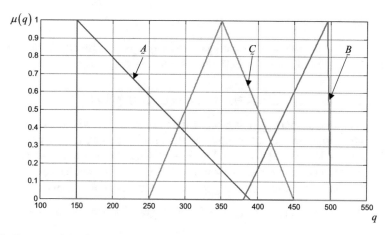

Fig. 3 Fuzzy models of uncertain demands

In the next step it is necessary to solve the Traveling Salesperson Problem for the total number of nodes N using one of the heuristic algorithms [27, 45], such as sweep-algorithm, tabu search algorithm, genetic algorithm, bio-geography optimization algorithm, simulated anneling, ant colony optimization algorithm or others. For example, Clarke and Write savings algorithm is based on calculating value s_{ij} for each pair of nodes (i,j):

$$s_{ij} = L_{0i} + L_{0j} - L_{ij}, (i = 1 \dots N; j = 1 \dots N; i \neq j), \tag{40}$$

The result of TSP solving depends on the number of nodes and on the selected heuristic. For example, for 35 nodes $\{12, 13, 14, \dots, 45, 46\}$ with coordinates, presented in Fig. 1 [41], and central depot with coordinates ($x_0 = 40$, $y_0 = 40$) TSP solutions are:

(a) for Clarke and Write saving algorithm:

[0 − 42 − 41 − 43 − 23 − 24 − 18 − 25 − 31 − 38 − 14 − 19 − 35 − 13 − 15 − 20 − 37 − 36 − 21 − 28 − 22 − 33 − 16 − 44 − 32 − 39 − 40 − 12 − 17 − 26 − 46 − 34 − 27 − 29 − 45 − 30 − 0]

and the total length of this Hamiltonian circuit is 386,98;

(b) for sweep algorithm:

[0 − 38 − 14 − 35 − 19 − 46 − 34 − 13 − 27 − 15 − 20 − 45 − 29 − 37 − 36 − 30 − 21 − 28 − 22 − 42 − 43 − 41 − 33 − 23 − 16 − 24 − 44 − 18 − 17 − 32 − 40 − 25 − 39 − 12 − 31 − 26 − 0]

and the total length of this Hamiltonian circuit is 486,80;

(c) for ant colony algorithm:

$[0 - 26 - 17 - 12 - 40 - 32 - 44 - 16 - 33 - 30 - 29 - 45 - 27 - 13 - 15 - 20 - 37 - 36 - 21 - 28 - 22 -$
$42 - 41 - 43 - 23 - 24 - 18 - 25 - 39 - 31 - 38 - 14 - 19 - 35 - 46 - 34 - 0]$

and the total length of this Hamiltonian circuit is 343,74.

The procedure of route planning is based on the used TSP-solution (Hamiltonian circuit) which was created at the previous step. It is necessary to find nodes which should be included in the corresponding route taking into account possibilities for service of each node with fuzzy demands and constraint tanker capacity.

Figure 4 shows the route planning procedure for 1-st node's demand q_1, Fig. 5 that for 2-nd node's demand q_2 and Fig. 6 applies for 3-rd node's demand q_3, where

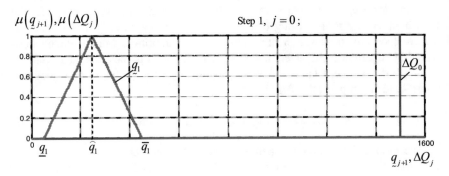

Fig. 4 Procedure of route planning (for 1-st node in TSP-solution)

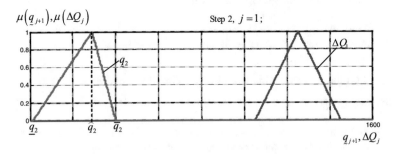

Fig. 5 Procedure of route planning (for 2-nd node in TSP-solution)

$\Delta Q_0 = \left(\Delta \underline{Q}_0, \Delta \hat{Q}_0, \Delta \bar{Q}_0 \right)$ is the initial value (Fig. 4) of cargo capacity of a tanker, which is presented as crisp number D in the style of triangular fuzzy number $D = \Delta D_0 = (D, D, D)$;

$\Delta Q_1 = \left(\Delta \underline{Q}_1, \Delta \hat{Q}_1, \Delta \bar{Q}_1 \right)$ is remaining cargo capacity on tanker (Fig. 5) after including 1-st node to the route, $\Delta Q_1 = \Delta Q_0 - Q_1$;

$\Delta Q_2 = \left(\Delta \underline{Q}_2, \Delta \hat{Q}_2, \Delta \bar{Q}_2 \right)$ is remaining cargo capacity (Fig. 6) on tanker after including 2-nd node to the route, $\Delta Q_2 = \Delta Q_1 - Q_2$;

$\Delta Q_3 = \left(\Delta \underline{Q}_3, \Delta \hat{Q}_3, \Delta \bar{Q}_3 \right)$ is remaining cargo on tanker after including 3-rd node to the route, $\Delta Q_3 = \Delta Q_2 - q_3$.

All first three nodes (Figs. 4, 5, and 6), a sequence of which was taken from TSP solution, are included in the planning route because fuzzy values of remaining cargo on the tanker are higher than the fuzzy values of corresponding demands (Steps 1, 2, 3).

The conflict situation always appears during the route planning process for CVRP on the corresponding step and it is necessary to make a decision about including or excluding the corresponding "conflict" node in the route.

Let's denote, that the conflict situation for the $(j+1)$-th port-applicant appears on condition that the highest value $\bar{q}_{j+1}$ of the fuzzy demand $q_{j+1} = \left(\underline{q}_{j+1}, \hat{q}_{j+1}, \right.$ $\left. \bar{q}_{j+1} \right)$ is higher than the lowest value $\left(D - \sum_{s=1}^{j} \bar{q}_s \right)$ of the fuzzy remaining tanker's cargo quantity $\Delta Q_j = \left(D - \sum_{s=1}^{j} \bar{q}_s, D - \sum_{s=1}^{j} \hat{q}_s, D - \sum_{s=1}^{j} \underline{q}_s \right)$, or in other words, in the situation when

$$\underline{q}_{j+1} < \left(D - \sum_{s=1}^{j} \bar{q}_s \right).$$

Figure 7 shows the conflict situation during route planning procedure for 4-th node (Step 4). During the planning procedure for the current route (Figs. 4, 5 and 6), fuzzy values of remaining tanker's cargo $\Delta \underline{Q}_j = \left(D - \sum_{s=1}^{j} \bar{q}_s, D - \sum_{s=1}^{j} \hat{q}_s, \right.$ $\left. D - \sum_{s=1}^{j} \underline{q}_s \right)$ step by step (Steps 1, 2, 3) shift to the left. Triangular fuzzy number ΔQ_j is crossed with a triangular fuzzy number q_{j+1} (Fig. 7, $j = 3$) on the 4-th Step or placed to the left of it (as for fuzzy demand $q_{j+1}' = underline q_4'$ of $(j+1)$-th node, shown in Fig. 7 by dashed lines).

In the last case of intersection of fuzzy sets ΔQ_j and q_{j+1}, the possibility degree of demand implementation decreases with increasing degree of shifting triangular fuzzy number ΔQ_j to the left.

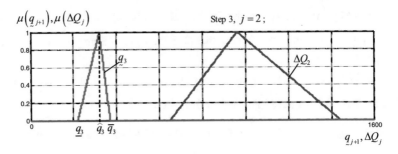

Fig. 6 Procedure of route planning (for 3-rd node in TSP-solution)

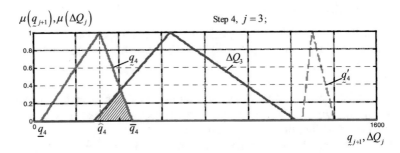

Fig. 7 The conflict situation in route planning procedure for 4-th node in TSP-solution

4.2 Fuzzy Decision-Making Based on Satisfaction Value

It is propose to make a decision about including considered port P_{j+1} in the current route (using developed fuzzy knowledge-based system—FKBS) based on the condition [16, 17]

$$\lambda_{j+1} \geq \lambda^*, \tag{41}$$

where λ^* is a desired satisfaction (preference) value; λ_{j+1} is a current value of satisfaction level. Such satisfaction level λ_{j+1} can be calculated at each serviced port (P_j) as possible level of the satisfaction of required order for the next serviced port (P_{j+1}) with its fuzzy demand q_{j+1}.

The developed decision-making algorithm based on fuzzy logic [16, 17] can be presented in the following way:

$$\lambda_{j+1} = FKBS\Big[FRB\Big(\underset{\sim}{x}_1, \underset{\sim}{x}_2, x_3\Big)\Big]. \tag{42}$$

In (42) we used the following notations:

- FKBS is a fuzzy knowledge–based system;
- FRB is a fuzzy rule base with the following structure of fuzzy rules. For example,

"**IF** (input signal x_1 is *Low*) **AND** (input signal x_2 is *Middle*) **AND** (input signal x_3 is *High*) **THEN** (output signal λ_j is *High*)";

- λ_{j+1} is a value of satisfaction level for each alternative decision-making;
- $x_1 = \tilde{q}_{j+1}/\Delta\tilde{Q}_i$ is the ration of fuzzy demands of the next $(j+1)$-th port $\tilde{q}_{j+1}$ to the fuzzy value of the remaining tanker cargo $\Delta\tilde{Q}_i$;
- $x_2 = \Delta\tilde{Q}_i/Q_i$—ration of fuzzy numbers $\Delta\tilde{Q}_i$ to tanker capacity Q_i;
- $x_3 = L_1/L_2$ is the ration of length L_1 to L_2 of two alternative routes R_1 and R_2 (R_1 is the route with 1st level of search of the next route candidate port and R_2 is the route of the 2nd level of search);
- $\Delta\tilde{Q}_i = \left(Q_i - \sum_{j=1}^{k} \bar{q}_j,\ Q_i - \sum_{j=1}^{k} \hat{q}_j,\ Q_i - \sum_{j=1}^{k} \underline{q}_j \right)$ is a fuzzy value of the remaining tanker cargo, where k is the number of served ships on the i-th route before the current decision; $\{Low, Middle, High\}$ is a set of the corresponding linguistic terms for input x_1, x_2, x_3 and output λ_{j+1} signals.

The characterized surface

$$Surf(x_1, x_3), x_2 = const,$$

of the fuzzy rule base FRB (42) for fixed input signal $x_2 = Middle$ is presented in Fig. 8.

If the condition (41) is correct for the 4-st node (Fig. 7), then this node will be included in the current planning route. Next node in the TSP-solution will be a first node in the next planning route and so on.

After planning all routes according to the proposed second fuzzy approach it is necessary to realize an additional optimization procedure by solving TSP for each separate route $R_s, (s = 1 \ldots r)$ that provides the minimization of the length of each separate route, as well as the total length of all routes $L_{\Sigma s}, (s = 1 \ldots r)$.

The final solution of the CVRPFD can be implemented for the practical realization of the respective bunkering program for r tankers.

Modelling results [16, 17] confirm the efficiency of the proposed fuzzy approach based on FKBS application as second alternative for solving CVRPFD.

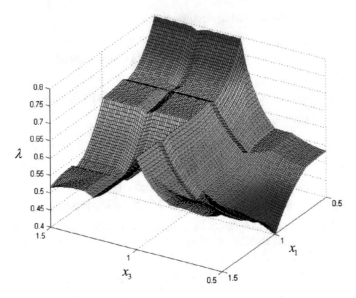

Fig. 8 Characteristic surface of FKBS (42) with fixed value of input signal x_2

5 Conclusions

The suggested theoretical approach, based on the fuzzy multi-criteria models and iterative multistage algorithm, allows to receive in a fuzzy context optimal solutions for the CVRPFD. Modelling results confirm the efficiency of the suggested multi-objective optimization approach as first alternative. Using FKBS is the base for the second alternative for efficiently solving CVRPFD. Nevertheless, successful implementation of the second fuzzy approach significantly depends on the choice of the desired satisfaction value λ^*. In future research work it is appropriate to extend the proposed models and algorithm for CVRPFD by solving time-windows problem.

Acknowledgements The authors gratefully acknowledge the support of this research work by the Ruhr University Bochum and Deutscher Akademischer Austauschdienst (DAAD), Germany, by awarding one of the author with the research 2000 fellowship and research 2010-2011 fellowship.

References

1. Christiansen, M., Fagerholt, K., Nygreen, B., Ronen, D.: Maritime transportation. In: Barnhart, C., Laporte, G. (eds.) Handbook in OR & MS, vo. 14, pp. 189–284. Elsevier (2007)
2. Gil-Aluja, J.: Investment in Uncertainty. Kluwer Academic Publishers, Dordrecht, Boston, London (1999)

3. Gil-Aluja, J.: Elements for a Theory of Decision in Uncertainty, vol. 32. Springer Science & Business Media (1999)
4. Gil-Aluja, J.: Fuzzy Sets in the Management of Uncertainty, vol.145. Springer Science & Business Media (2004)
5. Gil-Aluja, J.: The Interactive Management of Human Resources in Uncertainty, vol. 11. Springer Science & Business Media (2013)
6. Gil-Aluja, J.: Handbook of Management Under Uncertainty, vol. 55. Springer Science & Business Media (2013)
7. Gil-Aluja, J., Gil-Lafuente, A.M., Klimova, A.: The optimization of the economic segmentation by means of fuzzy algorithms. J. Comput. Opt. Econ. Finance (Nova Science Publishers) 1(3), 169–186 (2008)
8. Gil-Aluja, J., Gil-Lafuente, A.M., Merigó, J.M.: Using homogeneous groupings in portfolio management. Expert Syst. Appl. 38(9), 10950–10958 (2011)
9. Gil Aluja, J. (ed.): Les Universitats En El Centenari Del Futbol Club Barcelona. Estudis En L'Ambit De L'Esport, Proleg, Josef Lluis Nunez (1999)
10. Gil Lafuente, A.M.: Fuzzy Logic in Financial Analysis. Studies in Fuzziness and Soft Computing, vol. 175. Springer, Berlin (2005)
11. Gil-Lafuente, A.M., Zopounidis, C. (eds.): Decision Making and Knowledge Decision Support Systems, Lecture Notes in Economics and Mathematical Systems, vol. 675. Springer (2015)
12. Halvorsen-Weare, E.E., Fagerholt, K.: Routing and scheduling in a liquefied natural gas shipping problem with inventory and berth constraints. Ann. Oper. Res. (Springer) (2010)
13. Jamison, K.D., Lodwick, W.A.: Minimizing unconstraint fuzzy functions. Fuzzy Sets Syst. 103, 457–464 (1999)
14. Jamshidi, M., Kreinovich, V., Kacprzyk, J. (eds.): Advance Trends in Soft Computing. Series: Studies in Fuzziness and Soft Computing, vol. 312. Springer (2013)
15. Kauffman, A., Gil-Aluja, J.: Introduction of fuzzy sets theory to management of enterprises. Minsk, Higher School (1992). (in Russian)
16. Kondratenko, Y.P., Werners, B., Kondratenko, G.V.: Fuzzy models and algorithms for solving marine routing problem using values of statistical level. J. Model. Measur. Control AMSE Period. Ser. D 28(2), 47–59 (2007)
17. Kondratenko, G.V., Kondratenko, Y.P., Romanov, D.O.: Fuzzy models for capacitive vehicle routing problem in uncertainty. In: Proceeding of the 17th International DAAAM Symposium "Intelligent Manufacturing and Automation", Vienna, Austria, pp. 205–206 (2006)
18. Kondratenko, Y.P., Encheva, S.B., Sidenko E.V.: Synthesis of intelligent decision support systems for transport logistic. In: Proceeding of the 6th IEEE International Conference on Intelligent Data Acquisition and Advanced Computing Systems: Technology and Applications (IDAACS'2011), vol. 2, Prague, Czech Republic, Sept. 15–17, pp. 642–646 (2011)
19. Kondratenko, Y.P., Klymenko, L.P., Al Zu'bi, E.Y.M.: Structural optimization of fuzzy systems' rules base and aggregation models. Kybernetes 42(5), 831–843 (2013)
20. Kondratenko, Y.P., Kondratenko, N.Y.: Soft computing analytic models for increasing efficiency of fuzzy information processing in decision support systems. In: Hudson, R. (ed.) Decision Making: Processes, Behavioral Influences and Role in Business Management, pp. 41–78. Nova Science Publishers, New York (2015)
21. Kondratenko, Y., Kondratenko, V.: Soft computing algorithm for arithmetic multiplication of fuzzy sets based on universal analytic models. In: Ermolayev, V., et al. (eds.) Information and Communication Technologies in Education, Research, and Industrial Application. Communications in Computer and Information Science, vol. 469, ICTERI'2014, pp. 49–77. Springer International Publishing, Switzerland (2014)
22. Kondratenko, Y.P., Sidenko, Ie.V.: Decision-making based on fuzzy estimation of quality level for cargo delivery. In: Zadeh, L.A., Abbasov, A.M., Yager, R.R., Shahbazova, S.N., Reformat, M.Z. (eds.) Recent Developments and New Directions in Soft Computing. Studies in Fuzziness and Soft Computing, vol. 317, pp. 331–344. Springer International Publishing, Switzerland (2014)

23. Kondratenko, Y., Klymenko, L., Yemelyanov, V., Datsy, O., Koretskiy, N., Gil Lafuente, J., Luciano, E.V., Molina, L.A., Reverter, S.B., Merigo Lindahl, J.M., Klimova, A., Moro, L.S.: Explorando Nuevos Mercados: Ucrania. Real Academia de Ciencias Economicas y Financieras, Monograph. Directora Anna Maria Gil Lafuente. Barcelona (2012)

24. Kondratenko, Y.P., Klymenko, L.P., Sidenko, Ie.V.: Comparative analysis of evaluation algorithms for decision-making in transport logistics. In: Jamshidi, M., Kreinovich, V., Kazprzyk, J. (eds.) Advance Trends in Soft Computing, Studies in Fuzziness and Soft Computing, vol. 312, pp. 203–217. Springer (2014)

25. Kondratenko Y.P., Al Zubi, E.Y.M.: The optimisation approach for increasing efficiency of digital fuzzy controllers. In: Annals of DAAAM for 2009 & Proceeding of the 20th International DAAAM Symposium "Intelligent Manufacturing and Automation", pp. 1589–1591. DAAAM International, Vienna, Austria (2009)

26. Kondratenko, Y.P., Korobko, O.V., Kondratenko, V.Y., Kozlov, O.V.: Optimization models and algorithms of multistage processes of liquid cargoes transportation for computer DSS. In: Armborst, K., Degel, D., Lutter, P., Pietschmann, U., Rachuba, S., Shultz, K., Wiesche, L. (eds.) Management Science: Modelle und Methoden zur quantitativen Entscheidungsunterstutzung. Festschrift zum 60. Geburtstag von Brigitte Werners, pp. 241–270. Verlag Dr. Covac, Hamburg (2013)

27. Kondratenko, Y.P.: Optimisation Problems in Marine Transportation. Incidencia de las relaciones economicas internacionales en la recuperacion economica del area mediterranea. VI Acto Internacional celebrado en Barcelona el 24 de febrero de 2011, pp. 43–52. Real Academia de Ciencias Economicas y Financieras, Barcelona (2011)

28. Laporte, G.: The travelling salesman problem: an overview of exact and approximate algorithms. Eur. J. Oper. Res. **59**, 231–248 (1992)

29. Laporte, G.: The vehicle routing problem: an overview of exact and approximate algorithms. Eur. J. Oper. Res. **59**(3), 345–358 (1992)

30. Lodwick, W.A., Kacprzhyk, J. (eds.): Fuzzy Optimization. Studies in Fuzziness and Soft Computing, vol. 254. Springer, Berlin, Heidelberg (2010)

31. Merigó, J.M., Gil-Lafuente, A.M.: New decision-making techniques and their application in the selection of financial products. Inf. Sci. (2010). https://doi.org/10.1016/j.ins.2010.01.028

32. Merigo, J.M., Gil-Lafuente, A.M., Gil-Aluja, J.: Decision making with the induced generalized adequacy coefficient. Appl. Comput. Math. **2**(2), 321–339 (2011)

33. Merigó, J.M., Gil-Lafuente, A.M., Gil-Aluja, J.: A new aggregation method for strategic decision making and its application in assignment theory. Afr. J. Bus. Manag. **5**(11), 4033–4043 (2011)

34. Merigó, J.M., Gil-Lafuente, A.M.: The generalized adequacy coefficient and its application in strategic decision making. Fuzzy Econ. Rev. **13**, 17–36 (2008)

35. Merigo, J.M., Gil-Lafuente, A.M., Yager, R.R.: An overview of fuzzy research with bibliometric indicators. Appl. Soft Comput. **27**, 420–433 (2015)

36. Simon, D.: Evolutionary Optimization Algorithms: Biologically Inspired and Population-Based Approaches to Computer Intelligence. Wiley (2013)

37. Teodorovic, D., Pavkovich, G.: The fuzzy set theory approach to the vehicle routing problem when demand at nodes is uncertain. Fuzzy Sets Syst. **82**, 307–317 (1996)

38. Toth, P., Vigo, D. (eds.): The Vehicle Routing Problem. SIAM, Philadelphia (2002)

39. Tamir, D.E., Rishe, N.D., Kandel, A. (eds.): Fifty Years of Fuzzy Logic and Its Applications. Studies in Fuzziness and Soft Computing, vol. 326. Springer International Publishing, Cham, Switzerland (2015)

40. Werners, B.: An interactive fuzzy programming system. Fuzzy Sets Syst. **23**, 131–147 (1987)

41. Werners, B.: Interactive multiple objective programming subject to flexible constraints. Eur. J. Oper. Res. **31**, 342–349 (1987)

42. Werners, B., Drawe, M.: Capacitated vehicle routing problem with fuzzy demand. In: Verdegay, J.-L. (ed.) Fuzzy Sets Based Heuristics for Optimization, Studies in Fuzziness and Soft Computing, pp. 317–335. Berlin (2003)

43. Werners, B., Kondratenko, Y.P.: Tanker routing problem with fuzzy demands of served ships. Int. J. Syst. Res. Inf. Technol. **1**, 47–64 (2009)
44. Werners, B., Kondratenko, Y.P.: Tanker Routing Problem with Fuzzy Demand. Arbeitsberichte zur Unternehmensforschung Nr. 2001/04. Fakultät für Wirtschaftswissenschaft, Ruhr-Universität Bochum (2001)
45. Werners, B., Kondratenko, Y.P.: Fuzzy multi-criteria optimization for vehicle routing with capacity constraints and uncertain demands. In: Proceedings of the International Congress on Cost Control, Barcelona, Spain, 17–18 March 2011, pp. 145–159 (2011)
46. Werners, B.: Grundlagendes Operations Research: Mit Aufgaben und Lösungen, 2. Aufl., Berlin (2008)
47. Yager, R.R.: Golden rule and other representative values for intuitionistic membership grades. IEEE Trans. Fuzzy Syst. **23**, 2260–2269 (2015)
48. Yager, R.R.: On the OWA aggregation with probabilistic inputs. Int. J. Uncertain. Fuzziness Knowl. Based Syst. **23**(Suppl. 1), 143–162 (2015)
49. Zadeh, L.A.: Fuzzy sets. Inf. Control **8**, 338–353 (1965)
50. Zadeh, L.A., Abbasov, A.M., Yager, R.R., Shahbazova, S.N., Reformat, M.Z. (eds.): Recent Developments and New Directions in Soft Computing. Studies in Fuzziness and Soft Computing, vol. 317. Springer (2014)
51. Zadeh, L.A., Abbasov, A.M., Yager, R.R., Shahbazova, S.N., Reformat, M.Z. (eds.): Recent Developments and New Directions in Soft Computing Foundations and Applications. Studies in Fuzziness and Soft Computing, vol. 342. Springer, Berlin, Heidelberg (2016)

Retraction Note to: Towards the Convergence in Fuzzy Cognitive Maps Based Decision-Making Models

Leonardo Concepción, Gonzalo Nápoles, Isel Grau, Koen Vanhoof and Rafael Bello

Retraction Note to:
Chapter "Towards the Convergence in Fuzzy Cognitive Maps Based Decision-Making Models" in: C. Berger-Vachon et al. (eds.), *Complex Systems: Solutions and Challenges in Economics, Management and Engineering*, Studies in Systems, Decision and Control 125, https://doi.org/10.1007/978-3-319-69989-9_8

Due to an oversight the uncorrected version was accepted and printed, and hence this chapter had to be removed.

The retracted online version of this chapter can be found at
https://doi.org/10.1007/978-3-319-69989-9_8

© Springer International Publishing AG 2018
C. Berger-Vachon et al. (eds.), *Complex Systems: Solutions and Challenges in Economics, Management and Engineering*, Studies in Systems, Decision and Control 125, https://doi.org/10.1007/978-3-319-69989-9_32

E1

Printed in the United States
By Bookmasters